Toward Sustainable Agricultural Systems in the 21st Century

Committee on Twenty-First Century Systems Agriculture

Board on Agriculture and Natural Resources

Division on Earth and Life Studies

NATIONAL RESEARCH COUNCIL
OF THE NATIONAL ACADEMIES

THE NATIONAL ACADEMIES PRESS
Washington, D.C.
www.nap.edu

THE NATIONAL ACADEMIES PRESS   500 Fifth Street, N.W.   Washington, DC 20001

NOTICE: The project that is the subject of this report was approved by the Governing Board of the National Research Council, whose members are drawn from the councils of the National Academy of Sciences, the National Academy of Engineering, and the Institute of Medicine. The members of the committee responsible for the report were chosen for their special competences and with regard for appropriate balance.

This study was supported by a grant from the Bill & Melinda Gates Foundation and the W.K. Kellogg Foundation under Contract No. 48042 and P3005905. Any opinions, findings, conclusions, or recommendations expressed in this publication are those of the author(s) and do not necessarily reflect the views of the foundations that provided support for the project.

International Standard Book Number-13: 978-0-309-14896-2 (Book)
International Standard Book Number-10: 0-309-14896-0 (Book)
International Standard Book Number-13: 978-0-309-14897-9 (PDF)
International Standard Book Number-10: 0-309-14897-9 (PDF)
Library of Congress Control Number: 2010927922

Additional copies of this report are available from the National Academies Press, 500 Fifth Street, N.W., Lockbox 285, Washington, DC 20055; (800) 624-6242 or (202) 334-3313 (in the Washington metropolitan area); Internet, http://www.nap.edu.

Copyright 2010 by the National Academy of Sciences. All rights reserved.

Printed in the United States of America

THE NATIONAL ACADEMIES
Advisers to the Nation on Science, Engineering, and Medicine

The **National Academy of Sciences** is a private, nonprofit, self-perpetuating society of distinguished scholars engaged in scientific and engineering research, dedicated to the furtherance of science and technology and to their use for the general welfare. Upon the authority of the charter granted to it by the Congress in 1863, the Academy has a mandate that requires it to advise the federal government on scientific and technical matters. Dr. Ralph J. Cicerone is president of the National Academy of Sciences.

The **National Academy of Engineering** was established in 1964, under the charter of the National Academy of Sciences, as a parallel organization of outstanding engineers. It is autonomous in its administration and in the selection of its members, sharing with the National Academy of Sciences the responsibility for advising the federal government. The National Academy of Engineering also sponsors engineering programs aimed at meeting national needs, encourages education and research, and recognizes the superior achievements of engineers. Dr. Charles M. Vest is president of the National Academy of Engineering.

The **Institute of Medicine** was established in 1970 by the National Academy of Sciences to secure the services of eminent members of appropriate professions in the examination of policy matters pertaining to the health of the public. The Institute acts under the responsibility given to the National Academy of Sciences by its congressional charter to be an adviser to the federal government and, upon its own initiative, to identify issues of medical care, research, and education. Dr. Harvey V. Fineberg is president of the Institute of Medicine.

The **National Research Council** was organized by the National Academy of Sciences in 1916 to associate the broad community of science and technology with the Academy's purposes of furthering knowledge and advising the federal government. Functioning in accordance with general policies determined by the Academy, the Council has become the principal operating agency of both the National Academy of Sciences and the National Academy of Engineering in providing services to the government, the public, and the scientific and engineering communities. The Council is administered jointly by both Academies and the Institute of Medicine. Dr. Ralph J. Cicerone and Dr. Charles M. Vest are chair and vice chair, respectively, of the National Research Council.

www.national-academies.org

COMMITTEE ON TWENTY-FIRST CENTURY SYSTEMS AGRICULTURE

JULIA L. KORNEGAY, Chair, North Carolina State University, Raleigh
RICHARD R. HARWOOD, Vice Chair, Michigan State University *(Emeritus)*, East Lansing
SANDRA S. BATIE, Michigan State University, East Lansing
DALE BUCKS, Bucks Natural Resources Management, Elkridge, Maryland
CORNELIA BUTLER FLORA, Iowa State University, Ames
JAMES HANSON, University of Maryland, College Park
DOUGLAS JACKSON-SMITH, Utah State University, Logan
WILLIAM JURY, University of California, Riverside
DEANNE MEYER, University of California, Davis
JOHN P. REGANOLD, Washington State University, Pullman
AUGUST SCHUMACHER, JR., SJH and Company, Boston, Massachusetts
HENNING SEHMSDORF, S&S Homestead Farm, Lopez Island, Washington
CAROL SHENNAN, University of California, Santa Cruz
LORI ANN THRUPP, Fetzer Vineyards, Hopland, California
PAUL WILLIS, Niman Ranch Pork Company, Thornton, Iowa

Consultants

LAWRENCE ELWORTH, Center for Agricultural Partnerships, Asheville, North Carolina
C. CLARE HINRICHS, Pennsylvania State University, State College
SUSAN SMALLEY, Michigan State University, East Lansing

Editor

PAULA TARNAPOL WHITACRE, Full Circle Communications, LLC

Staff

EVONNE P.Y. TANG, Study Director
ERIN P. MULCAHY, Senior Program Assistant
JANET M. MULLIGAN, Research Associate
KAREN L. IMHOF, Administrative Assistant
ROBERTA A. SCHOEN, Board Director

BOARD ON AGRICULTURE AND NATURAL RESOURCES

NORMAN R. SCOTT, Chair, Cornell University, Ithaca, New York
PEGGY F. BARLETT, Emory University, Atlanta, Georgia
HAROLD L. BERGMAN, University of Wyoming, Laramie
RICHARD A. DIXON, Samuel Roberts Noble Foundation, Ardmore, Oklahoma
DANIEL M. DOOLEY, University of California, Oakland
JOAN H. EISEMANN, North Carolina State University, Raleigh
GARY F. HARTNELL, Monsanto Company, St. Louis, Missouri
GENE HUGOSON, Minnesota Department of Agriculture, St. Paul
KIRK C. KLASING, University of California, Davis
VICTOR L. LECHTENBERG, Purdue University, West Lafayette, Indiana
PHILIP E. NELSON, Purdue University, West Lafayette, Indiana
KEITH PITTS, Marrone Bio Innovations, Davis, California
CHARLES W. RICE, Kansas State University, Manhattan
HAL SALWASSER, Oregon State University, Corvallis
PEDRO A. SANCHEZ, The Earth Institute, Columbia University, Palisades, New York
ROGER A. SEDJO, Resources for the Future, Washington, D.C.
KATHLEEN SEGERSON, University of Connecticut, Storrs
MERCEDES VAZQUEZ-AÑON, Novus International, Inc., St. Charles, Missouri

Staff

ROBERTA A. SCHOEN, Director
RUTH S. ARIETI, Research Associate
CAMILLA YANDOC ABLES, Associate Program Officer
KAREN L. IMHOF, Administrative Assistant
KARA N. LANEY, Associate Program Officer
AUSTIN J. LEWIS, Senior Program Officer
ERIN P. MULCAHY, Senior Program Assistant
JANET M. MULLIGAN, Research Associate
KAMWETI MUTU, Research Associate
EVONNE P.Y. TANG, Senior Program Officer
PEGGY TSAI, Program Officer

Preface

Since the National Research Council published the report *Alternative Agriculture* in 1989, there has been a remarkable emergence of innovations and technological advances that are generating promising changes and opportunities for sustainable agriculture in the United States. At the same time, the agricultural sector worldwide faces numerous daunting challenges that will require innovations, new technologies, and new ways of approaching agriculture if the food, feed, and fiber needs of the global population are to be met.

This report, *Toward Sustainable Agricultural Systems in the 21st Century*, assesses the scientific evidence for the strengths and weaknesses of different production, marketing, and policy approaches for improving agricultural sustainability and reducing the costs and unintended consequences of agricultural production. It also evaluates the transferability of principles underlying farming systems and practices that could improve the sustainability of small-scale agricultural systems in less developed countries, with an emphasis on sub-Saharan Africa. The report includes case studies of different kinds of farms and farming systems in different regions of the United States that actively pursue the goal of sustainability and revisits some farms originally featured in *Alternative Agriculture*. We want to thank the farmers who so generously shared their expertise and experiences and to wish them well in their future farming endeavors. We also want to thank the consultants who conducted and documented the farmer interviews.

The study committee included 15 members with expertise in food production and agribusiness; crop, soil, and horticultural sciences; water-use and water-quality science; farming systems and agroecology; agricultural economics and social science; and federal farm, trade, international development, environmental, and regulatory policies (Appendix B). Two of the committee members are farmers. The committee also solicited information from a wide range of experts (Appendix C) with complementary expertise and experience. We are grateful for their willingness to give of their time and knowledge. During the development of the report, the committee held two workshops. The first focused on the state of the science on agricultural methods and systems for improving sustainability, and a second was on the lessons learned and transferability of agriculture practices and systems to improve sustainability of agriculture in developing countries. Two public committee meetings, in which other experts were invited to provide the committee with information on U.S. agricultural economics and policies, and their effect on farming systems, farmers' behavior, and the environment, were also held. Some of the committee members also attended the Sustainable Agricultural Research and Education (SARE) conference in Kansas City, Missouri, in 2008 to gather information.

Challenges that the committee immediately faced included understanding and interpreting the rapid changes and developing crises in the global economy and their effect on sustainable agriculture. For example, when the committee began its study, global price of crude fuel oil rose from about $75 per barrel to a peak of $147 in July 2008. This increase caused harmful reverberations across the global agriculture sector and shortages of corn, rice, and other food, especially in developing countries, and a significant increase in the demand for biofuels. It was immediately followed by the global economic crisis, which, among other impacts, restricted farmers' access to credit, lowered land values, and lowered prices for biofuels when fuel oil costs declined by half. On a more positive note, the committee faced a virtual cascade of new information and programs relating to sustainable agriculture, such as important new advances in science and in federal and state programs and policies. The new federal farm bill places greater emphasis on agricultural sustainability, organic agriculture, and renewable energy and fuels, and support is growing for regional and local food production systems.

The committee notes that although most farms have the potential and responsibility to contribute to different aspects of sustainability, U.S. agriculture needs both incremental and transformative changes to address the many challenges of the future. Incremental changes—such as pest-resistant varieties, conservation tillage, integrated pest management, and use of crop diversity including cover crops, crop rotations, and other biologically integrative technologies and practices—have been increasingly used in many regions, but have not yet been adapted to some fragile areas and to low-rainfall cropland. Transformative changes include the development of new farming systems that represent a dramatic departure from the dominant systems of present-day American agriculture and capitalize on synergies and efficiencies associated with complex natural systems and broader social and economic forces using integrative approaches to research and extension at both the farm and landscape levels. Examples include development and broad adoption of water-conserving production systems in areas of water shortage and overdraft, landscape-scale reduction of nutrient and other materials runoff from agricultural lands that contributes to major hypoxic zones, and assessment of the potential and cost for broad adoption of alternative animal production systems that address many environmental and social concerns of some dominant production systems.

The committee believes that its report identifies many of the most important challenges that U.S. agriculture faces today, but it is well aware that unforeseen threats as well as new opportunities could surface tomorrow. We hope that the sponsors of this study, the Bill & Melinda Gates Foundation and W.K. Kellogg Foundation, as well other groups and organizations, will find the report's conclusions and recommendation to be of value in their efforts to understand and develop sustainable agricultural systems that will meet the food, feed, fiber, and biofuel needs of a growing global population.

On behalf of the committee, we would like to express our thanks and appreciation to Robin Schoen, director of the Board of Agriculture and Natural Resources (BANR), and

Evonne Tang, the senior program officer responsible for our study. Without their planning, organization, and editing expertise, this large and complex report would have been impossible. We also want to thank all the BANR study staff for their support and assistance with our meetings and in preparing the final report.

Julia L. Kornegay, *Chair*
Richard R. Harwood, *Vice-Chair*
Committee on Twenty-First Century
Systems Agriculture

Acknowledgments

This report has been reviewed in draft form by persons chosen for their diverse perspectives and technical expertise in accordance with procedures approved by the National Research Council's Report Review Committee. The purpose of this independent review is to provide candid and critical comments that will assist the institution in making its published report as sound as possible and to ensure that the report meets institutional standards of objectivity, evidence, and responsiveness to the study charge. The review comments and draft manuscript remain confidential to protect the integrity of the deliberative process. We wish to thank the following individuals for their review of this report:

P. Stephen Baenziger, University of Nebraska-Lincoln
Jon T. Biermacher, The Samuel Noble Foundation, Inc.
Juliet Christian-Smith, Pacific Institute for Studies in Development, Environment, and Security
Michael DeFelice, Pioneer Hi-Bred
Thomas Dobbs, South Dakota State University
Michael Doyle, University of Georgia
Simeon Ehui, The World Bank
Peter Gleick, Pacific Institute for Studies in Development, Environment, and Security
Temple Grandin, Colorado State University
Gary Hirshberg, Stoneyfield, Inc.
Terry Howell, U.S. Department of Agriculture Agricultural Research Service
R. Cesar Izaurralde, Joint Global Change Research Institute
Fred Kirschenmann, Iowa State University
Max Pfeffer, Cornell University
Keith Prasse, University of Georgia
William Raun, Oklahoma State University
Andrew Thulin, California Polytechnic State University

Although the reviewers listed above have provided many constructive comments and suggestions, they were not asked to endorse the conclusions or recommendations, nor did they see the final draft of the report before its release. The review of this report was overseen by Drs. R. James Cook and Harley W. Moon. Appointed by the National Research Council, they were responsible for making certain that an independent examination of this report was carried out in accordance with institutional procedures and that all review comments were carefully considered. Responsibility for the final content of this report rests entirely with the authoring committee and the institution.

Contents

EXECUTIVE SUMMARY 1

SUMMARY 3

1 UNDERSTANDING AGRICULTURAL SUSTAINABILITY 15
    Purpose of this Report, 16
    Farming and Agriculture, 17
        General Definitions, 17
        Farming Practices and Systems, 19
            Conventional and Industrial Agricultural Systems, 19
            Ecologically Based Farming Systems, 20
            A Farming Systems Continuum, 20
    Agricultural Sustainability, 23
        Defining Sustainability Goals, 23
            Objectives, 24
            Important Qualities of Systems That Move Toward Greater Sustainability, 26
                Robustness, 26
                Scales, 27
                Synergies and Tradeoffs, 28
    Setting Priorities for Agricultural Sustainability, 29
        Who Decides?, 29
        The Role of Science in Facilitating Sustainability, 31
    Indicators of Sustainability, 32
        Characteristics and Types of Indicators, 33
        Relating Indicators to Sustainability Goals and System Attributes, 35
        Interpreting Indicators, 36
        Integrating Diverse Indicators in Holistic Assessments, 36

Summary, 37
Organization of the Report, 38
References, 39

2 A PIVOTAL TIME IN AGRICULTURE 43
   A Brief History of U.S. Agriculture, 45
   U.S. Agriculture Today, 48
   Challenges to U.S. Agriculture in the 21st Century, 54
      Increasing Demand on U.S. Agriculture, 54
      Natural Resource Scarcity, 55
         Land Availability, 55
         Water Availability, 59
         Climate Change, 59
      Environmental Degradation, 61
         Water Quality, 61
         Air Quality, 63
         Soil Quality, 64
         Reduced Genetic Diversity, 66
      Economic Concerns, 67
         Farm Sector Profitability and Rising Input Costs, 67
         Loss of Mid-Sized Commercial Family Farms, 69
      Social Concerns, 71
         Labor Concerns, 71
         Food Quality and Safety, 71
         Food Security Concerns, 72
         Animal Welfare Concerns, 73
         Community Well-Being, 73
         Community Health and Quality of Life, 74
   Systems Approach to Improving the Sustainability of Agriculture, 74
   Summary, 75
   References, 75

3 IMPROVING PRODUCTIVITY AND ENVIRONMENTAL SUSTAINABILITY
   IN U.S. FARMING SYSTEMS 83
   Soil Management, 85
      Conservation Tillage, 86
         Impact of Conservation Tillage, 86
            Physical Properties of Soil, 86
            Soil Organic Matter, 87
            Soil Microbial Activity and Diversity, 87
            Soil Erosion, 88
            Sediment Loading and Water Quality, 88
            Air Quality, 89
            Energy Use, 90
         Disadvantages of Conservation Tillage, 92
         Adoption of Conservation Tillage, 93
      Cover Cropping, 94
         Impact of Cover Cropping, 94
            Productivity, 94

Soil Quality, 94
Water Quality, 95
Water Use, 95
Nutrient Management, 95
Weeds, Insects, and Diseases, 96
Disadvantages of Cover Cropping, 96
Adoption of Cover Cropping, 96
Crop and Vegetation Diversity Management, 97
Crop Rotations, 99
Impact of Crop Rotations, 100
Productivity, 100
Soil Health, 100
Air Quality, 101
Water Use, 101
Disadvantages of Crop Rotations, 101
Adoption of Crop Rotations, 101
Intercropping, 103
Impact of Intercropping, 103
Productivity, 103
Nutrient Management, 104
Disadvantages of Intercropping, 104
Cultivar Mixtures, 104
Impact of Cultivar Mixtures, 104
Disadvantages of Cultivar Mixtures, 104
Management of Noncrop Vegetation, 105
Impact of Noncrop Vegetation on Biodiversity, 105
Disadvantages of Noncrop Vegetation, 105
Plant Breeding and Genetic Modification of Crops, 106
Molecular Markers and Genetic Engineering in Cultivar Development, 108
Impact, 109
Disadvantages, 109
Adoption, 110
Water-Use Management, 110
Irrigation Scheduling, 111
Impact of Irrigation Scheduling, 112
Adoption of Irrigation Scheduling, 112
Gravity Systems, 112
Disadvantages of Gravity Systems, 113
Adoption of Gravity Systems, 113
Sprinkler Irrigation, 113
Impact of Sprinkler Irrigation, 113
Adoption of Sprinkler Irrigation, 114
Trickle or Drip Irrigation, 114
Impact of Trickle or Drip Irrigation, 114
Adoption of Trickle or Drip Irrigation, 114
Regulated Deficit Irrigation, 115
Impact of Regulated Deficit Irrigation, 115
Disadvantages of Regulated Deficit Irrigation, 115
Water Reuse, 115

Disadvantages of Water Reuse, 116
Adoption of Water Reuse, 116
Small Dams, 116
Impact of Small Dams, 116
Disadvantages of Small Dams, 117
Water Quality Management, 117
Drainage Water Management Systems, 117
Impact of DWM on Productivity and Water Quality, 118
Disadvantages of DWM, 118
Adoption of DWM, 119
Wetlands, 119
Impact of Wetlands, 120
Nutrient Loading, 120
Pesticides, 120
Disadvantages of Wetlands, 120
Adoption of Wetlands, 121
Buffers, 121
Impact of Buffers, 121
Nutrient Loading, 121
Pesticides, 122
Fecal Coliform Bacteria, 122
Disadvantages of Buffers, 122
Nutrient Management, 122
Mass Balances for Nutrient Management, 123
Soil and Tissue Sufficiency Tests, 125
Nutrient Management Plans and Best Management Practices, 126
Nutrient Inputs, 126
Legumes, 126
Impact of Legumes, 126
Yield, 126
Nutrient Availability, 127
Disadvantages of Legumes, 127
Animal Manure, 127
Impact of Animal Manure, 127
Soil Quality, 127
Energy Use, 128
Disadvantages of Animal Manure, 128
Dietary Modification to Adjust Manure Composition, 128
Adoption of Animal Manure Application, 129
Compost, 129
Impact of Compost, 130
Productivity, 130
Soil Quality, 130
Disease Suppression, 130
Energy Use, 130
Disadvantages of Compost, 131
Precision Agriculture, 131
Impact of Precision Agriculture, 132
Nutrient Use, 132

Water Quality, 133
Adoption of Precision Agriculture, 133
Nanotechnology-based Applications, 133
Anaerobic Digestion with Biogas Recovery of Animal Manure, 134
Impact of Anaerobic Digestion, 134
Disadvantages of Anaerobic Digestion, 135
Adoption of Anaerobic Digestion, 135
Weeds, Pests, and Disease Management in Crops, 135
Managing the Crop–Weed–Disease–Pest Complex, 137
The Evolution of Integrated Pest Management, 138
Use of Disease and Insect Resistant Cultivars in IPM, 138
Arthropod Pest Management, 139
Pathogen Management, 141
Biofumigation, Organic Amendments, and Anaerobic Disinfestation Techniques, 142
Advances in Nematode Management, 143
Advances in Weed Management, 144
Designing Systems for Holistic Management of the Crop–Weed–Disease–Pest Complex, 147
Evaluation of Adoption, Effectiveness, and Future Challenges of IPM and Ecological Pest Management, 148
Future Role of Pesticides in IPM, 149
Managing Efficiency of Animal Production Systems, 150
Animal Breeding, 151
Disadvantages of Animal Breeding, 152
Animal Nutrition, 152
Impact of Nutritional Strategies, 153
Disadvantages of Nutritional Strategies, 153
Adoption of Nutritional Strategies, 153
Animal Welfare, 154
Housing, 155
Qualitative Diet Restriction, 156
Environmental Enrichment, 157
Research Needs, 157
Animal Health, 158
Alternatives to Subtherapeutic Antibiotics, 158
Impact of Alternatives to Antibiotics, 159
Disadvantages of Alternatives to Antibiotics, 160
Animal Identification, 160
Summary, 161
Soil Management, 161
Crop and Vegetation Diversity Management, 161
Water Use and Quality Management, 162
Nutrient Management, 162
Weed, Pest, and Disease Management in Crops, 163
Animal Housing, Nutrition, Health, and Breeding, 163
References, 164

4 ECONOMIC AND SOCIAL DIMENSIONS OF THE SUSTAINABILITY OF FARMING PRACTICES AND APPROACHES 189

Economic Security of Sustainable Farming Systems, 189
    Economic Security at the Farm Level, 191
        Economics of Production Practices That Can Improve Sustainability, 192
            Conservation Tillage, 192
            Crop Rotations, 193
            Cover Cropping, 193
            Crop Nutrient Management Strategies, 194
            Conservation Best Management Practices, 194
            Precision Agriculture for Nutrient Management, 195
            Integrated Pest Management, 196
        Business and Marketing Diversification Strategies, 196
            Value-Trait Marketing, 197
            Direct Marketing, 198
            Agritourism and Fee Hunting, 200
            Off-Farm Income, 201
        Quality of Life and Sustainable Farming System, 201
    Socioeconomic Aspects of Sustainability at the Community Level, 202
        Farm Labor Conditions and Security, 202
        Community Economic Security, 203
            Farming Practices for Improving Sustainability and Community Economic Security, 203
            Civic Agriculture, Local Foods, and Community Economic Security, 205
        Community Well-Being, 206
    Food Security, Safety, Quality, and Other Socioeconomic Dimensions, 207
        Satisfying Human Food, Feed, and Fiber Needs, 207
        Sustainable Agriculture and Food Access, 207
        Food Safety, 208
            Bacterial Pathogens in Natural Fertilizers and Irrigation Water, 208
            Fungal Pathogens, 209
            Pesticide Residue, 209
        Food Quality and Nutritional Completeness, 210
        Next Generation of Farmers, 210
    Summary, 211
    References, 212

5 EXAMPLES OF FARMING SYSTEM TYPES FOR IMPROVING SUSTAINABILITY 221

Organic Cropping Systems, 222
    Principles and Practices of Organic Farming, 223
    Impact on Productivity and Environmental Sustainability, 224
        Yield, 224
        Nutrient Cycling and Soil Quality, 226
        Water Quality, 226
        Weeds, 227
        Greenhouse-Gas Emissions, 228
    Economic Impact, 228
    Social Impact, 230

CONTENTS

    Labor Practices, 230
    Food Adequacy, 231
    Food Quality and Nutritional Completeness, 232
    Community Well-Being, 233
Alternative Livestock Production Systems, 233
  Integrated Crop–Livestock Systems, 234
  Management-Intensive Rotational Grazing Systems, 235
    Environmental Impact of MIRG Systems, 235
      Soil Quality and Soil Erosion, 236
      Carbon, Greenhouse Gas, and Nutrient Dynamics, 237
      Biodiversity, 239
      Economic Performance of MIRG Systems, 239
    Social Performance of MIRG Systems, 240
      Labor Practices, 240
      Impact on Human Nutrition and Health, 240
  Low-Confinement Integrated Hog-Producing Systems, 241
    Forces of Change in the Hog Sector, 241
    Guiding Principles, 242
    Environmental Impact of Low-Confinement Hog Systems, 244
      Nutrient Cycling, Odor Control, and Greenhouse-Gas Emissions, 244
      Landscape Diversity, Soil Quality, and Soil Erosion, 245
    Economic Impact of Low-Confinement Systems, 246
      Farm Operations, 246
      Marketing, 247
    Social Impact of Low-Confinement Hog Systems, 248
      Labor Use and Working Conditions, 248
      Meat Quality, 248
      Public Reaction, 249
    Summary, 249
Perennial Agriculture Systems, 249
  Perennial Grain System, 250
    Impact, 251
  Perennial Grasses for Biofuels, 251
    Impact on Food Security, 251
    Environmental Impact, 252
    Economic Impact, 252
Gaps in Existing Science at the Systems Level, 253
  Design Within Systems Types, 253
  Holistic Comparisons Between Farming Systems Types, 255
Biogeophysical Landscape-Level Sustainability Analysis and Planning, 255
Summary, 258
References, 259

6   DRIVERS AND CONSTRAINTS AFFECTING THE TRANSITION TO
    SUSTAINABLE FARMING PRACTICES   271
    Agricultural Markets as Contextual Factors, 272
      Concentration in the Agrifood System, 272
        Markets for Farm Inputs, 273
        Markets for Products, 274

Emerging Markets, 275
    Changes in Consumer Preferences, 275
    Sustainability Initiatives, 276
    Organic Food Markets, 276
    Direct-Sales Markets, 278
        Farmers' Markets and Farm Stands, 279
        Community Supported Agriculture, 279
        Farm to Institutions, 280
Grades, Standards, and Certification Labels, 280
    Grades and Standards, 281
    Sustainable Agriculture Standards, Certification, and Eco-Label Programs, 282
    Marketing Institutions for Mid-Sized Commercial Farmers: Branding, 283
Emerging Markets for Ecosystem Services, 285
    Payment for Environmental Services: Beneficiary Pays, 286
    Cap-and-Trade, 289
    Offsets or Conservation Credit Trading, 289
    Role of Valuation of Ecosystem Services, 291
Public Policy as a Contextual Factor, 291
    The Food, Conservation and Energy Act of 2008, 291
        Commodity Support Programs, 292
        Crop Insurance and Disaster Payments, 294
        Conservation Programs, 295
        Nutritional Assistance Programs, 297
        Trade Policies, 298
    Energy Policy, 299
    Environmental Regulation, 301
        Clean Air Act, 301
        Clean Water Act, 301
        Food Quality Protection Act, 302
        Food Safety Guidelines and Standards, 303
        Endangered Species Act, 304
    Water Use Policies, 304
        Surface Water, 305
        Ground Water, 306
        Conjunctive Use, 307
    Animal Welfare Regulations, 307
Knowledge Institutions as Contextual Factors, 307
    Publicly Funded Agricultural Research and Extension, 308
        Land-Grant Universities, 308
        U.S. Department of Agriculture, 308
        Distribution of Federal Funds for Agricultural Research, 309
        Broadening Review of Public Competitive Grant Programs, 311
    Private Sector Agricultural Research, 312
    Division of Labor Between Public and Private Agricultural Research, 314
    Expanding Beyond Productivity Research, 314
        Federal Sustainable Agriculture Research Programs, 316
        State and Civil Society Support for Sustainable Agriculture, 317
        University Sustainable Agriculture Programs, 318
        Cooperative Extension, 318

CONTENTS   xxi

        Farmer Participation and Innovation in Research and Development, 320
        Structuring Systems Research for Improving Agricultural Sustainability, 322
    Stakeholders and Social Movements, 323
        A Brief History of Agricultural Stakeholders and Social Movements, 323
    Diversity of Farmer Responses to Contexts, 326
        Local Conditions and Farm Sustainability, 327
        Farm and Farmer Characteristics and the Use of Sustainable
        Agricultural Practices, 328
            Farm Characteristics, 328
            Implications for the Adoption of Farming Systems for Improving
            Sustainability, 329
        Farmer Knowledge, Skills, and Perceptions, 331
        Farmer Values, Goals, and Perceptions, 332
    Summary, 333
        Markets, 333
        Public Policies, 334
        Knowledge Institutions, 335
        Stakeholders and Social Movements, 336
        Diversity of Farmer Responses, 336
    References, 337

7   ILLUSTRATIVE CASE STUDIES   351
    Follow-Up of the Case Studies Featured in *Alternative Agriculture*, 353
    Status of the Farms, 353
    Commonalities Among the Farms, 355
    • Mormon Trail Farm, 357
        Farming Philosophy, 357
        Management Features, 357
            Crop Rotations and the Soil, 357
            Livestock, 358
        Learning Networks, 360
        Use of Government Programs, 360
        Trial of Organic Production, 361
        Benefits from the Biofuel Industry, 361
        Summary and Future Outlook, 361
    • Ferrari Farms, Inc., 362
        Farming Philosophy, 362
        Management Features, 362
            Crops, 362
            Pest Management, 362
            Fertility Management, 363
            Labor Management, 363
            Equipment, 363
            Marketing, 363
        Learning Networks, 364
        Performance Indicators, 364
        Key Changes, 364
        Challenges, 365
        Summary and Future Outlook, 365

- Brookview Farm, 366
    - Farming Philosophy, 366
    - Management Features, 366
        - Crop Management, 366
        - Weed Management, 367
        - Fertility Management, 367
        - Livestock, 367
        - Environmental Management, 367
        - Marketing, 368
    - Learning Networks, 369
    - Performance Indicators, 369
    - Key Changes, 370
    - Challenges, 370
    - Summary and Future Outlook, 370
- Lundberg Family Farms, 371
    - Farming Philosophy, 371
    - Management Features, 371
        - Operations, 371
        - Crop Management, 371
        - Weed, Pest, and Disease Management, 372
        - Fertility Management, 372
        - Energy Use, 373
        - Environmental Management, 373
        - Labor Management, 373
        - Marketing, 373
    - Learning Networks, 374
    - Performance Indicators, 374
    - Key Changes, 375
    - Challenges, 375
    - Summary and Future Outlook, 376
- Pavich Family Farms, 377
    - Farming Philosophy, 377
    - Learning Networks, 377
    - Key Changes, 377
    - Challenges, 378
    - Summary and Future Outlook, 378
- Thompson Farm, 380
    - Farming Philosophy, 380
    - Management Features, 380
        - Crops, 380
        - Weed and Pest Management, 380
        - Livestock, 381
        - Equipment, 381
        - Labor, 381
        - Marketing, 381
    - Learning Networks, 381
    - Performance Indicators, 382
    - Key Changes, 383
    - Challenges, 383

CONTENTS   xxiii

   Summary and Future Outlook, 383
- Green Cay Farm and Green Cay Produce, 385
  - Farming Philosophy, 385
  - Management Features, 385
    - Crops, 385
    - Pest Management, 386
    - Fertility Management, 386
    - Marketing, 386
    - Labor, 386
  - Learning Networks, 386
  - Key Changes, 387
  - Challenges, 388
  - Summary and Future Outlook, 388

New Case Studies, 390
 Production Challenges, 392
  Soil Management, 392
  Weed, Pest, and Disease Management, 393
  Water Management, 393
  Energy Management, 393
  Management of Livestock, 394
 Socioeconomic Issues, 395
  Economic Viability, 395
  Marketing, 395
  Labor, 395
 Information Sources and Knowledge of Production, 396
 Government Programs and Policies, 397
 Lessons Learned, 397
- Bragger Farm, 402
  - Background and History, 402
  - Farm Production System, 402
    - Land, 402
    - Soils and Fertility, 403
    - Crops, 404
      - Farming Practices, 404
      - Pest Management, 405
      - Weed Management, 405
      - Equipment and Buildings, 405
    - Livestock Enterprises, 406
      - Dairy, 406
      - Beef Cattle, 407
      - Heifers, 407
      - Pullets, 407
      - Labor, 407
    - Manure and Nutrient Management Issues, 408
    - Other Land Enterprises, 408
  - Natural Resources, Energy, and Climate Change, 409
  - Marketing, Business Management, and Financials, 409
    - Marketing and Business Management, 409
    - Use of Federal and Conservation Programs, 410

- Social and Community Considerations, 410
  - Social and Community Interactions, 410
    - Farm Succession, 411
- Risks, Challenges, and Changes, 411
- Sustainability, 411
- Observations and Conclusions, 412
- Radiance Dairy, 413
  - Background and History of the Farm, 413
  - Farm Production System, 414
    - Pasture Management, 414
    - Livestock, 415
    - Herd Health, 416
      - Livestock Waste, 417
      - On-Farm Dairy Processing, 417
    - Labor, 418
    - Farm Equipment, 418
  - Natural Resources, Energy, and Climate Change, 418
    - Water and Air Issues, 418
    - Energy and Carbon Concerns, 419
  - Marketing, Business Management, and Financials, 419
    - Marketing, 419
    - Certifications, 420
    - Finance and Business Management, 420
  - Social and Community Considerations, 421
  - Risks, Challenges, and Changes, 421
  - Observations and Conclusions, 421
- Straus Family Creamery, 423
  - Background and History, 423
  - Farm Production System, 424
    - Farm Production, 424
      - Herd Management, 424
      - Pasture and Silage, 425
      - Fertility and Nutrient Management, 425
      - Pest Management Concerns, 425
    - Creamery, 426
      - Production, 426
      - Product Line, 426
      - Packaging, 426
      - Plant Procedures and Issues, 427
    - Labor: Farm and Creamery, 427
    - Further Business, Marketing, and Financial Considerations, 428
    - Organic Certification, 428
  - Natural Resources, Energy, and Climate Change, 428
    - Energy, 428
    - Water, 429
    - Waste, 429
    - Local Environment, 429
  - Distribution and Markets, 430
  - Social and Community Considerations, 431

- Sustainability, 432
  - Risks, Challenges, and Changes, 432
- Full Belly Farm, 433
  - Background and History, 433
  - Farm Production System, 434
    - Planting and Rotations, 435
    - Tillage, 435
    - Weed Management, 436
    - Pest Management, 436
    - Animals, 436
    - Nutrient Management, 437
    - Equipment, 437
  - Natural Resources, Energy, and Climate Change, 438
    - Energy, 438
    - Biodiversity, 438
    - Water, 438
  - Marketing, Business Management, and Financials, 438
    - Marketing, 438
    - Pricing, 439
    - Finances, 440
  - Social and Community Considerations, 440
    - Labor, 440
    - Internships, 440
    - Community Outreach and Connections, 441
    - Further Community Considerations, 441
  - Risks and Challenges, 442
    - Supply and Farmer Cooperation, 442
    - Ripples from Food Safety Incidents in Larger Food System, 442
    - Transitions into Farming, 442
    - Research, 443
    - Government Programs, 443
  - Observations and Conclusions, 443
- Peregrine Farm, 445
  - Background and History, 445
  - Farm Production System, 446
    - Soils and Fertility Management, 447
    - Weed, Pest, and Disease Management, 447
    - Animals, 447
    - Labor, 448
    - Equipment, 448
  - Marketing, Business Management, and Financials, 448
    - Marketing, 448
    - Certifications, 448
    - Business Management, 449
    - Finance, 449
  - Social and Community Considerations, 449
    - Markets as Community, 449
    - Outreach, 450
    - Government Programs, 450

        Learning and Obtaining Information, 450
        Food Safety, 450
        Labor Practices and Mentoring Workers, 450
    Natural Resource Issues, Energy, and Climate Change, 451
        Water, 451
        Energy and Recycling, 451
        Climate Change, 451
    Risks, Challenges, and Changes, 451
    Sustainability, 451
    Observations and Conclusions, 452
- Stahlbush Island Farms, 453
    Background and History, 453
    Farm Production System, 455
        Soils and Fertility, 455
        Rotations, 456
        Weed Management, 456
        Pest Management, 456
        Rainfall and Irrigation, 457
        Equipment, 457
    Natural Resource, Energy, and Climate Change, 457
        Water, 457
        Wildlife and Biodiversity, 458
        Energy, 458
    Marketing, Business Management, and Financials, 458
        Audits and Certifications, 459
    Social and Community Considerations, 460
        Labor and Staffing, 460
        Community Support, Service, and Recognition, 460
    Sustainability, 460
    Research and Policy Concerns, 461
        Research, 461
        Policy Concerns, 461
    Observations and Conclusions, 462
- Goldmine Farm, 463
    Background and History, 463
    Farm Production System, 464
        Soils and Fertility Management, 464
        Cropping System and Yields, 465
        Production Practices, 465
        Pest and Disease Management, 466
        Livestock, 466
        Equipment, 467
    Natural Resources, Energy, and Climate Change, 468
    Marketing, Business Management, and Financials, 468
        Marketing and Organic Certification, 468
        Financial, 469
        Risk Management and Insurance, 469
    Social and Community Considerations, 469
        Labor, 469

Community Involvement, 470
Observations on Access to Organic Food, 470
Research and Policy Concerns, 470
Farm Programs, 470
Research Participation and Needs, 470
Sustainability, 471
Farm Transition Issues, 471
Labor as a Limiting Factor, 471
Observations and Conclusions, 472
- Rosmann Family Farm, 473
    Background and History, 473
    Farm Production System, 474
        Crops, 474
        Planting, 474
        Yields, 474
        Inputs, 474
        Livestock, 475
            Hogs, 475
            Cattle, 475
            Poultry, 476
        Pest Management, 476
        Pasture Management, 476
        Equipment, 477
        Labor, 477
        Nutrient Management, 477
    Natural Resources, Energy, and Climate Change, 478
        Water, 478
        Energy and Carbon Concerns, 478
    Marketing, Business Management, and Financial, 478
        Marketing, 478
        Certifications, 479
        Finance and Business Management, 479
    Social and Community Considerations, 479
        Federal Farm Programs, 480
    Risk, Challenges, and Changes, 481
    Observations and Conclusions, 481
- Zenner Farm, 482
    Background and History, 482
    Farm Production System, 483
        Soils and Growing Conditions, 483
        Crops and Rotations, 483
        Fertility Program, 484
        Direct Seeding, 484
        Disease and Pest Management Issues, 485
    Natural Resources and Wildlife Concerns, 486
    Marketing, Business Management, and Financials, 486
        Financials, 488
    Social and Community Considerations, 488
        Labor, 488

                Learning, 488
                Community Relations and Service, 489
            Risks, Challenges, and Changes, 490
                Research Needs, 490
                Transportation, 490
                Farm Transition Concerns, 491
        Government Programs and Policy Involvement, 491
        Observations and Conclusions, 492

8   SUSTAINABLE AGRICULTURE IN SUB-SAHARAN AFRICA: "LESSONS LEARNED" FROM THE UNITED STATES   493
    The Importance of Context, 494
        Evolving Agriculture in Sub-Saharan Africa, 494
        Lessons Learned from the Green Revolution, 494
        A Second Green Revolution, 495
    Long-Term Evolution Towards Sustainability in Sub-Saharan Africa, 496
    Considerations of U.S. "Lessons" Learned, 497
        Transferability of Agricultural Practices for Improving Sustainability, 497
    Summary, 514
    References, 515

9   CONCLUSIONS AND RECOMMENDATIONS   519
    What is Sustainable Agriculture?, 520
        Defining Sustainable Agriculture, 520
        Measuring Progress Toward Sustainability, 521
    Toward Agricultural Sustainability in the 21st Century, 521
    Incremental Approach to Improving U.S. Agricultural Sustainability, 522
    Transformative Approach to Improving U.S. Agricultural Sustainability, 524
        A Systems Approach to Agricultural Research, 527
        Key Drivers of Change: Markets and Federal and Local Policies, 531
    Relevance of Lessons Learned to Sub-Saharan Africa, 532
    In Closing, 533

APPENDIXES
A   Statement of Task   537
B   Biographical Sketches   539
C   Presentations to the Committee on 21st Century Systems Agriculture   545
D   Follow-up of the 1989 Case Studies Featured in *Alternative Agriculture* Report:
    Topics of Discussion During Telephone Interview   549
E   Dairy Farms: Topics of Discussion During On-Farm Interview   551
F   Grain Farms: Topics of Discussion During On-Farm Interview   559
G   Specialty-Crop Farms: Topics of Discussion During On-Farm Interview   565

Executive Summary

U.S. agriculture has had an impressive history of productivity that has resulted in relatively affordable food, feed, and fiber for domestic purposes and increases in agricultural exports. Fewer farmers are producing more food and fiber on about the same acreage, while input and energy use per unit output has decreased over the last 50 years. Despite these tremendous advances, U.S. farmers are facing the daunting challenges of meeting the food, feed, and fiber needs of the nation and of a growing global population and of contributing to U.S. biofuel production, under the constraints of rising production costs, increasingly scarce natural resources, and climate change. Agriculture is at a pivotal stage in terms of meeting societal demands for products while improving sustainability.

This report of the National Research Council Committee on Twenty-First Century Systems Agriculture reviews the state of knowledge on farming practices, technologies, and management systems that have the potential to improve the environmental, social, and economic sustainability of agriculture, and it discusses the tradeoffs and risks that might occur if more farms were to adopt those practices, technologies, and systems. The report also identifies knowledge gaps and makes recommendations for future actions to improve agricultural sustainability.

DEFINING AGRICULTURAL SUSTAINABILITY

Improving sustainability is a process that moves farming systems along a trajectory toward meeting various socially determined sustainability goals as opposed to achieving any particular end state. Agricultural sustainability is defined by four generally agreed-upon goals:

- Satisfy human food, feed, and fiber needs, and contribute to biofuel needs.
- Enhance environmental quality and the resource base.
- Sustain the economic viability of agriculture.
- Enhance the quality of life for farmers, farm workers, and society as a whole.

The sustainability of a farming practice or system could be evaluated on the basis of how well it meets various societal goals or objectives. To be sustainable, a farming system needs to be sufficiently productive, robust (that is, be able to continue to meet the goals in the face of stresses and fluctuating conditions), use resources efficiently, and balance the four goals.

TOWARD AGRICULTURAL SUSTAINABILITY IN THE 21ST CENTURY

All farms have the potential and responsibility to contribute to different aspects of sustainability. However, the scale, organization, enterprise diversity, and forms of market integration associated with individual farms provide unique opportunities or barriers to improving their ability to contribute to global or local food production, ecosystem integrity, economic viability, and social well-being. Dramatic and continuous improvement in agricultural sustainability will require long-term research, education, outreach, and experimentation by the public and private sectors in partnership with farmers.

The committee proposes two parallel and overlapping efforts to ensure continuous improvement in the sustainability performance of U.S. agriculture: incremental and transformative. The incremental approach would be directed toward improving the sustainability performance of all farms, irrespective of size or farming system type, through development and implementation of specific sustainability-focused practices, many of which are the focus of ongoing research and with varying levels of adoption. Most, if not all, farms have adopted some practices for improving sustainability, but such methods have not been adapted to all environments, and none of the practices has reached its full potential for adoption. Continuous research, extension, and experimentation by researchers and farmers are necessary to provide the toolkit necessary for farmers to adapt their systems to the changing environmental, social, market, and policy conditions to ensure long-term sustainability.

Research has to address multiple dimensions of sustainability and explore agroecosystems properties if systemic changes in farming systems are to be pursued. Therefore, the incremental approach to improving agricultural sustainability needs to be complemented by a transformative approach that would dramatically increase integrative research by bringing together multiple disciplines to address key dimensions of sustainability simultaneously beyond the agroecological dimension. The transformative approach would apply a systems perspective to agricultural research to identify and understand the significance of the linkages between farming components and how their interconnectedness and interactions with the environment make systems robust and resilient over time.

KEY DRIVERS OF CHANGE: MARKETS AND FEDERAL AND LOCAL POLICIES

The decisions of farmers to use particular farming practices and their ability to move toward increasingly sustainable farming systems are influenced by many external forces, including science, knowledge, skills, markets, public policies, and their own values, resources, and land tenure arrangements. Although market, policy, and institutional contexts are important drivers of the trajectory of U.S. agriculture, the response of individual farmers to the incentives and disincentives created by market conditions and policy contexts can be diverse. Efforts to promote widespread adoption of different farming practices and systems for improving sustainability will require an understanding of how variability among individual, household, farm, and regional-level characteristics affect farmers' response to incentives and disincentives.

Summary

Agriculture is facing daunting challenges. Not only are farmers expected to produce adequate agricultural products at affordable prices to meet the food, fiber, feed, and biofuel needs of a rising global population, but also they are expected to do so under conditions of rising production costs and increasingly scarce natural resources and climate change. Growing awareness of unintended impacts associated with some agricultural production practices has led to heightened societal expectations for improved environmental, community, labor, and animal welfare standards in agriculture. The question arose as to whether U.S. agriculture can meet those challenges and expectations in a sustainable way.

The National Research Council (NRC) convened the Committee on Twenty-First Century Systems Agriculture to assess scientific evidence for the strengths and weaknesses of different production, marketing, and policy approaches to improving the sustainability of American agriculture (Box S-1). The committee was asked to discuss how the lessons learned in U.S. agriculture could be relevant to agriculture in different regional and national settings, specifically sub-Saharan Africa. This report provides an update to the 1989 report *Alternative Agriculture*[1] on knowledge gained about the productivity and economics of different practices and systems at increasing levels of complexities (from the level of individual components in a farm, to a whole farm, to a regional level) and their impacts on community and social well-being. It includes case studies on several farms featured in the previous report and an additional set of case studies on farms selected by this committee. This report emphasizes the properties of complex agricultural systems and the interactions, synergies, and tradeoffs that affect the performance and robustness of agricultural practices and farming systems. Targeted public policy can encourage the development and implementation of farming systems designed to improve sustainability.

---

[1] NRC (National Research Council). 1989. Alternative Agriculture. Washington, D.C.: National Academy Press.

> **BOX S-1**
> **Statement of Task**
>
> The National Research Council Committee on Twenty-First Century Systems Agriculture was tasked to:
>
> 1. Provide an overview of the current state of U.S. agriculture in the domestic and world economies, and describe major challenges to farmers and problems in agricultural production related to the environmental, social, and economic sustainability of agriculture.
> 2. Review the state of knowledge on farming practices and management systems that can increase the environmental, social, and economic sustainability of agriculture.
> 3. Identify factors that influence the adoption of farming practices and systems that contribute to the goals of increasing agricultural sustainability.
> 4. Provide an update to the 1989 report's methodology to compare the productivity and economics of different systems and practices at levels of increasing complexity (from the level of individual components in a farm, to a whole farm, to a regional level).
> 5. Describe and analyze several case studies (including some from the 1989 report) that illustrate farming practices and management systems that pursue greater agricultural sustainability. Include general information about the operation, features of the management systems being used, and indicators of productivity, environmental, and financial performance. For case studies from the 1989 report, include a retrospective review of the past performance and the evolution of decision making by those producers over time.
> 6. Recommend research and development needs for advancing a systems approach to farming in the United States, and suggest ways to strengthen federal policies and programs related to improving agricultural production.
> 7. Evaluate the transferability of principles underlying farming systems and practices that could improve sustainability of different agricultural settings, and develop supportable conclusions and recommendations to improve the sustainability of agriculture under different natural, economic, and policy conditions in different regional or national settings.
>
> This study is supported by the Bill & Melinda Gates Foundation and the W.K. Kellogg Foundation.

DEFINING AGRICULTURAL SUSTAINABILITY

Sustainability has been described as the ability to meet core societal needs in a way that can be maintained indefinitely without significant negative effects. Accordingly, development of a sustainable agricultural production system requires defining the core societal needs from agriculture, a process that will require a collective vision of what the future characteristics of agriculture should be. Such vision is heavily contested and unresolved in the United States at present. Although developing a widely accepted vision is outside the scope of this study, the committee identified four generally agreed-upon goals (each of which has a set of specific objectives; see Chapter 1) that help define sustainable agriculture:

- Satisfy human food, feed, and fiber needs, and contribute to biofuel needs.
- Enhance environmental quality and the resource base.
- Sustain the economic viability of agriculture.
- Enhance the quality of life for farmers, farm workers, and society as a whole.

Sustainability is best evaluated not as a particular end state, but rather as a process that moves farming systems along a trajectory toward greater sustainability on each of

the four goals. As such, agricultural sustainability is a complex, dynamic, and political concept that is inherently subjective in that different groups in society place different emphasis on each of the four goals. Progress toward the four goals will require robust systems that adapt, evolve, and continue to function in the face of stresses and fluctuating conditions, are productive, use resources efficiently, and balance the four goals within enterprises or farms across and at all scales. The pursuit of sustainability is not a matter of defining sustainable or unsustainable agriculture, but rather is about assessing whether choices of farming practices and systems would lead to a more or less sustainable system as measured by the four goals. Improving sustainability will require identification of key metrics and indicators that can measure progress toward goals, together with monitoring and collecting data and using adaptive management.

The committee concluded that **if U.S. agricultural production is to meet the challenge of maintaining long-term adequacy of food, fiber, feed, and biofuels under scarce or declining resources and under challenges posed by climate change, and to minimize negative outcomes, agricultural production will have to substantially accelerate progress towards the four sustainability goals. Such acceleration needs to be undergirded by research and policy evolution that are designed to reduce tradeoffs and enhance synergies between the four goals and to manage risks and uncertainties associated with their pursuit.**

SCIENTIFIC FOUNDATION FOR IMPROVING SUSTAINABILITY

Science—including biophysical and social sciences—is essential for understanding agricultural sustainability. Science generates the knowledge needed to predict the likely outcomes of different management systems and expands the range of alternatives that can be considered by farmers, policy makers, and consumers.

Although all farms have the potential and responsibility to contribute to different aspects of sustainability, the scale, organization, enterprise diversity, and forms of market integration associated with different farms provide unique opportunities or barriers to improving their ability to contribute to the four goals. Therefore, **the committee proposes two parallel and overlapping approaches to ensure continuous improvement in the sustainability performance of U.S. agriculture: incremental and transformative. The incremental approach is an expansion and enhancement of many ongoing efforts that would be directed toward improving the sustainability performance of all farms, irrespective of size or farming systems type, through development and implementation of specific sustainability-focused practices, many of which are the focus of ongoing research and with varying levels of adoption.**

The transformative approach aims for major improvement in sustainability performance by approaching 21st century agriculture from a systems perspective that considers a multiplicity of interacting factors. It would involve:

- Developing collaborative efforts between disciplinary experts and civil society to construct a collective and integrated vision for a future of U.S. agriculture that balances and enhances the four sustainability goals.
- Encouraging and accelerating the development of new markets and legal frameworks that embody and pursue the collective vision of the sustainable future of U.S. agriculture.
- Pursuing research and extension that integrate multiple disciplines relevant to all four goals of agricultural sustainability.

- Identifying and researching the potential of new forms of production systems that represent a dramatic departure from (rather than incremental improvement of) the dominant systems of present-day American agriculture.
- Identifying and researching system characteristics that increase resilience and adaptability in the face of changing conditions.
- Adjusting the mix of farming system types and the practices used in them at the landscape level to address major regional problems such as water overdraft and environmental contamination.

INCREMENTAL APPROACHES TO IMPROVING SUSTAINABILITY OF U.S. AGRICULTURE

The proposed expanded incremental approach would include focused disciplinary research on production; environmental, economic, and social topics; and policies (such as expanded agricultural conservation and environmental programs) to improve the sustainability performance of mainstream agriculture. For example, large livestock farms in the United States produce the majority of the nation's meat and dairy products. Similarly,

---

BOX S-2
Examples of Practices That Contribute to Sustainability

Production Practices

- **Conservation (or reduced) tillage systems** have contributed to reducing water-caused soil erosion and surface runoff of nutrients, chemicals, and crop residues. Increased understanding of how conservation or reduced tillage can work with different crops and soil types to reduce energy use and labor has led to increased adoption of those practices.
- **Cover cropping** provides ground cover to protect soil and provide other services, including maintaining soil organic matter, providing nutrients to subsequent crops (green manures), trapping excess nutrients in the soil profile following harvest of the primary crop, and preventing leaching losses (catch crops). However, cover crops are not widely planted because of technical, economic, and environmental limitations.
- **Crop diversity, including rotations, intercropping, and using different genetic varieties** can contribute to improving soil quality and managing pests and diseases and is particularly important in the management of organic cropping systems. Although incorporation of diversity in cropping systems has increased in some regions, it fluctuates widely with commodity prices.
- **Traditional plant breeding and modern genetic engineering (GE) techniques** have resulted in crop varieties with increased yields, pest and disease resistance, enhanced water-use and nutrient-use efficiencies, and other important traits. GE has the potential to contribute a number of solutions for problems, but new varieties need to be tested rigorously and monitored carefully by objective third parties to ensure environmental, economic, and social acceptability and sustainability before they are released for planting.
- Many technologies for **efficient water use** such as metering, improved distribution of high-pressure water, and low-pressure, directed-use systems offer promise to address water scarcity. **Water reuse** is another strategy for addressing water scarcity, but the biological and chemical quality of the reclaimed water requires careful monitoring. **Best management practices (BMPs)**, including nutrient management planning, surface and subsurface drainage management, field buffer strips, riparian area management, and livestock manure management, have been developed to mitigate the runoff of agricultural nutrients and chemicals into the nation's surface and ground waters. Effectiveness of BMPs at the watershed scale depends, in part, on the coordinated actions of all farms in a watershed.
- **Soil and plant tissue tests, nutrient management plans, and precision agriculture technologies** help farmers to increase their farms' productivity, input-use efficiency, and economic returns by reducing unnecessary use of agricultural fertilizers, pesticides, or water. Experimental and field studies suggest that impacts and economic benefits of those practices and tools can be variable across time and space.

a large portion of corn and soybean are produced on highly mechanized grain farms that specialize in producing a small number of crops and rely heavily on purchased farm inputs to provide crop nutrients and to manage pest, disease, and weed problems. Most, if not all, farms have adopted some practices for improving sustainability. Some farms, including large farms illustrated in the report's case studies, are highly integrated, but such methods have not been adapted to all environments, and none of the practices has reached its full potential for adoption. Each of those production systems has fostered high productivity and low costs, but many have led to some serious negative social and environmental outcomes (or externalized production costs) that could hinder agriculture's progress toward improved sustainability. The negative outcomes have led to policy changes and publicly funded research programs explicitly designed to address those concerns. Efforts to improve sustainability of mainstream production system might be incremental in nature, but could have significant benefits given the dominance of those production systems in U.S. agriculture.

Since the publication of the report *Alternative Agriculture*, research has increased understanding of how different farming practices maintain or increase productivity while enhancing natural resources and addressing environmental concerns. Box S-2 lists examples

- **Use of manure, compost, and green manure**, as often used in organic systems, can reduce the need for synthetic fertilizer and hence reduce the energy used for fertilizer production. Many farms featured as case studies in this report successfully use **on-farm inputs to maintain soil fertility** and to insulate themselves from fluctuations in costs of synthetic fertilizer. Published studies, however, show variable results as to whether systems using commercial fertilizers or systems using cover crop-based or animal manure-based nutrient management have higher profits. Those studies often do not include environmental costs and benefits.
- **Integrated pest management** (IPM) research has identified promising options for improving soil suppressiveness and inducing crop resistance to some diseases and pests, in addition to classical biological and ecological pest management. The need to study weeds, diseases, pests, and crops as an interacting complex has been recognized. Adoption of IPM has been reasonable for some crops, but overall IPM use is lagging despite its potential for reducing chemical use.
- **Genetic improvement of livestock** can contribute to improving sustainability by increasing feed use efficiency and by selecting traits to improve animal health and welfare. Improvements in feed conversion through genetics, nutrition, and management have reduced manure and nutrient excretion per unit animal product produced and reduced land required for production.

Business and Marketing Strategies

- **Diversification of farm enterprises** can provide multiple income streams for farm operations. Producing a range of crops and animal products can enhance the stability and resilience of farm businesses and can decrease the volatility of farm income. However, studies that analyze the financial impacts of enterprise diversification under real-world conditions are sparse.
- An increasing number of farmers are raising their farm-level income by increasing the value of their products through **sales to niche markets** (such as organic or health-food markets) or by selling their products directly to consumers **(direct sales)** to obtain a larger proportion of the consumers' dollar spent on the product and to gain control over the prices they get for their products.

Practices for Improving Community Well-Being

- **Diverse farm systems**, **diversified landscapes** that include non-crop vegetation, and **farming practices that improve water quality** can contribute to community and social well-being. Some direct marketing strategies, such as **direct sales at farmers' markets, community-supported agriculture, farm-to-school programs, and agritourism** connect farmers to the community and can contribute to community economic security, but those sales strategies lack underpinning research and extension.

of farming practices that can contribute to sustainability goals. Performance and adoption of many of those practices could be further improved by additional biophysical, social, and economic research.

Research on the economic and social dimensions of agricultural sustainability complementary to the research on productivity and environmental sustainability is scarce despite its importance in providing farmers with knowledge to design systems that balance different sustainability goals and improve overall sustainability. Studies on economic and social sustainability are complicated by the fact that economic viability is influenced by market and policy conditions and that social acceptability of farms is influenced by the behavior of key actors (including farmers and consumers) and the values of community members. The lack of information on the economic viability of practices and approaches for improving environmental and social sustainability and on how markets and policies influence the economics of those practices could be a barrier to their wide adoption. Nonetheless, the case studies of 15 operating farms in this report illustrate the feasibility of the farming practices and approaches and how they are combined with business and marketing to result in farming systems that balance the four sustainability goals.

Continuous research, extension, and experimentation by researchers and farmers are necessary to provide the toolkit necessary for farmers to adapt their systems to changing environmental, social, market, and policy conditions to ensure long-term sustainability. Examples of high-priority areas of research are listed in Box S-3. Because research to develop practices and approaches for improving environmental sustainability and to qualify or quantify their economic and social impacts does not result in a marketable product for industry, it is generally not attractive for private-sector investment. Therefore, such research would have to rely on public funding and institutions, farmer organizations, and civil society sectors.

> RECOMMENDATION: The U.S. Department of Agriculture and state agricultural institutions and agencies should continue publicly funded research and development (R&D) of key farming practices for improving sustainability to ensure that R&D keeps pace with the needs and challenges of modern agriculture. They should increase support for research that clarifies the economic and social aspects of the many current and potential technologies and management practices and that addresses issues of resilience and vulnerability in biophysical and socioeconomic terms.

TRANSFORMATIVE APPROACH TO IMPROVING SUSTAINABILITY OF U.S. AGRICULTURE

The transformative approach to improving agricultural sustainability would dramatically increase integrative research by bringing together multiple disciplines to address key dimensions of sustainability simultaneously beyond the agroecological dimension. It would apply a systems approach to agriculture that could result in production systems and an agricultural landscape that are a significant departure from the dominant systems of present-day agriculture. This approach would facilitate development of production approaches that capitalize on synergies, efficiencies, and resilience characteristics associated with complex ecosystems and their linked social, economic, and biophysical systems. It would emphasize integrating information about productivity, environmental, economic, and social aspects of farming systems to understand their interactions and address issues of resilience and vulnerability to changing climate and economic conditions. Moreover, integration would include expanded attention to the role and development of new markets,

> **BOX S-3**
> **Examples of High-Priority Research in an Incremental Approach to Improving Sustainability**
>
> **Productivity and Environmental Research**
> - Assessment of the effectiveness of cover crops in providing ecosystem services such as biological control of agricultural pests, weed suppression, and nutrient and water retention.
> - Assessment of water reuse systems, surface and subsurface drainage systems, and advanced livestock waste management systems that improve the effectiveness of wetlands, enhance water quality and water conservation, and reduce greenhouse-gas emissions.
> - Comparative study of greenhouse-gas emissions and nutrient balances associated with different field management practices for animal wastes and other organic amendments such as green manures and organic mulches and composts.
> - Research and development of non-chemical alternatives (for example, biological control, biofumigation, induced resistance, and soil suppressiveness) for managing weeds, pests, and diseases as a complex.
> - Research that identifies ecosystem benefits from changing agricultural practices, such as planting buffer strips or hedgerows, reducing tillage, and using best management practices, at multiple scales.
>
> **Socioeconomic Research**
> - Assessment of how production practices might affect food attributes (such as pesticide residue, taste, nutritional quality, and food safety).
> - Assessment and comparison of the costs of different production practices and combinations of practices under different policy and market contexts.
> - Research to document and analyze the economic sustainability of direct marketing—for example, to review financial and labor returns to such marketing strategies as sales at farmers' markets, community-supported agriculture, and farm-to-school programs.
> - Research to document and analyze labor benefits, practices, and their trends in agriculture and their effects on farm profitability.
>
> **Policy Research**
> - Research to improve understanding of the intended and unintended consequences of federal farm, food, and environmental policies that can affect the use of agricultural practices for improving sustainability.

new policies and new approaches to research and development that are likely to sustain a systems-oriented agriculture. Options include development of appropriate price signals or incentives to farmers who seek to improve the sustainability of their farms across all four dimensions of sustainability and policies that are less likely to produce unintended consequences in one area of sustainability while addressing another area. Attention to production system types different from the dominant types, such as integrated crop and livestock systems, non-confinement livestock production systems, or highly diversified cropping systems that reduce reliance on purchased inputs, is desirable because the dominant system types might limit the range of technical or managerial possibilities in the pursuit of greater sustainability. Examples suggested in the report include organic cropping systems, low-confinement livestock systems, management intensive rotational grazing, enhanced local food systems, developing perennial grain systems, regional planning and implementation of farming system changes that reduce water overdraft and environmental loading leading to larger hypoxic zones, and planning for production of new products such as cellulosic biofuels.

Interdisciplinary Systems Research

Most agricultural research in the United States is conducted to address specific problems. Although a disciplinary approach to agricultural research is necessary to build a foundation of scientific knowledge about how each component of a farming system works, it often undervalues the importance of interconnections and functional relationships between different components of the farming system. A systems approach is necessary to identify and understand the significance of linkages between farming components and other aspects of the environment and economy so that a robust system that takes advantage of synergies and balances tradeoffs can be designed.

With some farming system types, the combination of adverse environmental loading, human health and social welfare issues, impacts on local economies, food safety issues, or animal welfare concerns might give impetus to dramatic departures from mainstream agricultural practices. Organic cropping, integrated livestock and crop systems, management-intensive rotational grazing, and low-confinement hog systems are used as examples of potential departures from mainstream agriculture in this report. Those systems illustrate how complementarities and synergies of resource use and containment at the systems level, if managed well, can generate positive outcomes on ecological and social environments. Organic cropping systems integrate many practices for improving sustainability listed earlier. The approach is driven by a philosophy for using biological processes to achieve high soil quality, control pests, and provide favorable growing environments for productive crops, and by the prohibition of most synthetically produced inputs. The alternative livestock production systems were designed to enhance environmental quality, animal welfare, and social acceptability. However, few systematic research studies assess the ability of confined animal systems and other alternatives to address public concerns of production efficiency, food safety, environmental impacts or risks, animal welfare, and labor conditions. Although the farming types that the committee uses as illustrations of systems approaches represent a small proportion of farms in the United States as of 2010, they serve as valuable demonstrations and provide data on how sustainability could be improved.

Ultimately, it will be more effective to structure farms and agricultural systems toward balancing the four sustainability goals at the outset rather than to address unintended consequences through piecemeal "technological fixes." **To pursue systemic changes in farming systems, research and development have to address multiple dimensions of sustainability (productivity, and environmental, economic, and social sustainability) and to explore agroecosystem properties, such as complex cropping rotations, integrated crop and livestock production, and enhanced reliance on ecological processes to manage pests, weeds, and diseases (recognizing their interconnectedness and interactions with the environment), that could make systems robust and resilient over time.**

Despite the need for research to balance and further enhance the four sustainability goals of agriculture, the majority of public research funding is targeted to improving productivity and reducing production costs. Only one-third of public research spending is devoted to exploring environmental, natural resource, social, and economic aspects of farming practices. **The report *Alternative Agriculture* emphasized the importance of a systems approach to agricultural research 20 years ago, yet the proportion of long-term systems agricultural research remains small.**

Examples of transformative systems type studies include:

- Holistic comparison of existing organic, conventional, and innovative farming systems in different environments to assess how each system performs and balances overall system efficiencies and resilience with environmental and social impacts.

- Holistic comparison of the ability of confined animal systems and other alternatives to address production efficiency, food safety, environmental impacts or risks, animal welfare, and labor conditions.
- Policies and legal frameworks that provide appropriate pricing and incentives to encourage the balancing of the four sustainability objectives and enhance system resilience and adaptability under dynamic conditions.

RECOMMENDATION: Federal and state agricultural research and development programs should aggressively fund and pursue integrated research and extension on farming systems that focus on interactions among productivity, environmental, economic, and social sustainability outcomes. Research should explore the properties of agroecosystems and the interdependencies between biophysical and socioeconomic aspects of farming systems, and how they could make the systems robust and resilient over time.

Application of a systems approach to agriculture is not limited to the farm level. The collective and potentially synergistic effects of agricultural systems at a landscape or community scale have gained recognition. However, the scientific foundation for and data needed to develop a landscape approach for improving sustainability of agriculture is sparse. Research suggests that the distribution of farm types and farming activities across a landscape could be designed to achieve greater productivity, resistance, and resilience and improve the sustainability of local and regional agricultural systems that support personal and community well-being. In addition, effective public policy tools that are politically viable and effective in shaping patterns of the agricultural practices or land use at the landscape level are needed. No single agricultural landscape pattern is likely to work in every location; effective landscape patterns would have to be tailored to local conditions and meet particular community needs. **Although a landscape approach to agricultural research could inform the design of agroecosystems to maximize synergies, enhance resilience, and inform what policies would be useful in influencing collective actions, programs to encourage such research do not exist.**

Examples of transformative landscape-scale research include:

- Develop systems type mixes, patterns, and technologies for landscape diversity that maintain economic output while reducing overall water use.
- Develop systems type mixes and technologies to reduce nitrogen, phosphorus, and pesticide losses to downstream fragile water bodies, particularly in source regions responsible for hypoxia.
- Develop tools for modeling of systems and patterns for multipurpose economic, aesthetic, and environmental impacts to enhance community well-being and assist in planning, local policy, market identification, and farmer decision making.
- Develop policies and legal frameworks that encourage cooperative watershed landscape and ground water management across field and farm boundaries.
- Generate landscape design options to increase resilience and adaptability to changing conditions using a combination of the above approaches.

RECOMMENDATION: The U.S. Department of Agriculture should partner with the National Science Foundation, the U.S. Environmental Protection Agency, key land-grant universities, and farmer-led sustainable agricultural organizations to develop a long-term research and extension initiative that aims to understand

the aggregate effects of farming at a landscape or watershed scale and to devise, encourage, and support the development of collective institutions that could enhance environmental quality while simultaneously sustaining economic viability and community well-being.

Returns on research investments could be increased by incorporating farmer knowledge. Much of the technical and managerial innovation in sustainable agriculture has occurred through on-farm innovation and experimentation. Engaging farmers as partners with scientists in innovation, development, extension, and outreach processes could produce effective and long-lasting technology adaptation and adoption. In addition, farmers' networks and farmer-to-farmer mentoring programs can help spread knowledge gained from research and help adapt such knowledge to farmers' local conditions.

RECOMMENDATION: The U.S. Department of Agriculture and other federal and state agencies that support agricultural research should encourage researchers to include farmer-participatory research or farmer-managed trials as a component of their research. Those agencies should strengthen initiatives for participatory education and peer-to-peer partnerships that could enhance information exchange and enhance farmers' adoption of new practices and approaches for improving sustainability of agriculture.

Efforts to engage farmers and citizens in research and outreach to improve agricultural sustainability will require institutional support. Cooperative Extension programs at the state and regional levels can play a critical role as facilitators and catalysts for fostering interaction among the various stakeholders and for providing educational programs and access to current information.

Key Drivers of Change: Markets and Federal and Local Policies

Other than available science, knowledge, and skills, the decisions of farmers to use particular farming practices and their ability to move toward more sustainable farming systems are influenced by many external forces such as markets, public policies, and their own values, resources, and land-tenure arrangements. Those structural constraints are in turn influenced by the efforts of broad social movements and organized interest groups that have different perspectives about how agriculture should be organized and how food should be produced and distributed.

Growing interests by consumers in food produced using practices perceived as "environmentally friendly," or that address a particular social concern (such as animal welfare) have led to development of value-trait markets. Similarly, sustainability initiatives in large food retailers open up new markets for food products that are produced using certain practices or farm system types that improve sustainability. Those emerging markets can motivate farmers to transition to farming systems that balance and meet multiple sustainability goals. The use of marketing tools, such as certification and branding of products produced using particular farming practices and systems that increase sustainability, can enhance the value of those farm products and contribute to environmental, social, and economic sustainability of the farm.

The impact of public policies aimed at moving agriculture along the sustainability trajectory has been mixed. Some scholars attribute a decrease in the diversity of cropping systems, increases in the use of external farm inputs, and extensive hydrologic modification of landscapes in part to commodity support payments because these supports provide a strong incentive for farmers to focus on planting program crops by monocropping and to maximize yields per dollar of cost (that is, to focus on only two of the four sustainability goals). Risk management policies can affect sustainability initiatives because some crop insurance products carry substantial subsidized premium structures that can potentially encourage farmers to grow monocrops, which could increase the vulnerability of highly erodible soils and reduce system resilience. Conservation programs are a mechanism for encouraging adoption of particular farming practices, but they are voluntary programs, often with a small proportion of farms participating. Public programs designed to increase incentives to move toward greater sustainability are increasing, but they remain a small portion of the federal and state agricultural policy portfolio.

Although market, policy, and institutional contexts are important drivers of the trajectory of U.S. agriculture, the response of individual farmers to the incentives and disincentives created by market conditions and policy contexts can be diverse. Efforts to promote widespread adoption of different farming practices and systems for improving sustainability will require an understanding of how variability among individual, household, farm, and regional-level characteristics affect farmers' response to incentives and disincentives. The scientific research to date is inadequate to assess the full impacts of current and proposed policy frameworks.

> **RECOMMENDATION: Because of the critical importance of macro-structural or institutional drivers of farmer behavior, the U.S. Department of Agriculture should increase investment in empirical studies of the ways that current and proposed market structures, policies, and knowledge institutions provide opportunities or barriers to expanding the use of farming practices and systems that improve various sustainability goals so that the department can implement changes in policies and institutions that are identified as effective to meeting those goals.**

Transformation of the agriculture sector will not occur overnight. It will take long-term research and experimentation by the public and private sectors in partnership with farmers. The two parallel approaches to improving sustainability proposed by the committee would ensure incremental improvement toward sustainability, while long-term systemic changes in agricultural systems are being pursued.

RELEVANCE OF LESSONS LEARNED TO SUB-SAHARAN AFRICA

When considering the relevance of lessons learned in the United States to sub-Saharan Africa, it is important to recognize key differences between the two regions. African farmers produce a wide variety of crops using diverse farming systems across a range of agroecological zones. Most systems are rain-fed, and many soils are severely depleted of nutrients. External inputs are expensive. High transportation costs and lack of infrastructure often inhibit access to outside resources and markets. Specific management approaches need to be developed in that context. Nonetheless, the concepts of sustainability and many of the broad approaches presented in this report are relevant and concur with conclusions from some recent international reports. The committee concluded that:

- An interdisciplinary systems approach is essential to address the improvement and sustainability of African agriculture that recognizes the social, economic, and

policy contexts within which farming systems operate. Evolving systems would need to address all four sustainability goals and be adapted to local conditions.
- Research programs need to actively seek input and collaboration from farmers to ensure research being conducted and technologies tested are relevant to their needs.
- Women, who play a pivotal role in African agriculture, need to be provided with educational and training opportunities and be involved in the development of research agendas.
- Technologies are needed to address soil, water, and biotic constraints, but they have to be integrated with local ecological and socioeconomic processes. Use of locally available resources would have to be maximized and combined with judicious use of external inputs when necessary.
- Promising technologies and approaches include soil organic matter management, reduced tillage, integrated fertility management, water harvesting, drip irrigation, stress-resistant crop varieties, improved animal breeds, integration of crops and livestock, and use of global information systems for landscape and regional analysis and planning.
- Expanding market access will be essential to increase productivity and enhance livelihoods in rural Africa. Investing in rural infrastructure could improve access to local, regional, and international markets.

RECOMMENDATION: Agencies and charitable foundations that support research and development of sustainable agriculture in developing countries should ensure that funded programs emphasize a systems approach that reflects the need for adaptability of management strategies and technologies to dynamic local socioeconomic and biophysical conditions, and supports efforts to increase market access.

1

Understanding Agricultural Sustainability

Agriculture worldwide faces daunting challenges because of increasing population growth and changing food consumption patterns, natural resource scarcity, environmental degradation, climate change, and global economic restructuring. Yet, at the same time, there are unprecedented hopeful changes and opportunities for the future, including a remarkable emergence of innovations in farming practices and systems and technological advances that have generated promising results for improving agricultural sustainability and an increase in consciousness and concern by consumers about the sources of their food and how it is produced. One of the first comprehensive reports on the scope and importance of systems approaches to improve sustainability of agriculture was documented in the National Research Council report *Alternative Agriculture* (1989b). The report analyzed and described the economic and environmental results of agricultural practices that could improve sustainability being developed and used by a small subset of U.S. farmers in the second half of the 20th century, and it helped to legitimize an approach to agricultural-systems research that had previously been considered nonscientific. Many so-called alternative practices at that time (integrated pest management, no-till farming, and cover crop planting) are now used by some farmers in mainstream agriculture.

Despite the potential benefits of farming practices and systems that improve sustainability, their adoption is far less widespread than society might want. One reason for the low rate of adoption is because of social, economic, and policy incentives that discourage fundamental changes in farming systems. Another reason is that some of those practices have tradeoffs so that they might provide benefits in one aspect and negative consequences in another. The movement toward improved sustainability could be hampered by society's lack of common agreement on which objectives are the highest priority and how tradeoffs should be managed.

Since *Alternative Agriculture* was published, many changes have been made in how farmers farm. While incremental approaches offer movement toward sustainability, such approaches might not tackle fundamental problems. Systemic changes and multiple paths will also need to be pursued. Indeed it can be argued that sustainability is the process of constantly adapting farming or food systems to meet clearly articulated and desired outcomes in a robust and resource-efficient manner that also reflects social responsibility.

PURPOSE OF THIS REPORT

With the support of the Bill and Melinda Gates Foundation and the W.K. Kellogg Foundation, the National Research Council convened a committee to study the science and policies that influence the adoption of farming practices and management systems designed to reduce the costs and unintended consequences of agricultural production. (See Box 1-1 for the statement of task.) To address the statement of task, the committee solicited input from many experts in academia and federal agencies in a series of open meetings and workshops, in addition to drawing on members' expertise. Two sets of case studies were used to examine farming systems that address those concerns and to explore the factors that affect their implementation, economic viability, and success in meeting environmental and other goals of sustainability.

This report reviews the state of knowledge on farming practices, technologies, and management systems that have the potential to improve the environmental, social, and economic sustainability of agriculture and discusses the tradeoffs and risks that might present themselves if more farms were to adopt those practices, technologies, and systems. The report also identifies knowledge gaps in improving agricultural sustainability and makes recommendations for future actions aimed at improving agricultural sustainability.

BOX 1-1
Statement of Task

The National Research Council Committee on Twenty-First Century Systems Agriculture was tasked to:

1. Provide an overview of the current state of U.S. agriculture in the domestic and world economies, and describe major challenges to farmers and problems in agricultural production related to the environmental, social, and economic sustainability of agriculture.
2. Review the state of knowledge on farming practices and management systems that can increase the environmental, social, and economic sustainability of agriculture.
3. Identify factors that influence the adoption of farming practices and systems that contribute to the goals of increasing agricultural sustainability.
4. Provide an update to the 1989 report's methodology to compare the productivity and economics of different systems and practices at levels of increasing complexity (from the level of individual components in a farm, to a whole farm, to a regional level).
5. Describe and analyze several case studies (including some from the 1989 report) that illustrate farming practices and management systems that pursue greater agricultural sustainability. Include general information about the operation, features of the management systems being used, and indicators of productivity, environmental, and financial performance. For case studies from the 1989 report, include a retrospective review of the past performance and the evolution of decision making by those producers over time.
6. Recommend research and development needs for advancing a systems approach to farming in the United States, and suggest ways to strengthen federal policies and programs related to improving agricultural production.
7. Evaluate the transferability of principles underlying farming systems and practices that could improve sustainability of different agricultural settings, and develop supportable conclusions and recommendations to improve the sustainability of agriculture under different natural, economic, and policy conditions in different regional or national settings.

Sustainability in agriculture is a complex and dynamic concept that includes a wide range of environmental, resource-based, economic, and social issues. The committee's definition of sustainable farming does not accept a sharp dichotomy between unsustainable or sustainable farming systems because all types of farming can potentially contribute to achieving different sustainability goals and objectives. Ultimately, the committee believes that sustainability is best evaluated against a range of environmental, economic, and social goals that reflect the views of diverse groups in society. The most intense controversies about the relative sustainability of different farming practices or farming systems necessarily take place within the domain of politics because preferences for particular farming systems reflect the priorities of various stakeholders with different working definitions of sustainability.

Although a final assessment of the sustainability of any particular farming practice or system is a social and political act, the committee believes that public debates about improving the sustainability of U.S. agriculture need to be based within a good understanding of the existing scientific research. Science documents the performance and impacts of different agricultural practices and systems, predicts outcomes likely to result from the use of different systems, develops indicators to measure progress toward sustainability goals, and expands the range of technological tools and farming management approaches available. The issue of water quality illustrates the critical role science plays. Societal concerns about water quality led to passage of the Clean Water Act, which has resulted in various local, state, and federal guidelines that require use of "best management practices." Guidelines are provided for acceptable soil phosphorus upper limits, soil erosion rates, water conservation, tillage practices, pesticide use, and a wide array of other process-type practices, all of which were determined through extensive scientific research.

This chapter first defines key terms used in later chapters and then discusses concepts of agricultural sustainability. It identifies the boundaries, or scope, for the overall science assessment of sustainability. Ultimately, this report focuses on current scientific evidence about the performance of farming practices and systems that can contribute to moving U.S. farming systems along a trajectory toward meeting various sustainability goals. Indicators that can be used to provide quantitative assessment of progress toward sustainability are also discussed. The structure of the report is outlined at the end of this chapter.

FARMING AND AGRICULTURE

General Definitions

The terms "farming," "agriculture," and "farming systems" can comprise a diverse range of activities. The definitions of agriculture used in the National Research Council (NRC) report *Investing in Research: A Proposal to Strengthen the Agricultural, Food, and Environmental System* (NRC, 1989a) provide some useful definitions and underscore the potential breadth of farming-related activities and goals (Box 1-2).

Various crop and livestock enterprises on individual farms can interact in complex ways with one another and with their surrounding ecosystems (USDA-CSREES, 2007). Combinations of different activities can generate properties of system behavior that might not be understood or predicted by looking at each practice individually. That behavior is manifested at various scales ranging from the field, whole farm, landscape, watershed, and region. Moreover, every farm is embedded in a particular biophysical environment that provides opportunities and constraints for using different practices or management strategies, which shapes the impacts of farming activities on environmental, economic,

> **BOX 1-2**
> **General Definitions**
>
> **Agriculture** encompasses the entirety of the system that grows, processes, and provides food, feed, fiber, ornamentals, and biofuel for the nation. Agriculture includes the management of natural resources such as surface water and ground water, forests and other lands for commercial or recreational uses, and wildlife; the social, physical, and biological environments; and the public policy issues that relate to the overall system. All activities, practices, and processes of the public and private sectors involved in agriculture and forestry are contained within the system. (Adapted from **Investing in Research: A Proposal to Strengthen the Agricultural, Food, and Environmental System** [NRC, 1989a]).
>
> A **farm** is most correctly used to mean a single, identifiable operational unit that manages natural resources such as water, forests, and other lands to provide food, feed, fiber, ornamentals, energy, and a range of environmental and other services. "Any operation that sells at least one thousand dollars of agricultural commodities or that would have sold that amount of produce under normal circumstances" is considered a farm by the U.S. Department of Agriculture (USDA) (USDA-ERS, 2008). Every farm is embedded within a temporal and spatially dynamic context (environment) and interacts with the geophysical, biological, economic, and social variables of that environment. Farms employ a wide range of production techniques and strategies known as "**farming practices**." Farms also use marketing techniques and strategies.
>
> A **farming system** is the mix of crops or animal components, or some combination thereof in a farm, their arrangement over space and time within the farm, the resources and technologies used in their management, and the nature and effectiveness of hierarchical relationships both within the farm and with the ecological, social, economic, and political environments within which it operates. The farming system thus includes community linkages, market integration, labor relationships, and interaction with a wide array of other influencing factors.
>
> A **farming system type** can be defined by any commonalities that one might wish to specify—for example, rangeland, dryland, irrigated, dairy, field crop, high-value crop—and includes the broad range of context-specific farming systems within the type.
>
> **Food systems** and **agrifood systems** refer to the complex set of actors, activities, and institutions that link food production to food consumption. Studies of agrifood systems often use a "commodity chain" approach, where they examine the production, processing, trade, wholesaling, retailing, and consumption of particular commodities, as well as the upstream and downstream processes that connect the various links in the chain. These terms differ from farming system in that the primary focus is beyond the farm gate.
>
> **Systems agriculture** is used to define an approach to agricultural research, technology development, or extension that views agriculture and its component farming systems in a holistic way. The approach treats components and processes within and across hierarchical levels and scale with appropriate context, and gives major importance to interactions among them. The USDA Cooperative State Research, Extension, and Education Service (CSREES) defines its agricultural systems approaches as follows: "Agricultural enterprises—crop or livestock—deal with such concepts as labor supply, marketing, finances, natural resources, genetic stock, nutrition, equipment and hazards. Although it is possible to effectively manipulate each mechanism of successful farming individually, better results can often be obtained by treating the farming operation as a system. The interactions, then, among system components might be more important than how each component functions by itself. Treating production operations holistically offers greater management flexibility, provides for more environmentally and economically sound practices, and creates safer and healthier conditions for workers and for farm animals." SOURCE: USDA-CSREES (2007).

and social outcomes. This committee also recognizes that farms operate within a complex market, policy, and community context, and interactions with those institutions and social systems affect the performance of every type of agricultural system.

Because of the complex interactions at various scales, this committee has chosen to cast a relatively wide net in its assessment of the performance of different farming practices and systems. The ensuing chapters discuss the impact of individual farming practices not only as individual components of a particular farm enterprise, but also in the context of whole farming systems and as pieces of a larger farming landscape at the watershed, regional, or national scale. Although this report discusses how farmers might be affected by opportunities and constraints created by the larger agrifood system (see Box 1-2), a detailed assessment of the overall social, economic, and environmental performance of the food processing, distribution, and retailing sectors is beyond the scope of this report.

Farming Practices and Systems

At the time that the report *Alternative Agriculture* (NRC, 1989b) was released, the term "alternative agriculture" was commonly used to refer to farming approaches (most notably, organic farming) that appeared to be dramatically different from the dominant or "conventional" farming systems that characterized contemporary crop and livestock production in the United States. Today such alternatives are more likely to be referred to as "sustainable" farming, but the phrase is unfortunately ambiguous. Not only do farming enterprises reflect many combinations of farming practices, organization forms, and management strategies, but also all types of systems can potentially contribute to achieving various sustainability goals (for individual farmers and for resource use and environmental sustainability at the landscape scale).

At some level, the distinctions between what some call "conventional" agricultural systems and a range of alternative systems have some basis in empirical reality in the U.S. context. The characteristics and examples of practices associated with each system are summarized briefly in the following sections to clearly define those terms and how they are used in this report.

Conventional and Industrial Agricultural Systems

Conventional agricultural systems draw from a set of predominant farming practices in the United States, although conventional farms are diverse in what they produce and in the specific combination of practices they use. The size of farms using conventional production methods ranges from small to large. Most agricultural commodities in the United States are produced in conventional agricultural systems.

- *Conventional crop* production makes use of synthetic pesticides and herbicides, and supplements nutrients generated on the farm (manure) with synthetic fertilizer to maintain soil fertility. Fields are more frequently planted in few rotations of marketable crops than left fallow or planted with cover crops. Conventional corn, soybean, and cotton farms are increasingly planted with seeds that are genetically engineered to facilitate weed control or to reduce pest losses (and pesticide use).
- *Conventional animal* production varies depending on the species produced, the size of livestock inventories, and the amount of cropland or pasture available per animal unit. Animals might be housed in partial to full confinement structures, with beef cattle, sheep, and goats being less confined (for at least part of their life cycle), dairy cows being more confined than sheep and beef cattle, and conventional hogs and poultry most confined among livestock. Although beef cattle, sheep, and goat farms rely heavily on pastures, most conventionally raised dairy cows, hogs, and poultry receive virtually all their feed from harvested forage and grain crops raised on the farm or purchased from other farmers. Depending on the number of

animals and the resources of the producers, vaccines, antibiotics, medicated feeds, and growth hormones could be used in production. Animal manure is spread on fields or managed in lagoons, with excess liquid sprayed on fields as in the case of some hog farms.

- *Industrial crop and animal* production is a term that has come to be associated with operations that, generally speaking, are characterized by large size combined in some cases with a high degree of specialization. Producers of industrial scale are more likely to produce under contract with food processors and handlers. Animals are more likely to be grown in confined housing, with no pasture in the case of swine and poultry, or with portions of the animal life cycle on pasture or low confinement for dairy and beef animals. Feed is more likely to be a purchased input, rather than produced on the farm, than operations of smaller scale. Many of the larger operations likely have cropped land for liquid manure application, but loading limits over time rarely allow long-term stability for adequate disposal or low-rate application for efficient use of water and nutrients. Contract arrangements often are used for distant application of manure. Industrial farms typically operate at a scale that allows for more extensive division of labor and the use of capital intensive machinery and buildings. They more often rely on a hired workforce than do their smaller-scale counterparts.

Over the last few decades there has been a major effort to develop new management approaches and farming practices that not only improve the economic performance and productivity of conventional and industrial farming systems, but also prevent and mitigate their potentially negative effects on soil erosion and water quality, some of which also can improve the economic performance and productivity of conventional farming. Examples of practices used in these strategies are summarized in Box 1-3 and will be discussed in depth in Chapter 3.

Ecologically Based Farming Systems

Some farming approaches have been developed, at least in part, to respond to perceived problems associated with conventional farming. They represent a concerted departure from some of the key features of conventional farming such as the reliance on off-farm and synthetic inputs. Such approaches emphasize the use of natural processes within the farming system, often called "ecological" or "ecosystem" strategies, which build efficiency (and ideally resilience) through complementarities and synergies within the field, the farm, and at larger scales across the landscape and community. Examples of such systems include organic and biodynamic farming (Box 1-4), although "pure" forms of each system are difficult to identify. Like conventional farms, the specific practices used could vary widely from farm to farm, and often include combinations of the practices discussed in Box 1-3.

A Farming Systems Continuum

A dichotomy between "ecologically based" and "conventional" farming systems has been a component of public debates over the performance of the U.S. farming sector for several decades. The dichotomy has been a heuristic device that scientists use to explore the comparative performance of different farming systems. However, in reality, these terms are to be used with great caution because farming enterprises found in the United States clearly reflect a potentially infinite set of combinations of particular farming practices, organizational forms, and management strategies. (See the case studies in Chapter 7.)

BOX 1-3
Examples of Practices Designed to Improve Environmental Performance of Conventional Agriculture

- **Crop rotation**, which involves the successive planting of different crops on the same lands in sequential seasons to improve soil fertility and to avoid the build up of pathogens and pests that often occur when one species is continuously cropped. The most common rotations involve alternating production of corn and soybean.
- **Cover crops** as part of a crop rotation, which involves the planting of crop varieties that can potentially protect fields from soil erosion, suppress weeds, and enhance soil organic matter and nutrient levels.
- **Reduced-tillage** and **no-till practices**, in which a crop is planted directly into a seedbed not tilled since harvest of the previous crop. Instead of plowing soil and burying crop residues, no-till farmers minimize soil disturbance and leave residues on the surface of their fields after harvest.
- **Integrated pest management** (IPM), which involves the strategic use of complementary practices—including cultural, mechanical, biological, ecological, and chemical control methods—to keep pest levels below critical economic thresholds.
- **Precision farming** practices, which combine detailed spatial information about soil conditions and indicators of crop performance to target fertilization and other crop management practices where they are most needed.
- **Diversification of farm enterprises**, which helps increase biodiversity, control pests and diseases, and reduce risks from climatic and market volatility.
- **Other agricultural conservation best management practices (BMPs)**, which are recommended as part of federally funded conservation programs. BMPs include the use of **buffer or filter strips**, **riparian area access management**, **manure handling and management**, **nutrient management** planning, **wildlife habitat enhancement** within agricultural landscapes, **composting** to process agricultural wastes, and practices designed to increase **irrigation water use efficiency** (USDA-NRCS, 2009).
- The development of crops and animals that have **enhanced genetic resistance** to climatic extremes, pests, and other threats, often with the use of new genetic engineering tools.

BOX 1-4
Examples of Ecologically Based Farming Systems

Organic farming systems emphasize the use of renewable resources and the conservation of soil and water to enhance environmental quality for future generations. They typically rely on crop rotations, green manures, composts, naturally derived fertilizers and pesticides, biological pest controls, mechanical cultivation, and modern technology. Organic meat, poultry, eggs, and dairy products come from animals that are not given any antibiotics or growth hormones. Organic food is produced without the use of most conventional pesticides, fertilizers made with synthetic ingredients or sewage sludge, bioengineering, or ionizing radiation. Before a product can be labeled "organic" in the United States, a government-approved certifier inspects the farm where the food is grown to make sure the farmer is following all the rules necessary to meet USDA organic standards.

Biodynamic farming systems typically use the full range of organic production practices, but also use a series of eight soil, crop, and compost amendments, called preparations, made from cow manure, silica, and various plant substances. Biodynamic farming also places greater emphasis on (1) the integration of animals to create a closed nutrient cycle, (2) using an astronomical calendar to determine auspicious planting, cultivating, and harvesting times, and (3) an awareness of spiritual forces in nature. Biodynamic farmers view the soil and the whole farm as an integrated, living organism and self-contained individuality. More than a production system, biodynamic agriculture is a practice of living and relating to nature in a way that focuses on the health of the bioregion, landscape, soil, and animal, plant, and human life, and it promotes the inner development of each practitioner. The Demeter Association has certification programs for food and feed produced by strict biodynamic farming methods in different countries.

Moreover, the definitions that once clearly demarcated conventional from ecologically based farming systems have become muddled. One example is the "conventionalization" of the organic farming industry in the United States (Guthman, 2004), characterized by the entry of large-scale farms in the market for USDA-certified organic products. Most farms present examples of hybrid or intermediate stages on a continuum between the extremes of agricultural practices, and their adoption of various new practices has produced apparent gains in environmental performance (Keystone Center, 2009). Some of the "mixed farming systems" also have been given names (Box 1-5).

The committee concludes that no simple typology or set of categories can capture the complexity of the farming practices and systems used on diverse U.S. farms. The lack of a single accepted typology complicated the writing of this report. Because so much of the research literature is based on comparisons of particular farming practices, or of one or more of those stylized "farming systems," research findings are cited throughout the report using the categories described by the scientists who conducted the research. For this reason, this report cites organic farming systems more frequently than other ecologically based systems. The illustrative use of organic systems is not intended to imply that organic

---

BOX 1-5
Examples of Mixed Farming Systems

Conservation agriculture is a term used by the United Nation's Food and Agricultural Organization (FAO) to refer to the use of resource-conserving but high-output agricultural systems. According to the FAO, conservation farming typically involves the integrated use of minimal tillage systems, cover crops, and crop rotations (http://www.fao.org/ag/ca/).

Reduced- or low-input farming is based on a reduction of materials imported from outside the farm, such as commercially purchased chemicals and fuels. Low-input farming employs technologies and is structured in such a way that tightens flow loops and provides ecosystem services internal to the farm and field, and therefore reduces input use. Such internal resources include biological pest controls, solar or wind energy, biologically fixed nitrogen, and other nutrients released from green manures, organic matter, or soil reserves. Whenever possible, external resources are replaced by resources found on or near the farm. Many reduced-input or low-input farming systems are examples of integrated farming systems (see below).

Integrated farming system is a term commonly used in Europe to describe widely adopted production systems that combine methods of conventional and organic production systems in an attempt to balance environmental quality and economic profit. For example, integrated farmers build their soils with composts and green manure crops but also use some synthetic fertilizers; they use some synthetic or natural pesticides in addition to biological, cultural, and mechanical pest control practices.

Alternative livestock production systems refer to farms that use lower-confinement housing and rely more on pastures than conventional and industrial livestock farms. A common example in dairy farming is the use of **intensive rotational grazing practices** that involve the use of short duration, intensive grazing episodes followed by long rest periods that allow pastures or fields to recover.

Mixed crop-livestock farming is characterized by livestock enterprises where a significant fraction of the animal feed inputs are generated on cropland and pastures that are under the direct control of the livestock farmer. Those systems capitalize on the ability of the enterprise to use synergies between the crop and livestock enterprises to efficiently recycle nutrients, promote crop rotations, and insulate livestock farmers from price fluctuations in feed and input markets. They reflect the resurgence of traditional mixed crop-livestock farming systems that characterized most production units in the first half of the 20th century. On the other hand, the scale and sophistication of many 21st century mixed crop-livestock farms reflect the effects of new technologies, breed improvements, and greater awareness of environmental issues than their predecessors.

farming systems are more sustainable than other farming systems. Indeed, all farming system types have opportunities to improve in sustainability, and all farming system types could be unsustainable depending on their management and on environmental, social, and economic changes over time.

AGRICULTURAL SUSTAINABILITY

Defining Sustainability Goals

In its broadest sense, sustainability has been described as the ability to provide for core societal needs in a manner that can be readily continued into the indefinite future without unwanted negative effects. Most definitions of sustainability are framed in terms of three broad social goals: environmental, economic, and social health or well-being.[1] For example, a sustainable farming system might be one that provides food, feed, fiber, biofuel, and other commodities for society, as well as allows for reasonable economic returns to producers and laborers, cruelty-free practices for farm animals, and safe, healthy, and affordable food for consumers, while at the same time maintains or enhances the natural resource base upon which agriculture depends (USDA-NAL, 2007).

The legal definition of sustainable farming systems as defined in the Food, Agriculture, Conservation, and Trade Act (1990 Farm Bill and revised in 2007) is a useful starting point for identifying sustainability goals for the purposes of this report:

> an integrated system of plant and animal production practices having a site-specific application that will, over the long term: satisfy human food and fiber needs; enhance environmental quality and the natural resource base upon which the agricultural economy depends; make the most efficient use of nonrenewable resources and on-farm resources and integrate, where appropriate, natural biological cycles and controls; sustain the economic viability of farm operations; and enhance the quality of life for farmers and society as a whole.

This legal definition mingles a description of societal sustainability goals with strategies that can be used to achieve those goals (for example, "an integrated system which will use, where appropriate natural biological cycles and controls"; or "make the most efficient use of nonrenewable and on-farm resources"). This report makes a clear distinction between societal sustainability goals and the management systems used to pursue these goals. That distinction recognizes that the same goals can potentially be achieved through a range of different management and organizational approaches.

Modifying the Farm Bill definition slightly, the committee isolated four key societal sustainability goals that serve as the organizing principles for the remainder of this report (Figure 1-1):

- **Satisfy human food, feed, and fiber needs, and contribute to biofuel needs.**
- **Enhance environmental quality and the resource base.**
- **Sustain the economic viability of agriculture.**
- **Enhance the quality of life for farmers, farm workers, and society as a whole.**

The sustainability of a farming practice or system could be evaluated on the basis of how well it meets various societal goals or objectives. To be sustainable, a farming system needs

---

[1] In Europe, the three goals of sustainability are sometimes referred to as the 3Ps: people, prosperity, and planet, or, alternatively, as the "triple bottom line."

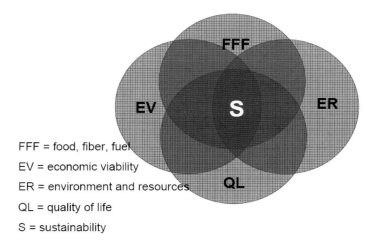

FIGURE 1-1 Sustainability goals used in this report. The area where the four goals overlap represents the highest sustainability in the continuum.

to be robust (that is, be able to continue to meet the goals in the face of stresses and fluctuating conditions; to adapt and evolve), be sufficiently productive, use resources efficiently, and balance the four goals. There are, however, often tradeoffs or synergies among the various goals and their related objectives, toward which sustainability is directed.

In the discussions that follow, the scientific evidence surrounding different farming practices or farming systems that illustrate their ability to further each of these four societal goals is discussed. The committee is not suggesting or implying that a farming practice would have to simultaneously accomplish each of these goals to be considered "sustainable." Rather, it recognizes and expects that combinations of practices used in a system will affect each of the four goals in different and often complicated ways. A sustainable system would balance and meet each of the four goals to a large extent.

Objectives

Each of the four sustainability goals consists of a large number of more specific objectives that represent different paths toward achievement of the goal. For example, the goal to "satisfy human food, feed, and fiber needs" requires managing farming systems in the aggregate so that there will be enough affordable food and fiber (including for energy production) in the future for all on the globe, although U.S. production would only play a part in overall global production. That goal has a long history and is a fundamental concern of all societies through time and can be summarized with the crucial question: *Will there be sufficient agricultural resources in the future?*

Achieving that goal will, at a minimum, require sufficient productivity (for example, the sheer volume of outputs produced from a given agricultural activity), farming practices that produce the outputs at a price that consumers can afford, and marketing and distribution systems to ensure that people have ready access to farm products. The concepts of productivity, affordability, and access represent specific objectives that are required to meet the overall goal.

Even relatively simple objectives can quickly become more complex. For example, agricultural productivity over time is influenced by the technologies that are available.

UNDERSTANDING AGRICULTURAL SUSTAINABILITY

Sometimes built capital, such as machinery or chemicals, can substitute for natural capital, such as natural soil fertility. Food productivity has risen over time in the United States, in part, because of the substitution of fertilizers for natural soil fertility. If the substitution is viewed as socially acceptable, if fertilizer is affordable and effective, and if its use is not accompanied by unwanted or detrimental side effects, then the loss of natural soil fertility as a result of a farming practice might be viewed as sustainable. If, however, fertilizer is viewed as a poor substitute for natural fertility, as having important unwanted side effects, or is thought to be unaffordable or ineffective in the future, then a farming system that results in losses of natural fertility of the soil will be viewed as unsustainable (Batie, 2008b).

Likewise, each of the other three goals comprise a number of specific objectives. The objectives listed in Box 1-6 are representative of various objectives associated with sustainability goals, and they are used to organize the review of the scientific literature in Chapters 3, 4, and 5. The specific objective of access to food is not discussed in this report because it is covered by another report, *The Public Health Effects of Food Deserts: Workshop Summary* (IOM and NRC, 2009). Quality and safety of food output is discussed in the context of its quality and safety at the farm gate. Food processing is beyond the scope of this report.

As the representative objectives listed in Box 1-6 illustrate, the question of the sufficient quantity of agricultural resources is not the only societal concern underlying calls for more sustainability. For example, the objectives listed with respect to environmental quality reflect a societal concern about the impacts of agriculture on the functional integrity of envi-

---

BOX 1-6
Representative Objectives Associated with Sustainability Goals

Satisfy human food, feed, and fiber, and contribute to biofuel needs
- Productivity of farming practice or system
- Quality and safety of food output
- Affordability of farm outputs
- Availability of farm outputs

Enhance environmental quality and quality of resource base
- Soil quality and health
- Water quality
- Air quality
- Biodiversity
- Animal health and welfare

Sustain the economic viability of agriculture
- Farm business profitability and viability
- Farm and household viability
- Farm labor economic security
- Community economic security

Enhance the quality of life for farmers, farm workers, and society as a whole
- Ensure that farm operators and their households are able to maintain an acceptable quality of life, including access to health and retirement benefits
- Protect the health and welfare of farmers, farm workers, and society
- Enhance community or social well-being from the surrounding agriculture, including access to local food, sustained provision of ecological services, and maintenance of attractive landscapes

ronmental resources and not just whether the environmental resources will be sufficient in quantity and quality for agricultural production. Similarly, the objectives about the welfare of farmers and animals are not always driven by concerns with respect to the sufficiency of agricultural resources; rather, they are ethical statements about how humans should treat each other and animals.

Important Qualities of Systems That Move Toward Greater Sustainability

Agricultural systems that move toward greater sustainability generally strive for several fundamental qualities. One of those qualities is to work with natural ecological and biogeochemical processes and cycles to maximize synergistic interactions and the beneficial use of internal resources, and to minimize dependence on external inputs. Another quality is to close nutrient, energy, and other resource cycles to the maximum extent feasible to reduce undesirable losses to the environment and additional waste disposal activities. Third, farmers, conventional or alternative, who work toward improved sustainability tend to understand and work with the social, cultural, and economic goals of people and institutions throughout the farm and food chain, which encourages synergistic relationships in the social and economic realm and increases the likelihood of desired outcomes emerging from investment of time and resources. The following chapters will illustrate different approaches being used to enhance these qualities in both the agronomic and socioeconomic aspects of U.S. farming systems.

Robustness

Farming is inherently a risky enterprise that requires constant adaptation to changes in environmental (for example, temperature, rainfall, wind), biotic (for example, prevalence of pests and diseases), as well as market (for example, commodity and input prices, consumer demand) and social conditions (for example, labor availability, policies). Farming practices or systems will differ in the extent to which they are vulnerable to different kinds of risks. With the advent of climate change and the corresponding increase in fluctuations and uncertainty in weather conditions, creating less vulnerable farming systems is of increased importance, because the capacity of crop insurance to cover greater crop losses will be limited and the potential for farming systems to fail might increase under climate change if appropriate adaptations are not made (Walker and Salt, 2006).

When thinking about **vulnerability**, two helpful concepts from ecosystem ecology are resistance and resilience. **Resistance** is the ability of a system to resist being dislodged from a stable condition by a disturbance such as some sort of system stressors and fluctuating conditions. In other words, resistance is the ability to resist change in functioning. **Resilience** has traditionally been regarded as the speed and extent at which stability returns to a system that is dislodged from a stable condition. New schools of thought emphasize that systems do not oscillate around a single stable state, but are highly dynamic and can shift between states depending on the extent of stresses affecting the system (Holling, 1996; Folke et al., 2002; Walker and Salt, 2006). Resilience, thus, is defined as "the capacity of the system to absorb a spectrum of shocks or perturbations and still retain and further develop the same fundamental structure, functioning and feedbacks" (Chapin et al., 2009). This report uses the new definition. Both resistance and resilience refer to the ability of a system, such as a farming system, to be able to function in the face of disturbances. At the landscape and community scales, resilience depends heavily on the diversity and types of farms and of their markets, as well as biodiversity (notably presence of perennial habitats). A related concept is **adaptability** (that is, the opposite of vulnerability), which reflects the ability of a system (biophysical or human) to evolve and change in response to long-term changes in the surrounding environment. The concepts of resistance, resilience, and adaptability apply equally to the properties of natural ecosystems and to social and institutional systems. The overall **robustness** of a farming system—that is, the ability of farming systems to withstand stresses, pressures, and changes in circumstances—will result from some mix of resistance, resilience, and adaptability.

Concepts of resistance and resilience have long been a cornerstone of ecological theory and other fields that deal with uncertainty and risk management (Holling, 1996, 2001; Walker and Salt, 2006; Nelson et al., 2007; Chapin et al., 2009). Although those concepts have been discussed in general terms as a desirable attribute of sustainable agricultural systems for over 20 years (Conway, 1987), little empirical research on the subject has been done for agricultural systems. Nonetheless, system resilience, resistance, and the ability to adapt in the face of climate change has garnered considerable attention in such fields as economics (for example, Brown and Lall, 2006; Goldstein, 2009), planning and engineering (for example, Fowler et al., 2003), and social science (for example, Adger, 2006; Ebi et al., 2006; Janssen et al., 2006). That kind of work has resulted in the development of a number of "resilience frameworks," modeling approaches, and other methods for determining resilience and resistance as well as mechanisms of adaptation (Turner et al., 2003; Eakin and Luers, 2006; Folke, 2006; Walker and Salt, 2006; Chapin et al., 2009). Furthermore, there is widespread agreement that an effective assessment of the long-term adaptive capacity of a given system requires linkages across multiple spatial scales and integration of biophysical, economic, and social considerations (Holling, 2001).

The ability to adapt to changing conditions is determined not only by the resilience and resistance of the biophysical system, but also by the capacity for self-organization and learning. The strength of social and institutional networks that support agriculture will play a pivotal role in the ability to adapt to climate change, increased variability in weather, and changing conditions. Considerable work, termed vulnerability science, is examining the relative vulnerability of different communities to climate change, and conceptualizing approaches to increase a community's ability to adapt to change that involves building social networks, appropriate institutional arrangements, and infrastructural capacity (Turner et al., 2003; Nelson et al., 2007). Agroecological and social systems are linked in current frameworks being used to address resilience, resistance, and adaptation (Turner et al., 2003; Eakin and Luers, 2006; Folke, 2006), and the linkage had been demonstrated by individual case studies (Milestad and Darnhofer, 2003; Robledo et al., 2004). Naylor (2008) provided a synthesis of resilience issues facing agricultural production systems globally and emphasized the important role that policies can play in supporting or detracting from creation of resilient systems. (See also Chapter 6.) A given farming system can therefore be robust when managed by farmers with access to adequate resources (for example, capital and labor) and where strong social networks and institutions are in place, but the system can be vulnerable when attempted by resource-poor farmers with a fragmented social system.

This report discusses how farming practices, management systems, and social organization can further various social objectives and goals. When possible, it examines evidence that those approaches help increase system robustness—that is, enhance the ability of a farm, farm household, or community to **resist** shocks (for example, market volatility, weather events, pest outbreaks), **adapt** or evolve in the face of changing conditions, and be **resilient** over the long term.

Scales
Assessing the sustainability of a farming system can become complex because of the importance of spatial and temporal scales. Initially, a farming system's sustainability might be evaluated differently at different spatial scales (such as at the farm, community, wa-

tershed, nation, or global level) and across different interconnected economic systems (including connections between food production, processing, and consumption). A good example is an assessment of the nutrient dynamics of farming systems. Annual cropping patterns might temporarily result in the deposition, uptake, or loss of various crop nutrients in a single season. Across a series of crop rotations, however, a field might be managed in a manner to present a relatively efficient and balanced nutrient budget over a whole rotation. Analysis of whole farm nutrient budgets on different types of farms can illustrate the ways that combinations of land type and availability, cropping patterns, and livestock feeding and manure management practices interact to create dramatically different nutrient outcomes depending on how producers manage their set of resources. At a larger scale, individual farms that are not able to use all their livestock manure nutrients efficiently can conceivably be organized under certain institutional settings to provide that resource to neighboring crop producers who can use them and effectively recycle the excess nutrients. Diversification of production systems and nonfarm habitat at a landscape or watershed level can greatly enhance the robustness of the system and reduce negative environmental impacts even if individual farms are still specialized and have limited diversification in land use (Santelmann et al., 2004; Boody et al., 2005).

Similar examples can also be found in the assessment of social and economic outcomes. A farm's separate crop or livestock enterprises might each produce positive or negative economic returns in a given year, yet the synergistic effects of the farm's combined enterprises might produce a different overall level of farm performance. It is common, for example, for dairy farmers to focus their management efforts on dairy herd performance, while using merely adequate (but perhaps not the best) management practices on their field crops.

Sustainability can also be assessed across different time scales, with potentially different results emerging across short-term and long-term time horizons. Those differences become important when implementing environmental policies based on performance and outcome-based indicators (discussed below). For example, the characteristics of perched water tables[2], and soils or stream sediments with high storage and release capacity for nutrients, will reflect the accumulation of chemicals or nutrients over many years. In those cases, effects of changed farming practices will not likely be detected until many years later. In those situations, means-based indicators based on the extent of use of different BMPs might be more appropriate as a policy tool than outcome measures such as nutrient concentrations in the water, at least over the short term. Time scale is an important factor because sustainability is a dynamic process moving along a trajectory toward meeting societal goals (that are also dynamic), as opposed to achieving some well-defined end point.

Synergies and Tradeoffs

Recognizing that there are multiple goals toward which sustainability can be examined (for example, generating food, feed, fiber, biofuel, environmental, economic, and social outputs), the complexity of the concept becomes readily apparent, and it is obvious why the topic has engendered much debate and contention. Each of the goals is made up of multiple aspects, and different goals can be mutually reinforcing (or synergistic) or present difficult tradeoffs among competing, socially desired outcomes. Synergies might create opportunities for potential win-win situations where pursuit of one outcome generates corollary benefits in other categories. Conflicts might result in tradeoffs, and the relative priority given to each goal will depend on the values of the stakeholders who are part of the decision-making process.

Tradeoffs can occur among different types of environmental impacts. For example,

---

[2]Water tables that occur above the regional water table as a result of an impermeable layer of soil or rock.

some practices designed to minimize negative impacts of farming practices on water quality can worsen problems with air quality. That point can be illustrated by the use of riparian zones and treatment wetlands. They can reduce nitrogen fluxes into surface waters in part by increasing rates of denitrification. However, the process of denitrification does not always result in the complete conversion of nitrate to nitrogen gas, in which case various potent greenhouse gases, nitrogen oxides, are produced; thus, a tradeoff exists between improving water quality and air quality (Crumpton et al., 2008). While riparian buffer strips are designed to reduce negative effects of crop farming on nearby water bodies and are beneficial in most extensive cropping systems, there are concerns that they provide habitats for wildlife that might defecate in the crop fields and contaminate vegetables and fruits that are consumed fresh (Atwill, 2008; Doyle and Erickson, 2008).

Contentious tradeoffs can also occur between environmental, social, and economic goals. Examples include production of food to feed a growing world population versus a desire to use production practices that protect soil, air, water, and biological resources and preserve some resources for nonfood production uses such as wildlife habitat. Efforts to use environmentally friendly practices or to improve the economic conditions of farmers or farm workers can sometimes increase production costs and possibly hinder access to affordable healthful food among low-income consumers. Opinions differ widely as to whether those goals necessarily are in direct conflict, or the extent of tradeoffs involved, but nonetheless balancing the different goals clearly has to be addressed.

Another potential tradeoff could be between the ability of a system to produce the outputs desired by society (for example, food, fiber, and fuel) and the resilience and resistance of that system. For example, diverse farming systems with multiple crops or integrated systems with livestock might be more able to sustain reasonable production and profit in the face of climatic or market volatility, but they might be less productive when measured by volume of production or by profits in "normal" or optimal years. However, the more variable and unpredictable conditions become, then the argument for trading some degree of maximum productivity, or efficiency, for greater stability becomes stronger (Walker and Salt, 2006). In the case of agriculture, that tradeoff may mean sacrificing the ability to achieve maximum yields and income in good years in return for a system that performs well over a wide range of conditions and is less likely to fail in bad years. Managing a system to achieve high yields clearly is an important component of sustainability, but maximizing one component can come at the expense of overall system resilience, which in turn reduces overall sustainability. (See Walker and Salt, 2006, for illustrative examples.) As used above, the term "efficiency" reflects efforts to maximize input use efficiency per unit production. In the rest of this report, the term "efficiency" is sometimes used with a similarly narrow definition (as in the case of the discussion on water use efficiency in Chapter 3). Other times, the term is used in the broad context of "systems efficiency" to reflect the notion of minimizing undesired outcomes (such as pollution and waste) from resource use while maximizing a wide group of desired outcomes (which could include production and support for important ecosystem services) and reducing the need for external inputs (which could be achieved by increasing nutrient cycling between animal and crop production).

SETTING PRIORITIES FOR AGRICULTURAL SUSTAINABILITY

Who Decides?

Any single farming system is unlikely to meet fully all of society's production, environmental, economic, and social goals and objectives. Indeed, it is most probable that meeting many of society's goals will require a mixture of many farming types and systems rather than the adoption of any one type. The societal choice about agricultural systems is a social choice about what types of agriculture are desirable and therefore what the future of agriculture ought to be. The debate over the wisdom of the "alternative" futures is made difficult by underlying value systems or philosophies of agriculture that produce competing opinions about that future. (See Box 1-7 for a discussion of contending philosophies of agriculture that underlie many of the societal disagreements about these goals.) The competing

---

BOX 1-7
Contending "Philosophies" of Agriculture

The philosopher Paul Thompson (2010) notes that one way to explain why debates over sustainable agriculture are so intense is that there are different perspectives as to what should be the objectives of agriculture and how agriculture should be structured. One view is termed **the industrial philosophy of agriculture**. According to this view, agriculture is just another sector of an industrial society where products are produced at the lowest cost possible and in a manner that provides sufficient food and fiber for society. The trend to fewer and larger commercial farms is not seen as a problem; rather, it is a way to capture economies of scale and lower the costs of food, fiber, and energy production. Indeed, advocates of industrial-scale agriculture view it important to export this structure to other nations to assure worldwide food sufficiency. Essentially, this view sees landscapes in terms of commodities the land can produce; thus, industrial philosophy puts great emphasis on increasing yields per acre or pounds of meat per animal. Although there are concerns within this philosophy about fairness to labor, the vitality of communities, animal welfare, and negative impacts on the environment, it is argued that those issues can be addressed without overhauling the industrial structure of agriculture.

Thompson terms a countervailing viewpoint as an **agrarian philosophy of agriculture** (sometimes called alternative or multifunctional agriculture) that views agriculture as having an important social function above and beyond its ability to produce food, feed, fiber, and biofuel. The social functions include providing positive ecological services and protecting ecological integrity and functioning. Because ecosystems place limits on what kind of farming can be continuously conducted, the agrarian philosophy believes that farming should not be conducted in such a way as to significantly harm ecological functioning; indeed, farming would restore ecosystems by recognizing the complex ecological relationships among plants, soils, and livestock. The agrarian philosophy questions whether the practices of industrial agriculture—with its heavy reliance on purchased inputs, particularly agricultural chemicals—are sustainable. Proponents of this view frequently advocate for reducing or eliminating those practices. Also, the agrarian philosophy frequently focuses on social sustainability: that is, the need for agriculture to support and be a part of rural communities. The large scale of industrial agriculture, and the perceived negative effects of consolidation of farms and ranches on diverse family farms, hence, is not conducive to sustaining rural communities. There is also concern about the effect of industrial agriculture on the welfare of agricultural workers and farm animals. The social sustainability concerns get reflected in calls for "fair trade" or for eating locally grown foods and "humanely produced" animal products.

The two contrary philosophies[1] illustrate that disagreements about agriculture's sustainability have much to do with differing perceptions on outcomes and the desirability of the outcomes produced by various ways to organize agricultural production. That is, there are different philosophical beliefs about what the agrifood system should do for us as a society; sustainability is a social goal (Thompson, 2010).

Others dispute that there are important differences between the visions of what agriculture should be, but they note that many goals do not result in as many conflicts between the outcomes of various systems as have been portrayed. For example, with respect to yields, systems that move toward increased sustainability are not necessarily small-scale, traditional agriculture, and they can be as productive as conventional and industrial systems (as illustrated by Stahlbush Island Farms, Goldmine Farm, and the Lundberg Family Farms in the case studies in Chapter 7). On the other hand, small-scale, diversified farms might be better associated with certain types of robust rural communities (as illustrated by Peregrine Farm in Chapter 7).

---

[1] Dobbs (presentation to the committee on March 27, 2008) and Josling (2002) also made this contrast.

expected reductions in nonpoint source pollution. On the other hand, outcome-based or effects-based indicators require some combination of direct measurement of reductions in runoff, leaching, levels of nutrients in surface and ground water, or outputs from simulation models.

In some situations, only means-based indicators are feasible and accessible to end users, particularly for large-scale monitoring for policy and regulatory contexts. Nonetheless, caution is to be used when relying on means-based indicators, particularly if changes measured in such means-based indicators (for example, BMP use) have a more complex relationship to landscape-scale outcomes than was present in the experimental conditions used to estimate their original impact coefficients. The most robust means-based indicators are those with simple causal links derived from a substantial body of evidence compiled from real-world conditions. A good example is the use of percentage of ground cover as an indicator of soil erosion potential, a relationship that has been demonstrated many times across different soil types, slopes, and climates.

There are dangers associated with the use of outcome-based indicators, notably when effects can be measured, but the causal links to what is actually responsible for causing the effects are not clear. That particular concern occurs when outcome-based indicators are used in a regulatory context or for awarding incentives. Using water quality as an example, changes in nutrient loadings in streams or lakes can be measured directly, but identifying the cause of the changes can be extremely difficult in the absence of detailed and comprehensive knowledge of surface or ground water hydrology, pollutant transport, and in-stream processing of pollutants in the region. Nutrient loadings in streams and lakes might reflect more of the effects of historical management than of the current one. An increase in nutrient levels in a section of a stream might not relate to the practices used on farms immediately adjacent to the stream, but rather to some combination of local input and loading from subsurface water that contain nutrients derived from a much larger area within the watershed. In that case, stream nutrient levels would not be a good indicator for forming the basis of farm-level regulatory decisions.

Means-based and outcome-based indicators can also be developed for measuring the impacts of farming systems on economic and social sustainability goals. However, there is typically a small scientific knowledge base to allow an analyst to link the use of particular production practices or agricultural system with a prediction for economic or social outcomes. For example, whether adoption of a particular farming practice or system is likely to increase the economic well-being of farmers and their households might be estimated by a means-based indicator such as the degree of adoption of a particular practice. However, to link well-being to that practice might require linking adoption to such factors as reduced costs of production per unit output. An outcome-based indicator might be to examine levels of net farm or household income among farms using the practice. Regardless of the indicator selected, complexities in the relationship between use of the practice and accomplishment of the ultimate outcome of improved economic well-being (such as variation in the management abilities or approaches of individuals, interactions among different cropping or livestock production activities on a farm, or different marketing outlets) make interpretation of each type of indicator challenging.

Relating Indicators to Sustainability Goals and System Attributes

The usefulness of indicators can be improved if they are designed to reflect performance in terms of a more complete set of systems attributes, including measures of productivity, resource-use efficiency, and robustness. For example, a commonly used indicator for the objective of high productivity is crop yield, typically average yield per acre. However a more complete indicator would also include assessment of system robustness (such as a measure of variability in yields over space and time, or the probability of yields falling below a certain threshold) and of resource-use efficiency of the system as measured by nutrient, fertility, water, and energy use expressed per unit crop yield. Similarly, measures of biodiversity could be improved by incorporating considerations of productivity (number of species supported and their population sizes), robustness (fluctuations in population sizes, and the number of extinctions and invasions over space and time), and system efficiency (for example, the amount of land and water used to support a given number of species and populations).

In the case of social goals, such as increasing community social and economic well-being, it would be valuable not only to examine productivity in terms of the ability of an agricultural system to produce desired outcomes (for example, increased net farm income, improved availability of affordable quality food, and decreases in income inequality), but also to pay attention to the robustness (stability of outcomes in the face of changing biophysical, market, and policy circumstances).

Interpreting Indicators

Indicator values have to be interpreted to be meaningful: that is, significance needs to be assigned to an indicator's numerical (or qualitative) value. Threshold values have not been established for many environmental indicators. Conclusions are typically drawn on the basis of whether numerical values are higher or lower than before, and rarely in terms of whether the differences are likely to be functionally significant. For example, measures of soil organic matter are a cornerstone of most sustainability and soil quality assessments, being seen as an integrative indicator for soil properties such as moisture-holding capacity, physical structure, and nutrient supply capacity. However, the numerical level that would be considered good, or what change in soil organic matter levels constitutes a significant functional change, has not been established (Loveland and Webb, 2003). In contrast, nutrient concentration thresholds for ecological integrity have been established for some freshwater lakes (Carpenter and Lathrop, 2008), but no such thresholds for ecological integrity currently exist for nitrate or phosphorus concentrations in freshwater streams. **Clearly, improving the understanding of the relationships between sustainability indicators and their functional significance is a priority for future work.**

Integrating Diverse Indicators in Holistic Assessments

Using even well-designed indicators still begs the question of how to make a holistic assessment of the relative sustainability of different systems given the multiple indicators that represent various sustainability goals and objectives. Even a single sustainability goal, such as enhancing environmental quality, contains many subobjectives, such as water quality, air quality, water use, and biodiversity conservation, each of which may be measured by multiple indicators. One practice is to combine individual indicators into an index, based on some additive (often weighted) procedure. A single index, however, obscures the values inherent in its calculation—which attributes are weighted more than others—and can be particularly problematic if the direction of change is positive for some measures but negative for others (Suter, 1993; Fisher et al., 2001). Despite that issue, reducing multiple indicators into a single indicator is done in some policy contexts, such as the use of an environmental benefits index to implement the Conservation Reserve Program of the Farm Bill.

An alternative is to evaluate system performance without creating a single number, in which case any tradeoffs or synergies might be identified. For example, the same suite of indicators is monitored over time across all countries and regions in EU, making it possible to look at the spatial distribution of relative importance of different sustainability issues and where trends are positive or negative (European Environment Agency, 2006). Data show that soil erosion and water overdrafts are most serious in the Mediterranean countries, whereas nitrate pollution is greatest in northern Europe. However, positive nitrogen balances are declining in most countries where the problems are most severe (for example, the Netherlands) and have decreased by 16 percent across the EU as a whole between 1990 and 2000. Other indicators used include greenhouse-gas emissions, changes in biodiversity, and landscape patterns. (See European Environment Agency [2006] for the complete list.) Those findings are then integrated into the various policy directives and used to evaluate program effectiveness.

Programs on indicators have been developed by other institutions, including the FAO (Tschirley, 1997), United Nations (United Nations, 2007), World Bank (Dumanski et al., 1998), USDA (USDA-ERS, 2003), various university programs in the United States (Aistars, 1999), and such nonprofit organizations as the International Institute of Sustainable Development (Pintér et al., 2005) and the Land Stewardship Project (Keeney and Boody, 2005). Although the proliferation of programs to develop indicators reflects the growing interest in quantification and evaluation of the sustainability of agriculture, there is not an alignment or consensus among the different organizations and scientists involved about the most appropriate or useful set of indicators. In other words, there are different interpretations, methods, and approaches for developing and using indicators. Collaborative efforts among different organizations to develop agreement about key indicators needed for measuring sustainability—and particularly performance outcomes—have emerged in the U.S. agriculture sector, such as Field to Market: The Keystone Alliance for Sustainable Agriculture (Keystone Center, 2009) and the Stewardship Index Initiative for Specialty Crops (Stewardship Index for Specialty Crops, 2009). Some university programs in the United States have also convened and attempted to collaborate in the development of indicators (for example, the Sustainable Agriculture Research and Education Program at University of California).

It should be noted, however, that the conclusions reached by the different methods can vary substantially even when applied to the same system (for example, van der Werf et al., 2007). Those differences emphasize the need for careful assessment of the assumptions behind each method and for more method comparison studies to be done.

SUMMARY

- Sustainable agriculture can involve a diverse number of possible farming practices or farming systems. The committee's definition of sustainable farming does not accept a sharp dichotomy between conventional and sustainable farming systems, not only because farming enterprises reflect many combinations of farming practices, organization forms, and management strategies, but also because all types of systems can potentially contribute to achieving various sustainability goals and objectives.
- Sustainability is a process that moves farming systems along a trajectory toward meeting societal defined goals, as opposed to any particular end state. As such, management systems will inevitably need to be adjusted to meet sustainability goals and objectives when circumstances and societal desires change.
- Sustainability of agriculture is a complex and dynamic concept. The committee rec-

ognizes that the concept is inherently subjective in that the goals will be viewed in different ways by various groups in society. Further, even with broad agreement for certain goals, the relative importance assigned to one goal over another will be highly contested. The definition of societal goals for sustainable agriculture emerges from the domain of politics, not science. Yet, science plays essential roles by generating knowledge to inform the political process, making predictions about outcomes likely to result from different management systems, answering specific questions when needed, and expanding the range of alternatives considered. Science also can supply the knowledge needed to develop new agricultural technologies.

- Finding ways to measure progress along a sustainability trajectory is an important part of an experimentation and adaptive management process. The rationale for selecting the indicators used to measure progress needs to be explicitly stated and justified since the choice is a political rather than a scientific question.
- It is important that indicators used are shown to be appropriate surrogates for the sustainability outcomes they are meant to represent, especially in the case of means-based indicators.
- Social indicators, in addition to environmental and economic measures, will help provide a more comprehensive assessment of movement toward sustainability goals; however, much more research is needed to develop appropriate social indicators because the development of such measures to date has been fledgling.

ORGANIZATION OF THE REPORT

Using the terms and the boundaries defined in this chapter, this report provides an overview of how U.S. agriculture has evolved over the years to the current state, and farmers' successes and the challenges they face (Chapter 2). The state of knowledge on farming practices and management systems that can increase the environmental and production (Chapter 3), social, and economic (Chapter 4) sustainability of agriculture are then discussed. However, individual farming practices are components of an agricultural system. Knowledge and understanding of the sum of the parts are important in designing, fine-tuning, and adapting the system to improve sustainability. Chapter 5 uses a few systems to illustrate how a collection of farming practices works in concert to improve sustainability and to illustrate some potential tradeoffs. In spite of the positive attributes of some farming practices and systems that can improve sustainability, they are not necessarily widely adopted by farmers. Chapter 6 highlights the importance of markets, policies, and research in shaping trajectories of farm change, and examines influences on farmers that make some of them more likely and able than others to change production practices or convert to new farming systems. The approaches that could improve environmental, social, and economic sustainability, their practical applications, and their ability to meet those goals are best illustrated in examples of real-life farms in Chapter 7. The committee used seven farms featured in the report *Alternative Agriculture* (NRC, 1989b) and nine additional farms as case studies to illustrate points made earlier in the report. Conclusions drawn in earlier chapters were used as the basis of the discussion on the lessons learned in U.S. agriculture that are relevant to agriculture in other regions in Chapter 8. A representative of the Gates Foundation specified that the foundation was most interested in the relevance of lessons learned to sub-Saharan Africa, which is the focus of Chapter 8. The committee's findings and recommendations for how to advance a systems approach to farming and to strengthen federal policies and programs related to improving agricultural production in the United States are summarized in Chapter 9.

REFERENCES

Adger, W.N. 2006. Vulnerability. *Global Environmental Change* 16:268–281.
Aistars, G.A. 1999. A life cycle approach to sustainable agriculture indicators. Available at http://css.snre.umich.edu/css_doc/Proceedings.PDF. Accessed on May 20, 2009.
Atwill, E.R. 2008. Implications of wildlife in *E. coli* outbreaks associated with leafy green produce. Paper read at 23rd Vertebrate Pest Conference, San Diego.
Batie, S. 2008a. Wicked problems and applied economics. *American Journal of Agricultural Economics* 9(5): 1176–1191.
Batie, S.S. 2008b. The sustainability of U.S. cropland soils. In Perspectives on Sustainable Resources in America, R.A. Sedjo, ed. Washington, D.C.: RFF Press.
Bell, S., and S. Morse. 2008. Sustainability Indicators: Measuring the Immeasurable. London: Earthscan.
Bockstaller, C., and P. Girardin. 2003. How to validate environmental indicators. *Agricultural Systems* 76(2): 639–653.
Boody, G., B. Vondracek, D.A. Andow, M. Krinke, J. Westra, J. Zimmerman, and P. Welle. 2005. Multifunctional agriculture in the United States. *BioScience* 55(1):27–38.
Brown, C., and U. Lall. 2006. Water and economic development: the role of variability and a framework for resilience. *Natural Resources Forum* 30 (4):306–317.
Carpenter, S.R., and R.C. Lathrop. 2008. Probabilistic estimate of a threshold for eutrophication. *Ecosystems* 11(4): 601–613.
Chapin, F.S., G.P. Kofinas, and C. Folke (eds.) 2009. Principles of Ecosystem Stewardship: Resilience-based Natural Resource Management in a Changing World. New York: Springer.
Collins, K. 2008. The role of biofuels and other factors in increasing farm and food prices—a review of recent developments with a focus on feed grain markets and market prospects. Available at http://www.foodbeforefuel.org/files/Role%20of%20Biofuels%206-19-08.pdf. Accessed on June 8, 2009.
Conway, G.R. 1987. The properties of agroecosystems. *Agricultural Systems* 24(2):95–117.
Crumpton, W.G., D.A. Kovacic, D.L. Hey, and J.A. Kostel. 2008. Potential of Restored and Constructed Wetlands to Reduce Nutrient Export from Agricultural Watersheds in the Corn Belt. St. Joseph, Mich.: ASABE.
Dobbs, T.L. 2008. Economic and policy conditions necessary to foster sustainable farming and food systems: U.S. policies and lessons from the European Union. Available at http://www.foodandsocietyfellows.org/publications.cfm?refID=104050. Accessed on August 6, 2009.
Doyle, M.P., and M.C. Erickson. 2008. Summer meeting 2007—the problems with fresh produce: an overview. *Journal of Applied Microbiology* 105(2):317–330.
Dumanski, J., E. Terry, D. Byerlee, and C. Pieri. 1998. Performance indicators for sustainable agriculture. Available at http://siteresources.worldbank.org/INTARD/864477-1112703179105/20434502/SustInd.pdf. Accessed on May 20, 2009.
Eakin, H., and A.L. Luers. 2006. Assessing the vulnerability of social-environmental systems. *Annual Review of Environment and Resources* 31:365–394.
Ebi, K.L., R.S. Kovats, and B. Menne. 2006. An approach for assessing human health vulnerability and public health interventions to adapt to climate change. *Environmental Health Perspectives* 114(12):1930–1934.
European Environment Agency. 2006. Integration of Environment into EU Agricultural Policy—the IRENA Indicator-Based Assessment Report. Copenhagen: European Environment Agency.
Fisher, W.S., L.E. Jackson, G.W. Suter, and P. Bertram. 2001. Indicators for human and ecological risk assessment: a US Environmental Protection Agency perspective. *Human and Ecological Risk Assessment* 7(5):961–970.
Folke, C. 2006. Resilience: the emergence of a perspective for social-ecological systems analyses. *Global Environmental Change—Human and Policy Dimensions* 16(3):253–267.
Folke, C., S. Carpenter, T. Elmqvist, L. Gunderson, C.S. Holling and B. Walker. 2002. Resilience and sustainable development: building adaptive capacity in a world of transformations. *Ambio* 31(5): 437–440
Fowler, H.J., C.G. Kilsby, and P.E. O'Connell. 2003. Modeling the impacts of climatic change and variability on the reliability, resilience, and vulnerability of a water resource system. *Water Resources Research* 39(8):1222.
Goldstein, B.E. 2009. Resilience to surprises through communicative planning. *Ecology and Society* 14(2):33. Available at http://www.ecologyandsociety.org/vol14/iss2/art33/. Accessed on March 3, 2010.
Guthman, J. 2004. Agrarian Dreams: The Paradox of Organic Farming in California. Berkeley: University of California Press.
Holling, C.S. 1996. Surprise for science, resilience for ecosystems, and incentives for people. *Ecological Applications* 6(3):733–735.
———. 2001. Understanding the complexity of economic, ecological, and social systems. *Ecosystems* 4:390–405.
IOM (Institute of Medicine) and NRC (National Research Council). 2009. The Public Health Effects of Food Deserts. Workshop Summary. Washington, D.C.: National Academies Press.

Janssen, M.A., M.L. Schoon, W.M. Ke, and K. Borner. 2006. Scholarly networks on resilience, vulnerability and adaptation within the human dimensions of global environmental change. *Global Environmental Change—Human and Policy Dimensions* 16(3):240–252.

Josling, T. 2002. Competing paradigms in the OECD and their impact on the WTO agricultural talks. In Agricultural Policy for the 21st Century, L. Tweeten and S.R. Thompson, eds. Ames: Iowa State Press.

Keeney, D., and G. Boody. 2005. Performance based approaches to agricultural conservation programs dealing with non-point source pollution, including utilization of the provisions of the conservation security program. Available at http://www.landstewardshipproject.org/mba/performance_based_policies.pdf. Accessed on August 10, 2009.

Keystone Center. 2009. Field to market: the Keystone Alliance for Sustainable Agriculture. Available at http://www.keystone.org/spp/environment/sustainability/field-to-market. Accessed on August 10, 2009.

Loveland, P., and J. Webb. 2003. Is there a critical level of organic matter in the agricultural soils of temperate regions: a review. *Soil & Tillage Research* 70(1):1–18.

Milestad, R., and I. Darnhofer. 2003. Building farm resilience: the prospects and challenges of organic farming. *Journal of Sustainable Agriculture* 22(3):81–97.

Naylor, R.L. 2008. Managing food production systems for resilience. In Principles of Natural Resource Stewardship: Resilience-Based Management in a Changing World, F.S. Chapin, G.P. Kofinas, and C. Folke, eds. New York: Springer.

Nelson, D.R., W.N. Adger, and K. Brown. 2007. Adaptation to environmental change: contributions of a resilience framework. *Annual Review of Environment and Resources* 32:395–419.

NRC (National Research Council). 1989. Investing in Research: A Proposal to Strengthen the Agricultural, Food, and Environmental System. Washington, D.C.: National Academy Press.

———. 1989b. Alternative Agriculture. Washington, D.C.: National Academy Press.

———. 1996. Understanding Risk: Informing Decisions in a Democratic Society. Washington, D.C.: National Academy Press.

———. 1999. Our Common Journey: A Transition Toward Sustainability. Washington, D.C.: National Academy Press.

Payraudeau, S., and H.M.G. van der Werf. 2005. Environmental impact assessment for a farming region: a review of methods. *Agriculture Ecosystems & Environment* 107(1):1–19.

Pielke, R.A., Jr. 2007. The Honest Broker: Making Sense of Science in Policy and Politics. New York: Cambridge University Press.

Pintér, L., P. Hardi, and P. Bartelmus. 2005. Sustainable Development Indicators: Proposals for a Way Forward. Winnipeg: International Institute for Sustainable Development.

Rigby, D., P. Woodhouse, T. Young, and M. Burton. 2001. Constructing a farm level indicator of sustainable agricultural practice. *Ecological Economics* 39(3):463–478.

Robledo, C., M. Fischler, and A. Patino. 2004. Increasing the resilience of hillside communities in Bolivia—has vulnerability to climate change been reduced as a result of previous sustainable development cooperation? *Mountain Research and Development* 24(1):14–18.

Santelmann, M.V., D. White, K. Freemark, J.I. Nassauer, J.M. Eilers, K.B. Vache, B.J. Danielson, R.C. Corry, M.E. Clark, S. Polasky, R.M. Cruse, J. Sifneos, H. Rustigian, C. Coiner, J. Wu, and D. Debinski. 2004. Assessing alternative futures for agriculture in Iowa, USA. *Landscape Ecology* 19(4):357–374.

Searchinger, T., R. Heimlich, R.A. Houghton, F. Dong, A. Elobeid, J. Fabiosa, S. Tokgoz, D. Hayes, and T-H. Yu. 2008. Use of U.S. croplands for biofuels increases greenhouse gases through emissions from land use change. *Science* 319:1238–1240.

Stewardship Index for Specialty Crops. 2009. Welcome to the homepage of the Stewardship Index for Specialty Crops. Available at http://www.stewardshipindex.org/. Accessed on August 10, 2009.

Suter, G.W. 1993. A critique of ecosystem health concepts and indexes. *Environmental Toxicology and Chemistry* 12(9):1533–1539.

Thompson, P.B. 2010. The Agrarian Vision: Sustainability and Environmental Ethics. Lexington: University of Kentucky Press.

Tokgoz, S., A. Elobeid, J. Fabiosa, D.J. Hayes, B.A. Babcock, T.H. Yu, F.X. Dong, and C.E. Hart. 2008. Bottlenecks, drought, and oil price spikes: impact on U.S. ethanol and agricultural sectors. *Review of Agricultural Economics* 30(4):604–622.

Tschirley, J.B. 1997. The use of indicators in sustainable agriculture and rural development: considerations for developing countries. In Sustainability Indicators, B. Moldan and S. Billharz, eds. New York: John Wiley and Sons.

> **BOX 2-1**
> **Farm Typology Developed by the U.S. Department of Agriculture Economic Research Service**
>
> The U.S. farm sector is so diverse that statistics summarizing the sector as a whole can be misleading. The USDA Economic Research Service (ERS) has developed a classification typology to identify relatively homogenous subgroups of U.S. farms. The typology is based largely on farm sales, organizational structure, and the operator's primary occupation. The farm classification developed by ERS focuses on the "family farm," or any farm organized as a sole proprietorship, partnership, or family corporation. Family farms exclude farms organized as nonfamily corporations or cooperatives and farms with hired managers.
>
> **Small Family Farms (sales less than $250,000)**
>
> - **Limited-resource.** Farms with gross sales less than $100,000 in 2003 and less than $105,000 in 2004. Operators of limited-resource farms must also have received low household income in both 2003 and 2004. Household income is considered low in a given year if it is less than the poverty level for a family of four, or it is less than half the county median household income. Operators may report any major occupation except hired manager.
> - **Retirement.** Small farms whose operators report they are retired (excludes limited-resource farms operated by retired farmers).
> - **Residential/lifestyle.** Small farms whose operators report a major occupation other than farming (excludes limited-resource farms with operators who report nonfarm work as their major occupation).
> - **Farming-occupation.** Farms whose operators report farming as their major occupation (excludes limited-resource farms whose operators report farming as their major occupation).
>   - **Low-sales.** Gross sales of less than $100,000.
>   - **Medium-sales.** Gross sales between $100,000 and $249,999.
>
> **Large-Scale Family Farms (sales of $250,000 or more)**
>
> - **Large family farms.** Farms with sales between $250,000 and $499,999.
> - **Very large family farms.** Farms with sales of $500,000 or more.
>
> **Nonfamily Farms**
>
> - **Nonfamily farms.** Farms organized as nonfamily corporations and cooperatives, as well as farms operated by hired managers. Also includes farms held in estates or trusts.
>
> SOURCE: USDA-ERS (2000).

The mid-sized family farms (sales between $100,000 and $500,000) are examples of the prototypical "family farm" that has captured much of the public imagination and public policy debates over the future of American agriculture (Browne et al., 1992). According to the 2007 census, these mid-sized farms represented just under 10 percent of all U.S. farms, produced 16.5 percent of all farm sales, and managed another quarter of the nation's farmland and nearly 30 percent of its cropland.

Small and mid-sized family farms together owned two-thirds of the total value of farmland, buildings, and equipment and managed roughly 60 percent of all U.S. farmland and cropland in 2007. Therefore, they will continue to play an important role in efforts to improve the environmental footprint of agriculture, and their experiences and activities will continue to shape the social and economic well-being of farm families and agricultural communities. Interestingly, the proportion of small and mid-size operations that have chosen to participate in federal land conservation programs is larger than that of

TABLE 2-1 Farm Typology Class and the Relative Contribution of Each Class to Various Farm Indicators in 2007

	Farm Typology Class						
	Small Family Farms (Sales < $250,000)					Large Family Farms (Sales $250,000+)	
Indicator	Limited-Resource Farms	Retirement Farms	Residential-Lifestyle Farms	Farming-Occup. (Sales < $100,000)	Farming-Occup. ($100,000–$249,999)	Large Family Farms ($250,000–$499,999)	Very Large Family Farms ($500,000+)	Non-Family Farms
	(Percent of U.S. Total)							
Farms	14.0	20.7	36.4	11.7	4.5	3.9	4.6	4.1
Value of Production[a]	0.9	2.3	3.7	2.2	5.8	10.7	54.3	22.9
Total Government Payments	3.3	9.9	12.1	7.2	11.0	16.1	33.1	7.4
Production Expenses	1.7	3.2	5.8	3.2	5.4	9.5	48.6	22.5
Net Cash Farm Income	−1.3	1.4	−1.2	0.8	7.4	13.9	60.5	18.4
Value of Farm Assets[b]	5.9	11.8	18.1	8.9	8.4	11.5	25.2	10.3
Hired Farm Workers	3.5	7.6	11.7	6.9	6.2	8.4	34.1	21.6
Farmland	4.6	9.7	13.1	9.5	11.3	13.3	22.9	15.6
Cropland	3.6	8.0	11.0	7.4	11.5	16.6	34.0	7.9
Irrigated Land	1.5	3.0	4.6	3.8	7.4	13.1	50.6	16.0
Conservation Program Land[c]	7.5	26.1	26.6	11.6	6.3	5.8	7.6	8.5
Crop Insurance Acres[d]	1.2	2.8	5.0	4.8	13.2	21.0	44.3	7.6
Organic Farms	16.4	12.1	27.3	19.6	8.1	4.6	4.9	7.2
Organically Certified Land	6.0	6.5	11.8	13.4	16.3	15.2	19.4	11.3
Organic Produce Sales	1.5	1.7	3.2	4.3	9.0	10.5	41.6	28.1

[a] Market value of agricultural product sales.
[b] Combined value of land, buildings, machinery, and equipment.
[c] Acres enrolled in Conservation Reserve Program, Wetlands Reserve Program, Conservation Reserve Enhancement Program, and farmable wetlands programs.
[d] Acres enrolled in crop insurance programs.

SOURCE: 2007 Census of Agriculture.

large operations. Eighty-four percent of all land in federal land conservation programs is managed by small and mid-sized farms. Small and mid-sized farms received 88 percent of U.S. total government payments for conservation programs in 2006 (Hoppe et al., 2008). In addition, 70 percent of organically certified land in the United States was managed by small and mid-sized farms in 2007 (although they accounted for only 30 percent of total organic product sales).

In contrast to the small and mid-sized farms, million-dollar farms—that is, those with annual sales of at least $1 million—accounted for nearly half of U.S. farm product sales in 2002, even though there were only about 35,000 of them. They represent only 2 percent of all U.S. farms (Hoppe et al., 2008). Most million-dollar farms were operated as family businesses, and many reflect joint operations that support multiple family members and households. These types of farms particularly dominate the value of U.S. production of high-valued specialty crops (72 percent), dairy products (59 percent), hogs (58 percent), poultry (55 percent), and beef (52 percent). In some crops, production is concentrated. For example "[d]ata on acres harvested [obtained] from the 2002 Census of Agriculture suggest that some specialty crops occur on a relatively small number of farms. For example, the 58 largest producers of head lettuce (out of 830 total producers) in 2002—each harvesting at least 1000 acres of the crop accounted for 65 percent of the total acreage in head lettuce. As another example, the 77 largest broccoli producers (out of 2,493 total producers)—each with at least 500 harvested acres of the crop—accounted for 69 percent of the total harvest acres" (Hoppe et al., 2008, p. 34).

Because of economies of size, and as illustrated in Figure 2-5, those large farms tend to have profit margins that give them a competitive edge when compared to similar, but smaller farms. The million-dollar farms can take better advantage than the small farms of technological changes, economic and financial innovations, business management principles, and coordination with suppliers and processors (Gray and Boehlje, 2007).

Relatively few of the million-dollar farms specialize in crops that are covered by Farm Bill commodity programs, although the 44 percent of these farms that did participate in

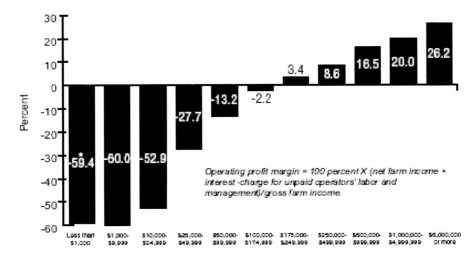

FIGURE 2-5 Operating profit margin, by sales class in 2006.
SOURCE: USDA, Economic Research Service, 2006 Agricultural Resource Management Survey, Phase III (as cited in Hoppe et al., 2008).

commodity programs received a total of 16 percent of all commodity program payments. The million-dollar farms account for 62 percent of all U.S. farm products produced under contracts with processors and other end buyers. Very large farms were somewhat less likely to participate in federal conservation programs than mid-sized farms. In 2006, 6 percent of total government conservation spending was distributed to the million-dollar farms (Hoppe et al., 2008).

Another important form of agricultural diversity in the United States becomes apparent when examining the acreage planted to various crops or used for livestock production. Efforts to address the sustainability of U.S. agriculture will need to confront the distinctive opportunities and challenges associated with production of different types of commodities. The most commonly raised commodities in U.S. agriculture are beef cattle, horses, and forages (each raised by more than a quarter of U.S. farms). However, the most economically important commodities—grains, poultry, dairy products, and specialty crops—are typically raised on a small fraction of U.S. farms (Table 2-2). Those commodities also represent the production systems that use most of the energy, fertilizers, agrichemicals, and hired labor in the United States. From a landscape perspective, most U.S. cropland is planted to

TABLE 2-2 Relative Importance of Different Commodities in U.S. Agriculture, 2002

Commodity Type	Percentage of U.S. Total		
	Farms Raising Commodity	Farm Sales from Commodity	U.S. Harvested Cropland
Livestock			
Beef cows	37.4	22.5	na
Horses	25.5	0.7	na
Sheep and goats	7.7	0.3	na
Poultry	4.6	11.9	na
Milk cows	4.3	10.1	na
Hogs and pigs	3.7	6.2	na
Crops			
Forages (all)	41.6	3.0	21.2
Grains and Oilseeds (any)	22.8	19.9	66.7
Corn grain	16.4		22.5
Soybean	14.9		23.9
Wheat	8.0		15.0
Corn silage	4.9		2.2
Oats	3.0		0.7
Barley	1.2		1.3
Rice	0.4		1.1
Fruit, Nuts, and Berries	6.2	6.9	1.9
Vegetables and Potatoes	3.0	6.4	3.0
Nursery/Greenhouse	2.6	7.3	0.3
Tobacco	2.7	0.8	0.1
Cotton	1.2	2.0	4.1

NOTES: Percent of farms raising each commodity = Number of farms reporting inventories of each livestock species or number of farms reporting acreage of each crop/Total number of farms in the United States.
Percent of U.S. farm sales by commodity = Sales of each commodity/Total U.S. farm sales.
Percent of U.S. harvested cropland = Percent of harvested acres in each crop/Percent of all U.S. harvested cropland.

corn, soybean, forage crops, and wheat. Efforts to significantly increase cropping diversity, change tillage practices, or reduce nonpoint source pollution from cropping activities will need to emphasize those commodity production systems.

The geography of U.S. agriculture is shaped by a range of biophysical, economic, and demographic factors that vary widely by region. Researchers at USDA demonstrate the landscape diversity by combining data on county-level farm characteristics with data on natural resource conditions, such as areas with similar physiographic, soil, and climatic traits. (For maps and definitions, see USDA-ERS, 2009c.) They identified nine major "farm resource regions" in the United States (Heimlich, 2000). Figure 2-6 describes these regions and highlights the importance of them, the combination of which accounts for almost half of U.S. farms, 60 percent of the value of production, and 44 percent of U.S. cropland. The three regions are the "heartland" region in the corn belt, where cash grain and cattle and hog production dominates; the "fruitful rim" along the Pacific coast, southern Texas, and Florida where large farms are concentrated and fruit, vegetable, nursery, and cotton production dominates; and the "northern crescent," a traditional dairy and cash grain region. Farm commodity systems and production practices often differ markedly across the various farm resource regions in the United States.

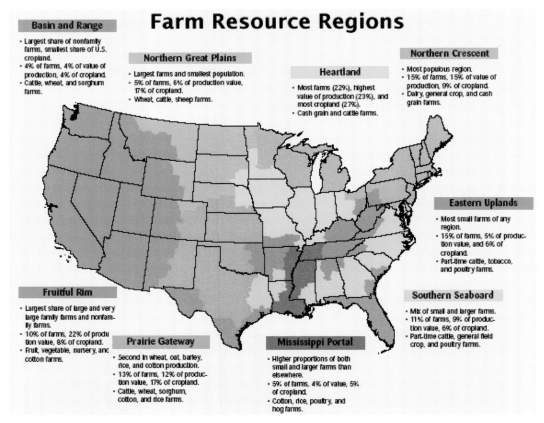

FIGURE 2-6 Farm resource regions in the United States.
SOURCE: USDA-ERS (Heimlich, 2000).

CHALLENGES TO U.S. AGRICULTURE IN THE 21ST CENTURY

The U.S. agricultural sector has evolved over time to meet the challenge of providing adequate food, feed, fiber, and landscape ornamental crops at acceptable consumer prices, but new challenges emerge. Demands on agriculture are not limited to food, feed, and fiber needs, but now include biofuel needs. As productivity in agriculture continues to increase, the natural resources used to support agriculture are being depleted or degraded. Such economic concerns as farm sector profitability and rising input costs and such social concerns as labor justice, food quality and safety, animal welfare, and community well-being are also becoming more prominent.

Increasing Demand on U.S. Agriculture

Agriculture faces the pressure of increasing demand for food, feed, fiber, and fuel as a result of population growth, changes in diet, and the emergence of the bioenergy industry. The world population is growing and reached 6.8 billion people in 2009—a 9 percent increase since 2000.

The U.S. population continues to increase at a similar rate as the world population, from 281 million people in 2000 to 307 million people in 2009 (U.S. Census Bureau, 2009). The total consumption of different food groups has been increasing as a result of a larger U.S. population (Lin et al., 2003). It is not merely the absolute population, however, that drives the demand for agricultural products, but rather population growth accompanied by income growth. As incomes grow, the composition of agricultural products demanded also changes. At low levels of intake of meat, milk, and eggs, an increase in consumption of these foods is known to be nutritionally beneficial because the biological value of protein in foods from animals is about 1.4 times that of foods from plants (CAST, 1999). Americans, however, are consuming on average more meat than the amount recommended by the federal dietary guideline (Wells and Buzby, 2008). About 3 pounds (lb) of grains is required to produce 1 lb of meat from any animal species (CAST, 1999). Therefore, a high level of meat consumption increases the demand on agriculture, because farmers have to produce feed for livestock, which in turn will provide meat for consumers (Pimentel and Pimentel, 2003).

In addition to meeting the demands of the domestic market, U.S. agriculture has to meet the demands of its foreign market. Agricultural exports are likely to increase in the coming years because of new demand from emerging markets (Gehlhar et al., 2007). Foreign demand for U.S. agricultural exports is largely determined by income level and the rate of economic growth. U.S. agricultural exports historically were largely dependent on such high-income markets as Canada, Japan, and the European Union. Slow income and population growth in Japan and the European Union have weakened demand for U.S. agricultural exports to those countries (Figure 2-7). Rising incomes in such emerging markets as Mexico and China have led to an increased proportion of U.S. agricultural export to those countries in the last 15 years (Gehlhar et al., 2007) and an overall increase in the value of U.S. agricultural exports (USDA-ERS, 2008).

In addition to meeting domestic and foreign demand for food, feed, and fiber and responding to incentives provided within federal legislation, U.S. agriculture has also been producing crops for the emerging bioenergy and biofuels market over the last 10 years. The corn grain ethanol and soybean biodiesel industries have opened a new market for corn and soybean. In the 2007 crop year (from September 2, 2007, to August 31, 2008), 8.2 billion gallons of ethanol were produced from 3 billion bushels of corn (NCGA, 2008) and 450 million gallons of biodiesel were produced from 275 million bushels of soybean (NBB, 2007). Those

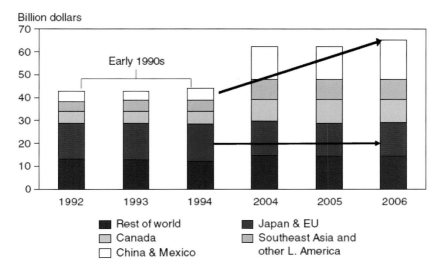

FIGURE 2-7 Shift in United States agricultural exports toward emerging markets.
SOURCE: USDA-ERS (Gehlhar et al., 2007).

quantities represent 23 percent of the year's corn harvest (up from 20 percent in 2006) and 17 percent of the year's soybean harvest (USDA-NASS, 2008). The biofuel market increases the demand for corn and soybean, which in turn raises the price of those commodities. Although higher commodity prices benefit some farmers, they increase production costs for others that use corn and soybean as part of animal feed (Westcott, 2007; Donohue and Cunningham, 2009). Moreover, increasing production of those commodities could displace other crops and have negative environmental and social consequences (Westcott, 2007; Donner and Kucharik, 2008; Pineiro et al., 2009).

The use of food commodities to produce biofuels has raised concern about competition between food versus fuel and the related impact of biofuel production on food prices. In the early years of corn grain ethanol production, there was little impact on the U.S. corn market because of the slow growth of the ethanol industry, higher corn yields, and large corn stocks (Baker and Zahniser, 2006). As corn grain ethanol production increases, however, the feedstock has to be either diverted from corn exports or come from higher production. Unless the conversion of grain to ethanol becomes more technically efficient, higher production can be achieved by diverting more land to corn production or growing corn more intensively. Both approaches could increase food prices and raise additional environmental concerns (Box 2-2). Although most economists agree that the emerging biofuel market contributes to higher food prices (FAO, 2008), its relative contribution has been widely debated and estimates vary from a few percentage points (Glauber, 2008) to as much as 39 percent of the increase in real prices (Collins, 2008; Rosegrant, 2008).

Natural Resource Scarcity

Land Availability

The long-term adequacy of agricultural land in the United States has been a continuing concern (Sampson, 1981). A study by the American Farmland Trust (2002) reported that

> **BOX 2-2**
> **Social, Economic, and Environmental Effects of Biofuels**
>
> Biofuel production in the United States has been considered an option to reduce U.S. reliance on oil because the feedstock for fuel production can be grown domestically. Technological developments and policy changes have contributed to the rapid growth in production of various agricultural crops that can be processed into liquid and gaseous fuels. A key factor was the passage of the Energy Policy Act in 2005 and the Energy Independence and Security Act (EISA) in 2007 by the U.S. Congress that mandates dramatically increased levels of biofuels consumption over the next 15 years.
>
> The biofuel industry has created a major new competitive use for U.S. grains. The portion of U.S. corn used to produce ethanol rose from 6 percent in the late 1990s to almost 25 percent in 2008. Growth of the industry, however, has slowed since 2008 partly because of the global economic recession. Production of corn grain ethanol and soybean biodiesel have unintended environmental, social, and economic consequences.
>
> **Influence on Food Security**
>
> As of 2009, the overwhelming bulk of biofuel produced in the United States was corn grain ethanol (Wescott, 2009), although some biodiesel is produced from soybean and animal fats. Growth in the use of corn for ethanol production has been associated with reduced availability of corn for livestock feeds (Doering, 2008). Although higher commodity prices benefit some farmers, they also increase production costs for others that use corn and soybean as part of animal feed (Westcott, 2007; Donohue and Cunningham, 2009). The use of dry distiller grain, a byproduct from corn ethanol production, as animal feed for beef and dairy cattle can only partly offset the adverse effects of increased feed costs, because every bushel of corn used for ethanol is two-thirds less of a bushel of corn for livestock feed (Miranowski et al., 2008). As a result, the use of commodity crops for fuel production or use of prime cropland to produce biofuel feedstock instead of food crops has raised concerns about "food versus fuel." Some estimate that, if the corn grain ethanol industry continues to expand, the positive effects of ethanol industry on rural economies will be offset by the negative economic effect on the livestock industry (Miranowski et al., 2008). In addition, many are concerned that using commodity crops for fuel might contribute to rising world food prices (Collins, 2008; FAO, 2008; Rosegrant, 2008 and references cited therein), although food prices are affected by many other factors, including energy costs (Dewbre et al., 2008). In the United States, retail prices for red meats, poultry, and eggs rose dramatically in 2007 and 2008 in response to many factors, including higher feed prices (Leibtag, 2008).
>
> **Influence on Environmental Sustainability**
>
> Biofuels are also viewed as a potential strategy to reduce greenhouse gas emissions from the transportation sector, because the exhaust emissions of carbon monoxide from burning biofuels are considerably lower compared to petroleum-based fuels (EPA, 2002; Ribeiro et al., 2007). Although corn grain ethanol and soybean biodiesel have positive net energy balances (that is, energy in the fuels produced exceeds the energy required to produce them) and have lower greenhouse-gas life-cycle emissions compared to petroleum-based fuel (Hill et al., 2006; Wang et al., 2007), the greenhouse-gas benefits might not be realized if forest or grassland is cleared elsewhere to grow crops displaced by corn and soybean in the United States (Fargione et al., 2008; Searchinger

urbanized land grew by about 47 percent from 1982 to 1997 even though the U.S. population only grew by about 17 percent during that period. Much of the growth of urbanized land was at the expense of high-productivity cropland. Although various studies and analyses have shown that current land use changes do not represent threats to the nation's total food production (USDA-NRCS, 2001), there is a growing concern that prime farmland[4] near urban areas is being lost to nonagricultural uses through development and that the

---

[4]Prime farmland, as defined by USDA, is land that has the best combination of physical and chemical characteristics for producing food, feed, forage, fiber, and oilseed crops and is also available for these uses.

et al., 2008). That is because carbon sequestration achieved by changing practices to reduce carbon on one landscape can be offset by increased carbon releases on other landscapes and result in a net increase in total carbon emission (Murray et al., 2004; IPCC, 2006).

Although difficult to document, it is plausible that the increasing demand and prices of corn and soybeans—in addition to expanding the acreage dedicated to corn and soybean—have also encouraged farmers to increase production by reducing crop rotations, planting marginal lands, reducing fallow, or returning acreage that was idled because of enrollment in the Conservation Reserve Program (Westcott, 2007), all of which might exacerbate any negative environmental and habitat effects associated with intensive corn production (Donner and Kucharik, 2008; Pineiro et al., 2009). If EISA's objective of 15 billion gallons of corn grain ethanol by 2022 is realized, for example, the average flux of dissolved nitrogen export by the Mississippi and Atchafalaya Rivers into the Gulf of Mexico is predicted to increase by 10 to 34 percent (Donner and Kucharik, 2008). That level of nutrient influx will greatly expand the hypoxia zone in the Gulf—where dissolved oxygen in the water is too low to support marine life. (See Box 2-4.)

Influence on Economic Sustainability

Expanding the biofuel industry can enhance farm-level economic security on some farms and decrease it on others. As mentioned earlier, increases in corn and soybean prices benefit those commodity farms, but can have adverse effects on farms that rely on grains for feed. Even for the commodity farmers, the demand of corn grain for ethanol production depends on several factors, including the price of ethanol and biodiesel relative to oil (NAS-NAE-NRC, 2009), the percentage of ethanol that can be blended in fuel, and the number of flex-fuel vehicles (that can use 85 percent ethanol) on the road (Westcott, 2009).

Because biomass is bulky and expensive to transport, biorefineries will likely be built in areas where biomass feedstock for fuel is abundant. Biomass feedstock production will likely attract biorefineries to rural communities (as in the case with corn grain ethanol and soybean biodiesel). Biorefineries, particularly those with local investment, provide some additional jobs in rural communities and potentially enhance economic vitality (Kleinschmit, 2007). However, some empirical evidence suggests that job gains have been overestimated (Swenson, 2009), and that there can be other mixed or negative social impacts in host communities (Selfa et al., 2009).

Encouraged by EISA, other government-provided incentives, and rising oil prices until 2008, facilities have been constructed to convert corn starch to ethanol and to convert primarily soybean oil and secondarily animal fats and waste cooking oils to biodiesel. The growth of the biofuels industry has slowed since 2008. The economic recession and the steep decline of crude oil prices in the last quarter of 2008 (from the peak of $147 per barrel on July 11, 2008, to about $35 per barrel at the end of the same year) resulted in many ethanol plants running below capacity and caused many to declare bankruptcy (Andrejczak, 2009; Knauss, 2009; The Associated Press, 2009; VeraSun Energy, 2009). According to one source, ethanol production in 2009 was estimated at approximately 10 billion gallons, far less than the previous capacity of 12.5 billion gallons (Krauss, 2009). Meanwhile, biodiesel plants continue to operate at 50 percent of capacity or less. The economic benefits brought by a biorefinery to local communities might be volatile because biofuels are new compared to petroleum-based fuels. The long-term stability of the biofuels market remains to be seen.

The ultimate environmental, social, and economic impacts of using corn and soybean grain for production of biofuels are under debate at this time. Another National Research Council study on the economic and environmental impacts of biofuels that is in progress will examine those issues in depth.

reversal of those developments is politically difficult and expensive (Klein and Reganold, 1997; Nizeyimana et al., 2001). In 2001, the total developed area in the contiguous United States was slightly more than 106 million acres (1 acre = 0.40 hectare), which included 33.2 million acres of agricultural lands—including cropland, pastureland, rangeland, and forest land—that were converted to developed uses (for example, large urban and built-up areas, small built-up areas, and rural transportation land) from 1982 to 2001 (Figure 2-8) (USDA-NRCS, 2003). The rate of prime farmland development increased from an average of 400,000 acres per year between 1982 and 1992 to more than 600,000 acres per year between 1992 and 2001 (Figure 2-9). Researchers (Reganold and Singer, 1984) have shown that ratios of

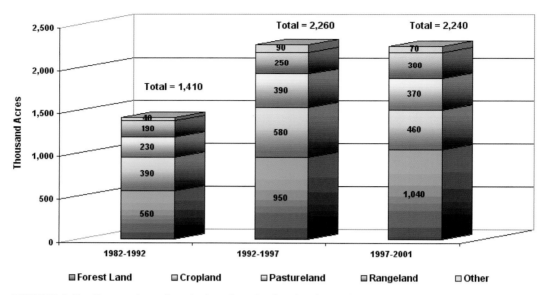

FIGURE 2-8 Conversion of agricultural and other lands to developed uses in the United States between 1982 and 2001.
SOURCE: USDA-NRCS (2003).

economic input to output for farming on prime farmlands are significantly lower than for non-prime farmlands. The economic return for farming on higher-quality soils tends to be better than farming on lower-quality soils and marginal lands for food production.

A related concern is that of pressure for nonfarm development in areas where many high-value specialty crops are grown. Much of the specialty fruit and vegetable production occurs in southern states such as Florida and California, which are experiencing the most

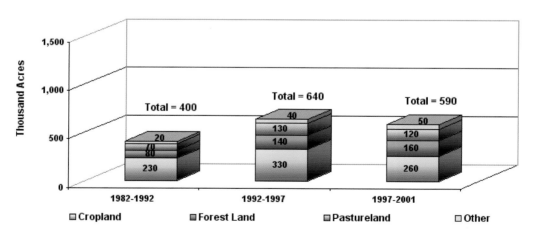

FIGURE 2-9 Conversion of prime farmland to developed uses in the United States between 1982 and 2001.
SOURCE: USDA-NRCS (2003).

rapid rates of land development in the country (Norris and Deaton, 2001). An American Farmland Trust study (2002) estimates that 86 percent of U.S. fruits and vegetables are produced in areas influenced by rapid urban development pressures. While the farms in those areas might benefit from close access to large population centers, they are competing for land and water with urban and industrial uses.

Water Availability

Agriculture accounts for 80 percent of consumptive water use[5] in the United States (USDA-ERS, 2004). Irrigated cropland is an important and growing component of the U.S. farm economy and was the largest use of freshwater by 2000 (Hutson et al., 2004). Although only 16 percent of the U.S. cropland is irrigated, that acreage is used for high-value crops that account for nearly half of the total U.S. crop sales (USDA-ERS, 2004).

Some regions of the United States with the greatest water shortages also have the most rapidly growing populations. California's population is expected to grow from 34 million to nearly 60 million from 2000 to 2050, mostly in the water-starved urban regions of southern California (California Department of Finance, 2007). The seven states sharing the Colorado River are projected to increase their populations by 47 percent from 2000 to 2030, significantly higher than the increase of 29 percent projected for the United States as a whole (U.S. Census Bureau, 2009).

Meanwhile, ground water is extensively pumped from many aquifers to provide water for domestic and agricultural use. When the rate of extraction exceeds the rate of recharge by natural processes, ground water is said to be in a state of overdraft, and water levels drop. Under prolonged overdraft conditions, the water level of an aquifer can fall to a depth where it is no longer economically feasible to pump and the resource becomes exhausted. The time required for natural recharge to return the water level to a depth practical for extraction can be considerable, and in the case of aquifers that were charged in previous climate cycles (so-called fossil aquifers such as the Ogallala), the resource is unrecoverable in any practical time frame (Box 2-3). The challenge to agriculture is to conserve and recycle water to extend the use of the fossil aquifers or adapt to dryland farming.

Although irrigation has become much more water and energy efficient, water scarcity will likely be a challenge for agriculture in years to come as the supply of water decreases and competing demands of water from other sectors increase. As the demand rises and supply declines, the price of water will likely increase, which in turn increases production costs of irrigated agriculture (Box 2-3). The Family Farm Alliance (2007) expressed concern that urban and industrial water demands will be met at the expense of domestic agricultural production.

Climate Change

Agriculture faces uncharted challenges posed by global climate change in the near future. Global surface temperature shows an increase of 0.78°C since the beginning of the 20th century (NRC, 2008); it is projected to increase by another 1.1–6.4°C by 2100 (IPCC, 2007). Climate models project an increase in carbon dioxide (CO_2) levels and temperature, changes in rainfall, and increased frequency in extreme weather events (for example, heat waves and heavy precipitation) in the mid-latitude to high-latitude regions (IPCC, 2007).

---

[5] Amount of withdrawn water lost to the immediate water environment through evaporation, plant transpiration, incorporation in products or crops, or consumption by humans and livestock.

> **BOX 2-3**
> **Overdrafting of the Ogallala Aquifer**
>
> The success of large-scale farming in areas lacking adequate precipitation and reliable perennial surface water depends heavily on pumping ground water for irrigation. In the Midwestern United States, ground water overdrafting of the Ogallala or High Plains aquifer presents a significant long-range problem. This aquifer is a large underground reservoir that encompasses portions of eight states ranging from South Dakota to Texas. About 27 percent of the total irrigated land in the United States overlies this aquifer system, which yields about 30 percent of the nation's ground water used for irrigation. The water is used to produce corn, wheat, cotton, alfalfa, and soybean; some of those crops are used to support about 40 percent of the U.S. supply of feedlot beef (USDA-ERS, 2005b). In addition, the aquifer system provides drinking water to 82 percent of the people who live within its boundary (Dennehy, 2000).
>
> Water-level declines started to occur in the Ogallala aquifer soon after the beginning of extensive ground water irrigation development following the end of World War II. By 1997, 13.7 million acres were being irrigated by ground water from the aquifer, accounting for some 20 percent of all U.S. irrigation. Because withdrawals exceed natural recharge, the water table has been declining (Colaizzi et al., 2009). By 1980, water levels in the Ogallala aquifer in parts of southwestern Kansas, New Mexico, Oklahoma, and Texas had declined more than 100 feet (Luckey et al., 1981). Some croplands in Texas have suspended the use of irrigation with ground water because of its expense relative to the value of the crops grown. In effect, the Ogallala aquifer is a nonrenewable resource, similar to a coal mine. The current pumping of groundwater for irrigation is permanently depleting (or mining) ground water quantities available for future uses (Kneese, 1986).

Although increases in CO_2 and temperature could benefit crop production, the interactive effects of CO_2, temperature, and rainfall might result in the opposite (Easterling et al., 2007). Because precipitation is the main driver of variability in the water balance over space and time, future changes in regional precipitation could have important implications for agricultural systems. Current climate models simulate a climate change-induced increase in annual precipitation in high and mid-latitudes (Carter et al., 2000). Precipitation extremes are predicted to increase in frequency, with more drought and flood occurrences. Rosenzweig et al. (2002) estimated that production losses as a result of excessive moisture could double in the United States under scenarios of heavy precipitation. Increased frequency of droughts is of particular concern in the arid Southwestern United States, where water resources are already stretched and the population is increasing rapidly, and in the southern regions of the Ogallala where shifts to dryland farming are expected.

Crop yields could be further decreased because of changes in the dynamic between crops and weeds, pests, and diseases as a result of climate change (Patterson et al., 1999). A study reported that C_3 weeds tend to benefit more from an increase in CO_2 than C_3 crops (Ziska, 2003). Research also suggests that the efficacy of glyphosate herbicide on weeds decreases with increasing CO_2 (Ziska et al., 1999; Ziska and George, 2004; Ziska and Goins, 2006). Interactions between CO_2 and temperature or CO_2 and precipitation have been recognized as key factors in determining plant damage (Easterling et al., 2007). For example, warming trends in the United States could increase winter survival of pests and hence proliferation (Diffenbaugh et al., 2008) and could lead to earlier spring activity. Weather extremes could increase the vulnerability of plants to pests and diseases and promote outbreaks. Those potential effects of climate change on U.S. agriculture are discussed in further detail in Hatfield et al. (2008).

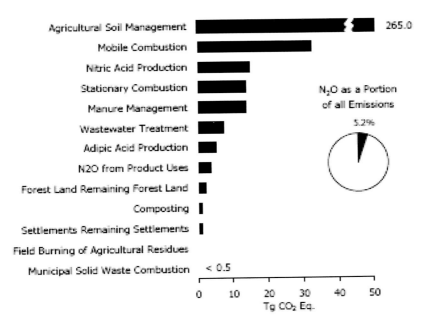

FIGURE 2-11 Nitrous oxide emission from different sources in the United States. SOURCE: EPA (2008).

2004). The soil loss through wind and water erosion has particularly long-term ecological and economical effects by reducing the overall productivity of agricultural lands and by impacting water quality, water supplies, navigation, and irrigation infrastructures. Chemical soil degradation includes salinization, nutrient depletion, acidification, contamination, and toxification from pesticides and excessive fertilization (NRC, 1993a).

Biological soil degradation includes a decline in biodiversity and soil carbon and an increase in soil-borne pathogens (Lal, 2004). Soil organisms contribute to the maintenance of soil quality and control many key processes, such as decomposition of plant residue and organic material, nitrogen fixation, and nutrient availability (Kennedy and Papendick, 1995). However, compared to physical and chemical soil degradation, little is known about how agricultural activities alter soil biological properties and how they affect soil functioning (Heimlich, 2003).

Modern agriculture's production of a few species of crops (and crops separated from livestock production), with limited rotations or crop diversity, runs counter to the natural tendency for more diversity that can result in high-quality soils. Conventional systems "lead, more often than not, to a decrease in soil quality, as indicated by the soil's ability to infiltrate and hold water, to maintain particle structure for optimal root habitat, and to hold and recycle nutrients. Less-than-optimal soil quality raises production costs in the long term, lowers production potential, and accentuates production variability" (Harwood, 1994, pp. 34–59).

One particular soil quality concern is that of salinization. In many arid zones where irrigation is practiced, salinization of soil can become a chronic problem. It is exacerbated by high water tables where water can wick upward to the surface or flow laterally to lower terrain and evaporate. Salinity poses a major challenge for farming in many arid areas relying on irrigation to grow crops. Even when relatively good quality water is used, salinity problems can arise whenever ground water levels are shallow enough to allow upward movement and evaporation at the soil surface. Enlightened management practices have to be used for irrigation water management under saline conditions to avoid land loss and crop damage. About 25 percent of irrigated cropland in the United States (14 million acres) is significantly affected by salts in soil and water (Hedlund and Crow, 1994). The worst salinity problems are in the productive San Joaquin Valley of California, which is experiencing loss of land and yield reductions from salinization due to high water tables (USDA-NRCS, 1997). Some 850,000 acres of the San Joaquin Valley's 2.5 million irrigated acres are affected by inadequate drainage and accumulating salts. The problem is exacerbated by the fact that the valley is a closed drainage basin with no current legal means for exporting drainage from the region. Land is lost each year due to surface accumulation of seepage from the higher regions to lower ones (San Joaquin Valley Drainage Program, 1990).

Saline soils are not limited to irrigated areas. Mineral weathering and dissolution of cretaceous shale occur over a large portion of the arid West. Saline seeps occur when water that exceeds plant requirements percolates unchecked below the root zone, then moves laterally downhill and emerges in a seep area. Those seeps frequently occur where farmers practice wheat fallow rotations. Seeps have affected about 500,000 acres of cropland in the Great Plains from Montana to Texas and some 2.5 million acres for all land uses (Brown et al., 1982).

Reduced Genetic Diversity

Large-scale farming systems, the need to feed large numbers of people, the globalization of markets, and loss of wild habitats as a result of land conversions have resulted in a dramatic reduction in crop genetic diversity throughout the world. The National Research Council documented those trends more than three decades ago in the report *Genetic Vulnerability of Major Crops* (1972). According to the Food and Agriculture Organization (FAO) of the United Nations (FAO, 1998), 75 percent of agricultural biological diversity was lost during the 20th century. Modern crop varieties have supplanted traditional varieties or landraces for over 70 percent of the world's corn, 75 percent of the Asian rice, and half of the wheat in Africa, Latin America, and Asia (Picone and Van Tassel, 2002).

The loss of genetic diversity in agriculture reduces the genetic material available for future use by farmers and plant breeders. In addition, genetic evolutionary processes that lead to the development of new genes and gene combinations might be curtailed. The increase in genetic uniformity within a crop can also lead to greater genetic vulnerability to pest and diseases (NRC, 1972, 1993b). More than the loss of a particular variety or landrace, however, the greatest concern is the irreversible loss of genes within plant gene pools that are critical for breeding. Much of the yield increase, resistance to biotic and abiotic constraints, and adaptation to poor soils, drought, or low temperatures in modern crops is a result of genes from traditional varieties. The ability of crops to adapt to future cropping systems and climate change will depend on access to genetic variation.

Not all breeding efforts, however, have led to reduced genetic diversity. The NRC study *Agricultural Crop Issues and Policies* (1993b) found that genetic diversity of U.S. wheat and corn has increased since the 1970s, in part because of efforts to breed in greater genetic diversity. However, the genetic uniformity of rice, beans, and many minor crops is still of concern. Since that study, the development and wide-scale adoption of genetically engineered crops (principally corn and soybean) have led to further concerns about the genetic homogenization of crops across large areas within the United States and worldwide. In addition, the potential contamination of landraces by transgenic varieties growing in the same region, as observed in maize landraces growing in farmers' fields in Mexico, is a concern (Pineyro-Nelson et al., 2009).

Similar to plant breeding, confinement livestock production systems have been associated with a decrease in the numbers of minor breeds and accelerated the development of genetically similar hogs, poultry, and beef and dairy cattle as a result of selective breeding by humans. Because large processing plants require a steady flow of uniform animals and bird types to achieve economies of scale (RTI International, 2007), uniformity in animals is achieved by controlling their genetics and length of feeding period (MacDonald and McBride, 2009). Genetic diversity is necessary for sustained genetic improvements of farm animals in the future and for rapid adaptation to changes in breeding objectives (Notter, 1999).

Much of the world's plant germplasm is stored in repositories in the United States in the USDA National Plant Germplasm System, the International Agricultural Research Centers, and other public and private collections. Likewise, animal germplasm is managed by the USDA National Animal Germplasm Program, which was initiated in 1999. Although there has been significant progress in collecting and conserving crop and animal germplasm, a number of problems remain; for example, collections were not adequately documented and seed and propagule viability is reduced as a result of improper storage (Plucknett, 1987; Blackburn, 2006). There has also been considerable discord on who owns, controls, and benefits from germplasm collections.

To slow or prevent the loss of crop genetic diversity worldwide and provide for the fair sharing of benefits arising from the use of genetic resources, a number of international agreements have been developed to encourage preservation of genetic diversity and to promote the exchange of germplasm. The most important is the International Treaty on Plant Genetic Resources for Food and Agriculture, which entered into force in June 2004. The treaty, of which the United States is a signatory, is a comprehensive international agreement that aims at guaranteeing food security through the conservation, exchange, and sustainable use of the world's plant genetic resources, as well as fair use and equitable benefit-sharing. It also recognizes farmers' rights to freely access genetic resources and to use and save seeds under national laws. The treaty implements a multilateral system of access to 64 of the most important food and forage crops essential for food security and interdependence for countries that ratified the treaty. Progress towards implementation of the treaty has been slowed by lack of consensus among the treaty parties on the value of particular genetic resources, and consequently many of the treaty's provisions are vague (Day-Rubenstein et al., 2005).

Economic Concerns

Farm Sector Profitability and Rising Input Costs

The estimated value of U.S. gross farm income has increased by 31 percent in real terms since 1970 (Figure 2-12). However, the aggregate value of net farm income received by farmers has not changed dramatically over the last 40 years. In essence, increases in gross farm receipts have generally been cancelled out by increases in the costs of production. For example, U.S. farms sold $324 billion in agricultural products in 2008 (up 65 percent from 1998 values in nominal dollars) but incurred $291 billion in production expenses, including $204 billion in purchased inputs (an increase of 57 percent and 73 percent since 1998, respectively). Much of the increase in purchased input costs was related to the rising prices

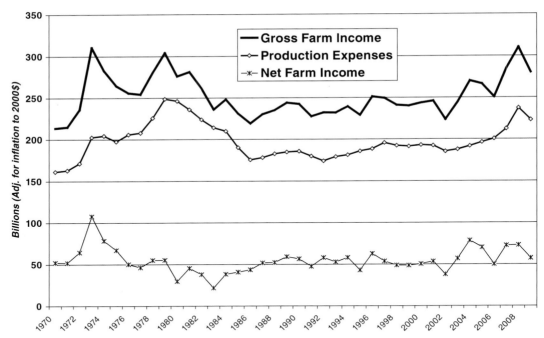

FIGURE 2-12 U.S. aggregate gross farm income, production expenses, and net farm income from 1970 to 2009 (expressed in constant 2000 dollars).
NOTES: Gross farm income includes estimated value of total farm production plus direct government payments. Production expenses include all purchased inputs and payments to stakeholders (payment to stakeholders includes hired labor, rental payments to nonoperator landlords, and payment of real estate interest and other interests. Net farm income is the result of subtracting production expenses from gross farm income.
DATA SOURCE: USDA-ERS (2009d).

of fuel and synthetic fertilizer, which were affected by crude oil prices that rose from an average price of about $12 per barrel in 1998 to $95 in 2008 (EIA, 2009). Annual farm product market gyrations are normal, but the potential effect of rising production costs, especially for fuel and fertilizers, could have a long-term impact on farm productivity, farm income, and increasing food prices.

Statistics on the aggregate profitability of the U.S. farm sector disguise considerable variation in the economic performance of individual farms. For example, in 2007, only 47 percent of all U.S. farms reported positive net farm income, a drop from 57 percent of all farms in 1987. Most farms that lost money were relatively small operations that relied principally on nonfarm sources of income. Most farms in the United States are essentially family businesses that rely mainly on farm family members for their labor force (Gasson and Errington, 1993; Hoppe et al., 2007), and the majority of farm families also gain income from off-farm work. Nonfarm work or transfer payments are commonly used to supplement income from the farm business. The proportion of farm operators who work off-farm increased from 44 percent in 1979 to 52 percent in 2004. The proportion of spouses working off-farm grew from 28 percent to 45 percent during the same period (Fernandez-Cornejo et al., 2007). The contribution of off-farm income to the total household income of U.S. farmers rose from about 50 percent in 1960 to more than 80 percent in 2004 (Fernandez-Cornejo et al., 2007).

Table 2-1 reported the percent of U.S. net cash farm income that was received by different types of farms. The data suggest that almost 80 percent of total U.S. net cash farm income went to farms with gross sales over $500,000 in 2007, while just 7 percent of total U.S. net farm income was shared by the 87 percent of farms that had gross sales under $250,000. Internal variations in the rate of return for different types of U.S. farms help explain why so many individual farmers report increasingly difficult economic stress at the same time that the performance of the overall farm sector appears to be relatively stable.

Producers of different commodities also report different levels of profitability. Table 2-3 reported data from the 2007 Census of Agriculture aggregated by the type of commodity that provides the majority of income to each farm. In general, the results suggest that crop farms generate higher average profit margins than do livestock operations. The worst performing commodity sectors are cattle feedlots and beef cattle ranching (which averaged less than a 10 percent return) and sheep, goat, aquaculture, and mixed livestock farms (which registered negative net income overall).

The profitability of many U.S. farms is partly determined by the level of federal government program payments. For example, in 2008, direct government payments totaled $12 billion, which accounted for 3.7 percent of total gross farm receipts and 13.9 percent of net farm income in the United States. Over the past 10 years, however, government payments have tended to be significantly higher than before, averaging roughly 7 percent of gross farm returns and 27 percent of net farm income.

The distribution of government payments across different segments of the farm sector varies widely (see Tables 2-1 and 2-3). According to data collected by the Census of Agriculture, only 38 percent of farms reported receiving government payments in 2007 (16 percent participated in conservation programs like the Conservation Reserve Program, Wetlands Reserve Program, and Conservation Reserve Enhancement Program; 2 percent received a loan from the Commodity Credit Corporation; and 31 percent participated in other federal farm programs). The farms that receive the bulk of government farm program payments raise cash grains that are eligible for non-recourse production loans, price supports, and other commodity programs. Government program payments are much more common among farms in the Midwest and Great Plains (where over two-thirds of farms in North Dakota, Iowa, Illinois, South Dakota, Nebraska, Minnesota, and Kansas received government program payments in 2007).

Loss of Mid-Sized Commercial Family Farms

Reduced and more volatile farm income can affect the economic vitality of individual farm households and rural communities.[7] Indeed, declining rates of return for individual farms have been linked to changes in the size structure of U.S. farms (Hallam, 1993; Gardner, 2002). Table 2-4 shows the percent of farms and farm sales that were produced by different-sized farms from 1997 to 2007. Mid-sized farm sales categories ($10,000–$249,999) are declining in importance, while small farms have increased in number and large farms

---

[7]Although farming activities are not restricted to rural areas (Jackson-Smith and Sharp, 2008), the economic interdependence between agriculture and community well-being is most pronounced in rural areas. In this report, the term "rural" refers to less densely populated towns or landscapes in the United States. However, researchers and federal agencies have used the term in multiple ways to reflect different analytical or program goals (Cromartie and Bucholtz, 2008).

TABLE 2-3 Distribution of Farm Income and Estimated Profit Margins by Farm Commodity Type

NAICS Farm Type (Majority of Income from Commodity)	Gross Farm Sales and Government Payments (Billion $)	% Total	Net Cash Farm Income from Operations (Billion $)	% Total	Crude Profit Margin (Net/Total)	Value of Government Payments (Billion $)	% Total	Crude Percent Return (Without Government Payments)
Crop Farms								
Oilseed and grain farming	78.6	25.8	28.7	38.4	36.5	4.052	50.7	33.0
Vegetable and melon farming	15.1	4.9	4.5	6.0	29.6	0.092	1.1	29.2
Fruit and tree nut farming	18.4	6.0	5.0	6.6	26.9	0.059	0.7	26.7
Greenhouse, nursery, and floriculture production	17.0	5.6	4.7	6.3	27.5	0.025	0.3	27.4
Other crop farming[a]	19.3	6.3	5.3	7.1	27.6	2.244	28.1	18.1
Specialty crop farms (combined)[b]	50.5	16.5	14.1	18.9	27.9	0.175	2.2	27.7
All crop farms	148.4	48.6	48.1	64.5	32.4	6.471	81.1	29.3
Livestock Farms								
Poultry and egg production	37.9	12.4	8.7	11.6	22.9	0.060	0.7	22.8
Dairy cattle and milk production	35.1	11.5	10.6	14.2	30.1	0.312	3.9	29.5
Cattle feedlots	30.4	9.9	3.0	4.0	9.8	0.103	1.3	9.5
Beef cattle ranching and farming	28.4	9.3	2.6	3.5	9.2	0.823	10.3	6.5
Hog and pig farming	18.2	6.0	2.9	3.9	16.0	0.109	1.4	15.5
Animal aquaculture and other animal production	6.4	2.1	-0.9	-1.3	-14.8	0.082	1.0	-16.3
Sheep and goat farming	0.6	0.2	-0.3	-0.4	-52.8	0.026	0.3	-59.9
All livestock farms	156.8	51.4	26.5	35.5	16.9	1.513	18.9	16.1
Total	305.2	100.0	74.6	100.0	24.4	7.984	100.0	22.4

[a]Includes tobacco, cotton, sugarcane, and other mixed crops.
[b]Subtotal that includes farms that receive most income from sale of vegetables, melon, fruit, treenut, greenhouse, nursery, and floriculture products.

SOURCE: 2007 Census of Agriculture.

fare and labor justice, but gaps in understanding remain. For example, how the collective actions of a number of farms could improve sustainability on a landscape scale is not well studied. Filling those gaps of understanding will require innovative new approaches, in particular in the realms of complex systems science and management as applied to agroecosystems, and a better understanding of economic and social outcomes of the farming approaches.

SUMMARY

U.S. agriculture has celebrated much success in the last 50 years as farmers continue to increase productivity on about the same acreage of farmland and increase energy efficiency in their production systems. However, agricultural sustainability is characterized by not only productivity and efficiency, but also by its impact on the environment and natural resource base, its economic vitality, and the quality of life of farmers and society as a whole. Although many farming practices, technologies, and approaches have improved one or two aspects of sustainability, they might have unintended negative effects on the other aspects of sustainability. As awareness on the importance of balancing the four sustainability goals increases, U.S. agriculture is at a pivotal point that can change the trajectory of farming toward improved sustainability by increasing understanding of the interactions and net impact of combinations of practices and approaches at the farm level and the collective actions of a number of farms on the landscape level.

REFERENCES

Arikan, O.A., W. Mulbry, and C. Rice. 2009. Management of antibiotic residues from agricultural sources: use of composting to reduce chlortetracycline residues in beef manure from treated animals. *Journal of Hazardous Materials* 164(2–3):483–489.

American Farmland Trust. 2002. Farming on the edge: sprawling development threatens America's best farmland. Washington, D.C.: Author.

Andrejczak, M. 2009. Pacific ethanol units joins others in bankruptcy court. *Market Watch*, May 18, 2009.

Aneja, V.P., W.H. Schlesinger, and J.W. Erisman. 2008. Farming pollution. *Nature Geoscience* 1(7):409–411.

Arcury, T.A., J.G. Grzywacz, D.B. Barr, J. Tapia, H.Y. Chen, and S.A. Quandt. 2007. Pesticide urinary metabolite levels of children in eastern North Carolina farmworker households. *Environmental Health Perspectives* 115(8): 1254–1260.

ASABE (American Society of Agricultural and Biological Engineers). 2005. Manure production and characteristics. *ASAE Standards* (March):20.

Baker, A., and S. Zahniser. 2006. Ethanol reshapes the corn market. *Amber Waves* 4(2):30–35.

Ball, E. 2005. Ag productivity drives output growth. Available at http://www.ers.usda.gov/AmberWaves/June 05/findings/AgProductivity.htm. Accessed on August 31, 2009.

Batie, S. 2008. The sustainability of U.S. cropland soils. In Perspectives on Sustainable Resources in America, R.A. Sedjo, ed. Washington, D.C.: RFF Press.

Berry, W. 2004. The Unsettling of America. 3rd edition. San Francisco: Sierra Club Books.

Blackburn, H.D. 2006. National animal germplasm program: challenges and opportunities for poultry genetic resources. *Poultry Science* 85:210–215.

Bricker, S.B., C.G. Clement, D.E. Pirhalla, S.P. Orlando, and D.R.G. Farrow. 1999. National estuarine eutrophication assessment: effects of nutrient enrichment in the nation's estuaries. Silver Spring, Md.: National Oceanic and Atmospheric Administration, National Ocean Service, Special Projects Office and the National Centers for Coastal Ocean Science.

Brown, P.L., A.D. Halvorsen, F.H. Siddoway, H.F. Mayland, and M.R. Miller. 1982. Saline-seep diagnosis, control, and reclamation. Available at http://www.wsi.nrcs.usda.gov/products/W2Q/downloads/Salinity/Saline_Seeps.pdf. Accessed on February 18, 2010.

Browne W.P., J.R. Skees, L.E. Swanson, P.B. Thompson, and L.J. Unnevehr. 1992. Sacred Cows and Hot Potatoes: Agrarian Myths in Agricultural Policy. Boulder, Colo.: Westview Press.

California Department of Finance. 2007. Population projections by race/ethnicity for California and its counties 2000–2050. Available at http://www.dof.ca.gov/html/DEMOGRAP/ReportsPapers/Projections/P1/P1.asp. Accessed on June 8, 2009.

Carter, T.R., M. Hulme, J.F. Crossley, S. Malyshev, M.G. New, M.E. Schlesinger, and H. Uomenvirta. 2000. Climate change in the 21st century: interim characterizations based on the new IPCC emissions scenarios. *The Finnish Environment* 433:148.

Cassman, K.G., and A.J. Liska. 2007. Food and fuel for all: realistic or foolish? *Biofuels Bioproducts & Biorefining-Biofpr* 1(1):18–23.

CAST (Council for Agricultural Science and Technology). 1999. Animal Agriculture and Global Food Supply. Ames, Iowa: Author.

Castanon, J.I.R. 2007. History of the use of antibiotic as growth promoters in European poultry feeds. *Poultry Science* 86(11):2466–2471.

CENR (Committee on Environment and Natural Resources). 2000. Integrated Assessment of Hypoxia in the Northern Gulf of Mexico. Washington, D.C.: National Science and Technology Council.

Chameides, B. 2009. Climate change: what is equivalent to "CO_2 equivalents"? Available at http://www.nicholas.duke.edu/thegreengrok/co2equivalents. Accessed on February 19, 2010.

Chander, Y., S.C. Gupta, K. Kumar, S.M. Goyall, and H. Murray. 2008. Antibiotic use and the prevalence of antibiotic resistant bacteria on turkey farms. *Journal of the Science of Food and Agriculture* 88(4):714–719.

Cochrane, W. 1979. The Development of American Agriculture: A Historical Analysis. St. Paul: University of Minnesota Press.

Colaizzi, P.D., P.H. Gowda, T.H. Marek, and D.O. Porter. 2009. Irrigation in the Texas High Plains: a brief history and potential reductions in demand. *Irrigation and Drainage* 58(3):257–274.

Collins, K. 2008. The role of biofuels and other factors in increasing farm and food prices. Available at http://www.foodbeforefuel.org/files/Role%20of%20Biofuels%206-19-08.pdf. Accessed on June 8, 2009.

Coronado, G.D., B. Thompson, L. Strong, W.C. Griffith, and I. Islas. 2004. Agricultural task and exposure to organophosphate pesticides among farmworkers. *Environmental Health Perspectives* 112(2):142–147.

Cox, N.A., M.E. Berrang, and J.A. Cason. 2000. Salmonella penetration of egg shells and proliferation in broiler hatching eggs—a review. *Poultry Science* 79(11):1571–1574.

Cromartie, J., and S. Bucholtz. 2008. Defining the "rural" in rural America. *Amber Waves* 6(3):28–34.

Das, R., A. Steege, S. Baron, J. Beckman, and R. Harrison. 2001. Pesticide-related illness among migrant farm workers in the United States. *International Journal of Occupational and Environmental Health* 7(4):303–312.

Day-Rubenstein, K., P. Heisey, R. Shoemaker, J. Sullivan, and G. Frisvold. 2005. Crop Genetic Resources: An Economic Appraisal. Washington, D.C.: U.S. Department of Agriculture Economic Resource Service.

Dennehy, K.F. 2000. 2009. High Plains regional ground-water study: U.S. Geological Survey fact sheet FS-091-00. Available at http://co.water.usgs.gov/nawqa/hpgw/factsheets/DENNEHYFS1.html. Accessed on August 12, 2009.

Dewbre, J., C. Giner, W. Thompson, and M. Von Lampe. 2008. High food commodity prices: will they stay? Who will pay? *Agricultural Economics* 39(3):393–403.

Diaz, R.J., and R. Rosenberg. 2008. Spreading dead zones and consequences of marine ecosystems. *Science* 321:926–929.

Dich, J., and K. Wiklund. 1998. Prostate cancer in pesticide applicators in Swedish agriculture. *The Prostate* 34:100–112.

Diffenbaugh, N.S., C.H. Krupke, M.A. White, and C.E. Alexander. 2008. Global warming presents new challenges for maize pest management. *Environmental Research Letters* 3(4).

Dobbs, T.L., and J.D. Cole. 1992. Potential effects on rural economies of conversion to sustainable farming systems. *American Journal of Alternative Agriculture* 7(1/2):70–80.

Doering, O. 2008. Biofuel Implications for Agriculture and the Environment. Presentation to the Committee on March 27, 2008, in Kansas City, Missouri.

Donner, S.D., and C.J. Kucharik. 2008. Corn-based ethanol production compromises goal of reducing nitrogen export by the Mississippi River. *Proceedings of the National Academy of Sciences* 105:4513–4518.

Donohue, M., and D.L. Cunningham. 2009. Effects of grain and oilseed prices on the costs of US poultry production. *Journal of Applied Poultry Research* 18(2):325–337.

Doyle, M.E. 2001. Alternatives to antibiotic use for growth promotion in animal husbandry. Available at http://fri.wisc.edu/briefs/antibiot.pdf. Accessed on July 14, 2009.

Doyle, M.P., and M.C. Erickson. 2008. Summer meeting 2007—the problems with fresh produce: an overview. *Journal of Applied Microbiology* 105(2):317–330.

Easterling, W.E., P.K. Aggarwal, P. Batima, K.M. Brander, L. Erda, S.M. Howden, A. Kirilenko, J. Morton, J.-F. Soussana, J. Schmidhuber, and F.N. Tubiello. 2007. Food, fibre and forest products. In Impacts, Adaptation and Vulnerability. Contribution of Working Group II to the Fourth Assessment Report of the Intergovernmental Panel on Climate Change, M.L. Parry, O.F. Canziani, J.P. Palutikof, P.J.V.D. Linden, and C.E. Hanson, eds. Cambridge: Cambridge University Press.

EIA (Energy Information Administration). 2009. Weekly all countries spot price FOB weighted by estimated export volume (dollars per barrel). Available at http://tonto.eia.doe.gov/dnav/pet/hist_xls/WTOTWORLDw.xls. Accessed on June 30, 2009.

EPA (U.S. Environmental Protection Agency). 2002. A Comprehensive Analysis of Biodiesel Impacts on Exhaust Emissions. Washington, D.C.: Author.

———. 2007. Wadeable Streams Assessment: A Collaborative Survey of the Nation's Streams. Washington, D.C.: Author.

———. 2008. Inventory of U.S. Greenhouse Gas Emissions and Sinks: 1990–2006. Washington, D.C.: Author.

Eskenazi, B., A.R. Marks, A. Bradman, K. Harley, D.B. Barr, C. Johnson, N. Morga, and N.A. Jewell. 2007. Organophosphate pesticide exposure and neurodevelopment in young Mexican-American children. *Environmental Health Perspectives* 115(5):792–798.

Family Farm Alliance. 2007. Water supply in a changing climate—the perspective of family farmers and ranchers in the irrigated west. Available at http://familyfarmalliance.clubwizard.com/IMUpload/FFA%20Report2.pdf. Accessed on February 18, 2010.

FAO (Food and Agriculture Organization). 1998. Biodiversity of food and agriculture. Available at http://www.fao.org/sd/epdirect/epre0040.htm. Accessed on February 19, 2010.

———. 2008. Soaring Food Prices: Facts, Perspectives, Impacts, and Actions Required. Rome: Author.

Fargione, J., J. Hill, D. Tillman, S. Polasky, and P. Hawthorne. 2008. Land clearing and the biofuel carbon debt. *Science* 319:1235–1238.

FDA (Food and Drug Administration). 2009. Pesticide residue monitoring program FY2004–2006. Available at http://www.fda.gov/Food/FoodSafety/FoodContaminantsAdulteration/Pesticides/ResidueMonitoringReports/ucm125183.htm. Accessed on July 27, 2009.

Fernandez-Cornejo, J., A. Mishra, R. Nehring, C. Hendricks, M. Southern, and A. Gregory. 2007. Off-farm Income, Technology Adoption, and Farm Economic Performance. Washington, D.C.: U.S. Department of Agriculture Economic Research Service.

Friedland, W., L. Busch, F.H. Buttel, and A.P. Rudy, eds. 1991. Towards a New Political Economy of Agriculture. Boulder, Colo.: Westview Press.

Frisvold, G.B. 2004. How federal farm programs affect water use, quality and allocation among sectors. *Water Resources Research* 40:W12SO15.

Fuglie, K.O., J.M. MacDonald, and E. Ball. 2007. Productivity Growth in U.S. Agriculture. Washington, D.C.: U.S. Department of Agriculture Economic Research Service.

Gardner, B.L. 2002. American Agriculture in the Twentieth Century: How It Flourished and What It Cost. Cambridge: Harvard University Press.

Gasson, R., and A. Errington. 1993. The Farm Family Business. Wallingford, UK: CAB International.

Gehlhar, M., E. Dohlman, N. Brooks, A. Jerardo, and T. Vollrath. 2007. Global Growth, Macroeconomic Change, and U.S. Agricultural Trade. Washington, D.C.: U.S. Department of Agriculture Economic Research Service.

Gilliom, R.J., J.E. Barbash, C.G. Crawford, P.A. Hamilton, J.D. Martin, N. Nakagaki, L.H. Nowell, J.C. Scott, P.E. Stackelberg, G.P. Thelin, and D.M. Wolock. 2006. The Quality of Our Nation's Waters: Pesticides in the Nation's Streams and Ground Water, 1992–2001. Reston, Va: U.S. Geological Survey.

Glauber, J. 2008. Statement of Joseph Glauber, Chief Economist before the Committee on Energy and Natural Resources, United States Senate, on June 12, 2008. Available at http://energy.senate.gov/public/_files/GlauberTestimony061208.pdf. Accessed on April 6, 2010.

Goolsby, D.A., W.A. Battaglin, G.B. Lawrence, R.S. Artz, B.T. Aulenbach, R.P. Hooper, D.R. Keeney, and G.J. Stensland. 1999. Flux and Sources of Nutrients in the Mississippi-Atchafalaya River Basin—topic 3. Silver Spring, Md.: National Oceanic and Atmospheric Administration, National Ocean Service, National Centers for Coastal Ocean Science.

Gray, A.W., and M.D. Boehlje. 2007. The industrialization of agriculture: Implications for future policy. Available at http://ageconsearch.umn.edu/bitstream/6712/2/wp070010.pdf. Accessed on February 19, 2010.

Hallam, A., ed. 1993. Size, Structure, and the Changing Face of American Agriculture. Boulder, Colo.: Westview Press.

Halloran, J.M., and D.W. Archer. 2008. External economic drivers and US agricultural production systems. *Renewable Agriculture and Food Systems* 23(4):296–303.

Halweil, B. 2007. Still No Free Lunch: Nutrient Levels in U.S. Food Supply Eroded by Pursuit of High Yields. Boulder, Colo.: The Organic Center.

Hart, J.F. 2003. The Changing Scale of American Agriculture. Charlottesville, Va.: University of Virginia Press.

Harwood, R.R. 1994. Managing the living soil for human well-being. In Environment and Agriculture: Rethinking

Development Issues for the 21st Century, S.A. Breth, ed. Morrilton, Ark.: Winrock International Institute for Agricultural Development.

Hatfield, J., K. Boote, P. Fay, L. Hahn, C. Izaurralde, B.A. Kimball, T. Mader, J. Morgan, D. Ort, W. Polley, A. Thomson, and D. Wolfe. 2008. Agriculture. Washington, D.C.: U.S. Climate Change Science Program.

Hedlund, J.D., and S. Crow. 1994. Draft Material on Salinity. Washington, D.C.: U.S. Department of Agriculture Natural Resource Conservation Service

Heimlich, R. 2000. Farm Resource Regions. Washington, D.C.: U.S. Department of Agriculture Economic Research Service.

———. 2003. Agricultural Resources and Environmental Indicators. Washington, D.C.: U.S. Department of Agriculture.

Hendrickson, M., and W. Heffernan. 2007. Concentration of agricultural markets. Available at http://www.nfu.org/wp-content/2007-heffernanreport.pdf. Accessed on December 11, 2009.

Hill, J., E. Nelson, D. Tilman, S. Polasky, and D. Tiffany. 2006. Environmental, economic, and energetic costs and benefits of biodiesel and ethanol biofuels. *Proceedings of the National Academy of Sciences* 103:11206–11210.

Hoppe, R.A., P. Korb, E.J. O'Donoghue, and D.E. Banker. 2007. Structure and Finances of U.S. Farms: Family Farm Report, 2007 Edition. Washington, D.C.: U.S. Department of Agriculture Economic Research Service.

Hoppe, R.A., P. Korb, and D.E. Banker. 2008. Million-Dollar Farms in the New Century. Washington, D.C.: U.S. Department of Agriculture Economic Research Service.

Horrigan, L., R. Lawrence, and P. Walker. 2002. How sustainable agriculture can address the environmental and human health harms of industrial agriculture. *Environmental Health Perspectives* 110(5):445–456.

Hutson, S.S., N.L. Barber, J.F. Kenny, K.S. Linsey, D.S. Lumia, and M.A. Maupin. 2004. Estimated Use of Water in the United States in 2000. Reston, Va.: U.S. Geological Survey.

IOM (Institute of Medicine) and NRC (National Research Council). 2009. The Public Health Effects of Food Deserts. Workshop Summary. Washington, D.C.: National Academies Press.

IPCC (Intergovernmental Panel on Climate Change). 2006. Guidelines for National Greenhouse Gas Inventories, Volume 4, Agriculture, Forestry and Other Land Use. Available at http://www.ipccnggip.iges.or.jp/public/2006gl/vol4.html. Accessed on April 24, 2009.

IPCC (Intergovernmental Panel on Climate Change). 2007. Climate Change 2007: Synthesis Report. Contribution of Working Groups I, II and III to the Fourth Assessment Report of the Intergovernmental Panel on Climate Change. Geneva: Author.

Jackson-Smith, D., and J. Sharp. 2008. Farming in the urban shadow: supporting agriculture at the rural-urban interface. *Rural Realities* 2(4):1–12.

Johnson, A.K. 2009. ASAS centennial paper: farm animal welfare science in the United States. *Journal of Animal Science* 87(6):2175–2179.

Keeling, L.J. 2005. Healthy and happy: animal welfare as an integral part of sustainable agriculture. *Ambio* 34(4–5):316–319.

Kemper, N., H. Färber, D. Skutlarek, and J. Krieter. 2008. Analysis of antibiotic residues in liquid manure and leachate of dairy farms in Northern Germany. *Agricultural Water Management* 95(11):1288–1292.

Kennedy, A.C., and R.I. Papendick. 1995. Microbial characteristics of soil quality. *Journal of Soil and Water Conservation* 50(3):243–248.

Kirner, L., and R. Kratochvil. 2006. The role of farm size in the sustainability of dairy farming in Austria: an empirical approach based on farm accounting data. *Journal of Sustainable Agriculture* 28(4):105–123.

Kirschenmann, F., G.W. Stevenson, F. Buttel, T.A. Lyson, and M. Duffy. 2008. Why worry about the agriculture of the middle. Pp. 3–22 in Food and the Mid-Level Farm, T.A. Lyson, G.W. Stevenson, and R. Welsh, eds. Cambridge, Mass.: MIT Press.

Klein, L.R., and J.P. Reganold. 1997. Agricultural changes and farmland protection in western Washington. *Journal of Soil and Water Conservation* 52(1):6–12.

Kleinschmit, J. 2007. Biofueling Rural Development: Making the Case for Linking Biofuel Production to Rural Revitalization. Durham: University of New Hampshire.

Knauss, T. 2009. Oswego County ethanol plant in bankruptcy. *The Post-Standard,* January 14, 2009.

Kneese, A. 1986. Water resource constraints: the case of the Ogallala aquifer. In The Future of the North American Granary: Politics, Economics, and Resource Constraints in North American Agriculture, C.F. Runge, ed. Ames: Iowa State University Press.

Koudela, M., and K. Petkikova. 2007. Nutritional composition and yield of endive cultivars—Cichorium endivia L. *Horticultural Science* 34(1):6–10.

———. 2008. Nutritional compositions and yield of sweet fennel cultivars—*Foeniculum vulgare* Mill. ssp *vulgare* var. *azoricum* (Mill.) Thell. *Horticultural Science* 35(1):1–6.

Krauss, C. 2009. Valero Energy, the oil refiner, wins an auction for 7 ethanol plants. Available at http://www.nytimes.com/2009/03/19/business/energy-environment/19ethanol.html. Accessed on February 19, 2010.

Kumar, K., S. C. Gupta, et al. (2005). Antibiotic uptake by plants from soil fertilized with animal manure. *Journal of Environmental Quality* 34(6):2082–2085.

Lal, R. 2004. Soil carbon sequestration impacts on global climate change and food security. *Science* 304.

Leibtag, E. 2008. Corn prices near record high, but what about food costs. *Amber Waves* 6(1):10–15.

Lin, B-H., J.N. Variyam, J. Allshouse, and J. Cromartie. 2003. Food and Agricultural Commodity Consumption in the United States: Looking Ahead to 2020. Washington, D.C.: U.S. Department of Agriculture Economic Research Service.

Lobao, L., and K. Meyer (2001). The great agricultural transition: crisis, change, and social consequences of twentieth century U.S. farming. *Annual Review of Sociology* 27:103–124.

Lobao, L., and C. Stofferahn. 2008. The community effects of industrialized farming: social science research and challenges to corporate farming laws. *Agriculture and Human Values* 25(2):219–240.

Lubowski, R.N., S. Bucholtz, R. Claassen, M. Roberts, J. Cooper, A. Gueorguieva, and R. Johansson. 2006. Environmental Effects of Agricultural Land Use. Washington, D.C.: U.S. Department of Agriculture Economic Research Service.

Luckey, R.R., E.D. Gutentag, and J.B. Weeks. 1981. Water-Level and Saturated-Thickness Changes, Predevelopment to 1980, in the High Plains Aquifer in Parts of Colorado, Kansas, Nebraska, New Mexico, Oklahoma, South Dakota, Texas, and Wyoming: U.S. Geological Survey Hydrologic Investigations Atlas HA-652, 2 sheets, scale 1:2,500,000.

Lusk, J.L., and B.C. Briggeman. 2009. Food values. *American Journal of Agricultural Economics* 91(1):184–196.

MacDonald, J.A., and W.D. McBride. 2009. The Transformation of U.S. Livestock Agriculture: Scale, Efficiency, and Risks. Washington, D.C.: U.S. Department of Agriculture Economic Research Service.

Mäder, P., D. Hahn, D. Dubois, L. Gunst, T. Alföldi, H. Bergmann, M. Oehme, R. Amadò, H. Schneider, U. Graf, A. Velimirov, A. Fließbach, and U. Niggli. 2007. Wheat quality in organic and conventional farming: results of a 21 year field experiment. *Journal of the Science of Food and Agriculture* 87(10):1826–1835.

Magkos, F., F. Arvaniti, and A. Zampelas. 2003. Organic food: nutritious food or food for thought? A review of the evidence. *International Journal of Food Sciences and Nutrition* 54(5):357–371.

McBride, W. 1997. Change in U.S. Livestock Production, 1969–92. Washington, D.C.: U.S. Department of Agriculture Economic Research Service.

Mead, P.S., L. Slutsker, V. Dietz, L.F. McCaig, J.S. Bresee, C. Shapiro, P.M. Griffin, and R.V. Tauxe. 1999. Food-related illness and death in the United States. *Emerging Infectious Diseases* 5(5):607–625.

Mench, J.A. 2008. Scientific Basis for Improving Animal Welfare. Presentation to the Committee on August 6, 2008, in Washington, D.C.

Methner, U., S. Alshabibi, and H. Meyer. 1995. Experimental oral infection of specific pathogen-free laying hens and cocks with *Salmonella enteritidis* strains. *Journal of Veterinary Medicine Series B-Zentralblatt Fur Veterinarmedizin Reihe B-Infectious Diseases and Veterinary Public Health* 42(8):459–469.

Miranowski, J., D. Swenson, L. Eathington, and A. Rosburg. 2008. Biofuel, the rural economy, and farm structure. Paper read at Farm Foundation Conference Transition to a Bioeconomy: Risk, Infrastructure and Industry Evolution, at Berkeley, California, on June 24, 2008. Available at http://www.farmfoundation.org/news/articlefiles/365-John%20Miranowski.pdf. Accessed on January 28, 2010.

Mississippi River/Gulf of Mexico Watershed Nutrient Task Force. 2001. Action Plan for Reducing, Mitigating, and Controlling Hypoxia in the Northern Gulf of Mexico. Washington, D.C.: Author.

Mitchell, L. 2001. Impact of consumer demand for animal welfare on global trade. In Changing Structure of Global Food Consumption and Trade, A. Regmi, ed. Washington, D.C.: U.S. Department of Agriculture Economic Research Service.

Murray, B.C., B.A. McCarl, and H.C. Lee. 2004. Estimating leakage from forest carbon sequestration programs. *Land Economics* 80(1):109–124.

NAS-NAE-NRC (National Academy of Sciences, National Academy of Engineering, and National Research Council). 2009. Liquid Transportation Fuels from Coal and Biomass: Technological Status, Costs, and Environmental Impacts. Washington, D.C.: National Academies Press.

NBB (National Biodiesel Board). 2007. National Biodiesel Board Annual Report 2007. Jefferson City, Mo.: Author.

NCGA (National Corn Growers Association). 2008. World of Corn. Chesterfield, Mo.: Author.

Nizeyimana, E.L., G.W. Peterson, M.L. Imhoff, H.R. Sinclair, Jr., S.W. Waltman, D.S. Reed-Margetam, E.R. Levine, and J.M. Russo. 2001. Assessing the impact of land conservation to urban uses on soils with different productivity levels in the USA. *Soil Science Society of America Journal* 65:391–402.

Nolan, B.T., K.J. Hitt, and B.C. Ruddy. 2002. Probability of nitrate contamination of recently recharged groundwaters in the conterminous United States. *Environmental Science & Technology* 36(10):2138–2145.

Nord, M., M. Andrews, and S. Carlson. 2008. Household Food Security in the United States, 2007. Washington, D.C.: U.S. Department of Agriculture Economic Research Service.

Norris, P., and B.J. Deaton. 2001. Understanding the demand for farmland preservation: implications for Michigan policies. In Staff paper 2001-18. East Lansing: Michigan State University, Department of Agricultural Economics.

Notter, D.R. 1999. The importance of genetic populations diversity in livestock populations of the future. *Journal of Animal Science* 77(1):61–69.

NRC (National Research Council). 1972. Genetic Vulnerability of Major Crops. Washington, D.C.: National Academy Press.

———. 1993a. Soil and Water Quality. Washington, D.C.: National Academy Press.

———. 1993b. Agricultural Crop Issues and Policies. Washington, D.C.: National Academy Press.

———. 2008. Understanding and Responding to Climate Change—Highlights of National Academies Reports. Washington, D.C.: The National Academies Press.

Oliver, S.P., D.A. Patel, T.R. Callaway, and M.E. Torrence. 2009. ASAS centennial paper: developments and future outlook for preharvest food safety. *Journal of Animal Science* 87(1):419–437.

Patterson, D.T., J.K. Westbrook, R.J.V. Joyce, P.D. Lingren, and J. Rogasik. 1999. Weeds, insects, and diseases. *Climatic Change* 43(4):711–727.

Pearce, N., and D. McLean. 2005. Agricultural exposures and non-Hodgkin's lymphoma. *Scandinavian Journal of Work, Environment & Health* 31(Suppl. 1):18–25.

Petherick, J.C. 2007. Spatial requirements of animals: allometry and beyond. *Journal of Veterinary Behavior-Clinical Applications and Research* 2(6):197–204.

Picone, C., and D. Van Tassel. 2002. Agriculture and biodiversity loss: industrial agriculture. In Life on Earth: An Encyclopedia of Biodiversity, Ecology, and Evolution, N. Eldredge, ed. Santa Barbara, Calif.: ABC-CLIO.

Pimentel, D., and M. Pimentel. 2003. Sustainability of meat-based and plant-based diets and the environment. *American Society for Clinical Nutrition* 78(3):660S–663S.

Pineiro, G., E.G. Jobbagy, J. Baker, B.C. Murray, and R.B. Jackson. 2009. Set-asides can be better climate investment than corn ethanol. *Ecological Applications* 19(2):277–282.

Pineyro-Nelson, A., J. Van Heerwaarden, H.R. Perales, J.A. Serratos-Hernandez, A. Rangel, M.B. Hufford, P. Gepts, A. Garay-Arroyo, R. Rivera-Bustamante, and E.R. Alvarez-Buylla. 2009. Transgenes in Mexican maize: molecular evidence and methodological considerations for GMO detection in landrace populations. *Molecular Ecology* 18(4):750–761.

Plucknett, D.L., N.J. Smith, J.T. Williams, and N.M. Anishetty. 1987. Genebanks and the World's Food. Princeton, N.J.: Princeton University Press.

Rabalais, N.N., R.E. Turner, and D. Scavia. 2002. Beyond science into policy: Gulf of Mexico hypoxia and the Mississippi River. *BioScience* 52:129–142.

Reganold, J.P., and M.J. Singer. 1984. Comparing farm production input/output ratios using two land classification systems. *Journal of Soil and Water Conservation* 39(1):47–53.

Ribaudo, M., and R. Johansson. 2006. Water quality: impacts of agriculture. In Agricultural Resources and Environmental Indicators, 2006 Edition / EIB–16. Washington, D.C.: U.S. Department of Agriculture Economic Research Service.

Ribeiro, N.M., A.C. Pinto, C.M. Quintella, G.O. da Rocha, L.S.G. Teixeira, L.L.N. Guarieiro, M.D. Rangel, M.C.C. Veloso, M.J.C. Rezende, R.S. da Cruz, A.M. de Oliveira, E.A. Torres, and J.B. de Andrade. 2007. The role of additives for diesel and diesel blended (ethanol or biodiesel) fuels: a review. *Energy & Fuels* 21(4):2433–2445.

Rosegrant, M.W. 2008. Biofuels and grain prices: impacts and policy responses. Available at http://beta.irri.org/solutions/images/publications/papers/ifpri_biofuels_grain_prices.pdf. Accessed on June 8, 2009.

Rosenzweig, C., F.N. Tubiello, R. Goldberg, E. Mills, and J. Bloomfield. 2002. Increased crop damage in the US from excess precipitation under climate change. *Global Environmental Change—Human and Policy Dimensions* 12(3):197–202.

RTI International. 2007. GIPSA livestock and meat marketing study. Volume 3: Fed cattle and beef industries. Available at http://archive.gipsa.usda.gov/psp/issues/livemarketstudy/LMMS_Vol_3.pdf. Accessed on July 24, 2009.

Russelle, M.P., M.H. Entz, and A.J. Franzluebbers. 2007. Reconsidering integrated crop-livestock systems in North America. *Agronomy Journal* 99:325–224.

Sampson, R.N. 1981. Farmland or Wasteland: A Time to Choose. Emmaus, Pa.: Rodale Press.

San Joaquin Valley Drainage Program. 1990. A Management Plan for Agricultural Subsurface Drainage and Related Problems on the Westside San Joaquin Valley. U.S. Department of Interior and California Resources Agency.

Sassenrath, G.F., P. Heilman, E. Lusche, G.L. Bennett, G. Fitzgerald, P. Klesius, W. Tracy, J.R. Williford, and P.V. Zimba. 2008. Technology, complexity and change in agricultural production systems. *Renewable Agriculture and Food Systems* 23(4):285–295.

Scavia, D., and K.A. Donnelly. 2007. Reassessing hypoxia forecasts for the Gulf of Mexico. *Environmental Science & Technology* 41(23):8111–8117.

Scavia, D., N.N. Rabalais, R.E. Turner, D. Justic, and W.J. Wiseman, Jr. 2003. Predicting the response of Gulf of Mexico hypoxia to variations in Mississippi River nitrogen load. *Limnology and Oceanography* 48(3):951–956.

Scavia, D., D. Justic, and V.J. Bierman. 2004. Reducing hypoxia in the Gulf of Mexico: advice from three models. *Estuaries* 27(3):419–425.

Schillinger, W.F., A.C. Kennedy, and D.L. Young. 2007. Eight years of annual no-till cropping in Washington's winter wheat-summer fallow region. *Agriculture Ecosystems & Environment* 120(2–4):345–358.

Schjonning, P., S. Elmholt, and B.T. Christensen. 2004. Managing Soil Quality: Challenges in Modern Agriculture. Cambridge: CABI Publishing.

Schnepf, R. 2004. Energy Use in Agriculture—Background and Issues. CRS Report for Congress. Washington, D.C.: Congressional Research Service.

Searchinger, T., R. Heimlich, R.A. Houghton, F. Dong, A. Elobeid, J. Fabiosa, S. Tokgoz, D. Hayes, and T.-H. Yu. 2008. Use of U.S. croplands for biofuels increases greenhouse gases through emissions from land use change. *Science* 319:1238–1240.

Selfa, T., L. Kulcsar, R. Goe, and G. Middendorf. 2009. Biofuels bonanza? Exploring community perceptions of the promises and perils of biofuels production. Available at http://sustainability.nationalacademies.org/pdfs/Selfa072909.pdf. Accessed on February 19, 2010.

Shapouri, S., S.R. Rosen, B. Meade, and F. Gale. 2009. Economic Research Service Food Security Assessment 2008–09 Outlook Report. Washington, D.C.: U.S. Department of Agriculture Economic Research Service.

Sharratt, B.S., and D. Lauer. 2006. Particulate matter concentration and air quality affected by windblown dust in the Columbia plateau. *Journal of Environmental Quality* 35:2011–2016.

Shoemaker, R., D. McGranahan, and W. McBride. 2006. Agricultural and rural communities are resilient to high energy costs. *Amber Waves* 4(2):16–21.

Shore, L.S., and A. Pruden, eds. 2009. Hormones and Pharmaceuticals Generated by Concentrated Animal Feeding Operations. New York: Springer.

Silbergeld, E.K., J. Graham, and L.B. Price. 2008. Industrial food animal production, antimicrobial resistance, and human health. *Annual Review of Public Health* 29:151–169.

Simpson, T.W., A.N. Sharpley, R.W. Howarth, H.W. Paerl, and K.R. Mankin. 2008. The new gold rush: fueling ethanol production while protecting water quality. *Journal of Environmental Quality* 37(2):318–324.

Strange, M. 1988. Family Farming: A New Economic Vision. Lincoln: University of Nebraska Press.

Swenson, D. 2009. Biofuels and rural development. Available at http://sustainability.nationalacademies.org/pdfs/Swenson.pdf. Accessed on February 19, 2010.

Sylvan, J.B., Q. Dortch, D.M. Nelson, A.F. Maier Brown, W. Morrison, and J.W. Ammerman. 2006. Phosphorus limits phytoplankton growth on the Louisiana shelf during the period of hypoxia formation. *Environmental Science & Technology* 40(24):7548–7553.

Teuber, M. 2001. Veterinary use and antibiotic resistance. *Current Opinion in Microbiology* 4(5):493–499.

The Associated Press. 2009. Owner of Nebraska ethanol plant files bankruptcy. July 20, 2009.

Tilman, D. 1999. Global environmental impacts of agricultural expansion: the need for sustainable and efficient practices. *Proceedings of the National Academy of Sciences* 96(11):5995–6000.

Tilman, D., K. Cassman, P.A. Matson, R. Naylor, and S. Polasky. 2002. Agricultural sustainability and intensive production practices. *Nature* 418:671–677.

Tucker, M., S.R. Whaley, and J.S. Sharp. 2006. Consumer perceptions of food-related risks. *International Journal of Food Science and Technology* 41(2):135–146.

Turner, R.E., N.N. Rabalais, E.M. Swenson, M. Kasprzak, and T. Romaire. 2005. Summer hypoxia in the northern Gulf of Mexico and its prediction from 1978 to 1995. *Marine Environmental Research* 59:65–77.

U.S. Census Bureau. 2009. The 2009 population estimate for the United States is 307,006,550. Available at http://factfinder.census.gov/servlet/SAFFPopulation?_submenuId=population_0&_sse=on. Accessed on February 11, 2010.

USDA-ERS (U. S. Department of Agriculture Economic Research Service). 2000. ERS Farm Typology for a Diverse Agricultural Sector. Washington, D.C.: Author.

———. 2004. Irrigation and water use briefing room. Available at http://www.ers.usda.gov/Briefing/WaterUse/. Accessed on June 5, 2009.

———. 2005a. Food CPI, price and expenditures: food expenditures by families and individuals as a share of disposable personal income. Available at http://www.ers.usda.gov/Briefing/CPIFoodAndExpenditures/Data/table7.htm. Accessed on May 13, 2006.

———. 2005b. Agricultural chemicals and production technology: sustainability and production systems. Available at http://www.ers.usda.gov/Briefing/AgChemicals/sustainability.htm. Accessed on July 30, 2009.

———. 2008. Value of U.S. agricultural exports by commodity group, 2003–07. Available at http://www.ers.usda.gov/Data/StateExports/2008/commx5yr.xls. Accessed on August 4, 2009.

———. 2009a. Value of U.S. agricultural trade, by calendar year. Available at http://www.ers.usda.gov/Data/Fatus/DATA/XMScy1935.xls. Accessed on February 11, 2010.

———. 2009b. A Preliminary analysis of the effects of HR 2454 on U.S. agriculture. Available at http://www.usda.gov/oce/newsroom/archives/releases/2009files/HR2454.pdf. Accessed on March 25, 2010.

———. 2009c. Agricultural resource management survey (ARMS): resource regions. Available at http://www.ers.usda.gov/Briefing/arms/resourceregions/resourceregions.htm. Accessed on August 4, 2009.

———. 2009d. Rural labor and education: farm labor. Available at http://www.ers.usda.gov/Briefing/LaborAndEducation/FarmLabor.htm. Accessed on June 30, 2009.

———. 2010. Agricultural productivity in the United States. Available at http://www.ers.usda.gov/Data/agproductivity/. Accessed on February 11, 2010.

USDA-NASS (U.S. Department of Agriculture National Agricultural Statistics Service). 2009. 2007 Census of Agriculture. Washington, D.C.: U.S. Department of Agriculture National Agricultural Statistics Service.

———. 2002. Historical highlights: 2002 and earlier census years. Available at http://www.agcensus.usda.gov/Publications/2002/Volume_1,_Chapter_1_US/st99_1_001_001.pdf. Accessed on April 13, 2010.

———. 2008. Quick stats. Available at http://www.nass.usda.gov/QuickStats/PullData_US.jsp. Accessed on October 6, 2008.

USDA-NRCS (U.S. Department of Agriculture National Resource Conservation Service). 1997. National Irrigation Guide. East Lansing, Mich.: U.S. Department of Agriculture Natural Resources Conservation Service.

———. 2001. 1997 National resources inventory summary report (revised December 2000). Available at http://www.nrcs.usda.gov/technical/NRI/1997/summary_report/report.pdf. Accessed on September 2, 2009.

———. 2003. National resources inventory 2001 annual NRI—urbanization and development of rural land. Available at http://www.nrcs.usda.gov/technical/NRI/2001/urban.pdf. Accessed on September 13, 2008.

———. 2006. Energy management. Conservation Research Brief No. 0608. Available at http://www.nrcs.usda.gov/feature/outlook/Energy.pdf. Accessed on February 6, 2010.

VeraSun Energy. 2009. VeraSun Energy Corporation launches chapter 11 case to enhance liquidity while it reorganizes. Available at http://www.verasun.com/Press/details.cfm?ID=161. Accessed on July 24, 2009.

Vesterby, M., and K.S. Krupa. 2001. Major uses of land in the United States, 1997. Washington, D.C.: U.S. Department of Agriculture Economic Research Service.

Vesterby, M., K.S. Krupa, and R.N. Lubowski. 2004. Estimating U.S. cropland area. *Amber Waves* 2(5):54.

Villarejo, D., D. Lighthall, D. Williams, A. Souter, R. Mines, B. Bade, S. Samuels, and S.A. McCurdy. 2000. Suffering in Silence: A Report on the Health of California's Agricultural Workers. Davis, Calif.: California Institute for Rural Studies

Vitousek, P.M., J.D. Aber, R.W. Howarth, G.E. Likens, P.A. Matson, D.W. Schindler, W.H. Schlesinger, and G.D. Tilman. 1997. Human alteration of the global nitrogen cycle: sources and consequences. *Ecological Applications* 7(3):737–750.

Wang, M., W. May, and H. Huo. 2007. Life-cycle energy and greenhouse gas emission impacts of different corn ethanol plant types. *Environmental Research Letters* 2:1–9.

Wells, H.F., and J.C. Buzby. 2008. Dietary Assessment of Major Trends in U.S. Food Consumption, 1970–2005. Washington, D.C.: U.S. Department of Agriculture Economic Research Service.

Westcott, P.C. 2007. Ethanol Expansion in the United States: How Will the Agricultural Sector Adjust? Washington, D.C.: U.S. Department of Agriculture Economic Research Service.

Wescott, P.C. 2009. Full throttle: U.S. ethanol expansion faces challenges down the road. *Amber Waves* 7(3): 28–35.

Wing, S., R.A. Horton, S.W. Marshall, K. Thu, M. Taiik, L. Schinasi, and S.S. Schiffman. 2008. Air pollution and odor in communities near industrial swine operations. *Environmental Health Perspectives* 116(10):1362–1368.

Wing, S., and S. Wolf. 2000. Intensive livestock operations, health, and quality of life among Eastern North Carolina residents. *Environmental Health Perspectives* 108(3):233–238.

Wirzba, N., ed. 2003. The Essential Agrarian Reader: The Future of Culture, Community, and the Land. Lexington: The University Press of Kentucky.

Ziska, L.H. 2003. Evaluation of yield loss in field sorghum from a C-3 and C-4 weed with increasing CO_2. *Weed Science* 51(6):914–918.

Ziska, L.H., and K. George. 2004. Rising carbon dioxide and invasive, noxious plants: potential threats and consequences. *World Resource Review* 16:427–447.

Ziska, L.H., and E.W. Goins. 2006. Elevated atmospheric carbon dioxide and weed populations in glyphosate treated soybean. *Crop Science* 46(3):1354–1359.

Ziska, L.H., J.R. Teasdale, and J.A. Bunce. 1999. Future atmospheric carbon dioxide may increase tolerance to glyphosate. *Weed Science* 47(5):608–615.

3

Improving Productivity and Environmental Sustainability in U.S. Farming Systems

The widespread implementation of management practices that improve productivity and environmental sustainability, along with new environmental policies and regulations in the last 20 years, have been effective in reducing many detrimental effects of agriculture. Research aimed at understanding how these management practices and engineering approaches work continue to provide additional tools for progress toward the sustainability goals outlined in Chapter 1. This chapter briefly discusses some of the management approaches and practices that are relevant to productivity and environmental sustainability and have an impact on agriculture's natural resource base (goals 1 and 2 in Chapter 1). Table 3-1 illustrates the relationships between the two sustainability goals and subgoals, management activities and specific practices that can be used to reach the goals, and a selection of potential indicators that are or could be used to assess progress toward specific goals.

Each section in this chapter discusses how specific practices can contribute to crop or livestock productivity and improve various aspects of environmental sustainability or enhance the quality of a resource. The extent to which the practices are adopted by farmers is discussed if data are available. However, a practice by itself might improve sustainability in relation to one goal but might have a negative effect on another; hence, the disadvantages of each practice are also discussed. A farm is a system that contains multiple interrelated elements, and the interrelationships between environmental conditions, management, and biological processes determine such outcomes as the environmental impact, efficiency, and resilience of the farm (Drinkwater, 2002). Some of the disadvantages of certain practices might be overcome if a complementary practice is used. In other words, the collective outcome of several agricultural practices would be different from simply adding the anticipated outcome of individual practices. Therefore, many in the scientific community have been adopting a "systems" approach, which emphasizes the connectivity and interactions among components and processes and across multiple scales, to understand and harness complex processes. "Systems agriculture" is an approach to agricultural research, technology development, or extension that analyzes agriculture and its component farming

TABLE 3-1 An Illustration of Activities and Practices Used to Achieve Agroecological Sustainability Goals and of Indicators for Evaluating Sustainability

Agroecological Sustainability Goals	Examples of Indicators	Activity	Examples of Practices
1. Satisfy human food, fiber, feed, and fuel needs			
a. Sustain adequate crop production	• Yield per unit area, yield per unit resource use (energy, water, and nutrients)	• Crop management	• Fertility, pest, and water management (see below for specifics). Plant breeding and genetic modification to improve yield and stress tolerance.
		• Plant breeding	• Crops bred for increased resistance to biotic and abiotic stresses, enhanced nutrient use efficiency, and yield stability
b. Sustain adequate animal production	• Production per unit land, production per animal, production per unit resource use (energy, water, nutrients), mortality, duration of productive life, conversion of feedstuff to human edible products, animal health	• Animal husbandry	• Use of local feedstuffs, careful use of resources (labor, water, energy), breeding for increasing feed efficiency, animal health and welfare, herd health management (disease prevention), improved housing environments, judicious use of antibiotics, waste management, manure applications to field, and advanced treatment technologies for manure
2. Maintain and enhance environmental quality and resource base			
a. Maintain or improve soil quality	• Soil nutrient levels, nutrient use efficiency	• Soil-fertility management	• Fertilizer and organic amendment application, use of soil and tissue tests, nutrient budget calculations
	• Soil organic matter content, microbial and macrofaunal populations and communities	• Organic-matter management	• Conservation tillage, organic amendments, composts, green manure
	• Soil physical structure such as bulk density, water-holding capacity, aggregate stability, porosity, water infiltration rate	• Organic-matter management	• Conservation tillage, organic amendments, compost, green manure
b. Maintain or improve water quality	• Fertility inputs, field or farm nutrient budget balances, nutrient, pesticide, and pathogen concentrations in water courses, leaching estimates, nutrient or water model outputs	• Soil-fertility management	• Use of nutrient budgets, use of slow release fertilizers and organic amendments, plant nutrient tissue tests, soil nutrient tests, manure disposal

TABLE 3-1 Continued

	• Ground cover, USLE[a], direct measures of nutrient, sediment and pesticide fluxes, area in cover crops or perennial vegetation, soil aggregate stability, water-holding capacity, porosity, water infiltration rate	• Crop-vegetation management, nutrient management, and erosion and runoff control	• Plant cover crops, use of organic amendments, soil and tissue tests, conservation tillage, mulches, grass waterways, buffer strips, riparian vegetation, treatment wetlands
c. Conserve freshwater supply	• Crop water use efficiency, water consumption, ground water overdraft, pumping rates	• Irrigation management	• Drip irrigation, irrigation scheduling based on evapotransporation or soil moisture
d. Reduce pesticide use	• Pest populations, natural enemy populations, weed biomass, percent weed cover, vegetation diversity, presence of perennial habitat	• Management of pest complex	• Integrated pest management practices, biological and ecological approaches, soil organic matter management, crop breeding
e. Conserve and enhance biodiversity	• Biodiversity estimates (for example, number of plant species, number of species within selected animal groups, habitat diversity, landscape complexity, and connectivity)	• Habitat management	• In-field insectaries, hedgerows, riparian vegetation, habitat corridors, natural habitat fragments

[a] Universal soil loss equation.

systems in a holistic way. Chapter 5 uses a few farming systems to illustrate how systems research is conducted and how the practices can work together to achieve multiple environmental, economic, and social sustainability goals.

The following sections, however, focus on a series of activities that constitute crop and animal production, and highlight particular practices that are seen, or have the potential, to enhance sustainability. The emphasis is on developments that have occurred over the last 20 years.

SOIL MANAGEMENT

Management of soil to improve sustainability is a complex matter that requires a thorough understanding of its physical, chemical, and biological attributes and their interactions. Proper soil management is a key component of sustainable agricultural production practices as it produces crops and animals that are healthier and less susceptible to pests and diseases. It provides a number of important ecosystem services, such as reduced nitrogen runoff and better water-holding capacity (NRC, 1993). Mismanagement of soil can result in physical, chemical, and biological degradation (Lal, 2004b), as discussed in Chapter 2. Soil management is critical to improving environmental sustainability of farming systems. Proper soil management practices aim to:

- Maintain or build up soil organic matter.
- Improve soil structure by increasing soil aggregates. The soil aggregates would in turn enhance water-holding capacity of soil.
- Minimize erosion. Reduction in wind erosion would improve air quality. Reduction in water and tillage erosion would improve water quality by reducing sediment loading.
- Enhance soil microbial activities and diversity.
- Reduce soil-borne pathogens.

Conservation Tillage

One of the most important changes in U.S. agriculture in the last 20 years has been the movement away from conventional tillage to conservation tillage. Conventional tillage, such as moldboard plowing, results in considerable disturbance of the soil and breaks down its aggregate structure. Because aggregation reduces soil density and helps to maintain a balance of air and water in the soil, disturbance by tillage that breaks aggregates apart can result in soil compaction and reduced oxygen levels. Although conventional tillage contributes to weed and pest control, it also destroys habitats or disrupts the life cycle of some beneficial organisms (for example, earthworms and microorganisms) and reduces soil organic matter in the surface layer.

Increased soil erosion as a result of intensive tillage is long recognized (NRC, 1989). Tillage erosion is the downslope displacement of soil through the action of tillage (Lindstrom, 2006) and results in soil loss on hilltops and soil accumulation at the base of slopes. Because water erosion tends to be more important at the base of slopes than at hilltop positions, tillage erosion tends to reinforce water erosion (Government of Manitoba, 2009) and thereby increases sediment runoff and sediment loading into surrounding surface water. Phosphorus, herbicides, and other contaminants that absorb readily to soil particles move with sediment into surface water. Phosphorus from agricultural fertilizers enriches the receiving bodies of water and can cause large blooms of algae.

Conservation tillage is an agricultural practice that reduces soil erosion and water runoff, increases soil water retention, and reduces soil degradation. Conservation tillage, including ridge-till, mulch-till, and no-till practices, is any tillage and planting system that leaves 30 percent or more of the soil surface covered by crop residues after planting to reduce soil erosion by water. No-till leaves 50 to 100 percent of the soil surface covered from harvest to planting, depending on the crop residue, because it uses specifically designed seed planters or drills to penetrate all remaining surface residues (Huggins and Reganold, 2008). Comparisons of conventional tillage practices to conservation tillage in corn, soybean, and winter wheat found that systems that use conservation tillage tend to use more herbicides for each crop, but less insecticides (USDA-ERS, 2005).

Impact of Conservation Tillage

Physical Properties of Soil

Soil under no-till management has been shown to have a higher proportion of water stable aggregates (Karlen et al., 1994a; Abid and Lal, 2008), and the aggregates have larger geometric mean diameter and mean weight diameter compared to chisel-plowed soil (Abid and Lal, 2008). The large aggregates contain finer soil textures that assist in retaining more water than small aggregates. Arshad et al. (1999) compiled data collected from two sites in northern British Columbia to ascertain the long-term effects of conventional tillage and no-till on soil components thought to be important in surface soil structural improvement. They observed that soil water retention was greater under no-till compared with conventional till without dramatically altering bulk density because of redistribution of pore size classes into more small pores and less large pores.

No-till and other conservation tillage systems can work in a wide range of climates, soils, and geographic areas. Continuous no-till is also applicable to most crops, with the notable exceptions of wetland rice and root crops, such as potatoes. However, no-till crop production on fine-textured, poorly drained soils can be problematic and often results in decreased yields. Yields of no-till corn, for instance, are often reduced by 5 to 10 percent on those kinds of soils, compared with yields with conventional tillage, particularly in northern regions. Because the crop residue blocks the sun's rays from warming the earth to the same degree as occurs with conventional tillage, soil temperatures are colder in the spring, which can slow seed germination and curtail the early growth of warm-season crops, such as corn, in northern latitudes (Huggins and Reganold, 2008).

Soil Organic Matter

The amount of organic matter in soil subject to conventional tillage has been compared to soil subject to conservation tillage or no-till in different locations. Dell et al. (2008) quantified the impacts of no-till and rye (*Secale cereale L.*) cover crops on soil carbon and physical properties. They found that the no-till fields had 50 percent more carbon particulate and mineral-associated pools in the upper 5 cm compared to conventional tillage. The sizes of the carbon pools below 5 cm in the two fields were similar. The stability of the soil aggregates is proportional to the carbon pool size. Another study by Motta et al. (2007) compared soil organic carbon at different depths of the soil in cotton fields subject to conventional tillage and no-till. They found that the no-till fields had much higher particulate organic carbon within the top 3 cm. Some scientists have questioned if substantial soil carbon sequestration can be accomplished by changing from conventional plowing to conservation tillage. Baker et al. (2007b) argued that soils were sampled to a depth of 30 cm or less in essentially all cases where conservation tillage was found to sequester carbon. In the few studies where sampling extended deeper than 30 cm, conservation tillage has shown no consistent accrual of soil organic carbon. Instead conservation tillage showed a difference in the distribution of soil organic carbon, with higher concentrations near the surface in conservation tillage and higher concentrations in deeper layers under conventional tillage. Blanco-Canqui and Lal (2008) assessed the impacts of long-term no-till and plow-based cropping systems on soil carbon sequestration in the top 60 cm of soils across Kentucky, Ohio, and Pennsylvania. They found that no-till farming increased organic carbon concentrations in the upper layers of some soils, but it did not store more organic carbon than plowed soils for the whole soil profile. In fact, total soil profile organic carbon was significantly higher in plowed-based soils in a number of the areas sampled. In another study, Christopher et al. (2009) found that the soil organic carbon pool in the whole soil profile (0–60 cm) was never greater in no-till than conventionally tilled fields across 12 contrasting but representative soils in the Midwestern United States and was actually lower in the no-till soils in some areas.

Soil Microbial Activity and Diversity

Bacteria, fungi, and nematodes are important in maintaining the physical structure of soil. In a study of soil quality with data collected following a long-term tillage study on continuous corn, Karlen et al. (1994a) found that plots managed using no-till practices have higher microbial activity and earthworm populations. Motta et al. (2007) also found higher microbial biomass in no-till cotton fields compared to conventional-till ones.

Soil Erosion

The greater the percentage of ground cover (residue or mulch), the lower is the soil loss ratio (Figure 3-1) due to water and wind. The soil loss ratio (SLR) is an estimate of the ratio of soil loss under actual conditions to losses experienced under the reference condition of clean-tilled continuous-fallow conditions (the reference condition). Leaving 30 percent of the soil surface covered with residue, as with conservation tillage, reduces erosion by half as compared with bare, fallow soil. Leaving 50 to 100 percent of the surface covered throughout the year, as no-till does, reduces soil erosion dramatically.

Montgomery (2007) looked at numerous studies on conventional (n = 448) and conservation (n = 47) agricultural systems and found an average net soil loss of 3.9 mm/yr under conventional agriculture and 0.12 under conservation agriculture that included conservation tillage, no-till methods, and terracing. Montgomery further examined 39 studies involving direct comparisons of soil erosion under conventional and no-till methods representing a wide variety of settings with different erosion rates and showed that no-till practices reduce soil erosion up to 1,000 times, enough to bring agricultural erosion rates into line with rates of soil production.

Sediment Loading and Water Quality

Agriculture is a major contributor to sediment pollution, primarily because of improper farming practices that increase soil erosion. Farming on steep slopes, excessive heavy tillage, and lack of conservation practices are principal causes. A number of studies document the effectiveness of conservation or no-till on reducing sediment in runoff. Blevins et al. (1990) compared the contributions of no-till, chisel-plow tillage, and conventional tillage systems used in corn production to sediment losses and surface runoff on a Maury silt loam. Over a four-year period, they measured soil losses of 20, 0.71, and 0.55 Mg/ha from conventional, chisel-plow, and no-till systems, respectively. Amounts of nitrate (NO_3^-), soluble phosphorus, and atrazine leaving the plots in surface runoff were greatest from conventional tillage and about equal from chisel-plow and no-till. Chichester and Richardson (1992) compared the effect of no-till and conventional chisel-till soil management on runoff

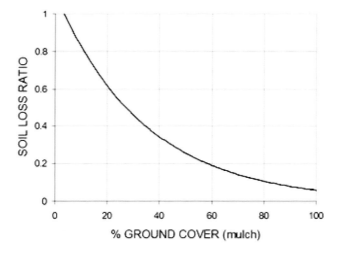

FIGURE 3-1 Soil loss ratio and percent ground cover.
SOURCE: McCarthy et al. (1993). Reprinted with permission from the University of Missouri Extension.

water volumes, sediment loss, and nitrogen and phosphorus loss from small watersheds on a clay soil. They found that runoff volume was not changed by tillage system, but sediment loss and nitrogen and phosphorus losses in runoff were far less, on average, from no-till than from chisel-till. Average annual quantities for sediment and nutrient losses were: 160 kg/ha and 1575 kg/ha for sediment, 3.8 kg/ha and 8.1 kg/ha for nitrogen, and 0.8 kg/ha and 1.5 kg/ha for phosphorus for no-till and chisel-till, respectively. Although erosion remains a significant problem in the United States, conservation and tillage changes have resulted in substantial improvements over the last 30 years. Soil erosion on cropland declined, as a result of changes in tillage practices and land retirement, from 3.1 billion tons per year in 1982 to 1.8 billion tons per year in 2001, while sheet and rill erosion dropped by almost 41 percent, and wind erosion dropped by 43 percent during the same time period (NRI, 2003).

Air Quality

With the advent of reduced and "zero" tillage in the past few decades made possible through the use of herbicides, releases of carbon dioxide (CO_2), nitrous oxide (N_2O), and particulate matter from agricultural soil have been reduced (Robertson et al., 2000; Madden et al., 2008). Reduced tillage reversed some of the soil carbon decline in surface soils. The impacts of tillage and different cropping systems on soil carbon (discussed earlier in this chapter) can be translated with reasonable accuracy into changes in CO_2 flux over time. When CO_2 flux is calculated and N_2O and methane (CH_4) fluxes are measured, the overall atmospheric impact of production systems can be assessed. Unfortunately, there are few production systems where such gaseous flux measurements have been done over a sufficient time span. One of the best sources of data comes from the Long Term Ecological Research (LTER) sites funded by the National Science Foundation (NSF). The LTER data in Box 3-1 are presented not to represent overall U.S. agricultural fluxes, but rather to show comparisons for the predominant gases between natural and managed systems, and the contribution of key management practices. The net greenhouse-gas emissions from agriculture in the United States were estimated to be about 50 g of CO_2 equivalent/m^2 per year (West and Marland, 2002). That estimate is comparable to the data presented in Table 3-2 in Box 3-1. A large database, several models for soil carbon accumulation, and ongoing research at the Natural Resources Ecology Laboratory in Colorado are focusing on the carbon accumulation potential of soils under different management (Easter et al., 2007).

Data from other long-term organic comparisons in California, Wisconsin, and Pennsylvania give similar effects for carbon sequestration. Studies conducted on finer-textured soils (most other than the Michigan trial) show higher levels of carbon sequestration and hence could be expected to show greater global warming mitigation potential. None of those studies monitored greenhouse gas over the long term. The bottom line is that agricultural systems can be designed and managed to compare favorably with natural ecosystems if moldboard plowing is eliminated in favor of either reduced or zero tillage.

Zero-till can also reduce emissions of particulate matter, especially if the practice is used with mulching or cover cropping. For example, in the dryland areas of wheat production on the eastern side of the Cascades and in portions of the Great Plains, clean fallow for moisture conservation has long been practiced. The fallowing leads to occasional high wind-blown soil erosion with increasingly unacceptable air quality problems (Sharratt and Lauer, 2006). New programs for crop rotation, weed and disease control, and reduced-till planting in the dryland wheat-growing area of eastern Washington are promising (Schillinger et al., 2007). Barley grown every other year seems to reduce rhizoctonia bare-patch area in wheat. Risk due to uncertain rainfall appears higher in crop rotation

> **BOX 3-1**
> **Tillage and Rotation Effects on Climate Change**
> **Greenhouse-Gas Emissions and Energy Use**
>
> The Kellogg Biological Station in southwestern Michigan is the only Long Term Ecological Research (LTER) site devoted to agricultural systems comparisons. It includes comparisons between agricultural and natural ecosystems in various stages of disturbance, making it unique. Greenhouse-gas flux measurements provide comparative data for agricultural systems and natural ecosystems at varying maturity stages (Robertson et al., 2000). Four corn–wheat–soybean rotations were replicated: (1) conventional chemical inputs and conventional (moldboard) plowing, (2) conventional inputs and zero tillage, (3) reduced chemical inputs, and (4) organic with no chemical inputs. Systems 3 and 4 included a winter legume cover crop following the corn and the wheat portions of the rotation. The trials were carried out on a silt loam soil that had been in continuous cultivation since the mid-1800s. Data are from the first eight years of the trial from 1991 to 1999.
>
> The net greenhouse warming potentials for the several agricultural systems and for comparison natural ecosystems in various stages of maturity are shown in Table 3-2. The annual crop rotations produced surface soil carbon changes consistent with those of other long-term trials in the U.S. corn belt. Conventional tillage with rotation had no change, indicating the long-term soil carbon equilibrium under conventional management with moldboard plowing that existed at the start of the experiment. No-till had the highest added carbon in soil and had negative CO_2 release (-110 g/m^2 per year of CO_2 equivalent). The low-input and organic systems were next. Perennial crops (alfalfa and poplar) had significantly higher carbon sequestration than annual crops. Natural communities added soil carbon depending on their length of time in development. The early succession treatment was kept in grasses and other herbaceous plants by annual mowing. That treatment was thus similar to a standard set-aside common to many farms enrolled in the U.S. Department of Agriculture (USDA) National Resource Conservation Service (NRCS) conservation programs. Such treatment has large carbon sequestration in the early decades following implementation, and therefore a significant greenhouse warming mitigation effect. Inputs calculated as CO_2 equivalents differed according to cropping system. Nitrogen fixation by legumes and from denitrificaton emits N_2O as a byproduct. It was roughly the same for the annual crop systems and for alfalfa in those systems. The biological pathways for nitrogen fixation and "leakage" and those for commercial fertilizer bioconversion differ, but the net effects on N_2O evolution are similar. Denitrification is higher when soils high in nitrogen experience waterlogging, producing low oxygen levels when soil temperatures are high. In LTER, the systems without fertilizer and with fewer legumes showed lower N_2O evolution. All systems oxidized CH_4 (removed it from the atmosphere), with the natural communities oxidizing slightly more than the commercial systems because of the canopy composition.

than in the traditional wheat-summer fallow. Modified tillage implements that undercut the root zone have promise. Some farmers in the area used no-till planting with rotations with success. The results are not ready for widespread adoption; continued research is essential. This research program appears to be a flagship program for low-rainfall cropping systems in the Pacific Northwest.

Energy Use

In 2006, no-till was practiced on 62.4 million acres of cropland in the United States and resulted in an annual savings of 243 million gallons of fuel for tillage (Table 3-3). The energy saving was estimated solely on the basis of reduced requirements for direct energy inputs for tillage. That estimate did not include the additional efficiencies gained from increased productivity as a result of increased soil quality as described above for enhancement of ecosystem services. When calculated for a 2100-acre Michigan corn–oats–soybean–wheat rotation farm, diesel fuel savings over conventional tillage would have been 28 percent for mulch-till, 27 percent for ridge-till, and 52 percent for no-till (USDA-NRCS, 2008a). In drier areas such as the western Corn Belt, returns are uncertain because of high variability in

TABLE 3-2 Relative Radiative Forcing Potential for Different Management Systems Based on Soil Carbon Sequestration, Agronomic Inputs, and Trace Gas Fluxes[a]

	CO$_2$ Equivalents of Change (g/m² per year)[b]						
Ecosystem Management	Soil C	N fertilizer	Lime	Fuel	N$_2$O	CH$_4$	Net Global Warming Period
Annual crops (corn–soybean–wheat rotation)[c]							
Conventional tillage	0	27	23	16	52	–4[d]	114
No-till	–110	27	34	12	56	–5	14
Low input with legume cover	–40	9	19	20	60	–5	63
Organic with legume cover	–29	0	0	19	56	–5	41
Perennial crops							
Alfalfa	–161	0	80	8	59	–6	–20
Poplar	–117	5	0	2	10	–5	–105
Successional communities							
Early successional	–220	0	0	0	15	–6	–211
Mid-successional (historically tilled)	–32	0	0	0	16	–15	–31
Mid-successional (never tilled)	0	0	0	0	18	–17	1
Late-successional forest	0	0	0	0	21	–25	–4

[a]Data source: Robertson et al. (2000).
[b]Results based on eight years of data (1991–1999), using IPCC (1996) conversion factors.
[c]Six replications of each for annual and perennial crops. Successional communities were nearby on similar soil types. Conventional and no-till treatments had full herbicide and fertilizer use. Low-input treatment used banded herbicides at low rates and low levels of nitrogen. Organic treatment had occasional lime input only, but no herbicides or fertilizer.
[d]Negative values represent a net CO$_2$ equivalent uptake, or a net reduction in greenhouse gases and a reduction in atmospheric radiative forcing.

Comparison of net effect on greenhouse-gas emissions showed that no-till had the least greenhouse-gas impact among the annual cropping systems. Conventional tillage and chemical use had the highest greenhouse-gas emission impact. The low-input system had lower greenhouse-gas impact than conventional tillage, but its yields were lower. The organic system had yields close to those of zero-till, followed by low input. Perennial systems and early succession communities had the most positive effects on reducing greenhouse-gas emissions.

rainfall. The accounting of long-term effects, including impact of increased surface organic matter and changed fertilizer requirements during the transition period, complicates total energy balance considerably. In U.S. studies, outputs are most often calculated as energy content of the harvested product. That measure of output complicates comparisons because

TABLE 3-3 Energy Savings and Production Potential from Conservation Practices and Measures in the United States

Conservation Practice	Conservation Measurement	Resource Savings		Energy Costs Reduction
		On-farm (per acre)	Total	Million $
Crop residue management	62.4 million acres of no-till (CTIC)	$11.70	243 million gallons	730
	Conversion of additional 50 million acres to no-till	$11.70	195 million gallons	585

SOURCE: USDA-NRCS (2006).

different crops in the rotation produce considerably different energy amounts, and their relative yields change dramatically over years; hence, long-term studies are needed for meaningful comparisons, which partly explains the paucity of such comparisons.

Disadvantages of Conservation Tillage

Potential problems with conservation tillage include weed control, soil crusting and compaction, flooding or poor drainage, delays in planting because fields are too wet or too cold, carryover of diseases or pests in crop residue, fewer options to work fertilizers and pesticides into the soil, new machinery requirements, increased risk of shifting weed populations that are resistant to specific herbicides, and the need for above-average farm management skills (Peigne et al., 2007; Huggins and Reganold, 2008). Because conservation tillage increases the size and prevalence of macropores in soil, there has been concern about increased leaching of pesticides to ground water in particular during heavy rainfall (Shipitalo et al., 2000). In some cases, tillage residues such as rye can have allelopathic effects on seed germination in other crops, especially when seeds are planted directly into recently killed rye residues or some mow-killed mulches (Mitchell et al., 2000). High carbon-to-nitrogen ratios in crop residues can also cause problems such as reduced nitrogen availability (Gebhardt et al., 1985; Troeh and Thompson, 2005; Baker et al., 2007).

Some of the problems mentioned above might be more prevalent in vegetable production systems than in field crops. Successful vegetable production with conservation tillage depends on careful crop selection. Crops that germinate quickly and grow rapidly in the first few weeks after planting are more competitive with weeds than crops that initially grow slowly. Cool-season vegetables perform better in spring no-till plantings than warm-season crops (Hoyt and Konsler, 1988). The availability of specialized equipment for planting horticultural crops in no-till systems can be a limitation, but large-seeded vegetables such as sweet corn, snap beans, and squash have been successfully planted with no-till planters designed for field corn or soybean, and no-till planters for planting cabbage, broccoli, and other vegetable transplants in no-till soils have been developed (Hoyt, 1999; Peet, 2008).

The impact of reduced tillage and no-till on rates of chemical use and on nutrient leaching has been mixed because it depends on whether herbicide and pesticide uses are increased as a result of reduced tillage and how nutrients and agricultural chemicals are applied (Lal, 1991; Daverede et al., 2003). There is, however, evidence that pesticide leaching and NO_3^- in drainage water is higher under no-till conditions because of movement through intact macropores (Isensee and Sadeghi, 1996; Stoddard et al., 2005). In addition, higher average concentration and load of soluble phosphorus have been found in runoff water of no-till systems compared to other tillage systems (McIsaac et al., 1995). Moldboard plowing has been shown to reduce nitrogen and phosphorus runoff by redistributing the nutrients into the soil profile (Gilley et al., 2007). Similarly, Garcia et al. (2007) and Quinke et al. (2007) proposed and demonstrated a promising strategy of tilling one-time only with a moldboard plow to reduce phosphorus in runoff, followed by no-till management. They observed a significant reduction in soluble phosphorus accumulation in runoff with no negative effects on soil quality or crop yield. Further research is needed on management of no-till systems to reduce negative water quality effects.

In organic farming systems, reduced tillage raises specific challenges because the use of herbicides to kill the preceding crop is prohibited. Nonetheless, the sparse research on reduced tillage methods (strip till, ridge till, or shallow tillage) has shown promising results (Schonbeck, 2009). The choice of crop rotation, cover crop, and cover crop management is critical. Winter-hardy cover crops that are amenable to no-till, no-herbicide management can be killed by mowing or rolling in early summer. Non-winter-hardy crops planted two to three months prior to the anticipated frost-kill date can be used to form in situ mulch and suppress winter and early spring weeds. Even with the use of managed cover crops, continuous no-till does not yet appear feasible under organic systems and more research is needed in this area. A high standard of management is required to successfully implement conservation tillage practices in organic systems, and the practices need to be tailored to local soil and site conditions (Kuepper, 2001; Peigne et al., 2007).

Adoption of Conservation Tillage

The passage of the Food Security Act by Congress in 1985 tied soil conservation practices to farmer eligibility for government-sponsored crop deficiency payments, crop loans, storage payments, federal crop insurance, and disaster payments. The overall purpose of the act was to remove incentives to produce crops on highly erodible land, and the program affected more than 125 million acres nationwide. In 1990, 26 percent of planted crop acreage was under conservation tillage practices; that number rose to 41 percent in 2004 (CTIC, 2004). Among the conservation tillage practices, no-till has been used on an increasing proportion of land (from 17 million acres in 1990 to 61 million acres in 2004; Figure 3-2).

Although weed control with conventional herbicides was successfully used on millions of acres of no-till (Derksen et al., 2002) before genetically engineered (GE) crop varieties with herbicide tolerance (HT) were introduced, GE corn, soybean, and crop varieties with HT might have further encouraged the adoption of conservation tillage practices, because

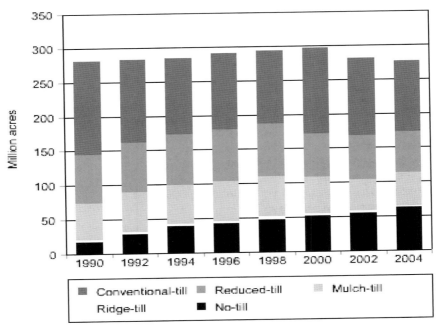

FIGURE 3-2 Area of cropland in the United States managed by different tillage systems from 1990 to 2004.
SOURCE: USDA-ERS (Sandretto and Payne, 2006).

they allow farmers to replace cultivation and tillage with chemical means of controlling weeds on those major crops. USDA survey data in 1997 showed that 60 percent of the acreage planted with HT soybean was under conservation tillage compared to about 40 percent of conventional soybean. By 2008, HT soybean varieties occupied more than 92 percent of the U.S. soybean acreage, HT cotton was grown on 68 percent of the total acreage, and HT corn on 63 percent of the acreage (USDA-ERS, 2009). However, HT crops are not a prerequisite for successful herbicidal weed control in conservation tillage because many farmers still grow non-GE crops successfully with conventional herbicides. Such practices as mulching, cover cropping, and crimping or rolling crop residues also can be used with conservation tillage to suppress weeds.

Cover Cropping

Cover cropping is the practice of using vegetative crops, such as clover or vetch, to prevent soil erosion, control weeds, and provide nitrogen to a subsequent crop. Cover crops grown in rotation between cash crops provide ground cover to protect the soil. They can also be used to provide other services, notably by being tilled into the soil to maintain soil organic matter and provide nutrients to subsequent crops (green manures) or being used to trap excess nutrients in the soil profile following harvest of the primary crop to prevent leaching losses (catch crops). Perennial cover crops can be used as ground covers in orchards.

Impact of Cover Cropping

Productivity

The impact of cover crops on yields can be difficult to quantify, but some studies have shown increased yields in cash crops when they are planted after certain cover crops. Sweeney and Moyer (1994) found that when hairy vetch or red clover were grown and then used as green manure, the yield of the sorghum crops in the eastern Great Plains immediately after was 79 to 131 percent higher compared to continuous grain sorghum. Summer cover crops have been shown to produce higher yields of conventionally grown and organically grown lettuce (Ngouajio et al., 2003) and of okra (Wang et al., 2006) compared to fallow. Preliminary results from a decade-long study in south central Colorado on cover crops and crop rotations show that the yield and quality of potatoes are 12 to 30 percent higher if they were planted after sudangrass was grown and plowed in as green manure, than if they were planted after wet fallow of the plot (Delgado et al., 2008). The ability of cover crops to replace or reduce the amount of chemical nitrogen fertilizer needed when used in combination has also been well established (Kramer et al., 2002; Cherr et al., 2006).

Soil Quality

Cover crops reduce soil erosion by wind and water, and therefore decrease particulate matter in the air and sediment runoff into surface water (Langdale et al., 1991). Cover crops also add to the soil organic matter pool (Sullivan, 2004). In turn, organic matter has a profound impact on soil quality as it enhances soil structure and fertility, increases water infiltration and storage, prevents surface crusting of the soil (Roberson et al., 1995), reduces the loss of nutrients and sediment in surface runoff, and reduces leaching losses of nutrients, especially nitrogen (Brady and Weil, 2008; Plaster, 2009). Decayed root channels of cover crops alleviate soil compaction problems. Williams and Weil (2004) found that soybean yields responded the most to the preceding cover crop at the test site that was most affected by drought and soil compaction, suggesting that the soybean plants used existing root channels to access subsoil water. Cover cropping has also been found to enhance soil microbial numbers and enzyme activities (Mullen et al., 1998; Steenwerth and Belina, 2008).

Water Quality

Cover crops increase soil biomass and therefore transpire more water, allow more rainfall to infiltrate into the soil, and decrease runoff and potential erosion to a greater extent than fallow (Dabney, 1998). Beyond taking up nutrients, cover crops also improve water quality by reducing erosion by protecting aggregates from the impacts of raindrops, reducing soil detachment and aggregate breakdown (Dabney et al., 2001).

Winter cover crops can reduce water flows, nitrate concentrations, and total nitrate load, particularly under some surface runoff or tile drainage landscapes. The effectiveness of cover crops in improving water quality varies with the growth of the cover crop, climatic conditions, and management of the main crop. More growth of the cover crop will result in greater reductions in nitrate leaching, but the growth of the cover crop can be limited by cold temperatures, water stress, nutrient availability, and delays in establishment. The lack of precipitation and soil freezing can greatly reduce NO_3^- leaching losses and thus reduce the impact of the cover crop. Reducing nitrogen fertilizer rates and applying nitrogen fertilizer closer to the time of crop uptake will also reduce losses from NO_3^- leaching and the impact of the cover crop (Kasper et al., 2008). Reductions in NO_3^- loadings because of rye or ryegrass cover crops range from 13 percent in Minnesota (Strock et al., 2004) to 94 percent in Kentucky (McCracken et al., 1994). Wyland et al. (1996) found that nitrate leaching was reduced by 65 to 70 percent in a broccoli system with a cover crop compared to a fallow rotation. In a meta-analysis of cover crop studies, Tonitto et al. (2006) found that over-wintering nonlegume and legume cover crops generally reduced nitrate leaching when compared to fallow fields.

Sharpley and Smith (1991) summarized research on the effect of cover crops on total phosphorus losses and found that reductions ranged from 54 (Yoo, 1988) to 94 percent (Pesant, 1987). They pointed out that the effects of cover crops on soluble phosphorus in runoff were variable and did not always result in reductions. Soluble phosphorus can be lost in runoff flowing over plant residues. However, some plant water use and infiltration can be expected, which would likely reduce the volume of runoff (Kasper et al., 2008).

Water Use

Cover crops can reduce evaporation from the soil surface. Baker et al. (2007) found that the introduction of a rye cover crop in a corn–soybean rotation in Minnesota lowered evaporation from soil because the rye and its straw residue reflected sunlight. A two-year corn–soybean rotation with a rye cover crop increased water use efficiency by nearly 35 percent compared to a traditional corn–soybean rotation (Baker et al., 2007). Winter cover crops can improve rainfall infiltration and enhance water storage in areas where rainfall occurs mostly in winter as short periods of heavy rain, such as the Sacramento Valley in California (Joyce et al., 2002). Cover crops are more suitable for humid and subhumid regions where precipitation is more reliable than for semiarid regions where precipitation is limited (Unger and Vigil, 1998).

Nutrient Management

Cover crops help support soil microbial communities, which break down organic matter and make nutrients available to subsequent crops. However, cover crop roots also increase nutrient accessibility. First, leguminous crops fix nitrogen through a symbiotic relationship with bacteria that live in root nodules. The bacteria convert nitrogen into ammonium (NH_4^+), which is accessible to the plants. Annual cover crops are generally seeded in the fall or winter and die at the end of the season. They are mowed several times during the growing season to add biomass, and hence nutrients, to the soil. At the end of the growing season, they can be plowed into the soil and used as green manure (Sullivan, 2003). If legumes are used in biodiverse rotations, they can provide at least a portion of nitrogen needs and, with longer rotations, all of the needs. With high nitrogen availability to the crop as needed for high yields, N_2O emissions appear inevitable (Dusenbury et al., 2008).

Weeds, Insects, and Diseases

Cover crops can suppress weeds by creating an environment too shady for weeds or by allelopathy[1] (Teasdale, 1998), provide habitats for beneficial insects and pests (Costello and Daane, 1998), and suppress diseases (Griffin et al., 2009). Those impacts will be discussed in detail in a later section of this chapter on weeds, pests, and disease management in crops.

Disadvantages of Cover Cropping

To include cover crops in a rotation, the growing season has to be long enough to establish both the main crop and cover crop for the rest of the year (Lu et al., 2000). The cover crops have to be selected carefully for several reasons. First, cover crops might use more water than cash crops in low-precipitation areas. Second, their common pests could affect the main and field crops (Lu et al., 2000). Third, if they are not managed properly, they could increase NO_3^- leaching (Moller et al., 2008). For cover crops used as green manure, the risk of NO_3^- leaching is much higher if the cover crops are plowed under in autumn than in winter (Moller et al., 2008) because of the effect of temperature on nitrogen mineralization. When used as a nitrogen source, the timing of nitrogen release from the cover crop can be difficult to predict, as it depends on weather conditions, the carbon-to-nitrogen ratio of the cover crop, and soil microbial activity. Furthermore, different legume species fix nitrogen at various levels, and nitrogen fixation is also dependent upon environmental conditions and soil microbial activity (Luna, 1998; Sullivan, 2003).

Cover crops can also improve pests' survival (Bugg and Waddington, 1994; Connell and Vossen, 2007). Studies have found mixed results as to whether cover crops reduce, increase, or have no effects on pest populations (Hanna et al., 2003; Hooks and Johnson, 2004; Wyland et al., 1996). Rothrock et al. (1995) found higher bacterial and fungal populations in a cropping system that includes hairy vetch than in one that includes winter fallow, but they did not observe significant differences in other cover crop treatments. The precise effect of cover crops on beneficial insect and pest complexes and on soil-borne diseases is likely to depend on several factors, including the composition of cover crops, the prevalence and types of pests and pathogens at the location, temperature, irrigation management, and tillage.

Adoption of Cover Cropping

Despite their benefits on soil and water quality, cover crops are not widely planted. The USDA Economic Research Service (ERS) has data on crop area with winter cover crops in

---

[1] Allelopathy broadly defined is the inhibition of one species by chemicals produced by another species. Examples of allelopathy or presumed allelopathy cited in this report pertain to the inhibition of one species of plant by chemicals produced by another species of plant or its residue. Allelopathy has also been a default diagnosis for injury to crops sometimes observed with crop residues and for the suppression of weeds with cover crops and green manures, whether or not a putative allelochemical is identified and shown to cause the crop injury or weed suppression.

1997 (Figure 3-3). Less than 10 percent of corn, soybean, cotton, and potato acreage include cover crops in the rotation (Padgitt et al., 2000). A more recent study examined whether cover crops were used in the U.S. Corn Belt by surveying 3,500 farmers in Illinois, Indiana, Iowa, and Minnesota (Singer et al., 2007). Although 96 percent of the farmers surveyed believe that cover crops are effective in controlling soil erosion and increasing soil organic matter, only 18 percent of the farmers surveyed have used cover crops. The low rate of adoption of cover cropping partly is due to the seeding costs, as discussed in Chapter 4, and the complexity of management (Cherr et al., 2006). About 56 percent of the respondents said that they would plant cover crops if cost-sharing were available (Singer et al., 2007).

CROP AND VEGETATION DIVERSITY MANAGEMENT

Crop diversity (that is, diversifying the types of crops grown and including different genetic varieties) is a method of managing risk on farms. Diversity on the farm can also be accomplished by integrating crop with livestock, a system discussed in Chapter 5. Numerous studies have documented the effect of crop diversity to reduce crop pest and

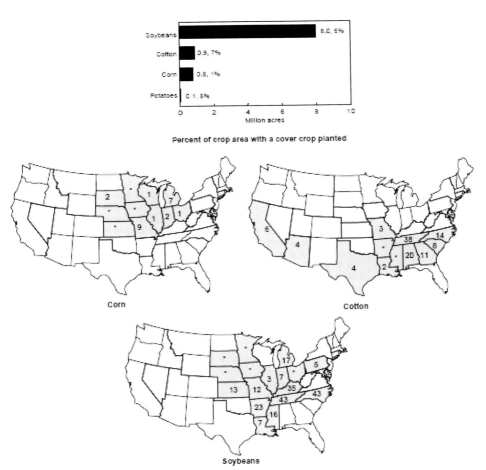

FIGURE 3-3 Crop area in the United States that was planted with winter cover crops in 1997.
SOURCE: USDA-ERS (Padgitt et al., 2000).

diseases, maintain soil fertility, and enhance water use (Power, 1990; Matson et al., 1997). However, the effects of diversity can be variable (Andow, 1991) depending on the kind of diversity present and the functional diversity (Moonen and Barberi, 2008; Shennan, 2008). Nonetheless, high levels of crop diversity continue to be the primary means of controlling risk in many subsistence farming systems throughout the world (Rhoades and Nazarea, 1999). Practices such as rotating crops, preserving genetic variety, planting crops together, incorporating cover crops, and managing noncropped land properly could increase the robustness and resilience of farming systems against the unpredictability of pest problems and against varying market conditions.

One of the premises of sustainable agriculture is that the tradeoff between higher productivity and loss of biodiversity is not inevitable (NRC, 1992; Thrupp, 1997). Increasing crop diversity has the potential to improve sustainability by achieving the following objectives (Box 3-2):

BOX 3-2
Agrobiodiversity and Its Relevance to Agricultural Sustainability

Biodiversity consists of genes, species, populations, and landscapes, along with the composition, structures, functions, and interactions that occur at each level of the ecosystem (Noss, 1990). Agricultural biodiversity, commonly known as agrobiodiversity, "encompasses the variety and variability of animals, plants and microorganisms which are necessary to sustain key functions of the agroecosystem, its structure and processes for, and in support of, food production and food security"(FAO, 1999).

Agrobiodiversity includes genetic resources, as well as domesticated and nondomesticated species and populations (within or outside of farming systems) that support food provision, including soil microorganisms, pollinators, and aquatic organisms. In addition to providing valuable crops and livestock species, biodiversity in agricultural systems performs many ecological services, including recycling of nutrients, pollination, management of organisms that are undesirable for agriculture, regulation of the local hydrological cycle and microclimate, and storage of carbon (Altieri, 1990; Thrupp, 1998).

The roles of biodiversity in agriculture can be defined in several ways, including: a) the **utilitarian** value (direct use) of components of biodiversity, such as medicinal values from particular species; b) **functional** values which biodiversity provides to support life and protect ecological integrity; c) **serendepic** or "option" value, which is the potential future value from particular species or genes for future generations; and d) the **intrinsic** value, which refers to cultural or aesthetic benefits (Swift et al., 2004). The **functional values** from agrobiodiversity services include formation of soil organic matter, nutrient cycling, useful watershed functions (for example, trapping sediment and mitigating runoff), and mitigation of pests and diseases—which are often known as ecosystem services. (See also Chapter 4 of this report for additional information on the valuation of ecosystem services.)

Although biodiversity in farming systems is increasingly recognized as a fundamental basis of sustained agricultural production and food security, biodiversity has been seriously eroded through the expansion of monocultural agriculture production systems and intensive use of agrochemicals, and other conventional patterns of agricultural development in many parts of the world. Commodity policies supporting monocultural production systems are among factors that have contributed to such losses internationally (Darymple, 1986; NRC, 1993; Thrupp, 1998). The decline of diversity of genetic resources, crop varieties, beneficial insects (including pollinators), soil and aquatic organisms, and other elements of biodiversity can seriously hinder sustainable production and can lead to irreversible biological losses (UNDP, 1995; FAO, 1996).

However, there is ample evidence to show multiple benefits of integrating, conserving, and enhancing biodiversity in agriculture at various levels and in a range of farming systems. The conservation of biodiversity is now recognized by scientists and practitioners as an important element of sustainable agriculture (UNDP, 1995; FAO, 1996). Numerous scientific studies and practical experiences have shown that biological diversity contributes to the resilience and stability of farming systems (UNDP, 1995; FAO, 1999; Swift et al., 2004).

More specifically, for example, recent studies reveal that soil biodiversity has a significant role in relation to soil

- Reduced pesticide and herbicide use.
- Improved resilience of the system to adverse environmental conditions.
- Greater conservation of biodiversity.
- Improved soil fertility and soil organic matter.

Crop Rotations

The environmental benefits of crop rotations are well documented (NRC, 1989). They include better control of weeds, pests, and diseases; increased soil moisture; increased availability of nutrients; and higher yields. Crop rotations can enhance accumulation of soil organic carbon. Including legumes in a rotation supplies symbiotically fixed nitrogen to the soil (Havlin et al., 1990). Studies have shown positive effects of crop rotations on soil microbial community composition, particularly mycorrhizae (Johnson et al., 1992).

---

health in agriculture. Each gram of soil can contain thousands or even millions of diverse microscopic organisms (Torsvik et al., 1994.) Although not generally visible to the human eye, "soil is one of the most diverse habitats on earth and contains one of the most diverse assemblages of living organisms" (Giller et al., 1997). Soil organisms incorporate plant and animal residues and wastes into the soil and digest them to create soil humus—the organic constituent that is important to good physical and chemical soil conditions, and they recycle carbon and mineral nutrients. The activities of soil organisms interact in a complex food web; the diverse soil organisms and their functions are valuable to both human societies and ecosystems (FAO, 2003). Although researchers have increased their knowledge about soil biodiversity and microbiology in farming systems, scientific understanding of the role of biodiversity in soil is still somewhat limited. Research has been constrained because of the tremendous diversity of soil organisms and by technical challenges. Yet, there appears to be great potential in this field for gaining insight for sustainable farming systems.

The following list contains examples of successful management practices that conserve or enhance agrobiodiversity at different levels and also have documented benefits towards increasing sustainability in many farming systems. Other examples illustrating the roles of biodiversity (at different levels) are mentioned within this chapter and in the case studies (Chapter 7) of this report.

- Crop Diversification Approaches
    - Temporal (crop rotation)
    - Spatial (polycultures, agroforestry, crop-livestock systems)
    - Genetic (multiple varieties within a farm)
- Recycling and Conservation of Soil Nutrients
    - Incorporating plant biomass (green manures, crop residues, mulches)
    - Reuse of nutrients and resources internal and external to farm (for example, tree litter)
    - Integration of diverse plants or organisms (for example, legume cover crops)
    - Strips of vegetation to prevent soil erosion
- Ecologically Based Integrated Pest Management
    - Natural biocontrol (conserving or enhancing natural control agents by eliminating broad-spectrum pesticides, by planting or conserving habitat that harbor beneficials, or by intercropping)
    - Introduction of imported biological control (augmentation)
    - Enhancing habitats and species in habitat surrounding and in farms
    - Diverse cropping or soil management methods
    - Using plants as natural pesticides
    - Use of nonhost plants that are used as a "decoy" crop to attract fungus or nematodes

More research is needed to better understand the functions and values of biodiversity at different levels in farming systems, and how much biodiversity, and of what kinds, is needed to achieve sustainability goals.

Impact of Crop Rotation

Productivity

Rotation has been shown to have beneficial effects on yields. Rotating corn with soybean can produce yield advantages of 5 to 30 percent compared to continuous corn (Lauer, 2007). In a rotation of spring wheat with field pea in North Dakota, gains of 9 to 11 bushels/ac were found in four of six years in spring wheat grain, while nitrogen gains in those years were 13–28 lbs/ac (Carr et al., 2006). Rotation length has been proven important to the productivity of alternative systems. When comparing continuous corn planting with two-year rotations of corn–alfalfa and corn–soybean and five-year rotations of corn–corn–oat–seedling alfalfa–alfalfa, corn–corn–corn–alfalfa–alfalfa, and corn–soybean–corn–oat with alfalfa seedling–alfalfa, Stanger and Lauer (2008) found that the two-year rotations did not improve grain yield trends, while the five-year rotations not only enhanced yields, but also decreased the need for nitrogen inputs. Peanut yields also increased in tandem with extended rotation lengths (Jordan et al., 2008). In an experiment that compared conventional, high-input, two-year rotations of corn and soybean with low-input and organic rotations, rotation length greatly affected productivity. When compared with conventional, high-input, two-year rotations of corn and soybean, two-year organic corn-soybean rotations produced only 70 percent of corn yields and 80 percent of soybean yields (Porter et al., 2003). Yields from two-year rotations provided with low levels of inputs fared somewhat better, averaging just under 90 percent of the conventional yields for both crops. However, when oats and alfalfa crops were added to expand the rotation to four years, corn yields from the organic system jumped to more than 90 percent of the high-input corn. Corn that received low levels of inputs almost equaled the conventional system's productivity. The four-year rotation did not impact organic soybean yields, but low-input soybean yields slightly exceeded those of the traditionally produced crop.

Corn seems more responsive than soybean to rotation length and crop diversity (Cavigelli et al., 2008; R.G. Smith et al., 2008). In a three-year Michigan study that examined the impact of rotation length and complexity on crop productivity, corn yields increased linearly with the addition of crops to the rotation system, even though no synthetic inputs were introduced. Corn yields in the most diverse rotation (corn–soybean–winter wheat with two cover crops per main crop) were more than 60 percent higher than corn in the two-year corn–soybean rotation that had no cover crops (R.G. Smith et al., 2008). In fact, corn yields in that system were more than 80 percent of average yield per hectare for Michigan corn. Soybean was less responsive than corn to rotation length and diversity; however, yields still increased approximately 30 percent and exceeded Michigan's average.

Soil Health

Crop rotation has also been shown to contribute to improved soil health. Studies have found organic carbon and nitrogen to be higher in rotation systems than in continuous soybean and continuous corn systems (Varvel, 1994). Rotation length, particularly the inclusion of forage crops, positively affected organic carbon (Karlen et al., 2006). Some studies have also demonstrated that soil microbial biomass is higher under rotation systems (Collins et al., 1992; Moore et al., 2000).

A comparison of soil quality between a conventional system with a two-year wheat–pea rotation and an organic system with a three-year wheat–pea–green-manure legume rotation in eastern Washington State showed significantly better soil quality and less soil erosion in the longer-rotation organic system (Reganold et al., 1987). Comparing the same farms for financial performance, Painter (1991) found that the conventional system achieved a

33 percent higher net return than the organic system, with both systems receiving government subsidies but no price premiums for the organic system. The main reason for this difference is that the shorter-rotation conventional system received greater wheat subsidy payments (wheat grown more often), even though the organic system reduced soil erosion and had potentially less environmental pollution from agrichemicals. Without government subsidies, Painter (1991) found the conventional system achieved a 10 percent higher net return. Thus, it is not surprising that farmers often do not adopt longer crop rotation systems because these systems reduce profitability or economic sustainability, even if they are more environmentally sustainable.

Air Quality

Complex crop rotations such as corn–corn–soybean–wheat with red clover underseeded can result in higher net returns, and might substantially lower greenhouse-gas emissions, than continuous corn (Meyer-Aurich et al., 2006).

Water Use

Soil water can also be affected by crop rotation. Bordovsky et al. (1994) found that changing a continuous cotton system to a cotton–wheat rotation increased soil water and improved yields. Soybean and corn rotations also improved water use efficiency, leading to increased root activity and yields (Copeland et al., 1993). Pala et al. (2007) showed that water use efficiency can depend on the type of rotation used; continuous wheat was the least efficient system for water use, but the types of crops built into the rotation improved yields based on how much water each used during its growing season.

Crop rotations can be designed to improve water use and to reduce saline seep. One example is the Triangle Conservation District Saline Seep Project in Montana. Local farmers are changing their land use and management over the water recharge area by switching to a flexible cropping system. The new system ensures that crops grown in sequence will use all available soil water, regardless of vagaries in the weather. The Saline Seep Program in the Central Rolling Red Plains area in Texas focuses on "salt" spots that hamper crop production in cultivated fields. Subsurface drains and deep-rooted vegetation that uses large amounts of available soil moisture are proven methods to reduce accumulations of salty water in shallow water tables (USDA-NRCS, 1997).

Disadvantages of Crop Rotations

Crop rotations require increased management skills because of the complexity involved in finding the right combination of crops to improve yields while also potentially reducing input expenses. As noted earlier, crop rotation patterns can potentially create water stress, thereby reducing yield and profitability (Pala et al., 2007). The unpredictability and variability of insect and disease pressures can also lower the economic incentive to pursue crop rotations. The economic aspects of crop rotations and cover crop use in rotations are discussed in Chapter 4.

Adoption of Crop Rotations

Most major crop production involves rotational cropping of some form, with the exception of cotton (Figure 3-4). The corn–soybean rotation is the most common system for corn and soybean. In the 10 major producing states, 80 percent of soybean acres and 75 percent of corn acres used that rotational system as of 2002.

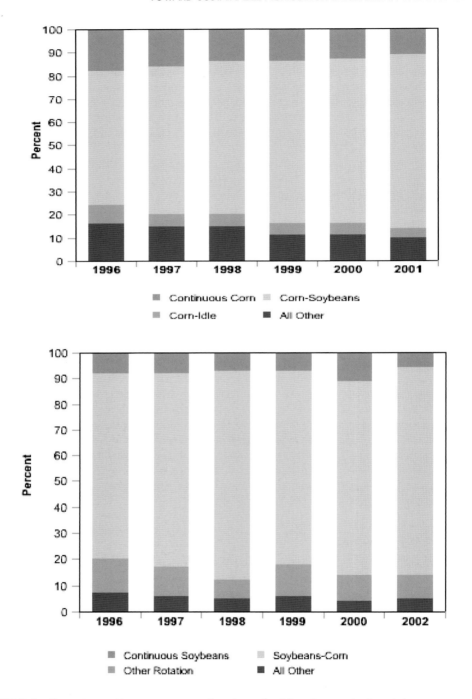

FIGURE 3-4 Cropping patterns on corn and soybean for 10 major producing states.
SOURCE: USDA-ERS (Sandretto and Payne, 2006).

Although the environmental benefits of crop rotations are well known, many farms specialize in a few crops so that production can be streamlined—that is, using the same planter, harvester, and marketing infrastructure for all crops (Cook, 2006). Rotations more complex than corn–soybean could be difficult to manage and not as profitable (Stanger et al., 2008). Other factors that might discourage farmers from adopting extensive crop rotations:

- Herbicide carryover.
- Farm rental arrangements.
- Increased management skills and information needed.
- Altered or new equipment to match changed farming practices.
- Additional storage units for wider variety of crops produced.
- Commodity prices and subsidies.

Intercropping

Intercropping is the agricultural practice of cultivating two or more crops in the same space at the same time (Andrews and Kassam, 1976). It is generally associated with the planting of two or more different food crop species in the same field, but it can also include different varieties of the same crop species. Intercropping systems are common in subsistence, small-scale farms in tropical areas as the practice increases crop genetic diversity and reduces the risk of crop loss. In the United States, intercropping is generally associated with small-scale, sustainable, and organic agricultural systems; it is much less common on large-scale mechanized farms.

Strip intercropping is the practice of growing two or more crops in strips that are wide enough that each can be managed separately, yet narrow enough that the strip components can interact. In theory, the interactions (physical, biological, ecological, and management) between the crop components enhance biomass yield and provide key ecological services such as nutrient cycling, biological pest control, and water and soil conservation. The challenge in strip cropping is to identify the correct assemblages of species to maximize their biological synergisms, while having compatible use of agricultural equipment and conservation tillage practices (Altieri and Nicholls, 1999).

Impact of Intercropping

Productivity

Research on strip intercropping in the United States has primarily been with corn and soybean, and the results have been mixed. One five-year study showed that corn yield increased when planted in strips with soybean, but soybean yields decreased (West and Griffith, 1992). Lesoing and Francis (1999) found that a maize, soybean, and grain sorghum strip intercrop produced up to 4 percent higher total yields than the individual crops in monoculture. A few technologically progressive farmers have successfully combined strip intercropping of HT soybean and corn with precision agriculture and the use of autoguidance farm equipment (C. Mitchell, presentation to the committee, August 5, 2008). Carr et al. (2004) suggested that intercropping forage with pea could enhance forage yield and quality. However, the best way to compare intercrop yields with monocrop yields is uncertain. That is, as posed by Shennan (2008), would it better to compare monocropping and intercropping systems that are similar (for example, similar densities and nutrient inputs), or would it be better to compare systems that are managed optimally?

Nutrient Management

Intercropping grain crops with legumes often produces a yield increase, as a result of transfer of nitrogen either from legume to nonlegume through root exudates or from transfer of residual nitrogen to a nonlegume crop that grows after the legume has been harvested (Vandermeer, 1995; Narwal, 2005). Corn that is intercropped with soybean instead of grown separately after soybean (or vice versa) has been shown to reduce nitrate leaching losses with subsurface drainage water (Kanwar et al., 2005).

Disadvantages of Intercropping

As mentioned above, the results of intercropping can be uncertain. Some experiments have shown no or inconsistent yield benefits (Hesterman et al., 1992; Pridham and Entz, 2008). Furthermore, if crop choices or timing differences in crop life cycles are not managed correctly, the two crops can compete with each other for water and nutrient resources with negative yield results (Brainard and Bellinder, 2004). One experiment found that intercropping was beneficial for the soil microbial community of sorghum but not of soybean, indicating the two crops competed with each other (Ghosh et al., 2006). Even with proper management, yields of intercrops can be easily influenced by growing conditions. Although growing conditions affect all agricultural systems, there is evidence to suggest that the complexity of intercropping can make that system more vulnerable to environmental stresses. Combined with the greater degree of management skills required to operate this system, yield uncertainty may hamper the adoption of intercropping.

Cultivar Mixtures

The preceding sections discuss diversity in the context of diversifying crop species. This section discusses diversity in the context of using mixtures of cultivars of the same species. In western agriculture, most crops are grown from uniform, genetically identical seeds or clonally propagated planting stock. Ecological principles, however, suggest that genetic diversity within species and cultivars can also increase fitness and productivity of the population (Hooper et al., 2005).

Impact of Cultivar Mixtures

Increasing the number of genetic varieties of a particular crop species in the same field can increase crop yield and improve resistance to diseases (Smithson and Lenne, 1996; Cowger and Weisz, 2008). Tilman et al. found that mixed species of native prairie grasses had higher productivity than monocultures (Reich et al., 2001; Tilman et al., 2006). Cultivar mixtures or blends have been shown to control powdery mildews and rusts of small grains (Mundt, 2002). Blends of wheat cultivars have been shown to have more stable yields than sole cultivars if the cultivars used in the mixtures had complementary disease resistance traits and similar growth and maturity characteristics (Pridham et al., 2007; Cowger and Weisz, 2008). Managing genetic diversity across farms and at the community level, in addition to the individual farm level, is important to managing crop performance and the risk of pest outbreaks (Hajjar et al., 2008).

Disadvantages of Cultivar Mixtures

Potential disadvantages of mixing cultivars include the added time and cost involved in the mixing, incompatibility of the varietal components in particular with regard to plant

TABLE 3-4 Crops on Which Transgenic Research Has Been Conducted

Abyssinian mustard	Chrysanthemum	Mulberry	Ryegrass
Alfalfa	Clover	Oat	Safflower
Anthurium	Coffee	Onion	Sorghum
Apple	Collard	Orange	Soybean
Arabidopsis	Cordgrass	Orchid	Spruce
Aspen	Cotton	Papaya	St. Augustine grass
Avocado	Cowpea	Pea	Strawberry
Bahiagrass	Creeping bentgrass	Peach	Sugarbeet
Banana	Cucumber	Peanut	Sugarcane
Barley	Eggplant	Pear	Sunflower
Bean	Elm	Pelargonium	Sweet potato
Beet	Field mustard	Peppermint	Sweetgum
Begonia	Flax	Perilla	Switchgrass
Bermudagrass	Gladiolus	Persimmon	Tall fescue
Birch	Grape	Petunia	Tobacco
Black nightshade	Grapefruit	Pine	Tomato
Blueberry	Guayule	Pineapple	Triticale
Broccoli	Indian mustard	Plum	Turf grass
Cabbage	Jumbay	Poplar	Walnut
Camelina	Kentucky bluegrass	Potato	Watermelon
Canola/Rape	Lettuce	Raspberry	Wheat
Carrot	Maize	Rhododendron	Wild Mustard
Cassava	Marigold	Rice	
Chestnut	Melon	Rose	
Chickpea	Mexican lime	Rye	

NOTE: Crops listed were identified as being modified transgenically through publications and field trials by J.K. Miller and K.J. Bradford, University of California, Davis. Reprinted with permission from K.J. Bradford.

Asia (NRC, 2008a). The research suggests that this technology may have potential to control viruses, bacteria, nematodes, and some insect pests in plants, and have applications for use against parasitic plants and fungi. It is also being investigated for switching off genes involved in the production of undesirable fatty acids in oilseeds. More research is needed to understand whether plants can discriminate between RNAi's that move between plants and pests, what size of RNA can be moved, and if an organism can develop resistance to RNA. However, if gene silencing techniques are successful, they have the potential to provide solutions to some of the more difficult pests and diseases in crops (NRC, 2008a).

Impact

Plant breeding has been improving yield through increased productivity, improved pest and weed resistance, and improved drought tolerance for decades. The impacts of biotechnology-derived crops are summarized by Johnson et al. (2008). Adoption of GE crops is associated with increased yield and decreased pesticide use in many cases (Fernandez-Cornejo and Caswell, 2006).

Disadvantages

Although GE crops have been widely adopted since their introduction in 1996 and are now grown on millions of acres worldwide, concerns about their effects on human health, the environment, and other aspects of sustainability persist. Lemaux (2008) reviewed a number of studies and concluded that there are no scientifically valid demonstrations that food safety issues of foods containing GE ingredients are greater than foods that do not contain them.

Some of the environmental concerns are not unique to cropping systems using GE. For example, the effect of GE crops on nontarget organisms is a concern, just as the effect of pesticides on nontarget organisms is a concern in conventional crop production (Marvier et al., 2007). Likewise, pests could evolve resistance to synthetic pesticides or transgene-derived proteins (Lemaux, 2009). Other issues include the potential of horizontal and vertical gene transfer from transgenic organisms to others (Pilson and Prendeville, 2004; O'Callaghan et al., 2005).

GE crops are banned in certified organic crop production systems, and there is significant resistance to GE crops in European and other countries. Hence, contamination of organic crops or crops intended for export to markets that do not accept GE crops is a concern.

Adoption

The percentage of acres planted with GE crops has been increasing in the United States (Figure 3-5) (USDA-ERS, 2009). Corn varieties with stacked genes (containing both HT and *Bt* genes) were grown on 46 percent of the total corn acreage in 2009 (USDA-ERS, 2009).

WATER-USE MANAGEMENT

To meet sustainability goals of conserving water resources, water management is critically important. The "drivers" for managing water use are the timing, intensity, and amount

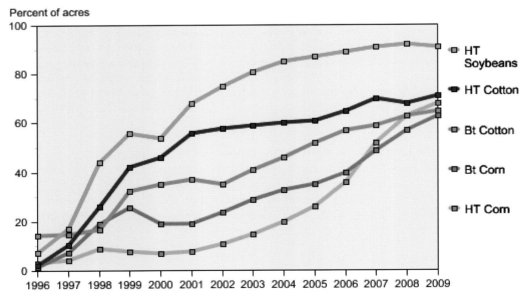

FIGURE 3-5 Adoption of GE crops in the United States.
SOURCE: USDA-ERS (2009).

of water being applied by precipitation, irrigation, or both for all agricultural lands. Those parameters in conjunction with evapotranspiration (ET), the amount of water that evaporates from the soil surface and transpires from the crop, determine the amount of excess water that drains from a field at any given location and time.

Because irrigation is the dominant form of water use, measures that improve the efficiency of water application and minimize water loss are most effective in conserving water and energy in regions facing limited supply (Table 3-5). Water-use efficiency is a complex subject, however, with many different definitions (Molden et al., 1998). From a systems perspective, comparing the amount of water withdrawn from a river or aquifer to the amount actually used beneficially by the crop might be most useful. Factors within the overall efficiency affecting performance are conveyance from the source to the farm, uniformity of application to the crop, and drainage losses following application. Globally, many systems perform poorly, resulting in only 30 to 50 percent of the water withdrawn actually being taken up by the crop (Faurés et al., 2007). However, water lost in that way from one farm might have a beneficial use elsewhere, so it is important to look at basinwide efficiency when estimating true water savings. In addition to managing water consumption, precipitation can be captured or water can be reused to improve the long-term sustainability of water use in agriculture.

Irrigation Scheduling

Quantitative irrigation scheduling methods rely on one of two approaches: soil or crop monitoring or a combination of both; or soil water balance computations. For the monitoring methods, the soil water content or matric potential is measured at several places in the field to decide when to irrigate. Methods based on plant measurements generally involve monitoring leaf water potential or canopy temperature. Soil water balance calculations require estimates of soil storage capacity, rooting depth, allowable depletion, and crop evapotranspiration to develop an irrigation schedule (Martin et al., 1990).

If direct monitoring of plant or soil water status is not possible, irrigation volume required for the crop could be estimated if evapotranspiration demand is known (Jensen et al., 1970).

TABLE 3-5 Energy Savings and Production Potential from Irrigation Water Management.

Conservation Measurement	Resource Savings		Energy Costs Reduction
	On-farm ($ per acre)	Total	Million $
Improving pumping system efficiency 10 percent on 16 million acres	15	80 million gallons	240
Conversion of medium pressure sprinkler system to low	40	560 Kwhr/acre	390
Conversion of high pressure sprinkler system to low	55	770 Kwhr/acre	120

SOURCE: USDA-NRCS (2006).

Impact of Irrigation Scheduling

Irrigation scheduling has been shown to significantly reduce water usage compared to traditional methods in a number of studies. For example, Mohammad and Al-Amoud (1993) achieved both higher wheat yields and a 25 percent decrease in water use when central pivot sprinkler irrigations were scheduled based on a calculation of evapotranspiration demand in Saudi Arabia. Clawson and Blad (1982) were able to reduce irrigation water additions to corn from 283 mm to 127 mm without significant yield reduction by precisely monitoring canopy temperature. An extension project in north central Nebraska also showed that irrigation scheduling reduced energy costs, applied less water, and led to higher harvested yields for center-pivot irrigated corn. Furthermore, there was in increase in annual return by $5.40 per hectare (Kranz et al., 1992).

Adoption of Irrigation Scheduling

A successful example of large-scale use of irrigation scheduling is the California Irrigation Management Information System (CIMIS), an integrated network of more than 125 automated active weather stations located throughout California. Specific weather parameters are collected on site and accessed daily by a computer at the Department of Water Resources. Reference Evapotranspiration (ETo) is calculated from this data and stored in a database along with the collected climatic data, where it can be accessed by Internet users. At present, approximately 6,000 registered CIMIS users from diverse backgrounds access the CIMIS computer directly (State of California, 2008).

Washington State University has also developed the Washington Irrigation Scheduling Expert (WISE) software and web-based information system to supplement scientific irrigation scheduling (SIS) used by farmers in the area. In 2002, half of the acreage being scheduled was for potato and fruit trees. Farmers surveyed said that energy savings and ensuring the quality of high-value crops were the main reasons for adopting the system (Leib et al., 2002).

In New Mexico's Mesilla Valley, pecan farmers have been slow to adopt new soil-based or climate-based irrigation scheduling technologies because these technologies require high in-season labor input. Kallestad et al. (2008) developed a simple, practical irrigation scheduling tool specifically for flood-irrigated pecan production using 14 years of archived climate data and model-simulated consumptive water use. Eventually, the hope is that these farmers will convert to using Internet databases as their main resource for climate and irrigation information (Kallestad et al., 2008).

Gravity Systems

Improving the uniformity of irrigation water application is one of the most effective means by which agriculture can save water. Nonuniform irrigations are wasteful, because water has to be added at rates greatly exceeding those needed by the parts of the field receiving the most water to avoid yield decreases on the parts that are receiving less water. Water is not pumped in gravity irrigation, but flows and is distributed by gravity. Gravity-flow systems distribute water laterally across the entire field or into furrows.

Various land treatments, system improvements, and water management measures have been developed to reduce water losses under gravity-flow systems. For example, precision laser-leveled irrigation is practiced on 3.7 million acres, mostly in the Southwest, Delta, and Northern Rockies regions. Improved gravity systems generally involve on-farm water conveyance upgrades that increase uniformity of applied water and reduce percolation losses and field runoff. Improved ditch systems, lined with concrete or another impervious substance, account for only 20 percent of gravity acres served by open ditches. Improved water management practices for gravity irrigation remains an area of significant growth potential, with many available technology or management improvements such as alternate row irrigation, furrow modification, tailwater reuse, or soil amendments not in widespread use (Schaible and Aillery, 2006).

Disadvantages of Gravity Systems

Water losses are comparatively high under traditional gravity-flow systems, with field application efficiencies typically ranging from 40 to 65 percent. However, improved gravity systems using laser-leveling and proper water management may achieve efficiencies of up to 80 to 90 percent (USDA-NRCS, 1997).

Adoption of Gravity Systems

Total acreage in gravity systems has declined by 26 percent since 1979, but still accounts for 44 percent of irrigated acreage nationwide, primarily in the Southwest, Central Rockies, Southern Plains, and Delta regions (USDA-NASS, 2010). Furrow application comprises about half of the acreage in gravity-flow systems, with border or basin or uncontrolled-flood application accounting for the remaining. Much of the uncontrolled flooding is used for hay and pasture production in the Northern and Central Rockies.

Sprinkler Irrigation

Sprinkler irrigation is a planned system in which water is applied by means of perforated pipes or nozzles operated under pressure to form a spray pattern (USGS, 2009b). Like gravity systems, pressurized systems improve irrigation uniformity. Pressurized systems include a variety of sprinkler and low-flow irrigation techniques to distribute water across a field. Low-energy precision application (LEPA) irrigation refers to methods by which water is delivered directly to the surface at very low pressure through drop tubes and orifice-controlled emitters, rather than spraying water into the air at moderate to high pressures. The applicators are generally attached to moving center pivot or linear advance lines to allow continuous advance over large areas. Lyle and Bordovsky (1981) initiated the concept on a lateral move system, although the vast majority of applications today are on center-pivot systems.

Impact of Sprinkler Irrigation

Field application efficiencies for properly designed and operated sprinkler systems range from 50 to 95 percent, with most systems achieving 75 to 85 percent (USDA-NRCS, 1997). Coates et al. (2006) have also worked on individual micro-sprinkler systems that regulate the amount of water each tree receives, which has promising application for orchard management.

The low-pressure center-pivot and linear-move systems combine high application efficiencies with reduced energy and labor requirements. In addition, center-pivot irrigation has been shown to improve the ground water contamination level, although it might not be the most economical irrigation system available at present (Kim et al., 2000).

Adoption of Sprinkler Irrigation

Acreage for all pressurized systems expanded from 19 million acres in 1979 to 30 million acres (57 percent of total irrigated acreage) in 2003, of which sprinkler systems alone accounted for 27 million acres. Acreage in sprinkler systems has continued to expand in recent years, with an increase of about 4 million acres from 2003 to 2008 (USDA-NASS, 2010). Center-pivot sprinkler systems accounted for roughly 79 percent of sprinkler acreage in 2003, increasing by nearly 13 million acres from 1979. Nearly two-thirds of the increase is attributable to net increases in irrigated area under sprinkler, while about one-third reflects the replacement of other sprinkler types with center-pivot systems. Low-pressure center-pivot systems account for 46 percent of center-pivot acreage and are especially popular in the Southern Plains where irrigation relies heavily on higher-cost groundwater pumping (Schaible and Aillery, 2006).

Trickle or Drip Irrigation

Trickle or drip irrigation applies water directly to the root zone of plants using applicators (for example, orifices, emitters, porous tubing, and perforated pipe) operated under low pressure. The applicators could be placed on or below the surface of the ground (USGS, 2009a). Shifting to trickle or drip irrigation has been the greatest strategic improvement in water-use efficiency and energy savings over the past three decades. Precision water application results in a significant conceptual and process-related change in energy use for those crops where it applies. Most orchards and vineyards are converting to these systems, and nearly all newly planted ones are using precision water application, as are a broad range of annual horticultural crops. The application tubes are placed in close proximity to the tree or vine of crop plants, and water is applied as needed, monitored by a host of newly engineered moisture and plant stress-sensing devices (Locascio, 2005).

Impact of Trickle or Drip Irrigation

Energy savings for systems changes are region- and crop-specific. In the central coast region of California, the better vegetable growers are saving upwards of 25 percent in water pumping, fertilizer, and herbicide costs by using subsurface drip irrigation technologies (California Energy Commission, 2008). The USDA-NRCS energy estimator (2008b) calculated, based on research reports for the costs of irrigation for orchards in the Michigan area, that the energy savings in a 100-acre orchard could be up to 20 percent by adding a flow meter, irrigation scheduling, and maintenance and upgrades to a basic diesel-powered sprinkler system. Use of a micro-irrigation system (after installation costs), with suggested management would reduce pumping costs from $488 per acre/year for a well-managed, well-maintained sprinkler system to $390 per acre/year, a 20 percent savings. Those savings are only for the direct cost of diesel fuel for pumping and do not include additional savings for fertilization, pest and disease control, and weed control. When switching from sprinkler to micro-systems the capital energy costs should be assigned, with repayment over time in a manner similar to payback of financial capital investments.

Adoption of Trickle or Drip Irrigation

Low-flow systems, including drip, trickle, and micro-sprinklers (with application efficiencies of 95 percent or greater) were used on 3 million acres in 2003, mostly for vegetables

Regulated Deficit Irrigation

Deficit irrigation refers to applying water below the crop's full evapotranspiration requirements so that it is allowed to withstand mild water stress (J.M. Costa et al., 2007). Studies have shown that deficit irrigation can be applied to various crops including cotton and potatoes with little or no negative effects on yield (Henggeler et al., 2002; Shock and Feibert, 2002). While deficit irrigation is not controlled, regulated deficit irrigation (RDI) subjects crops to moisture deficit during stress-tolerant growth stages to minimize negative effects of yield.

Impact of Regulated Deficit Irrigation

Regulated deficit irrigation could be applied to a variety of crops including grapes, pistachios, and stone fruits (Cooley et al., 2009). Cooley et al. (2009) estimated that RDI can reduce water use by 20 percent for almonds and pistachios and up to 47 percent for vineyards, but actual water savings depend on many factors including crop type, climatic conditions, and sensitivity of growth stages to stress.

RDI might improve some quality attributes but reduce quality of others. For example, Delicious apples from trees grown under RDI had higher levels of soluble solids but were smaller than apples from well-watered trees. Pistachios grown under RDI had significantly higher shell splitting at harvest (positive effect), but also lower fruit weight compared to pistachios grown without water deprivation (Goldhamer and Beede, 2004).

Disadvantages of Regulated Deficit Irrigation

Only some crops are suitable for regulated deficit irrigation, and growers need to have a clear understanding of crops' responses to water stress during different stages of growth and development and under different environmental conditions (Kirda, 2002; Cooley et al., 2009). For example, pistachios are particularly tolerant of stress during the shell-hardening phase, but not while they are in bloom or during the nut-filling stages (Cooley et al., 2009).

Water Reuse

A variety of sources of water of marginal quality can potentially be used to augment the supply of water for agriculture. Domestic waste water, if properly reclaimed, can serve a variety of uses beneficial to agriculture, either as a source of irrigation water or to free up high-quality water that was being used for an activity (such as landscaping) that can utilize reclaimed water without health risks to the public (Haruvy, 1997). At the present time, waste water provides only a small portion of the national water resource for agriculture. Saline water has some potential for augmenting water use in agriculture, primarily through reuse of drainage water on more salt-tolerant species or using cyclic rotations of good quality and saline water to grow a range of sensitive and more tolerant crops (Shennan et al., 1995). Those are especially attractive options for western San Joaquin Valley of California, where shallow saline ground water is affecting yields, and offsite disposal of drainage water collected by tile lines is prohibited (Benes et al., 2004).

Disadvantages of Water Reuse

The greatest concern with reuse is the biological and chemical quality of the reclaimed water. Removal of pathogens and anthropogenic chemicals is an important requirement for any wastewater treatment system. The reclaimed water would have to be monitored carefully to reduce chances of any harmful pollutants entering into an agricultural production system (Banin, 1999; Falconer et al., 2006).

Adoption of Water Reuse

In 1995, some 805,000 acre-feet per year of reclaimed water was used in the United States for irrigation, primarily in California and Florida (Solley et al., 1998). That amount was less than 2 percent of the total water reclaimed in the United States, most of which was released to streams or ground water.

Small Dams

The many adverse environmental consequences of large dams have spawned proposals for alternative means of water capture. A recent committee of the National Research Council recommended that managed underground storage and recovery should be seriously considered as a tool in a water manager's arsenal besides small surface water storage practices used extensively on agricultural lands (NRC, 2008b).

Impact of Small Dams

Small dams and other impoundments also provide temporary water storage of runoff from large storms, thereby reducing downstream flooding. Additional benefits include water storage for irrigation or livestock supply, municipal or industrial uses, and recreation, including fishing, boating, and wildlife habitat. USDA-NRCS estimates that these small dams yield an annual benefit of nearly $1.6 billion and prevent more than $700 million in damages annually through their control of flooding. These small watershed dams in the United States represent a $15 billion national infrastructure investment and beneficially impact hundreds of thousands of lives everyday (USDA-NRCS, 2008c).

NRCS has conservatively estimated the cost of rehabilitating small watershed dams to be between $500 million and $600 million. While the average rehabilitation cost per dam is approximately $242,000, local sponsors typically do not have sufficient resources to complete the necessary repairs to ensure the safety and critical functions of these small dams (USDA-NRCS, 2000).

Water quality impacts downstream of these small dams can be slight or significant, depending on sediment and nutrient inputs from upstream, water residence time in the impoundment, and whether surface (warmer) or deep (colder) water is released downstream. Sediments can be flushed or removed from behind these small dams during periods of dry-up or drought.

Some of the newer small dam designs have increased the flexibility in water storage capabilities, improved dam safety, and enhanced water quality and wildlife habitat benefits (Hanson et al., 2007; Hunt et al., 2008). Small dams have been endorsed by many federal and state agencies, the National Watershed Coalition, Ducks Unlimited, and many environmental groups (USDA-NRCS, 2008c).

Disadvantages of Small Dams

Excessive nutrients in slow-moving water that is impounded or stored from upstream agricultural uses can cause algae blooms, growth of aquatic plants to nuisance levels, and oxygen depletion because of organic matter decomposition. However, with proper management and maintenance, small dams can provide water storage and environmental benefits that outweigh the limitations (Lowrance et al., 2006).

Siltation and minimum storm water storage leading to spillway or dam failures are another concern. More than half of the small dams in the United States are now more than 40 years old and well beyond their original evaluated life (Hanson et al., 2007). Sediment pools have filled, and structural components have deteriorated on some of these dams. Public safety and environmental and social concerns will need to be addressed by rehabilitation or by using newer design and construction methods (Hanson et al., 2007).

WATER QUALITY MANAGEMENT

Reducing pollution of surface and ground water is a major goal for moving agriculture toward sustainability. As mentioned in Chapter 1, agricultural runoff and leaching contaminates ground water with agrichemicals and pollutes surface water as a result of sediment and nutrient runoff. Some of the most important landscape features affecting nutrient losses are surface drains (for example, waterways and drainage canals), mitigating features (for example, buffers and vegetative filter strips), or subsurface tile drainage, and whether those features are in place because they affect the relative volumes of surface runoff and subsurface drainage. On a smaller scale, the most important factors that determine the volume and timing of surface runoff are the rate of water infiltration in soil and rainfall. Factors that affect water infiltration in soil were discussed earlier in the context of soil management. Water-soluble pollutants move with water whereas those bound to soil particles move with sediment. Nutrient management is also key to protecting water quality, by ensuring that excess levels of nutrients do not build up in the soil and hence become vulnerable to runoff and leaching.

Drainage Water Management Systems

Drainage water management (DWM), often referred to as controlled drainage or water table management, is the practice in which the outlet from a conventional drainage system is intercepted by a water control structure that effectively functions as an inline weir. The drainage outlet's elevation is then artificially set at levels ranging from the soil surface to the bottom of the drains. Drainage water management systems are installed primarily to regulate drainage, thereby improve productivity, but they can be designed and managed to achieve additional environmental goals simultaneously (Evans et al., 1996). At the field scale, the drainage outlet can be set at or close to the soil surface between growing seasons to recharge the water table. The recharge temporarily retains soil water containing nitrate-nitrogen in the soil profile where it might be subjected to attenuating and nitrate transformation processes, depending on soil temperature and microbiological activity. In addition, it is possible to raise the outlet elevation after planting to help increase water availability to then-shallow plant roots, and to raise or lower it throughout the growing season in response to precipitation conditions. In some soils, water may even be added during very dry periods to reduce crop loss from drought, a related practice called subirrigation. Although there have been reported instances where subsurface DWM has resulted in reduced nitrate concentrations in the drainage outflow (from denitrification of the soil water within the soil profile), the general consensus is that the dominant process leading to reductions in nitrate loads is a reduction in drain outflow. With less water leaving the field through the tile drain, significantly less nitrate flows out of the drain, even with no change in nitrate concentration (Cooke et al., 2008).

Impact of DWM on Productivity and Water Quality

Researchers in North and South Carolina were among the first to recognize the potential of DWM to reduce nutrient losses from drained lands (Gilliam et al., 1979; Skaggs and Gilliam, 1981). They conducted field research and demonstration projects to determine the effectiveness of the method (Gilliam et al., 1978; Doty et al., 1985), developed design guidelines (Gilliam and Skaggs, 1986; Evans and Skaggs, 1989), and demonstrated the application of the method (Evans et al., 1990, 2000). The researchers in North Carolina continued to measure water quality effects associated with controlled drainage (Skaggs and Gilliam, 1981; Gilliam and Skaggs, 1986; Skaggs and Chescheir, 2003; Burchell et al., 2005) and began to improve a simulation model, called DRAINMOD or DRAINMOD-N, to predict water quality effects (Skaggs and Gilliam, 1981; Breve et al., 1997; Lou et al., 2000). DRAINMOD-N II was later developed to describe detailed nitrogen cycling, consider all forms of fertilizers and manures, and account for the carryover of nitrogen for different soils and plant organic matter (Youssef et al., 2004, 2005).

In Iowa, Kalita and Kanwar (1993) examined the effect of outlet control level on crop yield and nitrogen concentration. They observed a reduction in nitrate-nitrogen concentration for all outlet levels and an increase in crop yield for most. They also found, however, that it was possible to reduce yields by setting the outlet too close to the soil surface during the growing season. In Ohio, Cooper et al. (1991) reported increased soybean yields ranging from 23 to 58 percent where the DWM practice was used in combination with a subirrigation system in which additional water was added during most of the growing season. In field studies conducted using the DWM practice elsewhere in the United States, researchers have reported reductions in nitrate loading, ranging from 14 (Liaghat and Prasher, 1997) to 87 percent (Gilliam et al., 1997). A conservative estimate by consensus of researchers, extension specialists, and users is that the DWM practice can lead to a 30 to 40 percent reduction in nitrate loading in regions where appreciable drainage occurs in late fall, early spring, and winter seasons (ADMS, 2003).

Disadvantages of DWM

DWM has some practical limitations. Some existing drainage systems were not designed or configured in a way that improvements can be easily made; however, subsurface drainage systems can be retrofitted with all the equipment needed to efficiently operate and manage the DWM practice at a cost of less than $100 per hectare. Illinois farmers have made extensive changes that have ranged from $100 to $220 per hectare (Cooke et al., 2008). Costs of improvements in an existing surface drainage are often less than $30 per hectare, but could be more where additional land leveling, surface drains, buffers, or filter strips are being installed. Because DWM systems are typically managed during nongrowing season months or between crop rotations, there is little potential for yield loss.

Adoption of DWM

In 1985, USDA-ERS estimated that there were more than 13 million hectares of subsurface drainage in eight Midwestern states (USDA-ERS, 1987). Later, the amount of drained land in the entire Mississippi River Basin was estimated to have increased from about 2.5 to 30 million hectares over the past 100 years (Mitsch et al., 2001). In 1989, about 150,000 acres in eastern North Carolina had DWM systems installed (Evans et al., 1996). As more demonstrations and positive experiences are documented, farmers are beginning to combine DWM systems with other improved conservation and wetland practices to improve environmental quality and lessen some of the consequences of droughts and floods (Box 3-3).

Wetlands

A wetland is an ecosystem that depends on constant or recurrent, shallow inundation or saturation at or near the surface of the substrate (NRC, 1995). Because wetlands can be an effective method for removing a wide variety of water quality contaminants, including sediments, nitrogen, and phosphorus (Howard-Williams, 1985; Nixon and Lee, 1986; Kadlec and Knight, 1996; Reddy et al., 2005), the potential for using natural or constructed wetlands to clean up agricultural runoff has received considerable attention.

Emergent marshes provide significant potential for denitrification of nitrate and trapping of particular nutrients and can be effective in reducing sediment and other contaminant loadings associated with agricultural drainage (Reddy et al., 1999; Kovacic et al., 2000; Braskerud et al., 2005; Crumpton, 2005; Mitsch et al., 2005). In general, if wetlands are to serve as long-term sinks for nutrients, there has to be net storage in the system through accumulation and burial in sediments or net loss from the system, for example through denitrification (Crumpton et al., 2008). The processes involved in nitrogen transformation in wetlands are comparable to most types of aquatic systems and soils (Howard-Williams, 1985; Bowden, 1987; Reddy and Graetz, 1988; Crumpton and Goldsborough, 1998). Under anaerobic conditions, NO_3 can be converted to N_2O or nitrogen (N_2) by microorganisms via

---

BOX 3-3
Nitrate Loading from Agricultural Drainage into the Gulf of Mexico

A number of reviews and conferences on Gulf of Mexico hypoxia research and policies have highlighted the importance of agricultural drainage, the major pathway of nitrate loads in the Upper Midwest. For example, an economic study on two watersheds that used a constrained-optimization model to evaluate the cost-effectiveness of nitrogen-abatement policies (with explicit focus on drainage) showed that drained land dominates in nitrogen abatement and has substantially lower abatement costs relative to nondrained land. However, policies that remove drainage were not cost-effective. Furthermore, it was found that nutrient management, a policy strongly recommended by prior research, is relatively cost-ineffective as a means of abatement on nondrained (Petrolia and Gowda, 2006). Those two watersheds represent many of the watersheds that exist in the Upper Mississippi River Basin where surface and subsurface drainage enables farmers to produce high yields of corn and soybean based largely on proper drainage. If those farmers can improve both drainage and nutrient management, they should be able to improve their profitability and at the same time enhance the operation of a wetland used for improving environmental quality.

denitrification (Seitzinger, 1988). When wetlands are subjected to significant external nitrate loading, relatively high rates of denitrification are cited as the primary reason of nitrogen removal from drainage water (Crumpton et al., 2008).

There are three potential mechanisms by which wetlands reduce phosphorus in drainage water: deposition of sediment-bound phosphorus, sorption of dissolved phosphate, and accumulation of organic phosphorus in soil (Richardson, 1999). However, deposition of sediment-bound phosphorus is not considered long-term storage because future hydrologic events can re-suspend the sediment (Bruland and Richardson, 2006). Wetlands can mitigate pesticide contamination from agriculture in water by deposition of sediment-bound chemicals, sorption to wetland vegetation, or degradation (Reichenberger et al., 2007).

Impact of Wetlands

Nutrient Loading

In a constructed wetland used to treat dirty water from a dairy farm in Ireland, Mustafa et al. (2009) reported removal efficiency of 94 percent for suspended solids, 99 percent for ammonia-nitrogen, 74 percent for nitrate-nitrogen, and 92 percent for molybdate reactive phosphorus. However, the effectiveness of wetlands in reducing agricultural nutrient loads is influenced by a range of climatic and site-specific factors. Important factors related to wetland inputs include the timing and magnitude of nutrient and hydrologic loads to the wetland, the extent of surface and subsurface drainage, the concentrations of nutrients entering the wetland, and the chemical characteristics of the nutrients entering the wetland (for example, dissolved versus particulate fractions, nitrate versus ammonium and organic nitrogen, and liable refractory forms of phosphorus). Soil properties of wetlands, such as soil organic matter, exchangeable calcium, and oxalate extractable iron, are correlated to phosphorus sorption index. Therefore, the variability in the performance of wetlands in removing nitrogen and sequestering phosphorous can be expected (Crumpton et al., 2008). Research results over the last couple of decades clearly demonstrate that the design, operation, and maintenance of a wetland could be better understood and improved. Wetland restoration can be a promising approach particularly in heavily tile-drained areas like the Midwest (Crumpton et al., 2008). A restored wetland in Pennsylvania was shown to remove 65 percent of the nitrate load on average (Woltemade and Woodward, 2008).

Pesticides

Wetlands have been found to be effective in removing pesticides from water that passes through them (Reicherberger et al., 2007). Blankenberg et al. (2006) assessed the retention of four herbicides and three fungicides (fenpropimorph, linuron, metalaxyl, metamitron, metribuzin, propachlor, and propiconazole) commonly used on arable soil in Norway by two constructed wetlands. They observed pesticide retention of 3 to 67 percent. Munoz et al. (2009) observed sorption of chlorpyrifos on wetland vegetation, some of which was later degraded by sunlight. Similar to nutrient retention, the effectiveness of wetlands as a mitigation strategy for pesticides in water depends on several factors, including vegetation, properties of the pesticides, and width of the wetlands (Moore et al., 2007; Reicherberger et al., 2007).

Disadvantages of Wetlands

Loss of productive land and maintenance costs can be issues with use of wetlands for water treatment. Although wetland vegetation can sequester CO_2 from the atmosphere through photosynthesis, the CO_2 benefit can be offset by methanogenesis under anaerobic conditions and denitrification by soil microorganisms in the wetland soil (Crumpton et al., 2008). The carbon stored in the wetland soil can also be oxidized and emitted as CO_2 when the soil is drained (Crumpton et al., 2008).

Adoption of Wetlands

The widespread adoption of wetlands for nutrient reduction is not limited by science or engineering or by the availability of suitable land for large-scale wetland restoration; rather, the main obstacles are related to the scale of effort needed, cost, and policy and regulatory issues (Hey et al., 2004). The primary economic constraint associated with adoption of the practice is cost associated with wetland restoration and construction and with taking land out of production. These costs vary widely depending on the site characteristics and project size. Land costs are obviously higher for sites located on prime cropland than those located on marginal cropland or pasture, but these costs might be offset by lowering construction costs and, at least for nitrate, higher per acre rates of nutrient reduction (Crumpton et al., 2008).

Buffers

Buffers are small areas of permanent vegetation designed to manage environmental concerns (for example, nutrient and sediment runoff and pesticide contamination). Buffers can be planted at the edge of arable fields (hence, called edge-of-field buffers) or next to water resources to intercept pollutants (called riparian buffer). An edge-of-field buffer typically is a narrow trip of perennial vegetation. Riparian buffers can be designed to include trees, shrubs, native grasses and forbs, nonnative cool-season grasses, or some combinations of those to enhance ecosystem functions (for example, enhance surface and ground water quality, provide habitats for fish and wildlife, and reduce sediment transport) in specific habitats (Schultz et al., 2004).

Impact of Buffers

Nutrient Loading

Edge-of-field and riparian buffers have been shown to decrease nitrogen contamination of ground water (Lowrance et al., 2000; De Cauwer et al., 2006). Riparian forest and grass can reduce nitrate in shallow ground water near an upland area planted with row crops by up to 90 percent (Osborne and Kovacic, 1993).

The effectiveness of buffers in reducing nonpoint source phosphorus contamination is variable. One study showed that riparian zones can effectively limit the movement of phosphorus-enriched sediment and reduce dissolved phosphorus in contaminated ground water before the water reaches receiving bodies of water (Novak et al., 2002), while another showed no demonstrable effect (Snyder et al., 1998). Osborne and Kovacic (1993) observed that both forested and grass buffers in their study were less effective in reducing phosphorus concentrations in shallow ground water than nitrate concentrations. They also found that buffer vegetation could release phosphorus to ground water during dormant season.

Gypsum as a soil amendment for grassy buffer strips has been proposed as a strategy for enhancing buffer strips' effectiveness in reducing soluble phosphorus in surface runoff, particularly in land fertilized with manure (Watts and Torbert, 2009). Preliminary results show that it could be a useful strategy for reducing soluble phosphorus at the field edge.

Pesticides

Review of 14 publications revealed that edge-of-field buffer strips reduce pesticide load, but the efficiency varies widely (Reichenberger et al., 2007). Buffer strips reduce pesticide load mostly as a result of infiltration and sedimentation in the buffer strips. Grass strips were more effective than strips of crops or bare soil in reducing sediment loss and sediment-bound pesticides (Reichenberger et al., 2007). The effectiveness of riparian buffers in retaining pesticide has not been studied extensively, but Reichenberger et al. (2007) suggested that they are probably less than edge-of-field buffers. Surface runoff typically enters riparian vegetation as concentrated flow, which reduces the likelihood of pesticide retention by infiltration. Moreover, most riparian vegetation strips is too narrow or too sparse to be effective in reducing pesticide runoff.

Fecal Coliform Bacteria

Establishing edge-of-field buffer strips has been shown to be effective in reducing fecal coliform bacteria in runoff from pastureland amended with manure (Sullivan et al., 2007).

Disadvantages of Buffers

Although wider buffer strips tend to be more effective in nutrient removal, extending the width of the buffer strips takes land away from production (Hickey and Doran, 2004) and hence has economic implications. Edge-of-field buffers require active management to minimize unintended negative effects. Spontaneously developed plant communities in edge-of-field buffers might include weeds or noxious invasive species; therefore, it is better to sow the buffer vegetation (De Cauwer et al., 2008). The established buffer vegetation could contaminate the edge of the crop fields with weeds (Marshall and Moonen, 2002), and mowing and removal of cuttings might be necessary to reduce the risk of weed contamination (De Cauwer et al., 2008). Similar to edge-of-field buffers, grassy riparian buffers also require active management such as mowing (Lyons et al., 2000).

Like wetlands, the effectiveness of buffers in improving water quality depends on a number of factors, including hydrology, buffer vegetation, width of buffer, and climatic events (Dukes et al., 2002; Herring et al., 2006). For example, buffers are most effective in treating surface runoff that has slow, shallow, and diffuse flow (Lee et al., 2003). Flooding as a result of hurricanes, for example, can overwhelm a buffer's capacity to mitigate nutrient loading into water resources (Dukes et al., 2002).

NUTRIENT MANAGEMENT

The goals of a good nutrient management program are two fold: to provide sufficient nutrients for crop or animal growth throughout their life cycle, and to minimize negative impacts of nutrient losses on the environment. This section discusses the development of nutrient budgets to help manage fertility to balance inputs and desired outputs (products) and to minimize undesirable outputs (losses) into the environment. The different kinds of fertility inputs used in crop and pasture production (and issues involved with their use) subsequently are described. This section also provides examples of innovative ways of managing nutrient application (precision agriculture and nanotechnology) and the disposal and recycling of animal wastes.

A good on-farm nutrient management plan would aim to achieve the following goals to improve sustainability:

- Improve or maintain soil fertility.
- Minimize the use of off-farm nutrient inputs, especially synthetic fertilizer, thereby reducing energy used for fertilizer production.
- Ensure efficient use of nutrients, thereby reducing nutrient leaching and runoff and improving water quality.
- Ensure effective use and recycling of on-farm sources of nutrients.

In addition to providing the correct amounts of different nutrients for crop growth, it is equally important to synchronize the availability of the nutrients in the soil to meet the varying crop demands through the growing season. If the nutrient supply is not synchronized with the crop demand, then either the plants suffer nutrient stress (availability too low) or excess nutrients accumulate in the soil and are vulnerable to losses via leaching or as adsorbed nutrients on sediment lost with surface runoff (Crews and Peoples, 2005).

Use of organic nutrient sources requires their decomposition by soil organisms to convert the nutrients into plant-available forms. For example, the conversion of organic nitrogen in the soil to plant-available nitrate from fresh residue or existing soil organic matter is a two-step process mediated by soil microbes, first producing ammonium (mineralization) and then nitrate (nitrification). Environmental conditions such as soil moisture and temperature affect the rates of decomposition. The timing and rate of mineralization is also affected by the nature of the organic matter, notably its carbon-to-nitrogen ratio. A high carbon-to-nitrogen ratio (>20) leads to temporary immobilization of soil nitrate and ammonium, whereas a low carbon-to-nitrogen ratio (<20–25) leads to net mineralization. Nitrogen conversion is thus influenced by crop sequence, timing, the type and timing of nitrogen input, the soil microbial population, and the soil condition (Vigil et al., 2002). When nitrogen mineralization is brought into synchrony with crop needs and to minimize seasonal loss through nitrate leaching, crop growth and production per unit of nitrogen input (and the resultant energy balance) can be optimized (Fortuna et al., 2003). Along with the nutrient sequestration and release cycles, the timing and placement of fertilizer inputs can enhance cycling and uptake efficiency while reducing losses.

Applying nitrogen fertilizer in excess of that required also increases fertilizer costs, thereby reducing profits from crops. Crop quality and price can also suffer as a result of overfertilizing, and crops with parabolic yield response to fertilization might have a decrease in yield (Sibley et al., 2009).

Production of synthetic fertilizers is an energy-intensive process. In 2002, 490 trillion Btu was consumed in U.S. fertilizer production (Heller and Keoleian, 2000). For each kilogram of nitrogen fertilizer manufactured, transported, and stored, about 0.9–1.8 kg of CO_2 is emitted (Lal, 2004a). Use of manure, compost, and green manure can reduce the need for synthetic fertilizer and hence reduce the indirect energy use for the fertilizer production.

Mass Balances for Nutrient Management

Estimating nutrient mass balances (most commonly for nitrogen and phosphorus) can help producers develop a holistic approach to nutrient management by illustrating patterns of excessive or insufficient inputs for different nutrients over crop rotation cycles. Inputs can then be adjusted to obtain the correct nutrient balance. Mass balance calculations have been made at the field, farm (Haas et al., 2007), watershed (McIsaac and Hu, 2004), and national scales (Goodlass et al., 2003) to provide information on nutrient input excess or deficiency and implications for water quality. At the field scale, several studies (Drinkwater et al., 1998; Karlen et al., 1998; Jaynes et al., 2001; Webb et al., 2004) have shown agricultural practices and systems with higher nitrogen inputs compared to nitrogen outputs. All those studies use the conservation of mass to compute the N balance:

$$\text{Inputs} - \text{outputs} - \Delta \text{ soil residual mineral N} = \text{residual}$$

where output refers to nutrients that are removed by harvesting and nutrients lost to the environment.

Because of the difficulty in measuring all individual output pathways into the environment, and hence calculating the residual term, partial nutrient budgets are often used, especially for annual budgets, where the residual is assumed to be zero, and changes in soil mineral nitrogen may or may not be considered. The revised equation used commonly to estimate potential undesired losses from the field or farm is:

$$\text{Inputs} - \text{harvest outputs} = \text{potential loss into the environment}$$

Using nitrogen as an example, inputs of nitrogen consist of any fertilizers (synthetic or organic) applied, nitrogen contained in precipitation or irrigation water (wet deposition), and in dry deposition from particulate matter. All those components can be measured. If a legume is grown, nitrogen from fixation of atmospheric nitrogen will be another input, which can be estimated but with some uncertainty (Oenema et al., 2003). Outputs consist of harvested product removed from the field (easily measured) and losses into the environment. With nitrogen, these losses can be via leaching, surface runoff, and gaseous losses via denitrification or volatilization of ammonia. Considerable uncertainty is involved in measuring each of these pathways of loss (Oenema et al., 2003). Therefore it needs to be recognized that calculations of nutrient budgets should have uncertainty calculations accompany the numbers generated, although this is not commonly done (Oenema et al., 2003).

Nonetheless, partial budgets have been used as a way of comparing management systems in terms of potential nitrogen and phosphorus losses into the environment, and hence their potential impact on surface and ground water quality (Dechert et al., 2005; Drinkwater et al., 2008). In some cases, one or more of the losses to the environment might also be estimated, for example, leaching (Drinkwater et al., 1998), in which case the remaining nitrogen that is unaccounted for is assumed to be lost via some combination of denitrification, volatilization, and, over the long term, as additions to soil organic matter. In California, eight years of budget calculations illustrated the effects of organic, low-input, and conventional management on net balances of different nutrients. In that case, the budget included measures of changes in soil nutrient levels, thereby enabling an assessment of which systems were most efficient at retaining excess nutrients (Clark et al., 1998).

In the European Union, field- and farm-level nitrogen and phosphorus budgets are used as an indicator of sustainability as part of efforts to improve water quality (Ondersteijn et al., 2002; Ekholm et al., 2005). The main types of budget tools are: farm-gate that considers purchased inputs brought onto the farm versus loss of nutrients in products that are sold; soil surface budgets that measure inputs into the soil and removals via crop uptake and grazing; and soil system budgets that are more complex and take into account all nutrient inputs and outputs, including nutrient gains and losses within and from the soil (Oenema et al., 2003; Cherry et al., 2008).

In the Netherlands, regulations have required farmers to keep farm-gate nitrogen and phosphorus surpluses below a certain amount to meet water quality guidelines. In all, more than 50 different nutrient accounting systems are used among the European Union member states, with many using some type of farm-gate budgeting. In contrast, the Organisation for

Economic Co-operation and Development recognizes the gross soil surface balance as an effective agrienvironmental indicator (Goodlass et al., 2003; Cherry et al., 2008). Some argue that a standardized unified approach is needed (Oenema et al., 2003; Cherry, 2008), and the efficacy of the budget approach without further development and standardization is being questioned (Ondersteijn et al., 2002; Halberg et al., 2005). While budgeting using on-farm data provides a simple and readily communicable means of assessment, it does not currently consider the timing and transport aspects of loss and mitigation and assumes a direct causal relationship between potential and actual nutrient loss. The relationship between the nutrient surplus obtained from a farm-gate budget and actual losses into the environment varies with climate, topography, and other factors (Cherry et al., 2008). Nonetheless, in a number of examples, reductions in nutrient surplus at the farm-gate correlate well with reductions in leached losses or river nitrate levels (Cherry et al., 2008).

A variety of tools have been developed by different state extension systems, which are often built around a nutrient-credit system. Those tools are variants on a nutrient budget, where the idea is to work out how much fertilizer is needed to reach a predetermined potential yield. The amount of nitrogen required is therefore known. After credit is given for all other sources of nitrogen (such as nitrogen released from the soil as estimated by soil tests; any manure, compost, and other nutrients added; and nitrogen from incorporation of sod), the difference between the amount needed and the total credits indicates the rate of fertilizer to be applied. Many of the tools are available on the web and are interactive (Cornell University et al., 2009; USDA-ARS, 2009), enabling farmers to plug in information such as soil type and manure characteristics to calculate fertilizer needs.

While many management tools are developed for nitrogen, phosphorus is also a pressing concern especially for animal systems, where high phosphorus-content manure applied to crop fields can lead to excessive build up in the soil and losses not only via erosion and runoff, but also by leaching into the ground water (McDowell and Condron, 2004). Budget and other phosphorus management tools are also being developed (SERA-17, 2008). Other nutrient management tools, such as risk assessment and modeling, are also being developed, and they might be more effective than methods based on simple budgets (Cherry et al., 2008).

Soil and Tissue Sufficiency Tests

A great deal of work has been done over the past 20 years to keep refining soil and plant tissue sufficiency tests to help determine the level of fertilizer inputs necessary to support good crop growth. Soil tests are often carried out pre-planting or at early growth stages such as pre-sidedress, whereas plant tissue tests are often taken at multiple times during the season to allow for adjustments in later fertilizer applications. In addition, various crop canopy measures, such as leaf chlorophyll and canopy reflectance, are also used. There are excellent reviews on the topic that discuss the issues around soil and tissue testing and summarize the various tests developed for different crops (Schroder et al., 2000; Olfs et al., 2005; Zebarth et al., 2009).

The most effective test varies depending on the crop. For example, in one study, the best nitrogen test for maize was a pre-plant soil test; for barley, it was the mean stem NO_3^- content (measured across five phonological stages). Both tests showed strong linear relationships with yield (Montemurro and Maiorana, 2007). In the case of sugar beets, a petiole NO_3^- test was the best predictor of yield (Montemurro et al., 2006).

Soil tests have limitations, however, in that they do not take into account factors that affect the risk of actual loss from the field and of impacts on water quality. Those limitations led to the development of more complex measures, such as the phosphorus index, which includes some combination of soil test, rate, and application method for phosphorus from fertilizers and manure, soil erosion, runoff class, distance from surface water bodies, and irrigation erosion as inputs (Sonmez et al., 2009).

Nutrient Management Plans and Best Management Practices

Nutrient management plans are comprehensive plans for managing nutrients for crops and animals. Such plans are increasingly required to meet water quality guidelines. Many state extension services have developed tools to help farmers develop their plan. Typically, a plan incorporates some kind of soil testing, use of a budget or credit approach to determine input levels needed for a specified and realistic yield goal, and measurement of nutrient contents for all inputs including manure, composts, and use of other best management practices (BMPs). BMPs vary by regions but can include recommendations for methods and timing of fertility applications, use of specific soil or plant tissue tests, use of conservation buffers, use of cover crops, and use of conservation tillage. (See *An Introduction to Nutrient Management* [CTIC, 2007] for an example of BMPs for nitrogen and phosphorus.) Tools that focus on manure management are also available for dairy production (USDA-ARS, 2009). Furthermore, there are also efforts to coordinate nutrient management planning on a regional basis—for example, the Great Lakes Regional Water Program developed through partnerships with the USDA Cooperative State, Research, Education, and Extension Service and land-grant colleges and universities (The Great Lakes Regional Water Program, 2009).

Nutrient Inputs

The most commonly used fertility inputs in U.S. agriculture today are chemical fertilizers of different formulations. There is a very extensive literature on determination of recommended fertilizer input levels for different crops, together with various soil and tissue tests to help determine nutrient sufficiency during the growing season as discussed earlier. Split applications and slow release fertilizers can also help synchronize nutrient availability with crop demand (Chien et al., 2009; Sitthaphanit et al., 2009). This section, however, focuses on the three sources of nutrient inputs that can be generated on-farm and can enhance nutrient cycling—legumes, animal manure, and compost.

Legumes

Legumes form a symbiotic relationship with *Rhizobium*, root-nodule bacteria that fix atmospheric nitrogen to ammonium, and thus acquire nitrogen from the soil and the atmosphere. The fixed nitrogen is incorporated into legumes' biomass in the form of amino acids and proteins. Crop rotations that include actively fixing legumes can reduce nitrogen fertilizer needs because some of the fixed nitrogen is returned to the soil with incorporation of crop residue, and by direct release into the soil via root exudation and root death. As discussed in the earlier section on cover crops, leguminous cover crops can be used as green manures to improve soil fertility.

Impact of Legumes

Yield

Inclusion of legumes into rotation can be beneficial for subsequent crop yields. For example, regardless of the amount of fertilizer applied, grain yields following legume rotations are often 10 to 20 percent higher than continuous grain rotations (Heichel, 1987; Power, 1987). Similar grain yields have been achieved in studies that compare the use of legume or fertilizer as sources of nitrogen (Harris et al., 1994). In another study, a diverse rotation that included corn, soybean, wheat, and alfalfa led to higher grain nitrogen and sulphur, as compared to corn monoculture or a simple corn and soybean rotation. Furthermore, nitrogen application did not increase corn yield when it was grown in the diverse rotation and thus suggested the leguminous crops provided adequate nitrogen for the corn crop (Riedell et al., 2009). In some cases, better synchrony of nutrient availability and crop need also can be achieved by using a combination of legume residue and chemical fertilizer (Kramer et al., 2002).

Nutrient Availability

When legumes are used in rotation, they increase the nitrogen available in the soil (P. Smith et al., 2008; Sharifi et al., 2009) and reduce the need for commercial fertilizers. A number of studies suggest that legume residues can supply 36 to 266 kg ha^{-1} of nitrogen (as summarized in Christopher and Lal, 2007). The amount of nitrogen supplied depends on environmental conditions, the soil microbial biomass, management practices used (for example, tillage), and the legume species (Stute and Posner, 1993; Fageria and Baligar, 2005).

Disadvantages of Legumes

Nitrate leaching increases if leguminous crops or residues are incorporated into the soil in autumn (Moller et al., 2008), but leaching losses can be reduced substantially if a catch crop is grown during the autumn and winter that follow immediately before sowing subsequent spring wheat (Hauggaard-Nielsen et al., 2009). Similarly, if legumes are used as a winter cover crop, nitrate released following incorporation in the spring can be vulnerable to leaching losses (Moller et al., 2008). Such loss can be reduced by planting mixtures of legumes and nonlegumes to increase the carbon-to-nitrogen ratio of the residue (Cherr et al., 2006).

Animal Manure

Animal wastes in the form of raw manure are often used as a crop fertilizer or soil amendment. Substituting animal manure for synthetic fertilizer has the potential to improve carbon sequestration and reduce the fossil energy input required to produce synthetic fertilizer (Ceotto, 2005).

Impact of Animal Manure

Soil Quality

The application of animal manure to crops can provide multiple benefits to soil and crops when applied in appropriate quantities. Benefits include improved infiltration capacity (Boyle et al., 1989; Sullivan, 2004; Plaster, 2009) and increased soil carbon and nitrogen levels over the long term (Sommerfeldt et al., 1988).

Manure application also affects nutrient cycling in soil by providing carbon and other nutrients for microbial populations. For example, the application of chicken litter to Vertisol soil in Texas has been shown to result in higher microbial biomass carbon, nitrogen, and enzymatic activities compared to sites with no litter application (Acosta-Martinez and Harmel, 2006). Likewise, Larkin et al. (2006) observed that dairy and swine manure generally increase soil microbial populations. Dairy sludge in particular is nutrient-rich and high in organic matter (Ciecko et al., 2001). It stimulates the soil microbial respiration and enzymatic activity when added to soil (Jezierska-Tys and Frac, 2008, 2009).

Energy Use

It is more efficient to recycle nutrients within a farm system than to produce new fertilizers from fossil fuels. Fossil fuel energy use can be reduced when animal manure is used to fertilize crops instead of industrial nitrogen fertilizers (Ceotto, 2005). Concerns about global climate change have stimulated efforts to decrease agriculture's dependence on chemical fertilizers by using animal manure more efficiently.

Disadvantages of Animal Manure

Using animal manure requires more field labor (Karlen et al., 1995) and is more complicated than applying synthetic fertilizer. The nutrient content of manure depends on many factors including type and age of livestock, feed management, and manure storage. University extension provides guidance on manure sampling (Steinhilber and Salak, 2006; Martin and Beegle, 2009). Applying the appropriate amount of manure to meet the crops' nutrient requirements also requires knowledge of the mineralization patterns of the manure applied. However, nutrient release from applied manure depends on temperature, soil moisture, soil properties, manure characteristics, and microbial activity (Eghball et al., 2002). A common problem with using manure as a nutrient source is that application rates are usually based on the nitrogen needs of the crop. Some manure has about as much phosphorus as nitrogen, which exceeds the crops' uptake. The excess phosphorus often leads to a build up in the soil and subsequent loss into the environment and even leaching in extreme cases (McDowell and Sharpley, 2004). One solution is to adjust the manure rate to meet the phosphorus needs of the crop and to supply the additional nitrogen with fertilizer or a legume cover crop (Sullivan, 2004).

Manure from animal production operations will contain trace minerals (Petersen et al., 2007). Recommendations to include trace minerals in animal diets exist to meet metabolic needs, improve health, counteract elevated concentrations of interfering substances, and promote growth (NRC, 1980). The majority of trace minerals in livestock feed is excreted in feces and urine. Bioconcentration of trace elements will occur during manure storage as carbon, oxygen, hydrogen, and to some extent nitrogen are volatilized (Petersen et al., 2007). Another problem with the use of manure is microbial contamination, which will be discussed in Chapter 4.

Dietary Modification to Adjust Manure Composition

Animal diet can be adjusted to meet nitrogen and phosphorus requirements without much excess so that nutrients excreted in urine and feces are minimized (NRC, 2001). Changing the diet composition of poultry by adding crystalline amino acid supplements, adding enzymes such as phytase, an enzyme that improves mineral bioavailability (Lyberg et al., 2008), and lowering the protein and phosphorus contents can reduce the nitrogen, phosphorus, and other mineral contents in poultry manure and litter (Nahm, 2000; Plumstead et al., 2007). The addition of phytase to the diets of pigs has been found to reduce manure pH and lead to a decrease in ammonia losses from swine manure (Smith et al., 2004). Adjusting the dietary amino acids balance can reduce nitrogen excretion (Dourmad and Jondreville, 2007). Phase-feeding, which is feeding four or more diets to grower or finisher pigs, has also been found to reduce phosphorus excretion (Dourmad and Jondreville, 2007). Knowlton et al. (2007) showed that adding an exogenous phytase and cellulase enzyme formulation to diets for lactating cows reduced their fecal nitrogen and phosphorus excretion and fecal dry matter.

Dairy and beef cattle carry *E. coli* asymptomatically and shed it intermittently and seasonally in their feces (Bach et al., 2002). The presence of *E. coli* O157:H7 in manure could result in contamination of produce, soil, and water if the manure is applied as liquid. In cattle that are fed a grain ration, some starch could be passed to the hindgut without microbial degradation. Starch that is not degraded will be fermented in the hindgut where *E. coli* O157:H7 can use the sugars released from starch breakdown. Callaway et al. (2003) suggested that switching dairy cattle from a grain ration to forage could decrease *E. coli* O157:H7 populations in cattle. Gilbert et al. (2005) suggested that the type of dietary carbohydrate affects the fecal populations of *E. coli* in cattle. In experiments with small sample sizes of 6 or 30 cows, they observed significantly higher fecal *E. coli* populations in cattle fed with a finishing diet of grains compared to the ones given a finishing diet of roughage or roughage and molasses. Dietary manipulation has the potential to reduce nutrient and pathogen contamination in livestock manure, which in turn could mitigate some of the potential negative effects of using manure as natural fertilizers.

Adoption of Animal Manure Application

In 2006, animal manure was used on 16 million acres of U.S. cropland (about 5 percent) (MacDonald et al., 2009). USDA-ERS (Gollehon and Caswell, 2000) estimated that confined animals produced 1.23 million tons of recoverable manure nitrogen (collectible for spreading) in 1997, which was about 10 percent of total U.S. nitrogen consumption that year (USDA-ERS, 2008). Fertilizing crops with animal manure is not widely adopted because 52 percent of the harvested acres do not have livestock production at all (MacDonald et al., 2009). Those farms are not likely to use manure unless livestock or animal production facilities are nearby because transporting manure is costly.

Compost

Compost is a mixture of decaying organic material and can be made from farm manure, sewage sludge, agricultural residues, or food wastes. Composting has been defined as "an aerobic process of decomposition of organic matter into humus-like substances and minerals by the action of microorganisms combined with chemical and physical reactions" (Peigne and Girardin, 2004, pp. 46–47). Composting farm manure and other organic materials is a way to stabilize their nutrient content and create a product that is easier to handle than raw manure (DeLuca and DeLuca, 1997). Although compost is not as good a source of readily available plant nutrients as raw manure, a well-matured compost releases its nutrients slowly and thereby can minimize losses (although see also Evanylo et al., 2008). The raw materials used to produce compost and the conditions in which composting occur greatly affects the quality of the compost produced; hence, quality guidelines for commercial compost have been developed (Larney and Hao, 2007; Hargreaves et al., 2008). Although there are no U.S. national standards for commercial compost, California, for example, has quality criteria based on a series of tests including respiration, temperature, carbon-to-nitrogen ratio, visual and olfactory characteristics, seed germination, and a maturity index (California Integrated Waste Management Board, 2009).

Impact of Compost

Productivity

Composts can have favorable effects on crop productivity even when used alone, without chemical fertilizer as a supplement. For example, Delate et al. (2008) showed that growth and yields of peppers grown using a compost-based organic fertilizer can surpass those of conventionally grown peppers. In another study, yields of maize, wheat, and peppers grown with dairy leaf compost either equaled, and in the case of maize exceeded, conventionally grown crops (Hepperly et al., 2009). Furthermore, the compost treatment proved superior to both conventional synthetic fertilizer and raw dairy manure in building soil nitrogen and carbon, providing residual nitrogen for the subsequent unfertilized wheat crop, and in reducing nutrient losses via leaching. Olive pomace compost effectively replaced half of the mineral nitrogen fertilizer and gave equivalent yields of maize and barley to the highest rate of nitrogen fertilizer (Montemurro et al., 2006) and showed a similar nitrogen utilization efficiency. In contrast, supplemental fertilizer was needed to attain high yields of sweet corn in a vegetable rotation system, despite using high rates of compost application (Evanylo et al., 2008). The lower yield was due to mineralization of the compost occurring after the period of peak crop demand, and as a result higher levels of residual nitrate were left at harvest in the compost versus fertilizer treatments; use of a rye catch crop, however, prevented significant leaching losses.

Soil Quality

Use of composts adds carbon to the soil, increases soil organic matter, can increase nutrient availability, and improve soil moisture retention and water infiltration. For example, in an 18-year study, compost additions increased soil carbon by 16 to 27 percent and soil nitrogen by 13 to 16 percent (Hepperly et al., 2009). In another study, use of compost also led to higher soil organic carbon, decreased soil bulk density, and improved soil moisture retention relative to chemically fertilized plots (Jagadamma et al., 2009). In a field study in Spain, Gil et al. (2008) examined whether compost made from cattle manure combined with a nitrogen mineral fertilizer could substitute for conventional mineral fertilizer. Grain yields were similar across both treatments, but the soil in the field that received compost and mineral nitrogen had higher organic matter content, phosphorus, potassium, and sodium concentrations than the field that received conventional mineral fertilizer (Gil et al., 2008). Another study demonstrated that compost use led to both higher soil organic matter and higher soil water content than the control (Edwards and Burney, 2008). Evanylo et al. (2008) found that compost use affected bulk density, porosity, and water-holding capacity of the soil such that losses of nitrogen and phosphorus following a simulated rain event were greatly reduced (by over 70 percent relative to the fertilizer treatment), despite higher concentration of those nutrients in the runoff.

Disease Suppression

Compost can suppress a number of plant diseases and its contribution to disease suppression is discussed in a later section on weeds, pests, and disease management in crops.

Energy Use

Using composted manure can reduce energy consumption compared to using synthetic fertilizers (DeLuca and DeLuca, 1997), particularly if the organic materials for composting are wastes from on-farm sources. Using on-farm resources for composting improves nutrient recycling and eliminates transport costs of bringing in raw materials or compost from commercial suppliers.

Disadvantages of Compost

One major disadvantage is that composting can lead to significant losses of ammonia, CH_4, and N_2O to the atmosphere and contribute to greenhouse-gas emissions if the piles are too wet or if the carbon-to-nitrogen ratio is too low for quick retention of nitrogen compounds (IPCC, 2006). All production of well-finished compost depletes the carbon content of the starting materials by about 60 percent, with the released carbon going into the atmosphere as CO_2. For that reason, there has been a longstanding debate about the desirability of composting rather than direct application of manure or residues to the field and having decomposition occur in the field. However, Kirchmann and Bernal (1997) demonstrated that composting reduces the loss of CO_2 compared to a nondecomposed treatment when the calculation takes into account the loss during the treatment of the fresh material and the loss after its application to fields. Peigne and Girardin (2004, p. 52) concluded that "composting is responsible for a significant quantity of CO_2 emitted, but it is not a net source of CO_2 along the recycling chain of agricultural wastes." In contrast, compost is a net source of CH_4 and N_2O, but the amounts released depend largely on the raw materials used and other characteristics of the pile and how it is managed. Similarly, the magnitude of nitrate and phosphate losses by runoff and leaching during the composting process depends on the location of the piles, water additions, and whether the pile is covered (Peigne and Girardin, 2004).

Similar to the case of applied manure, nutrient mineralization from applied compost depends on the quality of compost, temperature, soil moisture, soil characteristics, and soil microbial communities so that the availability of nutrients from compost to plants varies (Eghball, 2002; Evanylo et al., 2008). Hence, it could be difficult to determine the appropriate amount of compost to apply to meet crops' needs. The carbon-to-nitrogen ratio has to be about 20:1 to ensure short-term nitrogen mineralization (Gaskell and Smith, 2007).

Some of the sources of material for composting might contain heavy metals (for example, sewage sludge and municipal waste). Compost from such materials could result in the accumulation of heavy metals in the soil and sometimes in the edible parts of vegetable plants. Metals released from composts might be leached out of the root zone and into ground water after irrigation or rainfall (Li et al., 2000).

Composting, with periodic heap turning, can inactivate some pathogens, thereby reducing the risk of microbial contamination. If appropriate practices are not followed, compost can contain plant and human pathogenic bacteria (Brinton et al., 2009). The time the compost pile is at a high temperature is the most important factor for eliminating pathogens (Noble et al., 2009). Studies have shown, however, that careful attention to ensure optimal time and temperature combinations can be effective at reducing enteric pathogens that pose a risk to humans (Heinonen-Tanski et al., 2006). Similarly, composting of biosolids (Class A stabilization) significantly reduced human pathogen levels as compared to class B stabilization and other treatments (Viau and Peccia, 2009).

Precision Agriculture

Precision agriculture can be broadly defined as "a management strategy that uses information technologies to bring data from multiple sources to bear on decisions associated with crop production" (NRC, 1997, p. 2). Precision agriculture presents farmers with the opportunity to use technologically advanced methods by which they can identify more efficient production practices. The pivotal technology in precision agriculture is the global positioning system (GPS) so that treatments applied during field operations can be related to localized requirements within a field. These technologies include real-time kinematics (RTK) GPS guidance that allows better seed and fertilizer placement and automated height adjustment for large-scale boom applicators. Yields, weeds, grid soil sampling, chemical (herbicide, insecticide, and fertilizer) use, and record keeping also can be monitored. The collected data can influence farmer decisions related to seeding, fertilizer and chemical applications, irrigation scheduling, and other farm input use and lead to economic savings on farm and reduced impact on the environment. In addition to large-scale production of crops such as corn, soybean, wheat, and barley, precision agriculture also is used in potato, onion, tomato, sugar beet, forages, citrus, grape, and sugarcane production systems (Zhang et al., 2002; Kach and Khosla, 2003).

One goal of precision agriculture is to reduce the input of nitrogen and phosphorus fertilizer into agricultural fields. There are various ways to achieve that goal, including monitoring crops' nutrient needs to determine the timing and amount of fertilizer application (Biermacher et al., 2009), using GPS technology to inject fertilizer in a precise location as needed instead of spraying an entire field indiscriminately (C. Mitchell, presentation to the committee on August 4, 2008; Smith, 2008), and balancing dairy cattle dietary phosphorus requirements precisely to reduce phosphorus concentration in manure (Ghebremichael et al., 2008).

Impact of Precision Agriculture

Nutrient Use

Geographic information system (GIS), GPS, and modeling technologies can be used to identify and simulate the spatial residual soil NO_3-N patterns (Delgado and Bausch, 2005). Sensor-based technologies have been developed to measure plant nitrogen and provide information for in-season nitrogen application (Osborne, 2007; Stroppiana et al., 2009). Those tools can provide information on soil nutrient levels much more quickly than soil tests. If nitrogen application is made on the basis of nitrogen reflectance index (fertilizer is applied when a certain proportion of crops have nitrogen reflectance index below a certain level), the in-season nitrogen application can be better synchronized with the crops' needs (Delgado and Bausch, 2005).

Clay Mitchell, an Iowa farmer, tested an RTK guidance system with sub-inch accuracy and found that corn planted in the center of the fertilized strip—made possible because of RTK autoguidance—yielded 245 bushels per acre. In contrast, corn planted 5 and 10 inches off the fertilized strip yielded 236 bushels per acre and 238 bushels per acre, respectively (Smith, 2008). Mitchell later used GPS-steering for his planter to eliminate slide slip on slopes and to ensure that the seeds were planted over the tilled strip. The precise planting enabled by GPS saved him up to 7 percent in seed costs. Stahlbush Farm, described in Chapter 7's case studies, also uses RTK-guided tractors. Bill Chambers, owner of Stahlbush Farm, said that those tractors improve the ease of operation at night, operate faster than nonguided tractors, use less fuel, and take less space to turn.

In addition to using precision agriculture to manage crop nutrients, a precision feed management (PFM) program has been proposed as a strategy to reduce phosphorus build up in soil by limiting feed and fertilizer purchases and by increasing high-quality home-grown forage production (Ghebremichael et al., 2007). The PFM program includes strategies that balance dairy cattle dietary phosphorus requirements precisely with actual intake, and strategies that improve on-farm forage production and utilization in the animal diet (Ghebremichael et al., 2008). The importance of precision feeding and cropland fertility management has improved the phosphorus balance of the dairy sector in New York State (Swink et al., 2009).

Water Quality

Water contamination can occur when inorganic or organic manure fertilizers are overapplied (Spalding and Exner, 1993; Jemison et al., 1994; Dinnes et al., 2002). Variable rate technology (VRT) methods of applying nutrients reduce leaching and improve water quality when compared to uniform application methods (Wang et al., 2003). VRT can be used to apply herbicide to areas of severe weed infestation (Thorp and Tian, 2004). Ghebremichael et al. (2007) also found that PFM reduces soluble phosphorus lost to the environment by 18 percent. Furthermore, adoption of the PFM system could result in a decrease of 7.5 kg per cow per year feed supplement for dietary mineral phosphorus and by 1.04 and 1.29 tons per cow per year for protein concentrates.

Adoption of Precision Agriculture

Adopters of precision agriculture are mostly large-scale farms in the Midwest (Whipker and Akridge, 2007) because small to medium-size producers see the initial cost, uncertain economic returns, and technology complexity as limiting factors (USDA-NIFA, 2009). Adoption varies from a few percent in some regions of the United States to 40 percent of tillable land in other regions, such as the sugar beet growing area of the Red River Valley in Minnesota and North Dakota (Robert, 2002). Surveys on precision agriculture have consistently found that age, attitude, and education of producers are correlated with adoption of precision agriculture (Robert, 2002; Walton et al., 2008).

Nanotechnology-based Applications

Nanotechnology is the manipulation or self-assembly of individual atoms, molecules, or molecular cluster into structures to create materials with unique characteristics. Nanotechnology generally is used when referring to materials with the size of 0.1 to 100 nanometers (or 1 billionth of a meter). Because of their ion exchange and reversible dehydration properties, nanotechnology-based soil amendments from naturally occurring minerals, such as zeolites, could be used as agents for the slow release of nitrogen and phosphorus fertilizers and to increase water retention. They could also be used to enhance the availability of micronutrients to absorb metal cations and reduce local concentrations of toxic substances that inhibit plant growth and nitrogen-fixing soil microbes (NRC, 2008a). Nanotechnology can potentially be used to improve herbicide application by providing better penetration through cuticles and tissues, allowing slow and constant release of the active substances and targeting delivery. However, issues of possible toxicity and scale and cost of production of nanoparticles and nanocapsules will have to be addressed before their widespread use (Perez-de-Luque and Rubiales, 2009).

Another major role for nanotechnology-enabled devices will be the increased use of autonomous sensors linked into GPS for real-time monitoring and precision farming. The nanosensors are distributed throughout the field where they monitor soil conditions and crop growth. Wireless sensors are being used in certain parts of the United States and Australia. Pickberry, a vineyard in California's Sonoma County, has installed wireless (Wi-Fi) systems with the help of the information technology company Accenture. The initial cost of setting up such a system is offset by the benefit of growing improved grape crops, which in turn produces improved wines that command a premium price. The use of such wireless networks is not restricted to vineyards (Joseph and Morrison, 2006).

Anaerobic Digestion with Biogas Recovery of Animal Manure

Treatment of animal manure with anaerobic digestion coupled with biogas recovery and use is one method for animal operators to reduce odors and pathogens in manures and generate biogas for energy at the same time. Anaerobic digestion requires the collection of fresh manure and lends itself to be a viable practice on larger animal and poultry facilities when animals are housed on a surface that can be scraped or flushed. Anaerobic digestion is a two-step process that requires microbial populations to digest organic material in the absence of oxygen (Balsam, 2006). In the first step, part of the volatile solids in manure is converted into fatty acids by acetogens (acid-forming bacteria). In the second step, the acids are converted to biogas (CH_4 and CO_2) by methanogens (CH_4-forming bacteria) in covered lagoons for liquid manure and plug flow or sequencing batch reactors for slurry manure. Biogas is captured from the enclosed area, transferred, and may be scrubbed. Biogas has most commonly been used to generate electricity and heat via internal or external combustion engines (EPA, 2002). More recently, it has been scrubbed and successfully injected into natural gas lines or pressurized to make compressed natural gas for use in vehicles. Research is ongoing in ways to convert biogas to energy in fuel cells. The biogas contains 60 to 70 percent CH_4, 30 to 40 percent CO_2, and trace amounts of hydrogen sulfide, ammonia, and sulfur-derived mercaptans (Balsam, 2006).

These alternatives for biogas use might reduce or maintain fuel or energy costs at an animal operation. New opportunities might be available to market greenhouse-gas emission reductions. The residual organic material can be dewatered or dried and used as animal bedding or as a soil amendment. Anaerobic digestion is beneficial for reducing odors and pathogens associated with manures.

Impact of Anaerobic Digestion

A few studies have examined the potential of using anaerobic digestion to stabilize swine and cattle manure slurry to recover biogas for energy generation. R.D. Costa et al. (2007) fed a laboratory-scale digester with 5 percent and 15 percent swine manure slurry and observed an average reduction of 58 percent total chemical oxygen demand and 85 percent dissolved oxygen demand. The CH_4 content of biogas ranged from 55 to 65 percent. The authors suggested that the stabilized sludge might be suitable for use as soil amendment for crops. Macias-Corral et al. (2005) used dairy manure and cotton gin waste in a two-phase anaerobic digestion system to assess the feasibility of producing CH_4 and soil amendment from mixed agricultural wastes. They obtained biogas that has 72 percent CH_4 and conducted nutrient analyses on the residuals to demonstrate that the residuals can be used as soil amendments.

Methane generation has received considerable attention, with the USDA, state research groups, and many alternative groups providing technology information and promoting its use. Methane gas has an energy equivalent of 600 Btu/ft^3, compared to 1,000 Btu/ft^3 for natural gas. The net energy contribution per day for wastes of selected animals, assuming 35 percent of gross to operate the digester, is swine, 1,500 Btu/day; dairy, 18,000 Btu/d; beef, 10,700 Btu/d; and poultry, 110 Btu/d per bird. Small numbers of such units have been constructed, although the technology is reasonably well developed and efficient. Biogas generation is attractive especially for larger animal units where waste recycling to the land is environmentally and socially sensitive. Ideally, an anaerobic digestion and biogas recovery system can convert manure from animal operations into energy.

Disadvantages of Anaerobic Digestion

The biogas includes some toxic gases (as discussed above) and has to be scrubbed and the toxic gases separated. The CH_4 produced is highly explosive if it comes into contact with atmospheric air at proportions of 6 to 15 percent CH_4 (Balsam, 2006). The anaerobic digestion and biogas recovery system requires a large capital investment and regular labor to maintain the system to ensure proper functioning (MacDonald et al., 2009). Installation of digesters historically occurs in the United States when federal or state funds are available to offset costs. The low adoption rate, coupled with a demolition or decommission rate 10 years later, can lead to a lack of the critical mass needed to establish a viable technical support service industry. The inability of operators to work with others with similar problems (lack of farmer-to-farmer interaction) contributes to low adoption rates (Morse et al., 1996). In airsheds where NO_x emissions are regulated, use of a combustion engine to generate electricity requires additional pollution emissions reduction technologies (catalytic converters) as the standard generators used in these systems may not meet regulatory mandates.

Adoption of Anaerobic Digestion

Large dairy and hog farms are more likely to adopt anaerobic digestion and biogas recovery systems; the economic costs and benefits play a role in adoption of the systems (MacDonald et al., 2009). The AgSTAR Program sponsored by the Environmental Protection Agency (EPA), USDA, and the U.S. Department of Energy is a voluntary program that encourages the use of CH_4 recovery (biogas) technologies at the confined animal feeding operations that manage manure as liquids or slurries. AgSTAR estimated that anaerobic digestion of animal wastes to produce CH_4 could be cost-effective on about 7,000 U.S. farms (National Sustainable Agriculture Information Service [ATTRA], 2006). Since its inception in 1994, the AgSTAR Program has been successful in encouraging the development and adoption of anaerobic digestion technology. The number of manure-operating digesters reached 140 in 2009 (Figure 3-6), and they collectively reduced direct greenhouse-gas emissions by about 800,000 tons of CO_2 equivalent in 2009 (Figure 3-7) (EPA, 2009). A few centralized combustion facilities also collect animal manure from nearby animal production facilities.

WEEDS, PESTS, AND DISEASE MANAGEMENT IN CROPS

Chemical herbicides, fungicides, and pesticides are often used to manage weeds, pests, and disease. However, societal concerns about pesticide exposure in rural communities and pesticide residues on food have increased (Harnly et al., 2005; Tucker et al., 2006; Ward et al., 2006). As described in Chapter 2, issues of pesticide contamination in the nation's surface and ground water supply are now well documented for major agricultural watersheds (Gilliom et al., 2006). Knowledge of the impacts of certain pesticides on wildlife is improving, and nonlethal effects caused by some pesticides, such as disruption of the endocrine systems of different organisms, are better known (Desneux et al., 2007).

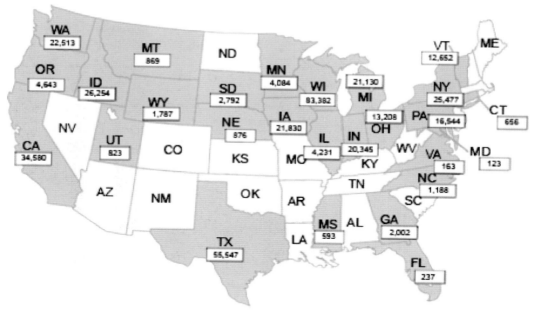

FIGURE 3-6 Number of operating manure digesters across the United States in 2009 and the amount of energy produced from the biogas generated. Numbers represent total annual energy production in MWh equivalent.
SOURCE: EPA (2009).

Impacts of pesticide on biodiversity are clear for some organisms but are typically compounded with loss of habitat, increased disturbance, and other attributes of intensive farming systems. This convergence of factors makes it difficult to determine the extent of direct impacts from pesticide use alone. One of the best-documented examples relates to the dramatic loss of native bees, representing the loss of a major ecological service (pollination) that is thought to increase size, quality, or stability of yields for 70 percent of the major crops produced globally (Ricketts et al., 2008) and is of paramount importance to the food supply (Allen-Wardell et al., 1998). Although difficult to pinpoint, the cause of pollinator decline is thought to be in part due to pesticide exposure, habitat loss, expansion of intensive agriculture, diseases, and parasites (Allen-Wardell et al., 1998; NRC, 2007; Rundlof et al., 2008; Black et al., 2009).

Continued reductions in pesticide use can reduce the potential of spray drift or leaching into ground water and potentially enhance biodiversity. In many situations, pesticide reduction could also result in decreased energy consumption (from reductions in production and transportation); however, any savings need to be balanced against any increases in tillage or decreases in productivity to determine overall energy use per unit of food or fiber produced.

To effectively manage the weed–disease–pest complex with reductions in, or elimination of, chemical use requires a suite of strategies (Shennan, 2008) that includes: breeding of crops that are pest and disease resistant and that are better able to compete with weeds in a given environment; use of different soil and crop management strategies; and diversification of crop rotations and noncrop vegetation.

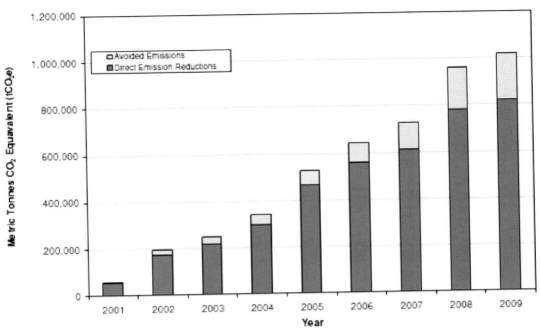

FIGURE 3-7 Reductions in greenhouse-gas emissions in 2009 as a result of the 140 operating manure digesters.
SOURCE: EPA (2009).

Managing the Crop–Weed–Disease–Pest Complex

For many decades the fields of weed, disease, and pest management focused primarily on chemical pest control materials targeted for the particular types of organisms in question, with little integration across the disciplines. Efforts to reduce chemical usage and develop more ecologically based approaches has called into question the value of this single disciplinary approach, as evidence of important interrelationships among each component of the crop–weed–disease–pest complex has emerged. Multitrophic interactions are known to occur in natural systems with important consequences, so it is not unexpected that they could play important roles in agroecosystems as well (Shennan, 2008). For example, foliar herbivory in grasslands has major consequences for the functioning of soil food webs (Wardle, 2006), and similarly, changes in soil food webs and nutrient dynamics also affect plant quality and attractiveness to herbivores (Awmack and Leather, 2002; Beanland et al., 2003). As a result, farm practices such as tillage, crop rotation, fertility inputs, and pesticides not only have direct effects on weed populations, disease incidence, and pest populations individually, but also important indirect effects mediated by other elements of the crop–weed–disease–pest complex. For example, changes in weed populations can provide new hosts for pests or pathogens increasing their severity; alternatively, they can provide refugia for beneficial arthropods and enhance soil suppressiveness to soil-borne pathogens, thus aiding biological control (Norris, 2005; Thomas et al., 2005; Wisler and Norris, 2005). Further, as in nature, crop plants are subjected to attack by more than one organism such that below-ground attack can influence responses to above-ground attack and vice versa, because of systemic induction of defense metabolism (Bruce et al., 2007). Finally, similar management techniques can be used to control more than one kind of pest, again arguing for interdisciplinary collaborations. Use of organic amendments to enhance soil suppressiveness is advocated to help manage fungal, bacterial, and nematode pest species (Alabouvette et al., 2006). Similarly, biofumigation through the incorporation of residues that are high in biocidal compounds, such as Brassica species, has the potential to control a range of soil pests (for example, nematodes) and pathogens (Matthiessen and Kirkegaard, 2006). The nature and outcomes of management interventions on the weed–disease–pest complex will be site- and organism-specific, making their study complex and challenging in terms of research design and statistical analysis (Kranz, 2005). Because the bulk of the literature is disciplinary studies, significant advances in each component of the complex are discussed, followed by a more holistic view of the impacts of different farming practices and systems.

The Evolution of Integrated Pest Management

Recently, the Food and Agriculture Organization (FAO) further revised its definition of integrated pest management (IPM) to reflect a continuing shift toward greater emphasis on ecologically based management, with application of pesticides seen as a last resort (W. Settle, presentation to the committee on January 14, 2009). However, others continue to use a narrower definition that implies a more primary role for improved pesticide use. The spectrum of definition is also reflected on the ground according to surveys of IPM practices among different farms in the United States and elsewhere. For some farmers, IPM means simply scheduling pesticide applications based on monitoring and established economic thresholds; others use more integrated IPM that combines a mix of cultural and biological control practices with or without pesticide use as a last resort (Shennan et al., 2001). The latter is sometimes referred to as ecological pest management (Shennan, et al., 2008) to distinguish it from improved management of pesticides. While IPM is most commonly associated with above-ground arthropod pest management, terms such as "integrated weed and disease management" are becoming more common, reflecting a similar increased attention to a diversified set of management approaches for these organisms. However, there are unique challenges for integrated management of soil pests and pathogens, and concepts of monitoring and use of thresholds are difficult to apply (Matthiessen and Kirkegaard, 2006).

While pesticides remain the primary method of pest control currently, increasing public and regulatory pressures for additional pesticide reductions are shifting the focus of research to development and implementation of nonchemical alternatives. The following section highlights promising new developments toward this goal.

Use of Disease and Insect Resistant Cultivars in IPM

Plant breeding has a crucial role in protecting crops against diseases and insect pests. For example, wheat cultivars are now grown with resistance to one or more of the following: stem rust, leaf rust, stripe rust, powdery mildew, soilborne mosaic, pseudocercosporella foot rot, Hessian fly, and greenbug (Cook, 2000). The transgenic Rainbow variety of papaya was engineered to resist papaya ringspot virus, which devastated the papaya industry in Hawai'i in the 1990s (Gonsalves et al., 2007). The use of resistant varieties is an important component of any IPM program. Development of resistant varieties depends on the availability of useful genes and the methods of selection, as discussed in the earlier an option for many row crops due to tillage and harvest requirements, it has potential for small grains and cover crops.

Weed seed bank dynamics are affected by tillage, residue management, and crop rotation, by reductions in annual seed input by weed suppression during the crop cycle, and by increasing seed predation in the soil (Menalled et al., 2001; Murphy et al., 2006). Reduced tillage and crop rotation were found to increase seed diversity, but reduce seed density in the soil by 80 percent over a six-year period (Murphy et al., 2006). Seed density also declined significantly in organic and reduced input systems relative to conventional and no-till systems in another study (Menalled et al., 2001), yet reductions in seed density do not always lead to reduced weed pressure during the crop cycle (Liebman and Davis, 2000).

Considerable progress has been made in understanding different mechanisms of weed suppression, and the ability to control weeds and reduce herbicide use will continue to improve as better combinations of approaches are developed and more competitive crop varieties become available. Linking weed and crop lifecycle models with models of seed bank dynamics could help design weed-suppressive rotations and management practices.

Designing Systems for Holistic Management of the Crop–Weed–Disease–Pest Complex

The preceding discussion lays out the elements necessary for effective management of the crop–weed–disease–pest complex. At the heart of any preventive management system is the maintenance of soil conditions that support the desired microbial, fungal, and nematode community assemblages that will suppress pathogenic fungi and nematodes, induce crop resistance responses, and reduce viable weed seed populations. Soil fertility needs to be managed for good crop growth, while at the same time ensuring that nutrient levels in the soil and plant tissue are not high enough to increase susceptibility to pathogens or attract higher levels of herbivory. In addition to good soil management, other key elements are the use of competitive and disease- or pest-resistant varieties, diversification of crop rotations to break pest cycles, and the inclusion of noncrop vegetation to provide habitat for natural enemies. Taken together, these strategies enable ecological interactions to occur that can greatly decrease the severity of pest, weed, and disease impacts, and potentially increase resilience of the production system to fluctuating conditions (Shennan, 2008).

The tools available to farmers to prevent outbreaks of pests, weeds, and disease are: varietal selection, crop rotation, tillage, fertility inputs, organic amendments, water management, and the provision of habitat diversity. Sufficient knowledge is now available to begin designing integrated management systems tailored to specific production systems using thoughtful combination of the practices discussed above.

Several factors and practices and their interactions are relevant to managing the weed–disease–pest complex. For example, crop rotation and organic matter management can enhance the populations of beneficial rhizobacteria and thus increase soil suppressiveness to diseases and nematodes (Welbaum et al., 2004). The effectiveness, however, can vary with soil types (Messiha et al., 2007). The effects of biological amendments (for example, biocontrol agents, microbial inoculants, mycorrhizae, and an aerobic compost tea) on disease suppressiveness appear to vary with crop rotations. It is possible that some crop rotations are better able than others to support populations of added beneficial organisms from amendments and enable more effective biological control (Larkin, 2008). Crop rotations can also be designed for improved weed control by using models of weed life cycles (Anderson, 2004) and can address other pest problems as highlighted by the following examples. Introduction of such tropical crops as American jointvetch (*Aeschynomene americana*) or castor (*Ricinus communis*) into the rotation with peanut and soybean was found to provide nematode control and increased peanut and soybean yields in Alabama (Rodríguez-kábana and Canullo, 1992). When corn is rotated with soybeans, it can reduce the need for insecticide by reducing the number of western corn rootworm larvae in the soil. The effectiveness of this management practice, however, has diminished over time because of a shift in the ovipositional behavior of the western corn rootworm (O'Neal et al., 1999; Schroeder et al., 2005).

Use of cover crops in rotations can similarly have multiple effects, and the choice of crop depends on which functions are seen as primary. For example, in addition to providing soil cover and cycling nutrients, cover crops can also suppress weed populations if appropriate species are chosen that shade and outcompete weeds or are allelopathic (Teasdale, 1998); provide habitats for beneficial insects and pests (Costello and Daane, 1998); or suppress certain diseases (Griffin et al., 2009).

Changing management practices, such as planting density and nitrogen fertility, can similarly have multiple effects. For example, the incidence of powdery mildew in no-till wheat depends upon a complex combination of nitrogen application rates, row spacing and seeding rates, and crop phenology (Tompkins et al., 1992). However, nitrogen fertility level and crop spacing also affect weed suppression (Liebman and Davis, 2000; Evans, 2003; Olsen et al., 2005a,b), and high nitrogen fertility can reduce soil suppressiveness to disease (Workneh et al., 1993) and increase crop palatability to arthropod herbivores (Staley et al., 2009).

Although information and management tools that can contribute to integrated management of specific pests or the whole weed–disease–pest complex are available, the challenge is determining which combination of tools to use to create the synergies necessary for effective control and to minimize negative interactions. Each system will need to be carefully tailored to the specifics of the location and context. Negative interactions might occur in some circumstances and need to be taken into account. For example, a cover crop that is desirable for one purpose, such as nitrogen fixation, may result in increased disease problems if the species is susceptible to key diseases in the area. Similarly, reduced tillage may have multiple benefits but still increase specific problems. Effects of reduced tillage on soil biota are somewhat predictable, favoring fungal food webs that are readily disrupted with soil disturbance, as well as higher AM fungal populations and increased seed predation (Shennan, 2008). This can be beneficial for disease suppressiveness and reductions in weed seed banks through increased seed predation and disease. Reduced tillage, however, also can reduce crop growth through poorer seed bed structure, cooler soil temperatures, and other factors (Triplett and Dick, 2008).

The complexity and unpredictability of the biotic interactions described above further reinforce the need to test ecological pest management tactics in systems-level field contexts (Shennan, 2008), as has been argued for biological control tactics for plant diseases (Alabouvette et al., 2006; Matthiessen and Kirkegaard, 2006) and use of green manures (Cherr et al., 2006). Further, monitoring outcomes when farmers adapt techniques for their individual contexts could provide important information on systems-level interactions. Putting a greater emphasis on participatory research approaches would combine farmer knowledge and experience with a generation of research information for subsequent meta-analysis and could increase the ability to predict when synergies or negative interactions are likely to occur in the field and adjust management accordingly.

Evaluation of Adoption, Effectiveness, and Future Challenges of IPM and Ecological Pest Management

While considerable research has been done to investigate more ecological approaches to pest management, little information is available on levels of adoption and effectiveness, even for arthropod IPM, which has a relatively long history. Indeed, few assessments are available. Comments range from the assertion that IPM has enjoyed significant success in the developed world (Way and van Emden, 2000) to a commentary (Devine and Furlong, 2007, p. 295) on its perceived failure: "It is worth noting that, despite the popularity of the IPM concept (reviewed by Kogan, 1998) there has been no decrease in overall insecticide usage, even in areas where that concept is very favorably viewed (for example, in the United Kingdom and California). If the success of the IPM concept is judged by reductions in the area of land sprayed by insecticides, then it has clearly failed."

It appears that inadequate datasets and lack of agreement on how to evaluate success explains such disparate views. Reductions in amounts used or acreage where pesticides are applied will be difficult to discern if highly aggregated data are used (for example, by country or region, across all types of pesticides, or across all crop types), because these data obscure any changes in pesticide use for a particular crop and pesticide combination for which an IPM system has been developed. Further, trends in application rates are also complicated by shifts in individual pesticides used, with replacement compounds often requiring lower application rates. In the absence of more nuanced analysis and adequate long-term data, it is inappropriate to draw broad conclusions about the impacts of IPM programs in the United States.

High levels of IPM use have clearly been documented for certain systems in the United States where growers are part of a network or organization that promotes sustainability as a goal (Warner, 2008). Likewise, there is evidence for use of IPM leading to reduction in pesticide use in developing countries, particularly where emphasis has been placed on farmer education through the Farmer Field School programs of organizations like the FAO. In their review, Van Den Berg and Jiggins (2007) looked at 14 studies that showed significant reductions of pesticide use (35 to 95 percent) in 13 of the studies and no effect in 1 study (although the design of the latter has been questioned). Many of the studies only measured immediate effects, however, and more long-term studies are needed.

Considerable progress has been made in development and commercialization of different kinds of augmentative biological control agents, yet sales of biocontrol products only account for 1 percent of total agricultural chemical sales (Fravel, 2005). A major barrier to use of biocontrol organisms is a lack of consistent and predictable levels of control under field situations. For example, in the case of antagonistic soil bacteria, variable results can be due to application problems (physiological state of the bacteria, timing, and dosage), or microclimate variation, differences in soil ecology, crop genotype, or weed community (Fravel, 1999, 2005; Sabaratnam and Traquair, 2002). Product registration can be a barrier, as in the case of AM fungal biocontrol agents (Whipps, 2004), and costs of formulating and producing mixtures may still be too high relative to chemical control options (Fravel, 2005). In addition, concerns about nontarget effects and ecological risks of microbial and other biocontrol agents are increasingly being voiced, particularly for those that are genetically modified (Wajnberg et al., 2001; Timms-Wilson et al., 2004). However, some argue that there is remarkable little evidence of negative side effects of biocontrol organisms, but this might be because of the lack of effort made to assess nontarget effects (Wajnberg et al., 2001).

Future Role of Pesticides in IPM

The future role for pesticides as part of integrated management of the pest complex was discussed in detail in the report *The Future Role of Pesticides in U.S. Agriculture* (NRC, 2000a). It is widely agreed that pesticides will continue to play an important role in many production systems, with continued attention to reducing nontarget effects by development of improved chemicals that have greater specificity, break down rapidly in the environment, and are less toxic to humans and other animals. That report (NRC, 2000a, p. 254) concluded that "the most promising opportunity for increasing benefits and reducing risks is to invest time, money, and effort into developing a diverse toolbox of pest-management strategies that include safe products and practices that integrate chemical approaches into an overall, ecologically based framework to optimize sustainable production, environmental quality, and human health." Pesticide use as part of an IPM system that combines cultural, physical, and biological strategies likely will continue, but there could be a shift to reduced pesticide use or using it as a last resort.

A number of priorities for future work emerge from the preceding discussion if a truly integrated pest, disease, and weed management system is to be developed. These involve more coordinated monitoring and data collection on the effectiveness and adoption of different strategies, as well as a shift to more integrated systems research on the crop–weed–disease–pest complex. Some suggested actions are:

- A coordinated effort to evaluate on-farm adoption and effectiveness of IPM and ecological pest management strategies, and track corresponding changes in pesticide use for specific cropping systems.
- Collection of field-based data to better understand the causes of variability in effectiveness of biocontrol organisms under different conditions; without this, adoption is unlikely to increase significantly.
- On-farm work to assess the ability of different kinds of vegetation diversification to increase indigenous biological control. If biological control advantages are well documented, growers will be more likely to diversify habitats within and around the farm, which will also provide other important ecological benefits.
- Field-based interdisciplinary work to increase the understanding of the effects of management practices and systems on dynamics among the whole weed–disease–pest complex.

Taken together those efforts will improve the ability to determine the adaptability and resilience of ecologically based management of the pest complex as compared to pesticide-based management.

MANAGING EFFICIENCY OF ANIMAL PRODUCTION SYSTEMS

Improvements in animal genetics, herd or flock management, and nutrition have resulted in increased conversion of animal feed to human-edible food and fiber of animal origin (Bull et al., 2008). Animals require nutrients for maintenance, growth, production, and reproduction. When nutrient intake is insufficient, the body prioritizes how nutrients are used with maintenance needs met first, followed by the other three categories in differing orders based on specific metabolic conditions. As an example, a female of reproductive age must meet maintenance requirements for regular estrus cycles, ovulation, and conception. Once pregnancy is established, the female requires additional nutrients for the developing fetus. After parturition, further nutritional needs exist to provide nutrients for the female to maintain herself, as well as sufficient nutrients to produce milk for her offspring. Improvements in genetics, management, and nutrition that allow for more closely targeted nutrient flow (inputs) to meet animal metabolic requirements will improve efficiency of feed conversion to animal product.

Feed conversion is the amount of feed required to produce one unit of product, where product can be eggs, meat, wool, or milk. As feed conversion efficiency improves, less feed is required per unit output, translating into a reduced need for farmland to grow feed inputs as well as reduced nutrient excretion (manure). Three key opportunities exist on livestock and poultry farms to improve feed conversion: genetics, nutrition, and management. This section discusses genetics and nutrition as approaches to improving efficiency of animal production systems.

Animal Breeding

Efforts in animal breeding have focused on traits that influence output, such as weight gain, feed efficiency, reproductive efficiency (Grosshans et al., 1994; Kelm et al., 2000), or meat quality (Bishop and Woolliams, 2004). Genetic improvement can contribute to improving sustainability by increasing feed utilization efficiency (Ward, 1999), by selecting traits to improve animal health and welfare (Star et al., 2008), and by reducing livestock's carriage of food-borne pathogens (Doyle and Erickson, 2006).

Quantitative genetics has been used in animal breeding for decades to select animals displaying desirable production traits. Broilers are used as an example in this section to discuss improvements associated with genetic versus nutrition alterations over time. Poultry breeders have developed broiler lines that exhibit increased growth rate and improved feed conversion efficiency, thereby reducing the time necessary for animals to reach market weight (with less feed inputs and less manure outputs). The challenge for scientists was to determine what fraction of the improvement resulted from improved genetics versus improved diet formulation. The Athens-Canadian random bred control (ACRBC) was established in 1957 by scientists at Agriculture Canada. That broiler line has been maintained genetically at the Southern Regional Poultry Breeding Laboratory (University of Georgia Department of Poultry Science, Athens, Georgia). The production traits of the ACRBC strain were evaluated in 1991 and found to have similar feed conversion as in 1957, which allowed comparison between old genetic lines (1957) and new genetic lines (2001) to evaluate feed conversion and growth weight (Havenstein et al., 2003a), carcass composition and yield (Havenstein et al., 2003b), and immune response (Cheema et al., 2003). Male or female birds were assigned from "old" (ACRBD) or "new" (Ross 208 broiler) genetic lines and on "old" (1957) or "new" (2001) feeding regimen. The Ross 308 broiler on the 2001 feed was estimated to have reached a body weight of 1,815 grams at 32 days of age with a feed conversion of 1.47, whereas the ACRBC on the 1957 feed would not have reached that body weight until 101 days of age with a feed conversion of 4.42. The shorter age to market as a result of improved feed conversion would require far less feed input (and associated land to grow the feed) to achieve similar product and have markedly less manure output. Comparisons of carcass weights of the Ross 308 on the 2001 diet versus the ACRBC on the 1957 diet showed they were 6.0, 5.9, 5.2, and 4.6 times heavier than the ACRBC at 43, 57, 71, and 85 days of age, respectively. Yields of hot carcass without giblets (fat pad included) were 12.3, 13.6, 12.2, and 11.1 percentage points higher for the Ross 308 than for the ACRBC at those ages. The yields of total breast meat and yields of saddle and legs for the Ross 308 were higher than for the ACRBC. The Ross 308 averaged more whole carcass fat than the ACRBC. Genetic selection for improved broiler performance has resulted in a decrease in the adaptive arm of the immune response but an increase in the cell-mediated and inflammatory responses (Cheema et al., 2003). The authors attributed 85 percent of the improvement in feed conversion and growth to genetics and 15 percent to nutrition.

Other tools in genetic improvement of livestock include genomics and transgenics. The development of the chicken, swine, and bovine genomic toolboxes provide "the needed platforms for developing whole-genome selection programs based on linkage disequilibrium for a wide spectrum of traits" (Green, 2009, p. 793). "Transgenic technology allows for the stable introduction of exogenous genetic information into livestock genomes" (Laible, 2009). USDA researchers have developed a transgenic dairy cattle that resist *Staphylococcus aureus*, a major mastitis pathogen (Donovan et al., 2005). The ability to identify genes that influence livestock production traits from genomic information complemented with transgenic technology could transform genetic improvement of livestock. Transgenic technology has not been applied in agricultural practices in the United States. As with genetic modification of crops, the use of transgenic livestock is a controversial topic and the potential risks, consumer acceptance, and the value of the product could be barriers to its development (Blasco, 2008; Laible, 2009).

As in the case with crops, the maintenance and use of genetic diversity in livestock will help manage the risks of animal production and improve resilience of animal production systems (Bishop and Woolliams, 2004). USDA established the National Animal Germplasm Program in 1999 to conserve livestock genetic resources (Blackburn, 2009), which is critical to future animal breeding efforts in the United States.

Disadvantages of Animal Breeding

Breeding animals for a specific trait could have unintended effects on animal health and welfare. Rauw et al. (1998) reviewed the undesirable behavioral, physiological, and immunological effects correlated to selection for high production efficiency in broilers, pigs, and dairy cattle. Star et al. (2008) suggested the concept of robustness in animal breeding, which emphasizes selection for individual traits of an animal that are relevant for health and welfare.

Animal Nutrition

To produce 1 lb of consumable meat takes about 4–18 lbs of meat (Rasby, 2007; Wulf, 2010). Research has increased efficiencies of converting food inputs to animal products. Incorporating research findings associated with basic chemistry and biochemistry has resulted in reduced needs for inputs and reduced nutrient excretion per unit animal product produced. The National Research Council has been releasing and updating reports on nutritional requirements of dairy cattle (NRC, 2001), beef cattle (NRC, 2000b), swine (NRC, 1998), poultry (NRC, 1994), and other animals since 1917. With improved understanding of nutrient requirements, animal diets can be managed to ensure that animals are provided adequate nutrients to meet the needs of maintenance, growth, reproduction, and lactation.

Animal and poultry science literature is rich in detailed studies conducted to improve production with less emphasis on determining output per unit product produced (meat animals) or per productive life of the animal (in the case of milk). Diets are often formulated with least-cost formulations where minimum constraints represent the NRC recommendations and maximum constraints identified (for select nutrients) identify caps for biochemical, palatability, or toxicity reasons. Those formulation programs typically do not include constraints associated with local environmental issues. As such, inclusion of byproduct feeds in diets to achieve least-cost formulation can be done and can reduce land requirements below those of grazing animals (Vandehaar and St. Pierre, 2006). However, the use of the byproducts might well exceed nutrient concentrations for sensitive nutrients. A recent review (CAST, 2002) identified advances through dietary modification to reduce excretion of nitrogen and phosphorus in food-producing animals. Formulation of diets to specific amino acid requirements (poultry and swine), inclusion of additives to improve bioavailability of nutrients (particularly phosphorus), and chemical modifications can result in no-loss in production or production gains, while reducing total amounts of specific nutrients excreted by animals per given unit of product produced. (See the earlier section on dietary modification for adjusting manure composition in this chapter.)

Impact of Nutritional Strategies

The improvements in feed conversion through genetics, nutrition, and management have significantly reduced manure and nutrient excretion per unit animal product produced and reduced land required for production. For example, the improvements in dairy cattle feed conversion is summarized by Bull et al. (2008). Broiler research identified that the time to reach market weight for a broiler in 1957 was 101 days with a feed requirement of 8.0 kg per broiler (Havenstein et al., 2003a; Havenstein, 2006). With improved genetics and feed, the same market weight was achieved in 2001 in 32 days with 2.68 kg of feed. In 2007, dairy cattle waste solids production was estimated to be less than half of the amount produced in 1950 (123 million lbs/day in 2000 versus 250 million lbs/day in 1950). Vandehaar and St. Pierre (2006) compared three types of dairy cattle (grazing, confined with no by-products, confined with byproducts) producing 5,000 kg/cow per year. Required land was 0.54, 0.66, and 0.30 ha/cow per year. Efficiency of land use improved within animal type (for example, confined and fed byproducts 76, 88, and 93 percent) as animal production increased (5,000; 10,000; 15,000 kg/herd per year). This efficiency was calculated as protein and energy yield per ha from dairy farming relative to the protein and energy yield from soybean and corn grown for direct human consumption. The comparative value for the grazing type cattle (5,000 kg/herd per year milk yield) was 43 percent.

Disadvantages of Nutritional Strategies

Improved genetics can be accomplished through incorporation of genetically improved lines (either through purchases or artificial insemination). Operations maintaining closed herds (that do not import live animals) rely on selection processes that might require longer time intervals for genetic improvement. Improvements in management can require costly infrastructure improvements or retraining of operators to achieve greater genetic potential. One great challenge associated with management and infrastructure improvements is the difficult nature of conducting controlled experiments to identify cost-effective alternatives to existing practices.

Nutritional strategies focus on more closely matching feed nutrient inputs to requirements of animals. The use of supplemental amino acids in poultry is based primarily on simple production economics (for example, least-cost, most-profitable production of meat and eggs) and is not specifically intended to decrease nitrogen excretion (reduce environmental costs). Least-cost diet formulation does not usually include decreasing nitrogen excretion because there has been little or no economic incentive to do so (CAST, 2002). In dairy cattle, the economic risk of underfeeding protein is greater than the risk of overfeeding protein (Vandehaar and St. Pierre, 2006) when environmental costs are not reviewed. As such, analysis of protein efficiency has not become an industry standard.

Adoption of Nutritional Strategies

The adoption of mitigation practices varies by species and herd or flock size. Animal operators often produce a product for a specific market. Often, broilers, turkeys, and swine are contracted out to meet a specified range of weights in a specified number of days. This provides the purchaser with a relatively uniform product for slaughter and further processing and delivers a more uniform and consistent product to the consumer. In some cases, specified diets are prescribed for specific production stages from the purchaser of the animal product. Adoption of nutritional strategies often results in a cost-savings or break-even situation (Powers and Angel, 2008). Most of the poultry industry routinely adds methionine and, in some cases, lysine to the diet so that lower concentrations of total protein and amino acids can be fed. Most of the industry also implements phase-feeding, but the number of diet changes may be less than optimal. A substantial part of the industry uses ideal protein to estimate more closely the amino acids requirements of older birds (CAST, 2002).

Increased attention and scrutiny on whole farm nutrient balance would provide additional opportunities for livestock and poultry nutrition consultants to focus on reducing nutrient excretion and potential sequestration of nutrients within facilities. For dairy animals, the key contributing factor responsible for excessive phosphorus supplementation is the prevailing belief that addition of phosphorus to diets will improve reproductive performance. Aggressive marketing of phosphorus supplements has contributed to unrealistic margins of safety in diet formulation programs. The emphasis in feeding is on maximizing animal production and profits, rather than on minimizing excretion of nutrients. Once animal producers understand the ramifications of nutrient accumulation (especially beyond regulatory thresholds), attention to dietary modification will increase. Nutritional strategies have achieved success in providing a partial solution for several of the prominent environmental issues (nitrogen, phosphorus, sulfur). Nutritional strategies can play an important role in reducing the environmental impact of animal production.

ANIMAL WELFARE

A recent report (Mench, 2008) summarized the scientific and social issues surrounding animal welfare. Research activities include areas of behavior (for example, natural behaviors, abnormal behaviors, and animal preferences, such as by Wemelsfeder and Farish, 2004; Smulders et al., 2006); physiology (for example, hormonal changes characteristic of stress, such as by Mormede et al., 2007); health (for example, pain, injury, and disease, such as by Webster and Cardina, 2004): and productivity (for example, growth rates and reproduction). Each of those measures has strengths and limitations, and it is generally agreed that there is no single indicator of good welfare and that multiple measures should therefore be evaluated. Animal-based outcome measures such as lameness, animal body condition score, sickness, and death losses are increasingly used to assess animal welfare (Grandin, 2010). The interpretation of the importance of those measures, however, is ultimately based on values and attitudes toward animals rather than on science. This section has a different format than the others in this chapter because animal welfare research is relatively new compared to research discussed earlier. Moreover, the impact of different practices aimed to improve animal welfare depends on the criteria used to measure welfare. Therefore, this section provides a brief overview of animal welfare, a few examples of research activities on animal welfare, and discussion of some of the controversies or tradeoffs. In a landmark report commissioned in the United Kingdom, the Brambell Committee (1965) identified five tenets of animal welfare:

- Freedom from hunger and thirst.
- Freedom from discomfort.
- Freedom from pain, injury, or disease.
- Freedom to express normal behavior.

phytate, and protein (Yang et al., 2009). Fiber-degrading enzymes break down nonstarch polysaccharides, such as cellulose, to sugars.
- Prebiotics. A prebiotic is a nondigestible food ingredient that beneficially affects the host by selectively stimulating the growth or activity of one or a limited number of bacteria in the gut (Gibson and Roberfroid, 1995). Prebiotics include fructo-oligosaccharides, mannan-oligosaccarhides, gluco-oligosaccharides, malto-oligosaccharides, stachyose, and oligochitosan. They can inhibit the growth of pathogens, promote digestion, and enhance immune response (Huang et al., 2007; Yang et al., 2009).
- Probiotics. A probiotic is "a preparation of or a product containing viable, defined microorganisms in sufficient numbers, which alter the microflora (by implantation or colonization) in a compartment of the host and by that exert beneficial health effects in this host" (Schrezenmeir and de Vrese, 2001, p. 362S). Probiotics maintain a beneficial population of microflora by competition exclusion—competing for substrate and attachment sites and producing antimicrobial metabolites to inhibit pathogens—and immune modulation. Microorganisms that have been used as probiotics include bacterial species, such as *Bacillus, Bifidobacterium, Enterococcus, Escherichia, Lactobacillus, Lactococcus,* and *Streptococcus,* yeast species, and mixed cultures (Yang et al., 2009).
- Immune modulators. Immune modulators are compounds that affect the working of the immune system and enhance resistance to disease. Those compounds include cytokines and unidentified components of spray-dried plasma.
- Organic acids. Organic acids have been widely used as food additives and preservatives for preventing food spoilage and prolonging shelf-life of perishable foods (Ricke, 2003). Organic acids have been suggested as a growth-promoter for livestock, but the mechanisms through which organic acids promote growth are not clear (Ricke, 2003). One potential mechanism is that organic acids reduce gastric pH and thus improve nutrient digestion (Doyle, 2001). The antimicrobial effects of the organic acids lead to beneficial effects, possibly by controlling bacterial populations in the intestinal tract of livestock (Doyle, 2001).

The search for antibiotic replacement has gained attention, likely because some countries banned the use of antibiotics in feed (Dibner and Richards, 2005). Data on adoption of those alternatives are not available to the committee's knowledge. None of the alternatives seem to be able to replace all the potential benefits of in-feed antibiotics and certainly cannot provide the same benefits as therapeutic antibiotics (Pettigrew, 2006).

Impact of Alternatives to Antibiotics

The alternatives to antibiotics discussed above have been shown to promote growth in livestock. Dietary enzymes can improve feed conversion and increase weight gain in pigs (Doyle, 2001). Supplementation of exogenous enzymes can improve growth rate in poultry by 2 to 3 percent, reduce incidence of sticky excreta, and improve litter conditions (Broz and Beardsworth, 2002; Yang et al., 2009).

Increased growth was observed in broilers treated with either an antibiotic growth promoter or a prebiotic compared to those that did not receive any supplement (Catala-Gregori et al., 2008). Yang et al. (2009) summarized the effects of different prebiotics on growth performance of broilers and they reported weight change ranging from −3 percent to 8 percent. The effect of prebiotics on feed conversion ratio of broilers ranges from −1 percent response to 6 percent. Some prebiotics resulted in growth-promoting effects similar to those of antibiotics (Huang et al., 2007; Li et al., 2008).

Probiotics added to feed for piglets protect them from intestinal pathogens and enhance nutrient uptake in their guts (Doyle, 2001). Doyle (2001) summarized research on the positive effects of probiotics, which include increased weight gain, reduced mortality, and improved feed efficiency in pigs, and improved growth and decreased incidence of diarrhea in piglets. In broilers, probiotics have been shown to reduce *Salmonella* colonization in broilers by 9 to 60 percent and enhance growth and reduce mortality (Yang et al., 2009).

Spray-dried porcine plasma protein could reduce mortality and diarrhea in piglets (Doyle, 2001). Another meta-analysis (van Dijk et al., 2001; Pettigrew, 2006) and a review of literature reported more than a 20 percent mean increase in growth rate of young pigs. In chickens, some cytokines can act as growth promoters by stimulating the immune system to ward off pathogens (Lowenthal et al., 2001). Cytokines can potentially be used as therapeutics and vaccine adjuvants (Hilton et al., 2002).

Organic acids have been shown to improve performance of weaned piglets, but the magnitude of performance depends on the acid used (Patanen and Mroz, 1999). Lactic acid seems to reduce gastric pH and coliforms in pigs consistently (Jensen, 1998). Other researchers showed evidence that organic acids improve the digestibility of proteins, minerals, and other nutrients (Doyle, 2001).

Disadvantages of Alternatives to Antibiotics

The response of poultry to alternatives to in-feed antibiotics depends on multiple factors including quality and quantity of feed, microbial status in the animal's gut, and the animal's age (Bedford, 2001; Doyle, 2001; Yang et al., 2009), so that improvements in feed efficiency and growth are not always observed. The benefits of prebiotics and probiotics also depend on the hygienic conditions of the farm, with benefits more readily observed under poor hygienic conditions (Doyle, 2001). The optimal dosage of prebiotics and probiotics could be difficult to determine because it depends on multiple factors including diet, species, age, stage of production of the animals, and hygiene status of the farm (Verdonk et al., 2005; Yang et al., 2009). Dosages that are too high could have negative effects on the gut flora and slow growth of the birds (Yang et al., 2009). Bacteria can develop acid resistance over time, similar to antibiotic resistance (Ricke, 2003), so that the benefits of in-feed organic acids likely will decrease over time.

Animal Identification

Most livestock on farm have some form of individual identification (USDA-APHIS, 2007, 2009). Livestock owners have different motives for establishing an identification system for their animals including evaluating product quality and genetic improvements, protecting their livestocks from loss or theft, and evaluating animal health and tracing back diseases (Golan et al., 2004; USDA-APHIS, 2007). Methods of identification include branding, tattooing, retina scanning, iris imaging, and tagging. Tags might have simple printed numbers, imbedded microchips, or machine-readable codes such as radio frequency identification. Increasingly, the animals are given individual identification that is linked to documentation of an individual's vaccination records, health history, breeding characteristics, and other process attributes. Some operations implement an animal identification system that allows traceability (Golan et al., 2004). Identification and recordkeeping systems used in the United States are summarized by Disney et al. (2001).

An animal identification system helps owners evaluate animal health, track disease in their own herds, and evaluate genetic improvements (USDA-APHIS, 2007). Beyond the farm scale, an animal identification and traceability system can help ensure that unhealthy animals will not contaminate healthy herds, and hence could prevent spread of animal diseases. An animal identification and traceability system would be useful for the control and eradication of animal products (Disney et al., 2001). For example, animal identification was an important element of the brucellosis eradication program in the United States (Golan et al., 2004).

SUMMARY

Chapter 3 summarizes how specific agricultural management practices and approaches can contribute to crop and livestock productivity and reduce some of the detrimental impacts on the environment. Because the practices are components of "agricultural systems," their interconnectivity and interactions are complex; a practice that by itself might improve sustainability in one aspect could have a negative effect in another. Hence, advantages and disadvantages of certain practices are discussed.

Soil Management

Proper soil management is a key component of sustainable agricultural production practices because it produces crops that are healthier and less susceptible to pests and diseases, and important ecosystem services, such as reduced nitrogen runoff and better water-holding capacity. Soil quality is a basic and critical starting point for robustness (including productivity) and resilience of all of agriculture. It is influenced by many factors, with one of the most critical being mechanical management and tillage.

- Conservation (reduced) tillage practices have been adopted on millions of acres of U.S. farmland during the past two decades, covering more than half of the acreage of corn, soybean, and cotton, with use in a wide range of agronomic and horticultural crops. Significant increases in soil quality have been nearly universal, and environmental loading has been markedly reduced. Research is needed to broaden crop coverage, solve problems in low-moisture areas, and increase the diversity of herbicides used while focusing on developing conservation tillage systems that would work with low-or-zero herbicide use, such as in organic agriculture.
- Cover crop use has seen a resurgence of interest and use in the past two decades, adding to landscape-level diversity, more effective nutrient containment, and improved soil quality, often in combination with conservation tillage. Enhanced research efforts are needed to identify improved varieties and to identify species for application in a wide range of crop production (both horticultural and agronomic) and biophysical conditions. Improved understanding of site specificity of cover crop performance is also needed, as noted by farmers in this report's case studies (see Chapter 7).

Crop and Vegetation Diversity Management

Biodiversity of both crops and animals at field, farm, and landscape scales is critical to soil quality, ecosystem function, pest and disease management, efficient nutrient flow at high rates with farm-and-field containment, and for farm and landscape-level productivity, robustness, and resilience. The following issues impact that diversity:

- Noncrop vegetation (including grassed waterways, buffer strips, and riparian vegetation) for protection of waterways and other environmentally sensitive areas, and for wildlife habitat is highly beneficial to ecosystem functioning.
- Ongoing genetic improvement of crop varieties through conventional and molecular-assisted technologies is critical for sustainability.

Water Use and Quality Management

Agricultural irrigation is the dominant form of water use; therefore, practices and applications that improve water application efficiency and minimize water loss are the most effective in conserving water and energy. Significant increases in population, expansion of housing, and a wide range of competing demands for water and land resources demand that agriculture responds to those pressures. Increasing efficiency and reducing major areas of hypoxia and other adverse environmental impacts are ways in which agriculture is meeting those challenges:

- Water-use efficiency has been increasing, driven both by increasing water scarcity and costs, through the use of farmer-assist models, and careful metering and low-pressure application technologies. Such savings are not nearly as widespread as they could be.
- Water reuse has been increasing, but with significant concern for maintaining quality. There will be a growing demand for recycling of tile drain water, both for water savings, but most importantly for reduction of loss of soluble nutrients and crop and animal residues.
- Small-scale dams and mini-watershed management approaches have considerable scope for improvement in some areas.
- Several regional and increasingly national-scale hypoxic zones could be more adequately addressed through reduction in nutrient and pesticide loading via more widespread and effective application of many of the technologies and practices mentioned in this chapter. Improved wetland management will be critical.

Nutrient Management

Nutrient loading at landscape and regional scales is increasingly critical as cropping intensity and animal densities increase at the same time that agriculture's share of environmental loading is reduced through social and regulatory pressures.

- Mass balances of nutrient flow at farm and landscape levels are highly relevant, particularly to large animal operations and regions of high animal census, regardless of the management systems used. Well-designed nutrient management plans would be useful for all production systems.
- Use of manure and of all nutrient inputs can enhance nutrient recycling on farms, but their use would have to be monitored carefully to ensure high nutrient uptake by plants and minimal nutrient loss to the environment. Compost use can have an important role in the recycling of plant and animal wastes and residues. Anaerobic digestion with biogas recovery can play a more important role in sensitive locations with many animal operations of various scales. Precision agriculture is another tool for nutrient management.

Weed, Pest, and Disease Management in Crops

Weed, pest, and disease management in crops has undergone significant improvement in the last two decades. Nowadays, information is available to better inform the use of ecological approaches that can bring about less environmental loading from pesticides. Most of the ecological approaches are based on use of multiple integrated practices that directly and indirectly affect pest population shifts and management, rather than enabling complete "control" of a particular organism. Because of the complex interactions among different components, holistic management of the crop–weed–disease–pest complex is needed. Although knowledge of interactions between soil and crop management and their effects on the crop–weed–disease–pest complex is improving, field-based research to address the applicability of manipulating those interactions in operating farming systems is sparse, but needed.

- The paucity of information in the United States on adoption of IPM practices and effects on pesticide use by crops has made it difficult to determine how effective IPM methods are in the fields, and by how much those methods can reduce pesticide use. Increased efforts are also needed to better understand how biodiversity in the farm landscape can enhance biological control.
- A number of promising avenues for pest and pathogen management are being pursued. They include the ongoing development of pest-resistant varieties, efforts to manipulate induced resistance responses, develop disease and pest-suppressive soils, and biofumigate through the use of plant residues to manage pathogens and nematodes. Those approaches deserve additional research attention with an emphasis on field testing under different conditions. Weed management requires a suite of approaches to reduce annual seed production and the preexisting seed bank. Methods such as crop rotations, soil tillage, and organic matter management; use of cover crops and other crops with allelopathic properties; plant spacing; and water management all can affect weed populations. The use of weed and crop lifecycle models in conjunction with seed bank models would inform the design of weed-suppressive cropping systems.

Animal Housing, Nutrition, Health, and Breeding

Most livestock in the United States, with the exception of beef cattle, are raised in large confinement facilities. There is major controversy over several aspects of those animal systems as the demands for animal products grow and the environmental and social dimensions of animal production come under increasing scrutiny.

- Animal housing and space allocation is one of the critical elements. Requirements and options differ widely with animal species. The issues are especially critical around pigs, cattle, and dairy animals where research on alternative housing and management systems highlight the interactions among space, animal health, environmental impacts, labor requirements, and worker safety. Research that characterizes and quantifies those interactions within the context of the increasing demands for air and water quality in animal-raising landscapes and changing global economies could be expanded.
- Interactions of the environmental variability and feed sources on the nutrition

of animals in mixed crop–animal and other alternative systems are not well understood.
- Animal health, diet, housing, and exercise interactions with animal immune levels and robustness leave significant areas for further research. Studies on how to reduce or eliminate routine use of antibiotics to maintain health without compromising productivity would be useful.
- If animals are to be raised with more rather than less space, greater exposure to environmental fluctuations, and fewer medications for disease and parasite control, then the breeding and selection criteria would change for many systems. Hence, continuous research on animal breeding is critical to improving agricultural sustainability.

REFERENCES

Abawi, G.S., and T.L. Widmer. 2000. Impact of soil health management practices on soilborne pathogens, nematodes and root diseases of vegetable crops. *Applied Soil Ecology* 15(1):37–47.

Abbasi, P.A., J. Al-Dahmani, F. Sahin, H.A.J. Hoitink, and S.A. Miller. 2002. Effect of compost amendments on disease severity and yield of tomato in conventional and organic production systems. *Plant Disease* 86(2):156–161.

Abid, M., and R. Lal. 2008. Tillage and drainage impact on soil quality. I. Aggregate stability, carbon and nitrogen pools. *Soil & Tillage Research* 100(1–2):89–98.

Acosta-Martinez, V., and R.D. Harmel. 2006. Soil microbial communities and enzyme activities under various poultry litter application rates. *Journal of Environmental Quality* 35(4):1309–1318.

Adediran, J.A., A.A. Adegbite, T.A. Akinlosotu, G.O. Agbaje, L.B. Taiwo, O.F. Owolade, and G.A. Oluwatosin. 2005. Evaluation of fallow and cover crops for nematode suppression in three agroecologies of south western Nigeria. *African Journal of Biotechnology* 4(10):1034–1039

ADMS (Agricultural Drainage Management Systems). 2003. Agricultural Drainage Management Systems task force-charter. Available at http://extension.osu.edu/~usdasdru/ADMS/charter.htm. Accessed on May 5, 2009.

Ahloowalia, B.S., M. Maluszynski, and K. Nichterlein. 2004. Global impact of mutation-derived varieties. *Euphytica* 135(2):187–204.

Alabouvette, C., C. Olivain, and C. Steinberg. 2006. Biological control of plant diseases: the European situation. *European Journal of Plant Pathology* 114(3):329–341.

Allen-Wardell, G., P. Bernhardt, R. Bitner, A. Burquez, S. Buchmann, J. Cane, P.A. Cox, V. Dalton, P. Feinsinger, M. Ingram, D. Inouye, C.E. Jones, K. Kennedy, P. Kevan, H. Koopowitz, R. Medellin, S. Medellin-Morales, G.P. Nabhan, B. Pavlik, V. Tepedino, P. Torchio, and S. Walker. 1998. The potential consequences of pollinator declines on the conservation of biodiversity and stability of food crop yields. *Conservation Biology* 12(1):8–17.

Altieri, A. 1990. Why study traditional agriculture? Pp. 551–564 in Agroecology, Carroll, C.R., J.H. Vandermeer, and P. Rosset, eds. New York: McGraw-Hill Publishing Company.

Altieri, A., and C.I. Nicholls. 1999. Biodiversity, ecosystem function, and insect pest management in agricultural systems. In Biodiversity in Agroecosystems, W.W. Collins and C.O. Qualset, eds. Boca Raton, Fla.: CRC Press.

Anderson, R.L. 2004. A planning tool for integrating crop choices with weed management in the Northern Great Plains. *Renewable Agriculture and Food Systems* 19(1):23–29.

Andow, D.A. 1991. Vegetational diversity and arthropod population response. *Annual Review of Entomology* 36:561–586.

Andrews, D.J., and A.H. Kassam. 1976. The importance of multiple cropping in increasing world food supplies. In Multiple Cropping, R.I. Papendick, A. Sanchez, and G.B. Triplett, eds. Madison, Wisc.: American Society of Agronomy.

Arshad, M.A., A.J. Franzluebbers, and R.H. Azooza. 1999. Components of surface soil structure under conventional and no-tillage in northwestern Canada. *Soil & Tillage Research* 53(1):41–47.

Awmack, C.S., and S.R. Leather. 2002. Host plant quality and fecundity in herbivorous insects. *Annual Review of Entomology* 47:817–844.

Bach, S.J., T.A. McAllister, D.M. Veira, V.P.J. Gannon, and R.A. Holley. 2002. Transmission and control of *Escherichia coli* O157:H7—a review. *Canadian Journal of Animal Science* 82(4):475–490.

Baker, J.M., T. Ochsner, and T.J. Griffis. 2007. Impact of a cover crop on carbon and water balance of corn/soybean systems. In The Fourth USDA Greenhouse Gas Conference, Baltimore, Md.

Bale, J.S., J.C. van Lenteren, and F. Bigler. 2008. Biological control and sustainable food production. *Philosophical Transactions of the Royal Society B-Biological Sciences* 363(1492):761–776.

Balsam, J. 2006. Anaerobic digestion of animal wastes: factors to consider. Available at http://www.attra.ncat.org/attra-pub/PDF/anaerobic.pdf. Accessed on September 7, 2009.

Banik, P., A. Midya, B.K. Sarkar, and S.S. Ghose. 2006. Wheat and chickpea intercropping systems in an additive series experiment: advantages and weed smothering. *European Journal of Agronomy* 24(4):325–332.

Banin, A. 1999. Recycling and reuse of wastewater for irrigation in the Mediterranean region: approaches, precautions and potentials. *Annali di Chimica* 89(7–8):479–488.

Bao, J., D. Fravel, G. Lazarovits, D. Chellemi, P. van Berkum, and N. O'Neill. 2004. Biocontrol genotypes of Fusarium oxysporum from tomato fields in Florida. *Phytoparasitica* 32(1):9–20.

Barnett, J.L. 2007. Effects of confinement and research needs to underpin welfare standards. *Journal of Veterinary Behavior–Clinical Applications and Research* 2(6):213–218.

Baumann, D.T., M.J. Kropff, and L. Bastiaans. 2000. Intercropping leeks to suppress weeds. *Weed Research* 40(4):359–374.

Baumann, D.T., L. Bastiaans, and M.J. Kropff. 2001. Competition and crop performance in a leek-celery intercropping system. *Crop Science* 41(3):764–774.

Baumann, D.T., L. Bastiaans, J. Goudriaan, H.H. van Laar, and M.J. Kropff. 2002. Analysing crop yield and plant quality in an intercropping system using an eco-physiological model for interplant competition. *Agricultural Systems* 73(2):173–203.

Baumgartner, J., T. Leeb, T. Gruber, and R. Tiefenbacher. 2003. Husbandry and animal health on organic pig farms in Austria. *Animal Welfare* 12(4):631–635.

Beanland, L., P.L. Phelan, and S. Salminen. 2003. Micronutrient interactions on soybean growth and the developmental performance of three insect herbivores. *Environmental Entomology* 32(3):641–651.

Bedford, M.R. 2001. The role of carbohydrases in feedstuff digestion. Pp. 319–336 in Poultry Feedstuffs: Supply, Composition, and Nutritive Value, J. McNab and K.N. Boorman, eds. Edinburgh, UK: CAB International.

Benes, S.E., P.H. Robinson, S.R. Grattan, D. Goorahoo, and V. Cervinka. 2004. Saline drainage water re-use systems for the Westside San Joaquin Valley of California: candidate forages, halophytes, and soil management. Paper read at Frontis Biosaline Agriculture workshop, June 27–30, 2004, at Wageningen University, the Netherlands.

Benitez, M.S., F.B. Tustas, D. Rotenberg, M.D. Kleinhenz, J. Cardina, D. Stinner, S.A. Miller, and B.B.M. Gardener. 2007. Multiple statistical approaches of community fingerprint data reveal bacterial populations associated with general disease suppression arising from the application of different organic field management strategies. *Soil Biology & Biochemistry* 39(9):2289–2301.

Beretti, M., and D. Stuart. 2008. Food safety and environmental quality impose conflicting demands on Central Coast growers. *California Agriculture* 62(2):68–73.

Berkelmans, R., H. Ferris, M. Tenuta, and A.H.C. van Bruggen. 2003. Effects of long-term crop management on nematode trophic levels other than plant feeders disappear after 1 year of disruptive soil management. *Applied Soil Ecology* 23(3):223–235.

Bertenshaw, C.E., and P. Rowlinson. 2008. Exploring heifers' perception of "positive" treatment through their motivation to pursue a retreated human. *Animal Welfare* 17(3):313–319.

Bhowmik, P.C., and Inderjit. 2003. Challenges and opportunities in implementing allelopathy for natural weed management. *Crop Protection* 22(4):661–671.

Biermacher, J.T., B.W. Brorsen, F.M. Epplin, J.B. Solie, and W.R. Raun. 2009. The economic potential of precision nitrogen application with wheat based on plant sensing. *Agricultural Economics* 40(4):397–407.

Bishop, S.C., and J.A. Woolliams. 2004. Genetic approaches and technologies for improving the sustainability of livestock productions. *Journal of the Science of Food and Agriculture* 84(9):911–919.

Black, S.H., M. Shepherd, M. Vaughan, C. LaBar, and N. Hodges. 2009. Yolo Natural Heritage Program pollinator conservation strategy. Available at http://www.xerces.org/wp-content/uploads/2010/01/yolo-nhp_pollinator-strategy_xerces.pdf. Accessed on February 28, 2010.

Blackburn, H.D. 2009. Genebank development for the conservation of livestock genetic resources in the United States of America. *Livestock Science* 120(3):196–203.

Blackshaw, R.E., J.R. Moyer, R.C. Doram, and A.L. Boswell. 2001. Yellow sweetclover, green manure, and its residues effectively suppress weeds during fallow. *Weed Science* 49(3):406–413.

Blanco-Canqui, H., and R. Lal. 2008. No-tillage and soil-profile carbon sequestration: an on-farm assessment. *Soil Science Society of America Journal* 72:693–701.

Blankenberg, A.G.B., B. Braskerud, and K. Haarstad. 2006. Pesticide retention in two small constructed wetlands: treating non-point source pollution from agriculture runoff. *International Journal of Environmental Analytical Chemistry* 86(3–4):225–231.

Blasco, A. 2008. The role of genetic engineering in livestock production. *Livestock Science* 113(2–3):191–201.

Blevins, R.L., W.W. Frye, P.L. Baldwin, and S.D. Robertson. 1990. Tillage effects on sediment and soluble nutrient losses from a maury silt loam soil. *Journal of Environmental Quality* 19(4):683–686.

Blok, W.J., J.G. Lamers, A.J. Termorshuizen, and G.J. Bollen. 2000. Control of soilborne plant pathogens by incorporating fresh organic amendments followed by tarping. *Phytopathology* 90(3):253–259.

Bonanomi, G., V. Antignani, C. Pane, and E. Scala. 2007. Suppression of soilborne fungal diseases with organic amendments. *Journal of Plant Pathology* 89(3):311–324.

Bordovsky, J.P., W.M. Lyle, and J.W. Keeling. 1994. Crop-rotation and tillage effects on soil-water and cotton yield. *Agronomy Journal* 86(1):1–6.

Bowden, R., J. Shroyer, K. Roozeboom, M. Claassen, P.Evans, B. Gordon, B. Heer, K. Janssen, J. Long, J. Martin, A. Schlegel, R. Sears, and M. Witt. 2001. Performance of wheat variety blends in Kansas. Available at http://www.oznet.ksu.edu/library/crpsl2/SRL128.pdf. Accessed on February 27, 2010.

Bowden, W.B. 1987. The biogeochemistry of nitrogen in freshwater wetlands. *Biogeochemistry* 4:313–348.

Boyle, M., W.T. Frankenberger, Jr., and L.H. Stolzy. 1989. The influence of organic matter on soil aggregation and water infiltration. *Journal of Production Agriculture* 2:290–299.

Brady, N.C., and R.R. Weil. 2008. The Nature and Properties of Soils, 14th edition. Upper Saddle River, N.J.: Prentice Hall.

Brainard, D.C., and R.R. Bellinder. 2004. Weed suppression in a broccoli-winter rye intercropping system. *Weed Science* 52(2):281–290.

Brambell, F.W.R. 1965. Report of the Technical Committee to Enquire into the Welfare of Animals Kept Under Intensive Livestock Husbandry Systems. London: Her Majesty's Stationery Office, London.

Braskerud, B.C., K.S. Tonderski, B. Wedding, R. Bakke, G.B. Blankenberg, B. Ulen, and J. Koshiaho. 2005. Can constructed wetlands reduce diffuse phosphorus loads to eutrophic water in cold temperature regions? *Journal of Environmental Quality* 34:2145–2155.

Breve, M.A., R.W. Skaggs, J.E. Parsons, and J.W. Gilliam. 1997. DRAINMOD-N: a nitrogen model for artificially drained soils. *Transactions of the ASABE* 40(4):1067–1075.

Brinton, W.F., P. Storms, and T.C. Blewett. 2009. Occurrence and levels of fecal indicators and pathogenic bacteria in market-ready recycled organic matter composts. *Journal of Food Protection* 72(2):332–339.

Broz, J., and P. Beardsworth. 2002. Recent trends and future developments in the use of feed enzymes in poultry nutrition. Pp. 345–362 in Poultry Feedstuffs: Supply, Composition and Nutrition Value, J.M. MacNab and K.N. Boorman, eds. Wallingford, Oxfordshire, UK: CABI.

Bruce, T.J., and J.A. Pickett. 2007. Plant defence signalling induced by biotic attacks. *Current Opinion in Plant Biology* 10(4):387–392.

Bruce, T.J.A., M.C. Matthes, J.A. Napier, and J.A. Pickett. 2007. Stressful "memories" of plants: evidence and possible mechanisms. *Plant Science* 173(6):603–608.

Bruland, G.L., and C.J. Richardson. 2006. An assessment of the phosphorus retention capacity of wetlands in the Painter Creek Watershed, Minnesota, USA. *Water Air and Soil Pollution* 171(1–4):169–184.

Buckler IV, E.S., and J.M. Thornberry. 2002. Plant molecular diversity and applications to genomics. *Current Opinion in Plant Biology* 5:107–111.

Bugg, R.L., and C. Waddington. 1994. Using cover crops to manage arthropod pests of orchards—a review. *Agriculture, Ecosystems & Environment* 50(1):11–28.

Bugg, R.L., F.L. Wackers, K.E. Brunson, J.D. Dutcher, and S.C. Phatak. 1991. Cool-season cover crops relay intercropped with cantaloupe—influence on a generalist predator, *Geocorpis-punctipes* (hemiptera, lygaeidae). *Journal of Economic Entomology* 84(2):408–416.

Bull, L.S., D. Meyer, J.M. Rice, O. D. Simmons, III. 2008. Recent changes in food animal production and impacts on animal waste management. Available at http://www.ncifap.org/bin/u/v/PCIFAP_FW_FINAL1.pdf. Accessed on April 13, 2010.

Burchell II, M.R., R.W. Skaggs, J.E. Parsons, J.W. Gilliam, and L.A. Arnold. 2005. Shallow subsurface drains to reduce nitrate losses from drained agricultural lands. *Transactions of the ASABE* 48(3):1079–1089.

Burger, H., M. Schloen, W. Schmidt, and H.H. Geiger. 2008. Quantitative genetic studies on breeding maize for adaptation to organic farming. *Euphytica* 163:501–510.

Caamal-Maldonado, J.A., J.J. Jimenez-Osornio, A. Torres-Barragan, and A.L. Anaya. 2001. The use of allelopathic legume cover and mulch species for weed control in cropping systems. *Agronomy Journal* 93(1):27–36.

California Energy Commission. 2008. Energy in agriculture program on-farm irrigation. Available at http://www.energy.ca.gov/process/agriculture/farm_irrigation.html. Accessed on May 8, 2009.

California Integrated Waste Management Board. 2009. Compost quality: performance requirements. Available at http://www.ciwmb.ca.gov/organics/products/Quality/Needs.htm. Accessed on September 3, 2009.

Callaway, T.R., R.O. Elder, J.E. Keen, R.C. Anderson, and D.J. Nisbet. 2003. Forage feeding to reduce preharvest escherichia coli populations in cattle, a review. *Journal of Dairy Science* 86:852–860.

Carr, P.M., R.D. Horsley, and W.W. Poland. 2004. Barley, oat, and cereal-pea mixtures as dryland forages in the Northern Great Plains. *Agronomy Journal* 96(3):677–684.

Carr, P.M., G.B. Martin, and R.D. Horsley. 2006. Impact of tillage and crop rotation on spring wheat yield: II. Rotation effect. Available at http://www.plantmanagementnetwork.org/pub/cm/research/2006/wheat2/. Accessed on June 26, 2009.

CAST (Council on Animal Technology). 2002. Animal Diet Modification to Decrease the Potential for Nitrogen and Phosphorus Pollution. Washington, D.C.: Author.

Catala-Gregori, P., S. Mallet, A. Travel, J. Orengo, and M. Lessire. 2008. Efficiency of a prebiotic and a plant extract alone or in combination on broiler performance and intestinal physiology. *Canadian Journal of Animal Science* 88(4):623–629.

Caton, B.P., T.C. Foin, and J.E. Hill. 1999. A plant growth model for integrated weed management in direct-seeded rice. III. Interspecific competition for light. *Field Crops Research* 63(1):47–61.

Cavigelli, M.A., J.R. Teasdale, and A.E. Conklin. 2008. Long-term agronomic performance of organic and conventional field crops in the Mid-Atlantic region. *Agronomy Journal* 100:785–794.

Ceccarelli, S., and S. Grando. 1991. Environment of selection and type of germplasm in barley breeding for low-yielding conditions. *Euphytica* 57(3):207–219.

Ceotto, E. 2005. The issues of energy and carbon cycle: new perspectives for assessing the environmental impact of animal waste utilization. *Bioresource Technology* 96(2)191–196.

Chable, V., M. Conseil, E. Serpolay, and F. Le Lagadec. 2008. Organic varieties for cauliflowers and cabbages in Brittany: from genetic resources to participatory plant breeding. *Euphytica* 164(2):521–529.

Chaney, W.E. 1998. Biological control of aphids in lettuce using in-field insectaries. Pp. 73–83 in Enhancing Biological Control: Habitat Management to Promote Natural Enemies of Agricultural Pests. Berkeley, Calif.: University of California Press.

Chase C.A., and O.S. Mbuya. 2008. Greater interference from living mulches than weeds in organic broccoli production. *Weed Technology* 22:280–285.

Cheema, M.A., M.A. Qureshi, and G.B. Havenstein. 2003. A comparison of the immune response of a 2001 commercial broiler with a 1957 randombred broiler strain when fed representative 1957 and 2001 broiler diets. *Poultry Science* 82(10):1519–1529.

Cherr, C.M., J.M.S. Scholberg, and R. McSorley. 2006. Green manure as nitrogen source for sweet corn in a warm-temperate environment. *Agronomy Journal* 98(5):1173–1180.

Cherry, K.A., M. Shepherd, P.J.A. Withers, and S.J. Mooney. 2008. Assessing the effectiveness of actions to mitigate nutrient loss from agriculture: a review of methods. *Science of the Total Environment* 406:1–23.

Chichester, F.W., and C.W. Richardson. 1992. Sediment and nutrient loss from clay soils as affected by tillage. *Journal of Environmental Quality* 21(4):587–590.

Chien, S.H., L.I. Prochnow, and H. Cantarella. 2009. Recent developments of fertilizer production and use to improve nutrient efficiency and minimize environmental impacts. *Advances in Agronomy* 102:267–322.

Christopher, S.F., and R. Lal. 2007. Nitrogen management affects carbon sequestration in North American cropland soils. *Critical Reviews in Plant Sciences* 26(1):45–64.

Christopher, S.F., R. Lal, and U. Mishra. 2009. Regional study of no-till effects on carbon sequestration in the Midwestern United States. *Soil Science Society of America Journal* 73:207–216.

Ciecko, Z., M. Wyszkowski, W. Krajewski, and J. Zabielska. 2001. Effect of organic matter and liming on the reduction of cadmium uptake from soil by triticale and spring oilseed rape. *Science of the Total Environment* 281(1–3):37–45.

Clark, M.S., H. Ferris, K. Klonsky, W.T. Lanini, A.H.C. van Bruggen, and F.G. Zalom. 1998. Agronomic, economic, and environmental comparison of pest management in conventional and alternative tomato and corn systems in northern California. *Agriculture, Ecosystems and Environment* 68:51–71.

Clawson, K.L., and B.L. Blad. 1982. Infrared thermometry for scheduling irrigation of corn. *Agronomy Journal* 74:311–316.

Coates, R.W., M.J. Delwiche, and P.H. Brown. 2006. Design of a system for individual microsprinkler control. *Transactions of the ASABE* 49(6):1963–1970.

Collier, T., and R. Van Steenwyk. 2004. A critical evaluation of augmentative biological control. *Biological Control* 31(2):245–256.

Collins, H.P., P.E. Rasmussen, and C.L. Douglas. 1992. Crop-rotation and residue management effects in soil carbon and microbial dynamics. *Soil Science Society of America Journal* 56(3):783–788.

Collins, K.L., N.D. Boatman, A. Wilcox, J.M. Holland, and K. Chaney. 2002. Influence of beetle banks on cereal, aphid predation in winter wheat. *Agriculture Ecosystems & Environment* 93(1–3):337–350.

Connell, J.H., and P.M. Vossen. 2007. Organic olive orchard nutrition. In Organic Olive Production Manual, P.M. Vossen, ed. University of California, Agriculture and Natural Resources.

Cook, R.J. 2000. Advances in plant health management in the twentieth century. *Annual Review of Phytopathology* 38:95–116.

———. 2006. Toward cropping systems that enhance productivity and sustainability. *Proceedings of the National Academy of Sciences USA* 103(49):18389–18394.

Cooke, R.A., G.R. Sands, and L.C. Brown. 2008. Drainage water management: a practice for reducing nitrate loads from subsurface drainage systems. In Final Report: Gulf Hypoxia and Local Water Quality Concerns Workshop. St. Joseph, Mich.: American Society of Agricultural and Biological Engineers.

Cooley, H., J. Christian-Smith, and P. Gleick. 2009. Sustaining California Agriculture in an Uncertain Future. Oakland, Calif.: Pacific Institute.

Cooper, R.L., N.R. Fausey, and J. G. Streeter. 1991. Yield potential of soybeans grown under a subirrigation/drainage water management system. *Agronomy Journal* 83(5):884–887.

Copeland, P.J., R.R. Allmaras, R.K. Crookston, and W.W. Nelson. 1993. Corn soybean rotation effects on soil-water depletion. *Agronomy Journal* 85(2):203–210.

Cordier, C., S. Gianinazzi, and V. Gianinazzi-Pearson. 1996. Colonisation patterns of root tissues by *Phytophthora nicotianae var parasitica* related to reduced disease in mycorrhizal tomato. *Plant and Soil* 185(2):223–232.

Cornell University, U.S. Department of Agriculture Agricultural Research Service, and The University of Vermont. 2009. Nitrogen management on dairy farms. Available at http://www.dairyn.cornell.edu/. Accessed on September 3, 2009.

Costa, J.M., M.F. Ortuno, and M.M. Chaves. 2007. Deficit irrigation as a strategy to save water: physiology and potential application to horticulture. *Journal of Integrative Plant Biology* 49(10):1421–1434.

Costa, R.D., C.R.G. Tavares, and E.S. Cossich. 2007. Stabilization of swine wastes by anaerobic digestion. *Environmental Technology* 28(10):1145–1151.

Costello, M.J., and K.M. Daane. 1998. Influence of ground cover on spider populations in a table grape vineyard. *Ecological Entomology* 23(1):33–40.

Cowger, C., and R. Weisz. 2008. Winter wheat blends (mixtures) produce a yield advantage in North Carolina. *Agronomy Journal* 100(1):169–177.

Crews, T.E., and M.B. Peoples. 2005. Can the synchrony of nitrogen supply and crop demand be improved in legume and fertilizer-based agroecosystems? A review. *Nutrient Cycling in Agroecosystems* 72(2):101–120.

Crumpton, W.G. 2005. Water Quality Benefits of Wetland Restoration: A Performance-Based Approach. Washington, D.C.: U.S. Geological Survey.

Crumpton, W.G., and L.G. Goldsborough. 1998. Nitrogen transformation and fate in prairie wetlands. *Great Plains Research* 8:57–72.

Crumpton, W.G., D.A. Kovacic, D.L. Hey, and J.A. Kostel. 2008. Potential of Restored and Constructed Wetlands to Reduce Nutrient Export from Agricultural Watersheds in the Corn Belt. St. Joseph, Mich.: American Society of Agricultural and Biological Engineers.

CTIC (Conservation Technology Information Center). 2004. Conservation tillage and other tillage types in the United States—1990–2004. Available at http://www2.ctic.purdue.edu/ctic/CRM2004/1990–2004data.pdf. Accessed on September 2, 2009.

———. 2007. An Introduction to Nutrient Management. West Lafayette, Ind.: Author.

D'Eath, R.B., B.J. Tolkamp, I. Kyriazakis, and A.B. Lawrence. 2009. "Freedom from hunger" and preventing obesity: the animal welfare implications of reducing food quantity or quality. *Animal Behaviour* 77(2):275–288.

Dabney, S.M. 1998. Cover crop impacts on watershed hydrology. *Journal of Soil & Water Conservation* 53:207–213.

Dabney, S.M., J.A. Delgado, and D.W. Reeves. 2001. Using winter cover crops to improve soil and water quality. *Communications in Soil Science and Plant Analysis* 32:1221–1250.

Darymple, D. 1986. Development and Spread of High Yielding Varieties in Developing Countries. Washington, D.C.: U.S. Agency for International Development.

Daverede, I.C., A.N. Kravchenko, R.G. Hoeft, E.D. Nafziger, D.G. Bullock, J.J. Warren, and L.C. Gonzini. 2003. Phosphorus runoff: effect of tillage and soil phosphorus levels. *Journal of Environmental Quality* 32(4):1436–1444.

Davis, A.S., and M. Leibman. 2001. Nitrogen source influences wild mustard growth and competitive effect on sweet corn. *Weed Science* 49(4):558–566.

Davros, N.M., D.M. Debinski, K.F. Reeder, and W.L. Hohman. 2006. Butterflies and continuous conservation reserve program filter strips: landscape considerations. *Wildlife Society Bulletin* 34(4):936–943.

Dawson, J., K. Murphy, and S. Jones. 2006. Evolutionary participatory wheat breeding in Washington State, USA. Paper read at Participatory Plant Breeding: Relevance for Organic Agriculture?, at La Besse, France.

De Cauwer, B., D. Reheul, I. Nijs, and A. Milbau. 2006. Effect of margin strips on soil mineral nitrogen and plant biodiversity. *Agronomy for Sustainable Development* 26(2):117–126.

———. 2008. Management of newly established field margins on nutrient-rich soil to reduce weed spread and seed rain into adjacent crops. *Weed Research* 48(2):102–112.

Dechert, G., E. Veldkamp, and R. Brumme. 2005. Are partial nutrient balances suitable to evaluate nutrient sustainability of land use systems? Results from a case study in Central Sulawesi, Indonesia. *Nutrient Cycling in Agroecosystems* 72(3):201–212.

Delate, K., C. Cambardella, and A. McKern. 2008. Effects of organic fertilization and cover crops on an organic pepper system. *HortTechnology* 18(2):215–226.

Delgado, J., S. Essah, M. Dillon, R. Ingham, D. Manter, A. Stuebe, and R. Sparks. 2008. Sustainable cover crop rotations with potential to improve yields, crop quality, and nutrient and water use efficiencies. Paper read at the annual meeting of the Soil and Water Conservation Society at Tucson, Ariz.

Delgado, J.A., and W.C. Bausch. 2005. Potential use of precision conservation techniques to reduce nitrate leaching in irrigated crops. *Journal of Soil and Water Conservation* 60(6):379–387.

Dell, C.J., P.R. Salon, C.D. Franks, E.C. Benham, and Y. Plowden. 2008. No-till and cover crop impacts on soil carbon and associated properties on Pennsylvania dairy farms. *Journal of Soil and Water Conservation* 63(3): 136–142.

DeLuca, T.H., and D.K. DeLuca. 1997. Composting for feedlot manure management and soil quality. *Journal of Production Agriculture* 10(2):235–241.

Derksen, D.A., R.L. Anderson, R.E. Blackshaw, and B. Maxwell. 2002. Weed dynamics and management strategies for cropping systems in the northern Great Plains. *Agronomy Journal* 94(2):174–185

Desneux, N., A. Decourtye, and J.M. Delpuech. 2007. The sublethal effects of pesticides on beneficial arthropods. *Annual Review of Entomology* 52:81–106.

Devine, G.J., and M.J. Furlong. 2007. Insecticide use: contexts and ecological consequences. *Agriculture and Human Values* 24(3):281–306.

Dibner, J.J., and J.D. Richards. 2005. Antibiotic growth promoters in agriculture: history and mode of action. *Poultry Science* 84(4):634–643.

Dinnes, D.L., D.L. Karlen, D.B. Jaynes, T.C. Kaspar, J.L. Hatfield, T.S. Colvin, and C.A. Cambardella. 2002. Nitrogen management strategies to reduce nitrate leaching in tile-drained Midwestern soils. *Agronomy Journal* 94(1):153–171.

Disney, W.T., J.W. Green, K.W. Forsythe, J.F. Wiemers, and S. Weber. 2001. Benefit-cost analysis of animal identification for disease prevention and control. *Revue Scientifique et Technique de L'Office International Des Epizooties* 20(2):385–405.

Donald, P.F., and A.D. Evans. 2006. Habitat connectivity and matrix restoration: the wider implications of agri-environment schemes. *Journal of Applied Ecology* 43:209–218.

Donovan, D.M., D.E. Kerr, and R.J. Wall. 2005. Engineering disease resistant cattle. *Transgenic Research* 14(5): 563–567.

Doty, C.W., J.E. Parsons, A.W. Badr, A. Nassehzadeh-Tabrizi, and R.W. Skaggs. 1985. Water table control for water resource projects on sandy soils. *Journal of Soil and Water Conservation* 40(4):360–364.

Dourmad, J.Y., and C. Jondreville. 2007. Impact of nutrition on nitrogen, phosphorus, Cu and Zn in pig manure, and on emissions of ammonia and odours. *Livestock Science* 112(3):192–196.

Doyle, M.E. 2001. Alternatives to antibiotic use for growth promotion in animal husbandry. Available at http://fri.wisc.edu/briefs/antibiot.pdf. Accessed on July 14, 2009.

Doyle, M.P., and M.C. Erickson. 2006. Reducing the carriage of foodborne pathogens in livestock and poultry. *Poultry Science* 85:960–973.

Drinkwater, L.E. 2002. Cropping systems research: reconsidering agricultural experimental approaches. *HortTechnology* 12(3):355–361.

Drinkwater, L.E., P. Wagoner, and M. Sarrantonio. 1998. Legume-based cropping systems have reduced carbon and nitrogen losses. *Nature* 396(19):262–265.

Drinkwater, L.E., M. Schipanski, S. Snapp, and L.E. Jackson. 2008. Ecologically-based nutrient management. Pp. 161–209 in Agricultural Systems: Agroecology and Rural Innovation for Development, S. Snapp, and B. Pound, eds. San Diego, Calif.: Academic Press.

Dukes, M.D., R.O. Evans, J.W. Gilliam, and S.H. Kunickis. 2002. Effect of riparian buffer width and vegetation type on shallow groundwater quality in the Middle Coastal Plain of North Carolina. *Transactions of the ASABE* 45(2):327–336.

Duncan, I.J.H. 1981. Animal rights animal-welfare: a scientist's assessment. *Poultry Science* 60(3):489–499.

Duncan, I.J.H., and M.S. Dawkins. 1983. The problem of assessing well-being and suffering in farm animals. Pp. 13–24 in Indicators Relevant to Farm Animal Welfare, D. Schmidt, ed. The Hague, Netherlands: Springer.

Dusenbury, M.P., R.E. Engel, P.R. Miller, R.I. Lemke, and R. Wallender. 2008. Nitrous oxide emissions from a northern great plains soil as influenced by nitrogen management and cropping systems. *Journal of Environmental Quality* 37:542–550.

Dyck, E., M. Liebman, and M.S. Erich. 1995. Crop-weed interference as influenced by a leguminous or synthetic fertilizer nitrogen source: 1. Doublecropping experiments with crimson clover, sweet corn, and lambsquarters. *Agriculture Ecosystems & Environment* 56(2):93–108.

Easter, M., K. Paustian, K. Killian, S. Williams, T. Feng, R. al-Adamat, N.H. Batjes, M. Bernoux, T. Bhattacharyya, C.C. Cerri, C.E.P. Cerri, K. Coleman, P. Falloon, C. Feller, P. Gicheru, P. Kamoni, E. Milne, D.K. Pal, D.S. Powlson, Z. Rawajfih, M. Sessay, and S. Wokabi. 2007. The GEFSOC soil carbon modeling system: a tool for conducting regional-scale soil carbon inventories and assessing the impacts of land use change on soil carbon. *Agriculture, Ecosystems & Environment* 122(1):13–25.

Edwards, L., and J. Burney. 2008. Effect of preceding crop management on crop yield and soil properties assessed using standard erosion plots. *Canadian Journal of Soil Science* 88(4):553–558.

Eghball, B. 2002. Soil properties as influenced by phosphorus- and nitrogen-based manure and compost applications. *Agronomy Journal* 94:128–135.

Eghball, B., B.J. Wienhold, J.E. Gilley, and R.A. Eigenberg. 2002. Mineralization of manure nutrients. *Journal of Soil and Water Conservation* 57(6):470–473.

Ekholm, P., E. Turtola, J. Gronroos, P. Seuri, and K. Ylivainio. 2005. Phosphorus loss from different farming systems estimated from soil surface phosphorus balance. *Agriculture Ecosystems & Environment* 110(3–4):266–278.

English-Loeb, G., M. Rhainds, T. Martinson, and T. Ugine. 2003. Influence of flowering cover crops on *Anagrus* parasitoids (Hymenoptera : Mymaridae) and *Erythroneura* leafhoppers (Homoptera : Cicadellidae) in New York vineyards. *Agricultural and Forest Entomology* 5(2):173–181.

EPA (U.S. Environmental Protection Agency). 2002. Managing manure with biogas recovery systems: improved performance at competitive costs. Available at http://www.epa.gov/agstar/pdf/manage.pdf. Accessed on January 6, 2009.

———. 2009. The AgSTAR Program. The accomplishments. Available at http://www.epa.gov/agstar/accomplish.html. Accessed on January 29, 2010.

Evans, R., J.W. Gilliam, and W. Skaggs. 1996. Controlled Drainage Management Guidelines for Improving Drainage Water Quality. Raleigh: North Carolina Cooperative Extension Service.

Evans, R.D. 2003. Mechanisms controlling plant nitrogen isotope composition. *Trends in Plant Sciences* 6:121–126.

Evans, R.O., and R.W. Skaggs. 1989. Design guidelines for water table management systems. *Applied Engineering in Agriculture* 5(4):539–548.

Evans, R.O., J.W. Gilliam, and R.W. Skaggs. 1990. Controlled Drainage Management Guidelines for Improving Water Quality. Raleigh: North Carolina Cooperative Extension Service.

Evans, R.O., J. Paul Lilley, R.W. Skaggs, and J.W. Gilliam. 2000. Rural Land Use, Water Movement, Coastal Water Quality. Raleigh: North Carolina Cooperative Extension Service.

Evanylo, G., C. Sherony, J. Spargo, D. Starner, M. Brosius, and K. Haering. 2008. Soil and water environmental effects of fertilizer-, manure-, and compost-based fertility practices in an organic vegetable cropping system. *Agriculture Ecosystems & Environment* 127(1–2):50–58.

Fageria, N.K., and V.C. Baligar. 2005. Enhancing nitrogen use efficiency in crop plants. *Advances in Agronomy* 88:97–185.

Falconer, I.R., H.F. Chapman, M.R. Moore, and G. Ranmuthugala. 2006. Endocrine-disrupting compounds: a review of their challenge to sustainable and safe water supply and water reuse. *Environmental Toxicology* 21(2):181–191.

Fall, N., U. Emanuelson, K. Martinsson, and S. Jonsson. 2008a. Udder health at a Swedish research farm with both organic and conventional dairy cow management. *Preventive Veterinary Medicine* 83(2):186–195.

Fall, N., K. Forslund, and U. Emanuelson. 2008b. Reproductive performance, general health, and longevity of dairy cows at a Swedish research farm with both organic and conventional production. *Livestock Science* 118(1–2):11–19.

FAO (Food and Agriculture Organization). 1996. Global Plan of Action for the Conservation and Sustainable Utilization of Plant Genetic Resources for Food and Agriculture. Rome: Author.

———. 1999. Sustaining agricultural biodiversity and agro-ecosystem functions. Available at http://www.fao.org/sd/epdirect/EPre0080.htm. Accessed on January 29, 2010.

———. 2003. Soil biodiversity. Available at http://www.fao.org/ag/agl/agll/soilbiod/soilbtxt.stm. Accessed on August 12, 2009.

Faurés, J.M., M. Svendsen, and H. Turral. 2007. Reinventing irrigation. In Water for Food, Water for Life: A Comprehensive Assessment of Water Management in Agriculture, D. Molden, ed. London and Colombo: Earthscan and IWMI.

Fernandez-Cornejo, J., and M. Caswell. 2006. The First Decade of Genetically Engineered Crops in the United States. Washington, D.C.: U.S. Department of Agriculture Economic Research Service.

Fortuna, A., R. Harwood, G. Robertson, J. Fisk, and E. Paul. 2003. Seasonal changes in nitrification potential associated with application of N fertilizer and compost in maize systems of southwest Michigan. *Agriculture, Ecosystems & Environment* 97:285–293.

Fossum, O., D.S. Jansson, P.E. Etterlin, and I. Vagsholm. 2009. Causes of mortality in laying hens in different housing systems in 2001 to 2004. *Acta Veterinaria Scandinavica* 51:3.

Fravel, D. 1999. Hurdles and bottlenecks on the road to biocontrol of plant pathogens. *Australasian Plant Pathology* 28(1):53–56.

Fravel, D.R. 2005. Commercialization and implementation of biocontrol. *Annual Review of Phytopathology* 43:337–359.

Freemark, K.E., and D.A. Kirk. 2001. Birds on organic and conventional farms in Ontario: partitioning effects of habitat and practices on species composition and abundance. *Biological Conservation* 101(3):337–350.

Frydman, E., F. Carrari, Y.S. Liu, A.R. Fernie, and D. Zamir. 2004. Zooming in on a quantitative trait for tomato yield using interspecific introgressions. *Science* 305:1786–1789.

Gahoonia, T.S., O. Ali, A. Sarker, M.M. Rahman, and W. Erskine. 2005. Root traits, nutrient uptake, multi-location grain yield and benefit-cost ratio of two lentil (*Lens culinaris*, Medikus.) varieties. *Plant and Soil* 272(1–2):153–161.

Gahoonia, T.S., O. Ali, A. Sarker, N.E. Nielsen, and M.M. Rahman. 2006. Genetic variation in root traits and nutrient acquisition of lentil genotypes. *Journal of Plant Nutrition* 29(4):643–655.

Gamez-Virues, S., R.S. Bonifacio, G.M. Gurr, C. Kinross, A. Raman, and H.I. Nicol. 2007. Arthropod prey of shelterbelt-associated birds: linking faecal samples with biological control of agricultural pests. *Australian Journal of Entomology* 46:325–331.

García, J.P., C.S. Wortmann, M. Mamo, R.A. Drijber, J.A. Quincke, and D. Tarkalson. 2007. One-time tillage of no-till: effects on nutrients, mycorrhizae, and phosphorus uptake. *Agronomy Journal* 99:1093–1103.

Gaskell, M., and R. Smith. 2007. Nitrogen sources for organic vegetable crops. *HortTechnology* 17(4):431–441.

Gebhardt, M.R., T.C. Daniel, E.E. Schweizer, and R.R. Allmaras. 1985. Conservation tillage. *Science* 230(4726):625–629.

Ghaouti, L., W. Vogt-Kaute, and W. Link. 2008. Development of locally-adapted faba bean cultivars for organic conditions in Germany through a participatory breeding approach. *Euphytica* 162(2):257–268.

Ghebremichael, L.T., P.E. Cerosaletti, T.L. Veith, C.A. Rotz, J.M. Hamlett, and W.J. Gburek. 2007. Economic and phosphorus-related effects of precision feeding and forage management at a farm scale. *Journal of Dairy Science* 90(8):3700–3715.

Ghebremichael, L.T., T.L. Veith, J.M. Hamlett, and W.J. Gburek. 2008. Precision feeding and forage management effects on phosphorus loss modeled at a watershed scale. *Journal of Soil and Water Conservation* 63(5):280–291.

Ghosh, P.K., M.C. Manna, K.K. Bandyopadhyay, Ajay, A.K. Tripathi, R.H. Wanjari, K.M. Hati, A.K. Misra, C.L. Acharya, and A.S. Rao. 2006. Interspecific interaction and nutrient use in soybean/sorghum intercropping system. *Agronomy Journal* 98(4):1097–1108.

Gibson, G.R., and M.B. Roberfroid. 1995. Dietary modulation of the human colonic microbiota—introducing the concept of prebiotics. *Journal of Nutrition* 125(6):1401–1412.

Gibson, K.D., A.J. Fischer, T.C. Foin, and J.E. Hill. 2003. Crop traits related to weed suppression in water-seeded rice (*Oryza sativa* L.). *Weed Science* 51(1):87–93.

Gil, M.V., M.T. Carballo, and L.F. Calvo. 2008. Fertilization of maize with compost from cattle manure supplemented with additional mineral nutrients. *Waste Management* 28(8):1432–1440.

Gilbert, R.A., N. Tomkins, J. Padmanabha, J.M. Gough, D.O. Krause, and C.S. McSweeney. 2005. Effect of finishing diets on *Escherichia coli* populations and prevalence of enterohaemorrhagic *E-coli* virulence genes in cattle faeces. *Journal of Applied Microbiology* 99(4):885–894.

Giller, K.E., M.H. Beare, P. Lavelle, A.M.N. Izac, and M.J. Swift. 1997. Agricultural intensification, soil biodiversity and agroecosystem function. *Applied Soil Ecology* 6(1):3–16.

Gilley, J.E., B. Eghball, and D.B. Marx. 2007. Nitrogen and phosphorus concentrations of runoff as affected by moldboard plowing. *Transactions of the ASABE* 50(5):1543–1548.

Gilliam, J.W., and R.W. Skaggs. 1986. Controlled agricultural drainage to maintain water quality. *Journal of Irrigation and Drainage Engineering* 112(3):254–263.

Gilliam, J.W., R.W. Skaggs, and S.B. Weed. 1978. An evaluation of the potential from using drainage control to reduce nitrate loss from agricultural fields to surface waters. Raleigh: North Carolina State University, Water Resource Research Institute.

———. 1979. Drainage control to diminish nitrate loss from agricultural fields. *Journal of Environmental Quality* 8(1):137–142.

Gilliam, J.W., D.L. Osmond, and R.O. Evans. 1997. Selected Agricultural Best Management Practices to Control Nitrogen in the Neuse River Basin, North Carolina. Raleigh: North Carolina State University.

Gilliom, R.J., J.E. Barbash, C.G. Crawford, P.A. Hamilton, J.D. Martin, N. Nakagaki, L.H. Nowell, J.C. Scott, P.E. Stackelberg, G.P. Thelin, and D.M. Wolock. 2006. The Quality of Our Nation's Waters: Pesticides in the Nation's Streams and Ground Water, 1992–2001. Reston, Va.: U.S. Geological Survey.

Golan, E., B. Krissoff, F. Kuchler, L. Calvin, K. Nelson, and G. Price. 2004. Traceability in the U.S. Food Supply: Economic Theory and Industry Studies. Washington, D.C.: U.S. Department of Agriculture Economic Research Service.

Goldhamer, D.A., and R.H. Beede. 2004. Regulated deficit irrigation effects on yield, nut quality and water-use efficiency of mature pistachio trees. *Journal of Horticultural Science & Biotechnology* 79(4):538–545.

Gollehon, N., and M. Caswell. 2000. Confined animal production poses manure management problems. *Agricultural Outlook* Sept:12–18.

Gonsalves, C., D.R. Lee, and D. Gonsalves. 2007. The adoption of genetically modified papaya in Hawaii and its implications for developing countries. *Journal of Development Studies* 43(1):177–191.

Gonyou, H.W. 1994. Why the study of animal behavior is associated with the animal-welfare issue. *Journal of Animal Science* 72(8):2171–2177.

Goodlass, G., N. Halberg, and G. Verschuur. 2003. Input output accounting systems in the European community—an appraisal of their usefulness in raising awareness of environmental problems. *European Journal of Agronomy* 20:17–24.

Government of Manitoba. 2009. Soil management guide. Available at http://www.gov.mb.ca/agriculture/soilwater/soilmgmt/fsm01s07.html. Accessed on August 12, 2009.

Grandin, T. 2010. Improving Animal Welfare: A Practical Approach. Wallingford, UK: CAB International.

Green, R.D. 2009. ASAS Centennial Paper: future needs in animal breeding and genetics. *Journal of Animal Science* 87(2):793–800.

Griffin, T.S., R.P. Larkin, and C.W. Honeycutt. 2009. Delayed tillage and cover crop effects in potato systems. *American Journal of Potato Research* 86(2):79–87.

Grosshans, T., O. Distl, G. Seeland, and J. Wolf. 1994. Estimation of individual cross-breeding effects on milk-production traits of the German-black-pied dairy-cattle using different genetic models. *Journal of Animal Breeding and Genetics-Zeitschrift Fur Tierzuchtung Und Zuchtungsbiologie* 111(5–6):472–492.

Guimarães, E.P., J. Ruane, B.D. Scherf, A. Sonnino, and J.D. Dargie. 2007. Marker-Assisted Selection, Current Status, and Future Perspectives in Crops, Livestock, Forestry, and Fish. Rome: FAO.

Guy, J. 2001. Environmental enrichment for broilers—will it prevent feather pecking? *World Poultry* 17(3).

Haas, G., C. Deittert, and U. Kopke. 2007. Farm-gate nutrient balance assessment of organic dairy farms at different intensity levels in Germany. *Renewable Agriculture and Food Systems* 22(3):223–232.

Hajjar, R., D.I. Jarvis, and B. Gemmill-Herren. 2008. The utility of crop genetic diversity in maintaining ecosystem services. *Agriculture Ecosystems & Environment* 123(4):261–270.

Halberg, N., H.M.G. van der Werf, C. Basset-Mens, R. Dalgaard, and I.J.M. de Boer. 2005. Environmental assessment tools for the evaluation and improvement of European livestock production systems. *Livestock Production Science* 96:33–50.

Hanna, R., F.G. Zalom, and W.J. Roltsch. 2003. Relative impact of spider predation and cover crop on population dynamics of Erythroneura variabilis in a raisin grape vineyard. *Entomologia Experimentalis Et Applicata* 107(3):177–191.

Hanson, G.J., L. Caldwell, M. Lobrecht, D. McCook, S. L. Hunt, and D. Temple. 2007. A look at the engineering challenges of the USDA small watershed program. *Transactions of the ASABE* 50(5):1677–1682.

Hargreaves, J., M.S. Adl, P.R. Warman, and H.P.V. Rupasinghe. 2008. The effects of organic amendments on mineral element uptake and fruit quality of raspberries. *Plant and Soil* 308(1–2):213–226.

Harnly, M., R. McLaughlin, A. Bradman, M. Anderson, and R. Gunier. 2005. Correlating agricultural use of organophosphates with outdoor air concentrations: a particular concern for children. *Environmental Health Perspectives* 113(9):1184–1189.

Harris, G.H., O.B. Hesterman, E.A. Paul, S.E. Peters, and R.R. Janke. 1994. Fate of legume and fertilizer N-15 in a long-term cropping systems experiment. *Agronomy Journal* 86(5):910–915.

Hartwig, N.L., and H.U. Ammon. 2002. 50th anniversary invited article: cover crops and living mulches. *Weed Science* 50(6):688–699.

Hartz, T.K., P.R. Johnstone, E.M. Miyao, and R.M. Davis. 2005. Mustard cover crops are ineffective in suppressing soilborne disease or improving processing tomato yield. *HortScience* 40(7):2016–2019.

Haruvy, N. 1997. Agricultural reuse of wastewater: nation-wide cost-benefit analysis. *Agriculture Ecosystems & Environment* 66(2):113–119.

Hauggaard-Nielsen, H., P. Ambus, and E.S. Jensen. 2001. Interspecific competition, N use and interference with weeds in pea-barley intercropping. *Field Crops Research* 70(2):101–109.

Hauggaard-Nielsen, H., S. Mundus, and E.S. Jensen. 2009. Nitrogen dynamics following grain legumes and subsequent catch crops and the effects on succeeding cereal crops. *Nutrient Cycling in Agroecosystems* 84(3): 281–291.

Havenstein, G.B. 2006. Performance changes in poultry and livestock following 50 years of genetic selection. *Lohmann Information* 41:30–37.

Havenstein, G.B., P.R. Ferket, and M.A. Qureshi. 2003a. Growth, livability, and feed conversion of 1957 versus 2001 broilers when fed representative 1957 and 2001 broiler diets. *Poultry Science* 82(10):1500–1508.

———. 2003b. Carcass composition and yield of 1957 versus 2001 broilers when fed representative 1957 and 2001 broiler diets. *Poultry Science* 82(10):1509–1518.

Havlin, J.L., D.E. Kissel, L.D. Maddux, M.M. Claassen, and J.H. Long. 1990. Crop rotation and tillage effects on soil organic carbon and nitrogen. *Soil Science Society American Journal* 54:448–452.

Heffner, E.L., M.E. Sorrells, and J.L. Jannink. 2009. Genomic selection for crop improvement. *Crop Science* 49(1): 1–12.

Heichel, G.H. 1987. Legumes as a source of nitrogen in conservation tillage systems. Pp. 29–35 in *The Role of Legumes in Conservation Tillage Systems*, J.F. Power, ed. Ankeny, Iowa: Soil Conservation Society of America.

Heinonen-Tanski, H., M. Mohaibes, P. Karinen, and J. Koivunen. 2006. Methods to reduce pathogen microorganisms in manure. *Livestock Science* 102(3):248–255.

Heller, M., and G. Keoleian. 2000. Life-Cycle Based Sustainability Indicators for Assessment of the U.S. Food System. Ann Arbor: University of Michigan Center for Sustainable Systems.

Hemsworth, P.H. 2003. Human-animal interactions in livestock production. *Applied Animal Behaviour Science* 81(3):185–198.

Henderson, I.G., J. Cooper, R.J. Fuller, and J. Vickery. 2000. The relative abundance of birds on set-aside and neighbouring fields in summer. *Journal of Applied Ecology* 37(2):335–347.

Henggeler, J.C., J.M. Enciso, W.L. Multer and B.L. Unruh. 2002. Deficit subsurface drip irrigation of cotton. In Deficit Irrigation Practices. Rome: Food and Agriculture Organization of the United Nations.

Hepperly, P., D. Lotter, C.Z. Ulsh, R. Seidel, and C. Reider. 2009. Compost, manure and synthetic fertilizer influences crop yields, soil properties, nitrate leaching and crop nutrient content. *Compost Science & Utilization* 17(2):117–126.

Herring, J.P., R.C. Schultz, and T.M. Isenhart. 2006. Watershed scale inventory of existing riparian buffers in northeast Missouri using GIS. *Journal of the American Water Resources Association* 42(1):145–155.

Hesterman, O.B., T.S. Griffin, P.T. Williams, G.H. Harris, and D.R. Christenson. 1992. Forage-legume small-grain intercrops–nitrogen–nitrogen-production and response of subsequent corn. *Journal of Production Agriculture* 5(3):340–348.

Hey, D.L., D.L. Montgomery, L.S. Urban, T. Prato, R. Zarwell, A. Forbes, M. Martell, J. Pollack, and Y. Steele. 2004. Flood Damage Reduction in the Upper Mississippi River Basin: An Ecological Means. Chicago, Ill.: The Wetlands Initiative.

Hickey, M.B.C., and B. Doran. 2004. A review of the efficiency of buffer strips for the maintenance and enhancement of riparian ecosystems. *Water Quality Research Journal of Canada* 39(3):311–317.

Hiddink, G.A., A.H.C. van Bruggen, A.J. Termorshuizen, J.M. Raaijmakers, and A.V. Semenov. 2005. Effect of organic management of soils on suppressiveness to *Gaeumannomyces graminis* var. *tritici* and its antagonist, *Pseudomonas fluorescens*. *European Journal of Plant Pathology* 113(4):417–435.

Hilton, L.S., A.G.D. Bean, and J.W. Lowenthal. 2002. The emerging role of avian cytokines as immunotherapeutics and vaccine adjuvants. *Veterinary Immunology and Immunopathology* 85(3–4):119–128.

Hoad, S., C. Topp, and K. Davies. 2008. Selection of cereals for weed suppression in organic agriculture: a method based on cultivar sensitivity to weed growth. *Euphytica* 163(3):355–366.

Hoffman, M.P., R.L. Ridgeway, E.D. Show, and J. Matteoni. 1998. Practical application of mass-reared natural enemies: selected case histories. Pp. 268–293 in Mass-Reared Natural Enemies: Application, Regulation, and Needs, R.L. Ridgeway, M.P. Hoffmann, M.N. Inscoe and C.S. Glenister, eds. Lanham, Md.: Entomology Society of America.

Hooks, C.R.R., and M.W. Johnson. 2004. Using undersown clovers as living mulches: effects on yields, lepidopterous pest infestations, and spider densities in a Hawaiian broccoli agroecosystem. *International Journal of Pest Management* 50(2):115–120.

Hooper, D.U., F.S. Chapin, J.J. Ewel, A. Hector, P. Inchausti, S. Lavorel, J.H. Lawton, D.M. Lodge, M. Loreau, S. Naeem, B. Schmid, H. Setala, A.J. Symstad, J. Vandermeer, and D.A. Wardle. 2005. Effects of biodiversity on ecosystem functioning: A consensus of current knowledge. *Ecological Monographs* 75(1):3–35.

Howard-Williams, C. 1985. Cycling and retention of nitrogen and phosphorus in wetlands: a theoretical and applied perspective. *Freshwater Biology* 15:391–431.

Hoyt, G.D. 1999. Tillage and cover residue affects on vegetable yields. *HortTechnology* 9:351–358.

Hoyt, G.D., and T.R. Konsler. 1988. Soil Water and Temperature Regimes Under Tillage and Cover Crop Management for Vegetable Culture. Paper read at 11th International Conference, ISTRO, at Edinburgh, Scotland.

Huang, R.L., Z.Y. Deng, C.B. Yang, Y.L. Yin, M.Y. Xie, G.Y. Wu, T.J. Li, L.L. Li, Z.R. Tang, P. Kang, Z.P. Hou, D. Deng, H. Xiang, X.F. Kong, and Y.M. Guo. 2007. Dietary oligochitosan supplementation enhances immune status of broilers. *Journal of the Science of Food and Agriculture* 87(1):153–159.

Huggins, D.R., and J.P. Reganold. 2008. No-till: the quiet revolution. *Scientific American* 299:70–77.

Hunt, S., D. Reep, K.C. Kadavy. 2008. RCC stepped spillways for Renwick Dam—a partnership in research and design. *Dam Safety Journal* 6(2):32–40.

Hyun, Y., M. Ellis, and R.W. Johnson. 1998. Effects of feeder type, space allowance, and mixing on the growth performance and feed intake pattern of growing pigs. *Journal of Animal Science* 76(11):2771–2778.

Inderjit, M. Kaur, and C.L. Foy. 2001. On the significance of field studies in allelopathy. *Weed Technology* 15: 792–797.

IPCC (Intergovernmental Panel on Climate Change). 2006. IPCC Guidelines for National Greenhouse Gas inventories: Agriculture, Forestry and Other Land Uses. Vol. 4. Geneva: Author.

———. 1996. Revised 1996 IPCC Guidelines for National Greenhouse Gas Inventories. Geneva: Intergovernmental Panel on Climate Change.

Isensee, A.R., and A.M. Sadeghi. 1996. Effect of tillage reversal on herbicide leaching to groundwater. *Soil Science* 161:382–389.

Jagadamma, S., R. Lal, and B.K. Rimal. 2009. Effects of topsoil depth and soil amendments on corn yield and properties of two Alfisols in central Ohio. *Journal of Soil and Water Conservation* 64(1):70–80.

Jay, M.T., M. Cooley, D. Carychao, G.W. Wiscomb, R.A. Sweitzer, L. Crawford-Miksza, J.A. Farrar, D.K. Lau, J. O'Connell, A. Millington, R.V. Asmundson, E.R. Atwill, and R.E. Mandrell. 2007. *Escherichia coli* O157:H7 in feral swine near spinach fields and cattle, central California coast. *Emerging Infectious Diseases* 13(12): 1908–1911.

Jaynes, D.B., T.S. Calvin, D.L. Karlen, C.A. Cambardella, and D.W. Meek. 2001. Nitrate losses in subsurface drainage as affected by nitrogen fertilizer rate. *Journal of Environmental Quality* 30:1305–1314.

Jemison, J.M., J.D. Jabro, and R.H. Fox. 1994. Evaluation of LEACHM: 1. Simulation of drainage, bromide leaching, and corn bromide uptake. *Agronomy Journal* 86(5):843–851.

Jensen. 1998. The impact of feed additives on the microbial ecology of the gut in young pigs. *Journal of Animal and Feed Sciences* 7:45–64.

Jensen, E., D.C.N. Robb, and C.E. Franzoy. 1970. Scheduling irrigations using climate-crop-soil data. *Journal of the Irrigation and Drainage Division* 96(1):25–38.

Jezierska-Tys, S., and M. Frac. 2008. Microbiological indices of soil quality fertilized with dairy sewage sludge. *International Agrophysics* 22(3):215–219.

———. 2009. Impact of dairy sewage sludge on enzymatic activity and inorganic nitrogen concentrations in the soils. *International Agrophysics* 23(1):31–37.

Johnson, N.C., P.J. Copeland, B.K. Crookston, and F.L. Pfleger. 1992. Mycorrhizae: possible explanation of yield decline with continuous corn and soybeans. *Agronomy Journal* 84:387–390.

Johnson, S.R., S. Strom, and K. Grillo. 2008. Quantification of the impacts on US agriculture of biotechnology-derived crops planted in 2006. Washington, D.C.: National Center for Food and Agricultural Policy.

Jordan, D.L., J.S. Barnes, T. Corbett, C.R. Bogle, P.D. Johnson, B.B. Shew, S.R. Koenning, W.M. Ye, and R.L. Brandenburg. 2008. Crop response to rotation and tillage in peanut-based cropping systems. *Agronomy Journal* 100(6):1580–1586.

Joseph, T., and M. Morrison. 2006. Nanotechnology in agriculture and food. Available at http://files.nanobio-raise.org/Downloads/nfnaf.pdf. Accessed on April 5, 2010.

Joyce, B.A., W.W. Wallender, J.P. Mitchell, L.M. Huyck, S.R. Temple, P.N. Brostrom, and T.C. Hsiao. 2002. Infiltration and soil water storage under winter cover cropping in California's Sacramento Valley. *Transactions of the ASAE* 45(2):315–326.

Kach, B., and R. Khosla, 2003. The role of precision agriculture in cropping systems. In Cropping Systems: Trends and Advances, A. Shrestha, ed. Binghamton, N.Y.: Food Products Press.

Kadlec, R.H., and R.L. Knight. 1996. Treatment Wetlands. Boca Raton, Fla.: Lewis Publishers.

Kalita, P.K., and R.S. Kanwar. 1993. Effect of water table management practices on the transport of nitrate-N to shallow groundwater. *Transactions of the ASABE* 36(2):413–422.

Kallestad, J.C., J.G. Mexal, T.W. Sammis, and R. Heerema. 2008. Development of a simple irrigation scheduling calendar for Mesilla Valley pecan growers. *HortTechnology* 18(4):714–725.

Kanwar, R.S., R.M. Cruse, M. Ghaffarzadeh, A. Bakhsh, D.L. Karlen, and T.B. Bailey. 2005. Corn-soybean and alternative cropping systems effects on NO_3-N leaching losses in subsurface drainage water. *Applied Engineering in Agriculture* 21(2):181–188.

Karban, R., and Y. Chen. 2007. Induced resistance in rice against insects. *Bulletin of Entomological Research* 97(4): 327–335.

Karlen, D.L., N.C. Wollenhaupt, D.C. Erbach, E.C. Berry, J.B. Swan, N.S. Eash, and J.L. Jordahl. 1994. Long-term tillage effects on soil quality. *Soil & Tillage Research* 32(4):313–327.

Karlen, D.L., M.D. Duffy, and T.S. Colvin. 1995. Nutrient, labor, energy, and economic evaluations of two farming systems in Iowa. *Journal of Production Agriculture* 8(4):540–546.

Karlen, D.L., A. Kumar, R.S. Kanwar, C.A. Cambardella, and T.S. Colvin. 1998. Tillage system effects on 15-year carbon-based and simulated N budgets in a tile drained Iowa field. *Soil & Tillage Research* 48:155–165.

Karlen, D.L., E.G. Hurley, S.S. Andrews, C.A. Cambardella, D.W. Meek, M.D. Duffy, and A.P. Mallarino. 2006. Crop rotation effects on soil quality at three northern corn/soybean belt locations. *Agronomy Journal* 98(3): 484–495.

Kasper, T.C., E.J. Kladivko, J.W. Singer, S. Morse, and D.R. Mutch. 2008. Potential and limitations of cover crops, living mulches, and perennials to reduce nutrient losses to water sources from agricultural fields in the Upper Mississippi River Basin. In Gulf Hypoxia and Local Water Quality Concerns Workshop. St. Joseph, Mich.: American Society of Agricultural and Biological Engineers.

Keeling, L.J. 2005. Healthy and happy: animal welfare as an integral part of sustainable agriculture. *Ambio* 34(4–5): 316–319.

Kelm, S.C., A.E. Freeman, and N.C.T. Comm. 2000. Direct and correlated responses to selection for milk yield: results and conclusions of regional project NC-2, "Improvement of dairy cattle through breeding, with emphasis on selection." *Journal of Dairy Science* 83(12):2721–2732.

Khanh, T.D., M.I. Chung, T.D. Xuan, and S. Tawata. 2005. The exploitation of crop allelopathy in sustainable agricultural production. *Journal of Agronomy and Crop Science* 191(3):172–184.

Kim, C.S., G.D. Schaible, and S.G. Daberkow. 2000. An efficient cost-sharing program to reduce nonpoint-source contamination: theory and an application to groundwater contamination. *Environmental Geology* 39(6): 649–659.

Kirchmann, H., and M.P. Bernal. 1997. Organic waste treatment and C stabilization efficiency. *Soil Biology & Biochemistry* 29(11–12):1747–1753.

Kirda, C. 2002. Deficit irrigation scheduling based on plant growth stages showing water stress tolerance. In Deficit Irrigation Practices. Rome: Food and Agriculture Organization of the United Nations.

Kloepper, J.W., C.M. Ryu, and S.A. Zhang. 2004. Induced systemic resistance and promotion of plant growth by *Bacillus* spp. *Phytopathology* 94(11):1259–1266.

Knowlton, K.F., M.S. Taylor, S.R. Hill, C. Cobb, and K.F. Wilson. 2007. Manure nutrient excretion by lactating cows fed exogenous phytase and cellulase. *Journal of Dairy Science* 90(9):4356–4360.

Kogan, M. 1998. Integrated pest management: historical perspectives and contemporary developments. *Annual Review of Entomology* 43: 243–270.

Kokalis-Burelle, N., N. Martinez-Ochoa, R. Rodriguez-Kabana, and J.W. Kloepper. 2002. Development of multi-component transplant mixes for suppression of *Meloidogyne incognita* on tomato (*Lycopersicon esculentum*). *Journal of Nematology* 34(4):362–369.

Kovacic, D.A., M.B. David, L.E. Gentry, K.M. Starks, and R.A. Cooke. 2000. Effectiveness of constructed wetlands in reducing nitrogen and phosphorus export from agricultural tile drainage. *Journal of Environmental Quality* 29:1262–1274.

Kramer, A.W., T.A. Doane, W.R. Horwath, and C. van Kessel. 2002. Combining fertilizer and organic inputs to synchronize N supply in alternative cropping systems in California. *Agriculture Ecosystems & Environment* 91(1–3):233–243.

Kranz, J. 2005. Interactions in pest complexes and their effects on yield. *Zeitschrift Fur Pflanzenkrankheiten Und Pflanzenschutz—Journal of Plant Diseases and Protection* 112(4):366–385.

Kranz, W.L., D.E. Eisenhauer, and M.T. Retka. 1992. Water and energy-conservation using irrigation scheduling with center-pivot irrigation systems. *Agricultural Water Management* 22(4):325–334.

Kratochvil, R.J., S. Sardanelli, K. Everts, and E. Gallagher. 2004. Evaluation of crop rotation and other cultural practices for management of root-knot and lesion nematodes. *Agronomy Journal* 96(5):1419–1428.

Kuepper, G. 2001. Pursuing Conservation Tillage Systems for Organic Crop Production. Fayetteville, Ariz.: National Center for Appropriate Technology.

Labrada, R. 2008. Allelopathy as a tool for weed management. *Allelopathy Journal* 22(2):283–287.

Laible, G. 2009. Enhancing livestock through genetic engineering—Recent advances and future prospects. *Comparative Immunology Microbiology and Infectious Diseases* 32(2):123–137.

Lal, R. 1991. Tillage and agricultural sustainability. *Soil & Tillage Research* 20(2–4):133–146.
———. 2004a. Carbon emission from farm operations. *Environmental International* 30:981–990.
———. 2004b. Soil carbon sequestration impacts on global climate change and food security. *Science* 304(5677): 1623–1627.
Landis, D.A., F.D. Menalled, A.C. Costamagna, and T.K. Wilkinson. 2005. Manipulating plant resources to enhance beneficial arthropods in agricultural landscapes. *Weed Science* 53:902–508.
Landis, J.N., J.E. Sanches, G.W. Bird, C.E. Edson, R. Isaacs, R.H. Lehnert, A.M.C. Schilder, and S.M. Swinton, eds. 2002 Fruit Crop Ecology and Management. Extension Bulletin E-2759. East Lansing: Michigan State University.
Langdale, G.W., R. L. Blevins, D.L. Karlen, D.K. McCool, M.A. Nearing, E.L. Skidmore, A.W. Thomas, D.D. Tyler, and J.R. Williams. 1991. Cover crop effects on soil erosion by wind and water. In Cover Crops for Clean Water, W.L. Hargrove, ed. Ankeny, Iowa: Soil and Water Conservation Society.
Larkin, R.P. 2008. Relative effects of biological amendments and crop rotations on soil microbial communities and soilborne diseases of potato. *Soil Biology and Biochemistry* 40:1341–1351.
Larkin, R.P., C.W. Honeycutt, and T.S. Griffin. 2006. Effect of swine and dairy manure amendments on microbial communities in three soils as influenced by environmental conditions. *Biology and Fertility of Soils* 43(1):51–61.
Larney, F.J., and X.Y. Hao. 2007. A review of composting as a management alternative for beef cattle feedlot manure in southern Alberta, Canada. *Bioresource Technology* 98(17):3221–3227.
Lauer, J.A. 2007. A retrospective on crop rotation research during the last century—what have we learned? In ASA-CSSA-SSSA International Annual Meeting, New Orleans, La.
Lee, K.H., T.M. Isenhart, and R.C. Schultz. 2003. Sediment and nutrient removal in an established multi-species riparian buffer. *Journal of Soil and Water Conservation* 58:1–7.
Leib, B.G., M. Hattendorf, T. Elliott, and G. Matthews. 2002. Adoption and adaptation of scientific irrigation scheduling: trends from Washington, USA as of 1998. *Agricultural Water Management* 55(2):105–120.
Lemaux, P.G. 2008. Genetically engineered plants and foods—a scientist's analysis of the issues (Part I). *Annual Review of Plant Biology* 59:771–812.
———. 2009. Genetically engineered plants and foods—a scientist's analysis of the issues (Part II). *Annual Review of Plant Biology* 60:511–559.
Leone, E.H., and I. Estevez. 2008. Economic and welfare benefits of environmental enrichment for broiler breeders. *Poultry Science* 87(1):14–21.
Lesoing, G.W., and C.A. Francis. 1999. Strip intercropping effects on yield and yield components of corn, grain sorghum, and soybean. *Agronomy Journal* 91(5):807–813.
Letourneau, D.K. 1998. Conserving biology, lessons for conserving natural enemies. In Conservation Biological Control, P. Barbosa, ed. San Diego, Calif.: Academic Press.
Letourneau, D.K., and S.G. Bothwell. 2008. Comparison of organic and conventional farms: challenging ecologists to make biodiversity functional. *Frontiers in Ecology and the Environment* 6(8):430–438.
Li, X.Q., L. Qiang, and C.L. Xu. 2008. Effects of supplementation of fructooligosaccharide and/or Bacillus subtilis to diets on performance and on intestinal microflora in broilers. *Archiv Fur Tierzucht—Archives of Animal Breeding* 51(1):64–70.
Li, Y.C., P.J. Stoffella, and H.H. Bryan. 2000. Management of organic amendments in vegetable crop production systems in Florida. *Soil and Crop Science Society of Florida Proceedings* 59:17–21.
Liaghat, A., and S.O. Prasher. 1997. Role of soil and grass strips in reducing nitrate-N pollution in subsurface-drained farmlands: Lysimeter results. *Canadian Water Resources Journal* 22(3):117–127.
Liebman, M., and A.S. Davis. 2000. Integration of soil, crop and weed management in low-external-input farming systems. *Weed Research* 40(1):27–47.
Lindstrom, M.J. 2006. Tillage erosion: description and process. In Encyclopedia of Soil Science, R. Lal, ed. New York: Taylor & Francis.
Livestock Conservation Institute. 1988. Environmental enrichment for confinement pigs. Paper read at Livestock Conservation Institute 1998 Annual Meeting, Kansas City, Mo.
Locascio, S.J. 2005. Management of irrigation for vegetables: past, present, and future. *HortTechnology* 15(3): 482–485.
Lou, W.R., R.W. Skaggs, and G.M. Chescheir. 2000. DRAINMOD modifications for cold conditions. *Transactions of the ASABE* 43(6):1569–1582.
Lowenthal, J.W. 2001. Therapeutic applications of cytokines—what can the chicken teach us? Pp. 1–7 in Current Progress on Avian Immunology Research, K.A. Schat, ed. Kennett Square, Pa: American Association of Avian Pathologists.

Lowrance, R., L.S. Altier, R.G. Williams, S.P. Inamdar, J.M. Sheridan, D.D. Bosch, R.K. Hubbard, D.L. Thomas. 2000. REMM: The riparian ecosystem management model. *Journal of Soil and Water Conservation* 55:27–34.

Lowrance R., T.M. Isenhart, W.J. Gburek, F.D. Shields, Jr., P.J. Wigington, Jr., and S.M. Dabney. 2006. Landscape Management Practices. Pp. 269–317 in Environmental Benefits of Conservation on Cropland: The Status of Knowledge, M. Schnepf and C. Cox, eds. Ankeny, Iowa: Soil and Water Conservation Society.

Lu, Y.C., K.B. Watkins, J.R. Teasdale, and A.A. Abdul-Baki. 2000. Cover crops in sustainable food production. *Food Reviews International* 16(2):121–157.

Luna, J. 1998. Multiple impacts of cover crops in farming systems. Available at http://ifs.orst.edu/pubs/multiple_impacts_cover_cro.html. Accessed on June 25, 2009.

Lyberg, K., H.K. Andersson, J.S. Sands, and J.E. Lindberg. 2008. Influence of phytase and xylanase supplementation of a wheat-based diet on digestibility and performance in growing pigs. *Acta Agriculturae Scandinavica Section a-Animal Science* 58(3):146–151.

Lyle, W., and J. Bordovsky. 1981. Low Energy Precision Application (LEPA) irrigation system. *Transactions of the ASABE* 24(5):1241–1245.

Lyons, J., S.W. Trimble, and L.K. Paine. 2000. Grass versus trees: managing riparian areas to benefit streams of central North America. *Journal of the American Water Resources Association* 36(4):919–930.

MacDonald, J.M., and W.D. McBride. 2009. The Transformation of U.S. Livestock Agriculture: Scale, Efficiency, and Risks. Washington, D.C.: U.S. Department of Agriculture Economic Research Service.

MacDonald, J.M., M.O. Ribaudo, M.J. Livingston, J. Beckman, and W. Huang. 2009. Manure Use for Fertilizer and for Energy Report to Congress. Washington, D.C.: U.S. Department of Agriculture Economic Research Service.

Macias-Corral, M.A., Z.A. Samani, A.T. Hanson, R. DelaVega, and P.A. Funk. 2005. Producing energy and soil amendment from dairy manure and cotton gin waste. *Transactions of the ASABE* 48(4):1521–1526.

Madden, N.M., R.J. Southard, and J.P. Mitchell. 2008. Conservation tillage reduces PM10 emissions in dairy forage rotations. *Atmospheric Environment* 42(16):3795–3808.

Marshall, E.J.R., and A.C. Moonen. 2002. Field margins in northern Europe: their functions and interactions with agriculture. *Agriculture, Ecosystems & Environment* 89(1–2):5–12.

Martin, D.L., E.C. Stegman, and E. Fereres. 1990. Irrigation scheduling principles. In Management of Farm Irrigation Systems. St. Joseph, Mich.: American Society of Agricultural Engineers.

Martin, J., and D. Beegle. 2009. Manure sampling for nutrient management planning. University Park: The Pennsylvania State University. Available at http://cropsoil.psu.edu/extension/facts/agfact69.pdf. Accessed on February 9, 2010.

Marvier, M., C. McCreedy, J. Regetz, and P. Kareiva. 2007. A meta-analysis of effects of Bt cotton and maize on nontarget invertebrates. *Science* 316:1475–1477.

Mason, H.E., and D. Spaner. 2006. Competitive ability of wheat in conventional and organic management systems: a review of the literature. *Canadian Journal of Plant Science* 86(2):333–343.

Matson, P.A., W.J. Parton, A.G. Power, and M.J. Swift. 1997. Agricultural intensification and ecosystem properties. *Science* 277(5325):504–509.

Matthiessen, J.N., and J.A. Kirkegaard. 2006. Biofumigation and enhanced biodegradation: opportunity and challenge in soilborne pest and disease management. *Critical Reviews in Plant Sciences* 25(3):235–265.

McBride, W.D. 1997. Change in U.S. Livestock Production, 1969–92. Washington, D.C.: U.S. Department of Agriculture Economic Research Service.

McCarthy, J.R., D.L. Pfost, and H.D. Currence. 1993. Conservation Tillage and Residue Management to Reduce Soil Erosion. University of Missouri Extension Publication G1650. Columbia: University of Missouri.

McCouch, S. 2004. Diversifying selection in plant breeding. *Plos Biology* 2(10):1507–1512.

McCouch, S., Y.X. L. Teytelman, K.B. Lobos, K. Clare, M. Walton, B. Fu, R. Maghirang, Z. Li, Y. Xing, Q. Zhang, M.Y. I. Kono, R. Fjellstrom, G. DeClerck, D. Schneider, S. Cartinhour, D. Ware, and L. Stein. 2002. Development and mapping of 2240 new SSR markers for rice (*Oryza sativa* L.). *DNA Research* 9:199–207.

McCracken, D.V., M.S. Smith, J.H. Grove, C.T. Mackown, and R.L. Blevins. 1994. Nitrate leaching as influenced by cover cropping and nitrogen source. *Soil Science Society of America Journal* 58(5):1476–1483.

McDowell, R.W., and L.M. Condron. 2004. Estimating phosphorus loss from New Zealand grassland soils. *New Zealand Journal of Agricultural Research* 47(2):137–145.

McDowell, R.W., and A.N. Sharpley. 2004. Variation of phosphorus leached from Pennsylvanian soils amended with manures, composts or inorganic fertilizer. *Agriculture Ecosystems & Environment* 102(1):17–27.

McEwen, S.A. 2006. Antibiotic use in animal agriculture: what have we learned and where are we going? *Animal Biotechnology* 7(2):239–250.

McIsaac, G.F., J.K. Mitchell, and M.C. Hirschi. 1995. Dissolved phosphorus concentrations in runoff from simulated rainfall on corn and soybean tillage systems. *Journal of Soil and Water Conservation* 50(4):383–388.

McIsaac, G.F., and X. Hu. 2004. Net N input and riverine N export from Illinois agricultural watersheds with and without extensive tile drainage. *Biogeochemistry* 70:251–271.

Menalled, F.D., K.L. Gross, and M. Hammond. 2001. Weed aboveground and seedbank community responses to agricultural management systems. *Ecological Applications* 11(6):1586–1601.

Menalled, F.D., P.C. Marino, S.H. Gage, and D.A. Landis. 1999. Does agricultural landscape structure affect parasitism and parasitoid diversity? *Ecological Applications* 9(2):634–641.

Mench, J.A. 2002. Broiler breeders: feed restriction and welfare. *Worlds Poultry Science Journal* 58(1):23–29.

———. 2008. Farm animal welfare in the USA: farming practices, research, education, regulation, and assurance programs. *Applied Animal Behaviour Science* 113(4):298–312.

Mench, J.A., J. Morrow-Tesch, and L.R. Chu. 1998. Environmental enrichment for farm animals. *Lab Animal* 27(3):32–36.

Messiha, N.A.S., A.H.C. van Bruggen, A.D. van Diepeningen, O.J. de Vos, A.J. Termorshuizen, N.N.A. Tjou-Tam-Sin, and J.D. Janse. 2007. Potato brown rot incidence and severity under different management and amendment regimes in different soil types. *European Journal of Plant Pathology* 119(4):367–381.

Meyer-Aurich, A., A. Weersink, K. Janovicek, and B. Deen. 2006. Cost efficient rotation and tillage options to sequester carbon and mitigate GHG emissions from agriculture in Eastern Canada. *Agriculture Ecosystems & Environment* 117(2–3):119–127.

Mikola, J., and P. Sulkava. 2001. Responses of microbial-feeding nematodes to organic matter distribution and predation in experimental soil habitat. *Soil Biology & Biochemistry* 33(6):811–817.

Millet, S., C.P.H. Moons, M.J. Van Oeckel, and G.P.J. Janssens. 2005. Welfare, performance and meat quality of fattening pigs in alternative housing and management systems: a review. *Journal of the Science of Food and Agriculture* 85(5):709–719.

Mitchell, J., C. Summers, T. Prather, J. Stapleton, and L. Roche. 2000. Potential allelopathy of sorghum-sudan mulch. *HortScience* 35:442.

Mitsch, W.J., J.W. Day Jr., J.W. Gilliam, P.M. Groffman, D.L. Hey, G.W. Randall, and N. Wang. 2001. Reducing nitrogen loading to the Gulf of Mexico from the Mississippi River Basin: strategies to counter a persistent ecological problem. *BioScience* 51:373–388.

Mitsch W.J., J.W. Day, L. Zhang, and R.R. Lane. 2005. Nitrate-nitrogen retention in wetlands in the Mississippi river basin. *Ecological Engineering* 24(4):267–278.

Mohammad, F.S., and A.I. Al-Amoud. 1993. Water conservation through irrigation scheduling under arid climatic conditions. *Agricultural Water Management* 24:251–264.

Molden, D.J., R. Sakthivadivel, C.J. Perry, C.J. de Fraiture, and W.H. Kloezen. 1998. Indicators for Comparing Performance of Irrigated Agricultural Systems. Colombo, Sri Lanka: International Irrigation Management Institute.

Moller, K., W. Stinner, and G. Leithold. 2008. Growth, composition, biological N-2 fixation and nutrient uptake of a leguminous cover crop mixture and the effect of their removal on field nitrogen balances and nitrate leaching risk. *Nutrient Cycling in Agroecosystems* 82(3):233–249.

Mols, C.M.M., and M.E. Visser. 2002. Great tits can reduce caterpillar damage in apple orchards. *Journal of Applied Ecology* 39(6):888–899.

Momma, N. 2008. Biological soil disinfestation (BSD) of soilborne pathogens and its possible mechanisms. *Japan Agricultural Research Quarterly* 42(1):7–12.

Montemurro, F., and M. Maiorana. 2007. Nitrogen utilization, yield, quality and soil properties in a sugarbeet crop amended with municipal solid waste compost. *Compost Science & Utilization* 15(2):84–92.

Montemurro, F., M. Maiorana, D. Ferri, and G. Convertini. 2006. Nitrogen indicators, uptake and utilization efficiency in a maize and barley rotation cropped at different levels and sources of N fertilization. *Field Crops Research* 99(2–3):114–124.

Montgomery, D.R. 2007. Soil erosion and agricultural sustainability. *Proceedings of the National Academy of Sciences USA* 104:13268–13272.

Moonen, A.C., and P. Barberi. 2008. Functional biodiversity: an agroecosystem approach. *Agriculture Ecosystems & Environment* 127(1–2):7–21.

Moore, J.M., S. Klose, and M.A. Tabatabai. 2000. Soil microbial biomass carbon and nitrogen as affected by cropping systems. *Biology and Fertility of Soils* 31(3–4):200–210.

Moore, M.T., C.M. Cooper, S. Smith, R.F. Cullum, S.S. Knight, M.A. Locke, and E.R. Bennett. 2007. Diazinon mitigation in constructed wetlands: influence of vegetation. *Water Air and Soil Pollution* 184(1–4):313–321.

Mormede, P., S. Andanson, B. Auperin, B. Beerda, D. Guemene, J. Malnikvist, X. Manteca, G. Manteuffel, P. Prunet, C.G. van Reenen, S. Richard, and I. Veissier. 2007. Exploration of the hypothalamic-pituitary-adrenal function as a tool to evaluate animal welfare. *Physiology & Behavior* 92:317–339.

Morrison, R.S., L.J. Johnston, and A.M. Hilbrands. 2007. The behaviour, welfare, growth performance and meat quality of pigs housed in a deep-litter, large group housing system compared to a conventional confinement system. *Applied Animal Behaviour Science* 103(1–2):12–24.

Morse, D., J.C. Guthrie, and R. Mutters. 1996. Anaerobic digester survey of California dairy producers. *Journal of Dairy Science* 79:149–153.

Motta, A.C.V., D.W. Reeves, C. Burmester, and Y. Feng. 2007. Conservation tillage, rotations, and cover crop affecting soil quality in the Tennessee valley: particulate organic matter, organic matter, and microbial biomass. *Communications in Soil Science and Plant Analysis* 38(19–20):2831–2847.

Mullen, M.D., C.G. Melhorn, D.D. Tyler, and B.N. Duck. 1998. Biological and biochemical soil properties in no-till corn with different cover crops. *Journal of Soil and Water Conservation* 53(3):219–224.

Mundt, C.C. 2002. Use of multiline cultivars and cultivar mixtures for disease management. *Annual Review of Phytopathology* 40:381–410.

Munoz, A.R., M. Trevisan, and E. Capri. 2009. Sorption and photodegradation of chlorpyrifos on riparian and aquatic macrophytes. *Journal of Environmental Science and Health Part B–Pesticides Food Contaminants and Agricultural Wastes* 44(1):7–12.

Murphy, D.V., E.A. Stockdale, P.R. Poulton, T.W. Willison, and K.W.T. Goulding. 2007. Seasonal dynamics of carbon and nitrogen pools and fluxes under continuous arable and ley-arable rotations in a temperate environment. *European Journal of Soil Science* 58(6):1410–1424.

Murphy, K., D. Lammer, S. Lyon, B. Carter, and S.S. Jones. 2005. Breeding for organic and low-input farming systems: An evolutionary-participatory breeding method for inbred cereal grains. *Renewable Agriculture and Food Systems* 20(1):48–55.

Murphy, S.D., D.R. Clements, S. Belaoussoff, P.G. Kevan, and C.J. Swanton. 2006. Promotion of weed species diversity and reduction of weed seedbanks with conservation tillage and crop rotation. *Weed Science* 54(1):69–77.

Mustafa, A., M. Scholz, R. Harrington, and P. Carroll. 2009. Long-term performance of a representative integrated constructed wetland treating farmyard runoff. *Ecological Engineering* 35(5):779–790.

Mutch, D.R., T.E. Martin, and K.R. Kosola. 2003. Red clover (*Trifolium pratense*) suppression of common ragweed (*Ambrosia artemisiifolia*) in winter wheat (*Triticum aestivum*). *Weed Technology* 17(1):181–185.

Nahm, K.H. 2000. A strategy to solve environmental concerns caused by poultry production. *Worlds Poultry Science Journal* 56(4):379–388.

Narwal, S.S. 2005. Allelopathy in sustainable agriculture. In Allelopathy: A Physiological Process with Ecological Implications, M.J. Reigosa, N. Pedrol, and L. Gonzalez, eds. Dordrecht, the Netherlands: Springer.

National Sustainable Agriculture Information Service (ATTRA). 2006. Anaerobic digestion of animal wastes: factors to consider. Available at http://www.attra.ncat.org. Accessed on April 7, 2010.

Neuenschwander, P., C. Borgemeister, and J. Langewald, eds. 2003. Biological Control in IPM Systems in Africa. Wallingford, UK: CAB International.

Ngouajio, M., M.E. McGiffen, and C.M. Hutchinson. 2003. Effect of cover crop and management system on weed populations in lettuce. *Crop Protection* 22(1):57–64.

Ni, H.W., and C.X. Zhang. 2005. Use of allelopathy for weed management in China—a review. *Allelopathy Journal* 15:3–12.

Nicholls, C.I., M. Parrella, and M.A. Altieri. 2001. The effects of a vegetational corridor on the abundance and dispersal of insect biodiversity within a northern California organic vineyard. *Landscape Ecology* 16(2):133–146.

Nixon, S.W., and V. Lee. 1986. Wetlands and Water Quality: A Regional Review of Recent Research in the United States on the Role of Freshwater and Saltwater Wetlands as Sources, Sinks, and Transformers of Nitrogen, Phosphorus, and Various Heavy Metals. Vicksburg, Miss.: Army Corps of Engineers, Waterways Experiment Station.

Noble, R., A. Dobrovin-Pennington, C. Wright, P.J. Hobbs, and J. Williams. 2009. Aerating recycled water on mushroom composting sites affects its chemical analysis and the characteristics of odor emissions. *Journal of Environmental Quality* 38(4):1493–1500.

Norris, R.E. 2005. Ecological bases of interactions between weeds and organisms in other pest categories. *Weed Science* 53(6):909–913.

Noss, R. 1990. Indicators for monitoring biodiversity—a hierarchial approach. *Conservation Biology* 4(4):355–364.

Novak, J.M., P.G. Hunt, K.C. Stone, D.W. Watts, and M.H. Johnson. 2002. Riparian zone impact on phosphorus movement to a Coastal Plain black water stream. *Journal of Soil and Water Conservation* 57(3):127–133.

NRC (National Research Council). 1980. Mineral Tolerance of Domestic Animals. Washington, D.C.: National Academy Press.

———. 1989. Alternative Agriculture. Washington, D.C.: National Academy Press.

———. 1992. Conserving Biodiversity—A Research Agenda for Developing Agencies. Washington, D.C.: National Academy Press.

———. 1993. Soil and Water Quality. Washington, D.C.: National Academy Press.
———. 1994. Nutrient Requirements of Poultry: Ninth revised edition. Washington, D.C.: National Academy Press.
———. 1995. Wetlands: Characteristics and Boundaries. Washington, D.C. National Academy Press.
———. 1997. Precision Agriculture in the 21st Century: Geospatial and Information Technologies in Crop Management. Washington, D.C.: National Academy Press.
———. 1998. Nutrient Requirements of Swine: 10th revised edition. Washington, D.C.: National Academy Press.
———. 2000a. The Future Role of Pesticides in U.S. Agriculture. Washington, D.C.: National Academy Press.
———. 2000b. Nutrient Requirements of Beef Cattle: Seventh revised edition. Washington, D.C.: National Academy Press.
———. 2001. Nutrient Requirements of Dairy Cattle: Seventh revised edition. Washington, D.C.: National Academy Press.
———. 2003. Air Emissions from Animal Feeding Operations: Current Knowledge, Future Needs. Washington, D.C.: National Academies Press.
———. 2007. Status of Pollinators in North America. Washington, D.C.: National Academies Press.
———. 2008a. Emerging Technologies to Benefit Farmers in Sub-Saharan Africa and South Asia. Washington, D.C.: National Academies Press.
———. 2008b. Prospects for Managed Underground Storage of Recoverable Water. Washington, D.C.: National Academies Press.
NRI (Natural Resources Institute). 2003. National Resources Inventory, 2003 Annual NRI Soil Erosion. Washington, D.C.: USDA National Resources Conservation Service.
O'Callaghan, M., T.R. Glare, E.P.J. Burgess, and L.A. Malone. 2005. Effects of plants genetically modified for insect resistance on nontarget organisms. *Annual Review of Entomology* 50:271–292.
O'Neal, M.E., M.E. Gray, and C.A. Smyth. 1999. Population characteristics of a western corn rootworm (Coleoptera : Chrysomelidae) strain in east-central Illinois corn and soybean fields. *Journal of Economic Entomology* 92(6):1301–1310.
Oenema, O., H. Kros, and W. de Vries. 2003. Approaches and uncertainties in nutrient budgets: implications for nutrient management and environmental policies. *European Journal of Agronomy* 20(1–2):3–16.
Olfs, H.W., K. Blankenau, F. Brentrup, J. Jasper, A. Link, and J. Lammel. 2005. Soil- and plant-based nitrogen-fertilizer recommendations in arable farming. *Journal of Plant Nutrition and Soil Science—Zeitschrift Fur Pflanzenernahrung Und Bodenkunde* 168(4):414–431.
Olsen, J., L. Kristensen, and J. Weiner. 2005a. Effects of density and spatial pattern of winter wheat on suppression of different weed species. *Weed Science* 53(5):690–694.
Olsen, J., L. Kristensen, J. Weiner, and H.W. Griepentrog. 2005b. Increased density and spatial uniformity increase weed suppression by spring wheat. *Weed Research* 45(4):316–321.
Ondersteijn, C.J.M., A.C.G. Beldman, C.H.G. Daatselaar, G.W.J. Giesen, and R.B.M. Huirne. 2002. The Dutch mineral accounting system and the European nitrate directive: implications for N and P management and farm performance. *Agriculture Ecosystems & Environment* 92(2–3):283–296.
Osborne, L.L., and D.A. Kovacic. 1993. Riparian vegetated buffer strips in water-quality restoration and stream management. *Freshwater Biology* 29(2):243–258.
Osborne, S.L. 2007. Utilization of existing technology to evaluate spring wheat growth and nitrogen nutrition in South Dakota. *Communications in Soil Science and Plant Analysis* 38(7–8):949–958.
Padgitt, M., D. Newton, R. Penn, and C. Sandretto. 2000. Production Practices for Major Crops in U.S. Agriculture, 1990–97. Washington, D.C.: U.S. Department of Agriculture Economic Research Service.
Painter, K. 1991. Does sustainable farming pay: a case study. *Journal of Sustainable Agriculture* 1:37–48.
Pala, M., J. Ryan, H. Zhang, M. Singh, and H.C. Harris. 2007. Water-use efficiency of wheat-based rotation systems in a Mediterranean environment. *Agricultural Water Management* 93(3):136–144.
Patanen, K.H., and Z. Mroz. 1999. Organic acids for performance enhancement in pig diets. *Nutrition Research Reviews* 12:117–145.
Peet, M. 2008. Conservation tillage methods. Available at http://www.ncsu.edu/sustainable/tillage/cons_til.html. Accessed on August 12, 2009.
Peigne, J., and P. Girardin. 2004. Environmental impacts of farm-scale composting practices. *Water Air and Soil Pollution* 153(1–4):45–68.
Peigne, J., B.C. Ball, J. Roger-Estrade, and C. David. 2007. Is conservation tillage suitable for organic farming? A review. *Soil Use and Management* 23(2):129–144.
Perez-de-Luque, A., and D. Rubiales. 2009. Nanotechnology for parasitic plant control. *Pest Management Science* 65(5):540–545.

Perfecto, I., J.H. Vandermeer, G.L. Bautista, G.I. Nunez, R. Greenberg, P. Bichier, and S. Langridge. 2004. Greater predation in shaded coffee farms: the role of resident neotropical birds. *Ecology* 85(10):2677–2681.

Pesant, A.R., J. L. Dionne, and J. Genest. 1987. Soil and nutrient losses in surface runoff from conventional and no-till corn systems. *Canadian Journal of Soil Science* 67:835–843.

Petersen, J., R. Belz, F. Walker, and K. Hurle. 2001. Weed suppression by release of isothiocyanates from turnip-rape mulch. *Agronomy Journal* 93:37–43.

Petersen, S.O., S.G. Sommer, F. Beline, C. Burton, J. Dach, J.Y. Dourmad, A. Leip, T. Misselbrook, F. Nicholson, H.D. Poulsen, G. Provolo, P. Sorensen, B. Vinneras, A. Weiske, M.P. Bernal, R. Bohm, C. Juhasz, and R. Mihelic. 2007. Recycling of livestock manure in a whole-farm perspective. *Livestock Science* 112(3):180–191.

Petrolia, D.R., and P.H. Gowda. 2006. Missing the boat: Midwest farm drainage and Gulf of Mexico hypoxia. *Review of Agricultural Economics* 28 (2):240–253.

Pettigrew, J.E. 2006. Reduced use of antibiotic growth promoters in diets fed to weanling pigs: dietary tools, part 1. *Animal Biotechnology* 17(2): 207–215.

Phillips, S.L., and M.S. Wolfe. 2005. Evolutionary plant breeding for low input systems. *Journal of Agricultural Science* 143:245–254.

Pilson, D., and H.R. Prendeville. 2004. Ecological effects of transgenic crops and the escape of transgenes into wild populations. *Annual Review of Ecology Evolution and Systematics* 35:149–174.

Plaster, E.J. 2009. Soil Science and Management. 5th edition. Clifton Park, N.Y.: Delmar.

Platt, J.O., J.S. Caldwell, and L.T. Kok. 1999. Effect of buckwheat as a flowering border on populations of cucumber beetles and their natural enemies in cucumber and squash. *Crop Protection* 18(5):305–313.

Plumstead, P.W., H. Romero-Sanchez, R.O. Maguire, A.G. Gernat, and J. Brake. 2007. Effects of phosphorus level and phytase in broiler breeder rearing and laying diets on live performance and phosphorus excretion. *Poultry Science* 86(2):225–231.

Poggio, S.L. 2005. Structure of weed communities occurring in monoculture and intercropping of field pea and barley. *Agriculture Ecosystems & Environment* 109(1–2):48–58.

Pope, R.O., and D.E. Stoltenberg. 1991. A Review of Literature Related to Vegetative Filter Strips. Ames: Iowa State University Press.

Porter, P.M., D.R. Huggins, C.A. Perillo, S.R. Quiring, and R.K. Crookston. 2003. Organic and other management strategies with two- and four-year crop rotations in Minnesota. *Agronomy Journal* 95(2):233–244.

Power, A.G. 1987. Plant community diversity, herbivore movement, and an insect-transmitted disease of maize. *Ecology* 68:1658–1669.

Power, A.G. 1990. Cropping systems, insect movement, and the spread of insect-transmitted diseases in crops. In Agroecology: Researching the Ecological Basis for Sustainable Agriculture, S.R. Gliessman, ed. New York: Springer-Verlag.

Powers, W., and R. Angel. 2008. A review of the capacity for nutritional strategies to address environmental challenges in poultry production. *Poultry Science* 87(10):1929–1938.

Pridham, J.C., and M.H. Entz. 2008. Intercropping spring wheat with cereal grains, legumes, and oilseeds fails to improve productivity under organic management. *Agronomy Journal* 100(5):1436–1442.

Pridham, J.C., M.H. Entz, R.C. Martin, and R.J. Hucl. 2007. Weed, disease and grain yield effects of cultivar mixtures in organically managed spring wheat. *Canadian Journal of Plant Science* 87(4):855–859.

Pullaro, T.C., P.C. Marino, D.M. Jackson, H.F. Harrison, and A.P. Keinath. 2006. Effects of killed cover crop mulch on weeds, weed seeds, and herbivores. *Agriculture Ecosystems & Environment* 115(1–4):97–104.

Pyrowolakis, A., A. Westphal, R.A. Sikora, and J.O. Becker. 2002. Identification of root-knot nematode suppressive soils. *Applied Soil Ecology* 19(1):51–56.

Pywell, R.F., K.L. James, I. Herbert, W.R. Meek, C. Carvell, D. Bell, and T.H. Sparks. 2005. Determinants of overwintering habitat quality for beetles and spiders on arable farmland. *Biological Conservation* 123(1):79–90.

Quincke, J.A., C.S. Wortmann, M. Mamo, T. Franti, R.A. Drijber, and J.P. García. 2007. One-time tillage of no-till systems: soil physical properties, phosphorus runoff, and crop yield. *Agronomy Journal* 99:1104–1110.

Rasby, R. 2007. Beef cattle production—frequently asked questions. Available at http://beef.unl.edu/FAQ/2007 07291.shtml. Accessed on February 2, 2010.

Rauw, W.M., E. Kanis, E.N. Noordhuizen-Stassen, and F.J. Grommers. 1998. Undesirable side effects of selection for high production efficiency in farm animals: a review. *Livestock Production Science* 56(1):15–33.

Rector, B.G. 2008. Molecular biology approaches to control of intractable weeds: new strategies and complements to existing biological practices. *Plant Science* 175(4):437–448.

Reddy, K.R., and D.A. Graetz. 1988. Carbon and nitrogen dynamics in wetland soils. In The Ecology and Management of Wetlands, D.D. Hook, ed. Portland, Ore.: Timber Press.

Reddy, K.R., R.H. Kadlec, E. Flag, and P.M. Gale. 1999. Phosphorus retention in streams and wetlands: a review. *Critical Reviews in Environmental Science and Technology* 29:83–146.

Reddy, K.R., R.G. Wetzel, and R.H. Kadlec. 2005. Biogeochemistry of phosphorus in wetlands. In Phosphorous: Agriculture and the Environment. Madison, Wis.: American Society of Agronomy, Crop Science Society of America, and Soil Science Society of America.

Reeves, D.W., A.J. Price, and M.G. Patterson. 2005. Evaluation of three winter cereals for weed control in conservation-tillage nontransgenic cotton. *Weed Technology* 19(3):731–736.

Reganold, J.P., L.F. Elliott, and Y.L. Unger. 1987. Long-term effects of organic and conventional farming on soil erosion. *Nature* 330:370–372.

Reich, P.B., D. Tilman, J. Craine, D. Ellsworth, M.G. Tjoelker, J. Knops, D. Wedin, S. Naeem, D. Bahauddin, J. Goth, W. Bengtson, and T.D. Lee. 2001. Do species and functional groups differ in acquisition and use of C, N and water under varying atmospheric CO2 and N availability regimes? A field test with 16 grassland species. *New Phytologist* 150(2):435–448.

Reichenberger, S., M. Bach, A. Skitschak, and H.G. Frede. 2007. Mitigation strategies to reduce pesticide inputs into ground- and surface water and their effectiveness; a review. *Science of the Total Environment* 384(1–3):1–35.

Renema, R.A., M.E. Rustad, and F.E. Robinson. 2007. Implications of changes to commercial broiler and broiler breeder body weight targets over the past 30 years. *Worlds Poultry Science Journal* 63(3):457–472.

Rhoades, R.E., and V. Nazarea. 1999. Local management of biodiversity in traditional agroecosystems. In Biodiversity in Agroecosystems, W.W. Collins and C.O. Qualset, eds. New York: CRC Press.

Richardson, C.J. 1999. The role of wetlands in storage, release, and cycling of phosphorus on the landscape: a 25-year retrospective. In Phosphorus Biogeochemistry in Sub-Tropical Ecosystems, K.R. Reddy, ed. Boca Raton, Fla.: CRC Press/Lewis Publishers.

Ricke, S.C. 2003. Perspectives on the use of organic acids and short chain fatty acids as antimicrobials. *Poultry Science* 82:632–639.

Ricketts, T.H., J. Regetz, I. Steffan-Dewenter, S.A. Cunningham, C. Kremen, A. Bogdanski, B. Gemmill-Herren, S.S. Greenleaf, A.M. Klein, M.M. Mayfield, L.A. Morandin, A. Ochieng, and B.F. Viana. 2008. Landscape effects on crop pollination services: are there general patterns? *Ecology Letters* 11(5):499–515.

Riedell, W.E., J.L. Pikul, A.A. Jaradat, and T.E. Schumacher. 2009. Crop rotation and nitrogen input effects on soil fertility, maize mineral nutrition, yield, and seed composition. *Agronomy Journal* 101(4):870–879.

Roberson, E.B., S. Sarig, C. Shennan, and M.K. Firestone. 1995. Nutritional management of microbial polysaccharide production and aggregation in an agricultural soil. *Soil Science Society of America Journal* 59(6):1587–1594.

Robert, P.C. 2002. Precision agriculture: a challenge for crop nutrition management. *Plant and Soil* 247(1):143–149.

Robertson, G.P., E.A. Paul, and R.R. Harwood. 2000. Greenhouse gases in intensive agriculture: contributions of individual gases to the radiative forcing of the atmosphere. *Science* 289(5486):1922–1925.

Roda, A.L., D.A. Landis, M.L. Coggins, E. Spandl, and O.B. Hesterman. 1996. Forage grasses decrease alfalfa weevil (Coleoptera: Curculionidae) damage and larval numbers in alfalfa-grass intercrops. *Journal of Economic Entomology* 89(3):743–750.

Rodríguez-kábana, R. and G.H. Canullo. 1992. Cropping systems for the management of phytonematodes. *Phytoparasitica* 20(3):211–224.

Rothrock, C.S., T.L. Kirkpatrick, R.E. Frans, and H.D. Scott. 1995. The influence of winter legume cover crops on soilborne plant-pathogens and cotton seedling diseases. *Plant Disease* 79(2):167–171.

Rundlof, M., H. Nilsson, and H.G. Smith. 2008. Interacting effects of farming practice and landscape context on bumblebees. *Biological Conservation* 141(2):417–426.

Sabaratnam, S., and J.A. Traquair. 2002. Formulation of a *Streptomyces* biocontrol agent for the suppression of *Rhizoctonia* damping-off in tomato transplants. *Biological Control* 23(3):245–253.

Salmon, S., L. Frizzera, and S. Camaret. 2008. Linking forest dynamics to richness and assemblage of soil zoological groups and to soil mineralization processes. *Forest Ecology and Management* 256(9):1612–1623.

Sandilands, V., B.J. Tolkamp, and I. Kyriazakis. 2005. Behaviour of food restricted broilers during rearing and lay—effects of an alternative feeding method. *Physiology & Behavior* 85(2):115–123.

Sandretto, C., and J. Payne. 2006. Chapter 4.2. Soil Management and Conservation. In Agricultural Resources and Environmental Indicators, 2006 Edition. Washington, D.C.: U.S. Department of Agriculture Economic Research Service.

Sarandon, S.J., and R. Sarandon. 1995. Mixture of cultivars: pilot field trial of an ecological alternative to improve production or quality of wheat (*Triticum-aestivum*). *Journal of Applied Ecology* 32(2):288–294.

Sato, K., P.C. Bartlett, R.J. Erskine, and J.B. Kaneene. 2005. A comparison of production and management between Wisconsin organic and conventional dairy herds. *Livestock Production Science* 93(2):105–115.

SCARM (Standing Committee of Agricultural and Resource Management Working Group). 2000. Review of layer hen housing and labelling of eggs in Australia. Available at http://www.daff.gov.au/__data/assets/pdf_file/0009/146709/synopsis.pdf. Accessed on September 3, 2009.

Schaible, G., and M. Aillery. 2006. Irrigation water management. In Agricultural Resources and Environmental Indicators, 2006 Edition/EIB-16. Washington, D.C.: U.S. Department of Agriculture Economic Research Service.

Schillinger, W.F., A.C. Kennedy, and D.L. Young. 2007. Eight years of annual no-till cropping in Washington's winter wheat-summer fallow region. *Agriculture Ecosystems & Environment* 120(2–4):345–358.

Schneider, R.W., ed. 1982. Suppressive Soils and Plant Disease. St. Paul, Minn.: The American Phytopathological Society.

Schonbeck, M. 2009. What is "organic no-till," and is it practical? Available at http://www.extension.org/article/18526. Accessed on August 12, 2009.

Schrezenmeir, J., and M. de Vrese. 2001. Probiotics, prebiotics, and synbiotics—approaching a definition. *American Journal of Clinical Nutrition* 73(2)361S–364S.

Schroder, J.J., J.J. Neeteson, O. Oenema, and P.C. Struik. 2000. Does the crop or the soil indicate how to save nitrogen in maize production? Reviewing the state of the art. *Field Crops Research* 66(2):151–164.

Schroeder, J.B., S.T. Ratcliffe, and M.E. Gray. 2005. Effect of four cropping systems on variant western corn rootworm (Coleoptera : Chrysomelidae) adult and egg densities and subsequent larval injury in rotated maize. *Journal of Economic Entomology* 98(5):1587–1593.

Schultz, R.C., T.M. Isenhart, W.W. Simpkins, and J.P. Colletti. 2004. Riparian forest buffers in agroecosystems—lessons learned from the Bear Creek Watershed, Central Iowa, USA. *Agroforestry Systems* 61(1):35–50.

Seitzinger, S.P. 1988. Denitrification in freshwater and coastal marine systems: ecological and geochemical significance. *Limnology and Oceanography* 33:702–724.

SERA-17. 2008. SERA-17: Organization to minimize phosphorus losses from agriculture. Available at http://www.sera17.ext.vt.edu/index.htm. Accessed on September 2, 2009.

Sharifi, M., B.J. Zebarth, G.A. Porter, D.L. Burton, and C.A. Grant. 2009. Soil mineralizable nitrogen and soil nitrogen supply under two-year potato rotations. *Plant and Soil* 320(1–2):267–279.

Sharpley, A.N., and S.J. Smith. 1991. Effects of cover crops on surface water quality. In Cover Crops for Clean Water: Proceedings from International Conference, W.L. Hargrove, ed. Ankeny, Iowa: Soil and Water Conservation Society of America.

Sharratt, B.S., and D. Lauer. 2006. Particulate matter concentration and air quality affected by windblown dust in the Columbia plateau. *Journal of Environmental Quality* 35:2011–2016.

Shennan, C. 2008. Biotic interactions, ecological knowledge and agriculture. *Philosophical Transactions of the Royal Society B: Biological Sciences* 363:717–739.

Shennan, C., C.L. Cecchettini, G.B. Goldman, and F.G. Zalom. 2001. Profiles of California farmers by degree of IPM use as indicated by self-descriptions in a phone survey. *Agriculture, Ecosystems and Environment* 84:267–275.

Shennan, C., S.R. Grattan, D.M. May, C.J. Hillhouse, D.P. Schachtman, M. Wander, B. Roberts, R.G. Burau, C. McNeish, L. Zelinski, and S. Tafoya. 1995. Feasibility of cyclic reuse of saline drainage in a tomato-cotton rotation. *Journal Environmental Quality* 24:476–486.

Shipitalo, M.J., W.A. Dick, and W.M. Edwards. 2000. Conservation tillage and macropore factors that affect water movement and the fate of chemicals. *Soil & Tillage Research* 53(3–4):167–183.

Shock, C.C., and E.B.G. Feibert. 2002. Deficit irrigation of potato. In Deficit Irrigation Practices. Rome: Food and Agriculture Organization of the United Nations.

Sibley, K.J., T. Astatkie, G. Brewster, P.C. Struik, J.F. Adsett, and K. Pruski. 2009. Field-scale validation of an automated soil nitrate extraction and measurement system. *Precision Agriculture* 10(2):162–174.

Singer, R.S., L.A. Cox, J.S. Dickson, H.S. Hurd, I. Phillips, and G.Y. Miller. 2007. Modeling the relationship between food animal health and human foodborne illness. *Preventive Veterinary Medicine* 79(2–4):186–203.

Sitthaphanit, S., V. Limpinuntana, B. Toomsan, S. Panchaban, and R.W. Bell. 2009. Fertiliser strategies for improved nutrient use efficiency on sandy soils in high rainfall regimes. *Nutrient Cycling in Agroecosystems* 85(2):123–139.

Skaggs, R.W., and G.M. Chescheir. 2003. Effects of subsurface drain depth on nitrogen losses from drained lands. *Transactions of the ASABE* 46(2):237–244.

Skaggs, R.W., and J.W. Gilliam. 1981. Effects of drainage system design and operation on nitrate transport. *Transactions of the ASABE* 24(1):934–940.

Smith, D. 2008. GPS sparks big changes. Available at http://www.allbusiness.com/agriculture-forestry/agriculture-crop-production-oilseed/10597000-1.html. Accessed on April 7, 2010.

Smith, D.R., P.A. Moore, B.E. Haggard, C.V. Maxwell, T.C. Daniel, K. VanDevander, and M.E. Davis. 2004. Effect of aluminum chloride and dietary phytase on relative ammonia losses from swine manure. *Journal of Animal Science* 82(2):605–611.

Smith, M.D., P.J. Barbour, L.W. Burger, and S.J. Dinsmore. 2005. Density and diversity of overwintering birds in managed field borders in Mississippi. *Wilson Bulletin* 117(3):258–269.

Smith, P., D. Martino, Z. Cai, D. Gwary, H. Janzen, P. Kumar, B. McCarl, S. Ogle, F. O'Mara, C. Rice, B. Scholes, O. Sirotenko, M. Howden, T. McAllister, G. Pan, V. Romanenkov, U. Schneider, S. Towprayoon, M. Wattenbach, and J. Smith. 2008. Greenhouse gas mitigation in agriculture. *Philosophical Transactions of the Royal Society B-Biological Sciences* 363(1492):789–813.

Smith, R.G., K.L. Gross, and G.P. Robertson. 2008. Effects of crop diversity on agroecosystem function: crop yield response. *Ecosystems* 11(3):355–366.

Smithson, J.B., and J.M. Lenne. 1996. Varietal mixtures: a viable strategy for sustainable productivity in subsistence agriculture. *Annals of Applied Biology* 128(1):127–158.

Smulders, D., G. Verbeke, P. Mormede, and R. Geers. 2006. Validation of a behavioral observation tool to assess pig welfare. *Physiology & Behavior* 89(3):438–447.

Snapp, S.S., S.M. Swinton, R. Labarta, D. Mutch, J.R. Black, R. Leep, J. Nyiraneza, and K. O'Neil. 2005. Evaluating cover crops for benefits, costs and performance within cropping system niches. *Agronomy Journal* 97(1):322–332.

Snyder, N.J., S. Mostaghimi, D.F. Berry, R.B. Reneau, S. Hong, P.W. McClellan, and E.P. Smith. 1998. Impact of riparian forest buffers on agricultural nonpoint source pollution. *Journal of the American Water Resources Association* 34(2):385–395.

Soliman, K.M., and R.W. Allard. 1991. Grain-yield of composite cross populations of barley—effects of natural selection. *Crop Science* 31(3):705–708.

Solley, W.B., R.R. Pierce, and H.A. Perlman. 1998. Estimated use of water in the United States in 1995: U.S. Geological Survey Circular 1200. Available at http://water.usgs.gov/watuse/pdf1995/pdf/circular1200.pdf. Accessed on January 30, 2010.

Sommerfeldt, T.G., C. Change, and T. Entz. 1988. Long-term annual manure applications increase soil organic matter and nitrogen, and decrease carbon to nitrogen ratio. *Soil Science Society of America Journal* 52:1668–1672.

Sonmez, O., G.M. Pierzynski, L. Frees, B. Davis, D. Leikam, D.W. Sweeney, and K.A. Janssen. 2009. A field-based assessment tool for phosphorus losses in runoff in Kansas. *Journal of Soil and Water Conservation* 64(3):212–222.

Spalding, R.F., and M.E. Exner. 1993. Occurrence of nitrate in groundwater—a review. *Journal of Environmental Quality* 22:392–402.

Staley, J.T., A. Stewart-Jones, G.M. Poppy, S.R. Leather, and D.J. Wright. 2009. Fertilizer affects the behaviour and performance of *Plutella xylostella* on brassicas. *Agricultural and Forest Entomology* 11(3):275–282.

Stanger, T.F., and J.G. Lauer. 2008. Corn grain yield response to crop rotation and nitrogen over 35 years. *Agronomy Journal* 100(3):643–650.

Stanger, T.F., J.G. Lauer, and J.P. Chavas. 2008. The profitability and risk of long-term cropping systems featuring different rotations and nitrogen rates. *Agronomy Journal* 100(1):105–113

Star, L., E.D. Ellen, K. Uitdehaag, and F.W.A. Brom. 2008. A plea to implement robustness into a breeding goal: Poultry as an example. *Journal of Agricultural & Environmental Ethics* 21(2):109–125.

State of California. 2008. California Irrigation Management Information System (CIMIS). Available at http://www.cimis.water.ca.gov/cimis/welcome.jsp. Accessed on August 12, 2009.

Steenwerth, K., and K.M. Belina. 2008. Cover crops enhance soil organic matter, carbon dynamics and microbiological function in a vineyard agroecosystem. *Applied Soil Ecology* 40(2):359–369.

Steinhilber, P., and J. Salak. 2006. Nutrient management—sampling manure for nutrient content. College Park: Maryland Cooperative Extension. Available at http://anmp.umd.edu/Pubs/NM-6.pdf. Accessed on February 9, 2010.

Stoddard, C.S., J.H. Grove, M.S. Coyne, and W.O. Thom. 2005. Fertilizer, tillage, and dairy manure contributions to nitrate and herbicide leaching. *Journal of Environmental Quality* 34(4):1354–1362.

Strock, J.S., P.M. Porter, and M.P. Russelle. 2004. Cover cropping to reduce nitrate loss through subsurface drainage in the northern U.S. Corn Belt. *Journal of Environmental Quality* 33:1010–1016.

Stroppiana, D., M. Boschetti, P.A. Brivio, and S. Bocchi. 2009. Plant nitrogen concentration in paddy rice from field canopy hyperspectral radiometry. *Field Crops Research* 111(1–2):119–129.

Sturz, A.V., B.R. Christie, and J. Nowak. 2000. Bacterial endophytes: potential role in developing sustainable systems of crop production. *Critical Reviews in Plant Sciences* 19(1):1–30.

Stute, J.K., and J.L. Posner. 1993. Legume cover crop options for grain rotations in Wisconsin. *Agronomy Journal* 85(6):1128–1132.

Sullivan, P. 2003. Overview of cover crops and green manures. Available at http://attra.ncat.org/attra-pub/PDF/covercrop.pdf. Accessed on June 24, 2009.

———. 2004. ATTRA, National Sustainable Agriculture Information Service. Available at http://www.attra.ncat.org/attra-pub/PDF/soilmgmt.pdf. Accessed on September 16, 2008.

Sullivan, T.J., J.A. Moore, D.R. Thomas, E. Mallery, K.U. Snyder, M. Wustenberg, J. Wustenberg, S.D. Mackey, and D.L. Moore. 2007. Efficacy of vegetated buffers in preventing transport of fecal coliform bacteria from pasturelands. *Environmental Management* 40(6):958–965.

Sweeney, D.W., and J.L. Moyer. 1994. Legume green manures and conservation tillage for grain–sorghum production on prairie soil. *Soil Science Society of America Journal* 58(5):1518–1524.

Swezey, S.L., D.J. Nieto, and J.A. Bryer. 2007. Control of western tarnished plant bug *Lygus hesperus* knight (Hemiptera : Miridae) in California organic strawberries using alfalfa trap crops and tractor-mounted vacuums. *Environmental Entomology* 36(6):1457–1465.

Swift, M., A.M. Izac, and M. Noordwijk. 2004. Biodiversity and ecosystem services in agricultural landscapes—are we asking the right questions? *Agriculture Ecosystems and Environment* 104:113–134.

Swink, S.N., Q.M. Ketterings, L.E. Chase, K.J. Czymmek, and J.C. Mekken. 2009. Past and future phosphorus balances for agricultural cropland in New York State. *Journal of Soil and Water Conservation* 64(2):120–133.

Szumigalski, A., and R. Van Acker. 2005. Weed suppression and crop production in annual intercrops. *Weed Science* 53(6):813–825.

Teasdale, J.R. 1998. Cover crops, smother plants, and weed management. In Integrated Weed and Soil Management, J.L. Hatfield, D.D. Buhler, and B.A. Stewart, eds. Chelsea, Mich.: Ann Arbor Press.

The Great Lakes Regional Water Program. 2009. The Great Lakes Regional Water Program. Available at http://www.uwex.edu/ces/regionalwaterquality/. Accessed on September 3, 2009.

Thies, C., and T. Tscharntke. 1999. Landscape structure and biological control in agroecosystems. *Science* 285(5429): 893–895.

Thomas, S.H., J. Schroeder, and L.W. Murray. 2005. The role of weeds in nematode management. *Weed Science* 53:923–928.

Thorp, K.R., and L.F. Tian. 2004. Performance study of variable-rate herbicide applications based on remote sensing imagery. *Biosystems Engineering* 88(1):35–47.

Thrupp, L.A. 1997. Linking Biodiversity and Agriculture: Challenges and Opportunities for Sustainable Food Security. Washington, D.C.: World Resources Institute.

———. 1998. Cultivating Diversity: Agrobiodiversity and Food Security. Washington, D.C.: World Resources Institute.

Tilman, D., J. Hill, and C. Lehman. 2006. Carbon-negative biofuels from low-input high-diversity grassland biomass. *Science* 314:1598–1600.

Timms-Wilson, T.M., K. Kilshaw, and M.J. Bailey. 2004. Risk assessment for engineered bacteria used in biocontrol of fungal disease in agricultural crops. *Plant and Soil* 266(1–2):57–67.

Tompkins, D.K., A.T. Wright, and D.B. Fowler. 1992. Foliar disease development in no-till winter wheat: influence of agronomic practices on powdery mildew development. *Canadian Journal of Plant Science* 72(3):965–972.

Tonitto, C., M.B. David, and L.E. Drinkwater. 2006. Replacing bare fallows with cover crops in fertilizer-intensive cropping systems: a meta-analysis of crop yield and N dynamics. *Agriculture Ecosystems & Environment* 112(1):58–72.

Torsvik, V., J. Goksøyr, F. L. Daae, R. Sørheim, J. Michalsen, and K. Salte. 1994. Use of DNA analysis to determine the diversity of microbial communities. Pp. 39–48 in Beyond the Biomass, K. Ritz, J. Dighton, and K. E. Giller, eds. London, UK: Wiley-Sayce Publications.

Traquair, J.A. 1995. Fungal biocontrol of root diseases—endomycorrhizal suppression of cylindrocarpon root-rot. *Canadian Journal of Botany* 73(S1):89–95.

Triplett, G.B., and W.A. Dick. 2008. No-tillage crop production: a revolution in agriculture! *Agronomy Journal* 100(3):S153–S165.

Troeh, F.R., and L.M. Thompson. 2005. Soils and Soil Fertility. Sixth edition. Ames, Iowa: Blackwell Publishing.

Tscharntke, T., R. Bommarco, Y. Clough, T.O. Crist, D. Kleijn, T.A. Rand, J.M. Tylianakis, S. van Nouhuys, and S. Vidal. 2008. Conservation biological control and enemy diversity on a landscape scale (Reprinted from *Biological Control* 43:294–309, 2007). *Biological Control* 45(2):238–253.

Tucker, M., S.R. Whaley, and J.S. Sharp. 2006. Consumer perceptions of food-related risks. *International Journal of Food Science and Technology* 41(2):135–146.

Underwood, W.J. 2002. Pain and distress in agricultural animals. *Journal of the American Veterinary Medical Association* 221(2):208–211.

UNDP (United Nations Development Program). 1995. Agroecology: Creating the Synergism for a Sustainable Agriculture. New York: Author.

Unger, P.W., and M.F. Vigil. 1998. Cover crop effects on soil water relationships. *Journal of Soil and Water Conservation* 53(3):200–207.

U.S. Congress, Office of Technology Assessment. 1995. Biologically Based Technologies for Pest Control. Washington, D.C., U.S. Government Printing Office.

USDA-APHIS (U.S. Department of Agriculture Animal and Plant Health Inspection Service). 2007. Dairy cattle identification practices in the United States, 2007. Available at http://nahms.aphis.usda.gov/dairy/dairy07/Dairy07_ID.pdf. Accessed on February 15, 2010.

———. 2009. Cattle identification practices on U.S. beef cow-calf operations. Available at http://www.aphis.usda.gov/vs/ceah/ncahs/nahms/beefcowcalf/beef0708/Beef0708_cattleID_infosheet.pdf. Accessed on February 15, 2010.

USDA-ARS (U.S. Department of Agriculture Agricultural Research Service). 2009. U.S. dairy forage research center. Available at http://ars.usda.gov/mwa/madison/dfrc. Accessed on September 3, 2009.

USDA-ERS (U.S. Department of Agriculture Economic Research Service). 1987. Farm Drainage in the United States: History, Status, and Prospects. Washington, D.C.: Author.

———. 2005. Agricultural chemicals and production technology—soil management. Available at http://www.ers.usda.gov/briefing/agchemicals/soilmangement.htm. Accessed on February 27, 2010.

———. 2008. U.S. consumption of nitrogen, phosphate, and potash, 1960–2007. Available at http://www.ers.usda.gov/Data/FertilizerUse/Tables/Table1.xls. Accessed on February 10, 2010.

———. 2009. Adoption of genetically engineered crops in the U.S.: extent of adoption. Available at http://www.ers.usda.gov/Data/BiotechCrops/adoption.htm. Accessed on August 6, 2009.

USDA-NASS (U.S. Department of Agriculture National Agricultural Statistics Service). 2010. Farm and Ranch Irrigation Survey (2008). Washington, D.C.: Author.

USDA-NIFA (U.S. Department of Agriculture National Institute of Food and Agriculture). 2009. Precision, geospatial, and sensor technologies—adoption of precision agriculture. Available at http://www.csrees.usda.gov/nea/ag_systems/in_focus/precision_if_adoption.html. Accessed on February 28, 2010.

USDA-NRCS (U.S. Department of Agriculture National Resource Conservation Service). 1997. National Irrigation Guide. East Lansing, Mich.: Author.

———. 2000. A report to congress on aging watershed infrastructure: an analysis and strategy for addressing the nation's aging flood control dams. Available at http://www.nrcs.usda.gov/Programs/WSRehab/wsr_main/AgingDams.pdf. Accessed on April 5, 2010.

———. 2006. Conservation resource brief: energy management. Available at http://www.nrcs.usda.gov/feature/outlook/Energy.pdf. Accessed on April 5, 2010.

———. 2008a. Energy consumption awareness tool: tillage. Estimator. Available at http://ecat.sc.egov.usda.gov/. Accessed on April 5, 2010.

———. 2008b. Energy consumption awareness tool—irrigation. Estimator. Available at http://ipat.sc.egov.usda.gov/. Accessed on April 5, 2010.

———. 2008c. Watershed projects authorized by the Watershed Protection and Flood Protection Act (PL 83-566): helping communities solve natural resource issues. Available at http://www.watershedcoalition.org/Watershed%20Benefits%203%2008.pdf. Accessed on April 5, 2010.

———. 2009. National conservation practice standards—NHCP. Available at http://www.nrcs.usda.gov/technical/Standards/nhcp.html. Accessed on February 8, 2009.

USGS (U.S. Geological Survey). 2009a. National Water-Quality Assessment Program. Available at http://water.usgs.gov/nawqa/. Accessed on August 6, 2009.

———. 2009b. Some irrigation methods. Available at http://ga.water.usgs.gov/edu/irquicklook.html. Accessed on July 17, 2009.

Vallad, G.E., and R.M. Goodman. 2004. Systemic acquired resistance and induced systemic resistance in conventional agriculture. *Crop Science* 44(6):1920–1934.

Van Den Berg, H. and J. Jiggins. 2007. Investing in farmers—the impacts of farmer field schools in relation to integrated pest management. *World Development* 35:663–686.

Van Bueren, E.T.L., P.C. Struik, and E. Jacobsen. 2002. Ecological concepts in organic farming and their consequences for an organic crop ideotype. *Netherlands Journal of Agricultural Science* 50(1):1–26.

Vandehaar, M.J., and N. St. Pierre. 2006. Major advances in nutrition: relevance to the sustainability of the dairy industry. *Journal of Dairy Science* 89:1280–1291.

Vandermeer, J. 1995. The ecological basis of alternative agriculture. *Annual Review of Ecology and Systematics* 26:201–224.

van Dijk, A.J., H. Everts, M.J.A. Nabuurs, R. Margry, and A.C. Beynen. 2001. Growth performance of weanling pigs fed spray-dried animal plasma—a review. *Livestock Production Science* 68(2–3):263–274.

Vanek, S., H.C. Wien, and A. Rangarajan. 2005. Time of interseeding of lana vetch and winter rye cover strips determines competitive impact on pumpkins grown using organic practices. *HortScience* 40(6):1716–1722.

Vargas-Ayala, R., and R. Rodriguez-Kabana. 2001. Bioremediative management of soybean nematode population densities in crop rotations with velvetbean, cowpea, and winter crops. *Nematropica* 31(1):37–46.

Varvel, G.E. 1994. Rotation and nitrogen-fertilization effects on changes in soil carbon and nitrogen. *Agronomy Journal* 86(2):319–325.

Recognizing those complexities, this chapter's overview will focus on the objectives associated with economic security as noted above: that is, farm business security, including production, marketing, and other diversification strategies, and quality-of-life issues that pertain to the sustainability of the operations (Figure 4-1).

Economic Security at the Farm Level

Strategies to improve economic security at the farm level include reducing production costs, increasing the value of farm products, and diversifying income streams. This section first addresses economic viability of different practices and systems associated with improving environmental performance of agriculture. It then discusses strategies for marketing and diversification that can improve economic security. The committee recognizes that those economic aspects are often interrelated with nonfinancial dimensions, which can also affect the sustainability of the farm business and are explored in later sections.

When the report *Alternative Agriculture* (NRC, 1989) was written, there was considerable skepticism that emerging alternative production systems—for example, organic farming, integrated pest management, or nonconfinement livestock farming—could be economically competitive with the dominant conventional farming practices. Since then, numerous case studies, enterprise-level and farm-level models, and farm accounting datasets have demonstrated that it is possible to realize economic gains and sometimes gain competitive advantages from the use of those alternative systems and other related practices. However, such gains and advantages are not guaranteed. (See the Sustainable Agriculture Research and Education website at www.sare.org for examples.) Rather than presenting a comprehensive summary of all of the relevant studies, illustrative examples are provided in this chapter to highlight key factors that influence economic outcomes for farm businesses.

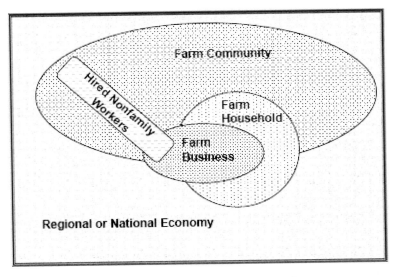

FIGURE 4-1 Levels of analysis for understanding economic security.

Economics of Production Practices That Can Improve Sustainability

This section highlights evidence related to the financial performance of some of the practices described in Chapter 3 for improving environmental sustainability—reduced-tillage systems, crop rotations, crop nutrient management strategies, and other conservation best management practices (BMPs). The committee notes that the financial performance of those practices depends not only on production costs, but also on the prices at which the products are sold. Production costs and prices are dynamic and depend on multiple factors including policies, market demand, and geographic location. Therefore, the illustrations of financial performance used by the committee are context dependent.

Conservation Tillage

As noted in Chapter 3, conservation tillage practices (including no-till and minimum tillage that disturbs 30 percent or less of the soil) have proven to be effective ways to reduce soil erosion. Conservation tillage has proven to be amenable to various scales of production, ranging from small to large operations, in a variety of crops. Huggins and Reganold (2008) noted the benefits and tradeoffs of no-till systems. The benefits include reduction of soil erosion, conservation of water, improvement of soil health, reduction in fuel and labor costs, reduction of sediment and fertilizer pollution in waterways, and sequestration of carbon. The tradeoffs are that transition to no-till from conventional tillage systems can be difficult, necessary equipment such as no-till seeders are expensive, no-till often increases reliance on herbicides, plants pests can shift in unexpected ways, more nitrogen fertilizer might be required initially, and increased ground cover might slow germination and reduce yields.

One of the main economic questions concerning no-till versus conventional tillage is whether the gains from reducing labor and fuel outweigh any reduction in yield. The economic results seem to vary by crop, region, and cropping system. Triplett and Dick (2008) surveyed several studies. In Iowa, conventional tillage had higher returns than no-till with continuous corn, but no-till systems were more profitable with a corn–soybean rotation. Yield stability was similar for the two systems. Similar results were obtained in another study conducted in Indiana and Ohio. In Mississippi, Martin and Hanks (2009) evaluated different types of tillage with crop rotations. Increasingly, farmers in that region are rotating cotton with corn. The highest net returns were found when cotton and corn rotations were combined with minimum tillage

In Washington State, under dryland conditions with low rainfall, a study noted the environmental advantages of the annual no-till rotation of winter wheat over the conventional tillage system with winter wheat and summer fallow in terms of the reduction in wind erosion and improvement in soil health (Schillinger et al., 2007). However, that study also found that the conventional tillage system was more profitable because of the lower yields associated with the no-till system. In terms of other small grains, Lankoski et al. (2006) found in Finland that no-till production of barley was more profitable than conventional tillage, but that conventional tillage systems were more profitable with wheat and oats. In Canada, Mohr et al. (2007) found that the highest returns for wheat–pea cropping systems were found on the "high soil disturbance seeding system" in the clay loam soil, but that the highest returns were found with the "low soil disturbance seeding system" in the loam soil.

There are other aspects to the economic decision to use no-till systems. Huggins and Reganold (2008) found that with the reduced labor associated with no-till systems, some farmers can almost double the acreage they farm with the same machinery complement. Others might pursue a better quality of life with their labor savings. Although yields might be reduced with no-till systems in the short term, the Food and Agriculture Organization (FAO) concluded that yields over time with the no-till systems are likely be higher because of the improved soil health (FAO, 2008).

Another important question is whether non-herbicide-based no-till or minimum tillage methods can be effective and economical in organic systems. Because the majority of no-till farming systems in the United States depend on herbicide inputs as a means of controlling weeds, some sustainability analysts (and organic advocates) perceive this dependence as both an ecological and economic concern. The use of organic methods and no-till agriculture is an emerging area of research. Both researchers and farmers in the United States and in other countries have developed organic methods for conservation tillage that do not incorporate the use of herbicides. Using mulches (such as straw or crop residues), putting a transparent plastic cover over soil to solarize it (Law et al., 2008), crimping, rolling or mowing weeds to reduce competition, and growing particular varieties of cover crops that can outcompete weeds are a few examples (Chase and Mbuya, 2008). The committee is not aware of any economic analyses on those practices.

Crop Rotations

Corn–soybean rotations have been shown repeatedly to have higher net returns than continuous corn rotations as a result of reduced production costs (less fertilizer and herbicide input), although tillage practices and management inputs can affect comparative net returns (Katsvairo and Cox, 2000). Moreover, corn–soybean rotations exhibited significantly less risk of serious income declines over a 14-year study period; part of the risk reduction was the result of diversification inherent in any rotation, although some came from positive yield interactions between the two crops. Olmstead and Brummer (2008) found that adding alfalfa to Iowa farmers' corn–soybean rotations can produce significant economic gains. They found that a simulated five-year rotation that included corn–soybean–oats/alfalfa–alfalfa–alfalfa would result in a 24 percent net income increase compared to a five-year rotation of corn–soybean–corn–soybean–corn, even if the row crops received farm support payments. However, they pointed out that commodity program incentives have served as a disincentive for producers to move toward forage crops in rotations. Zentner et al. (2002) found that including oilseed and pulse crops in rotations with grains contributed to higher and more stable net farm income in Canada.

Other studies have reported comparative economic disadvantages associated with some diversified crop rotations under current market conditions. For example, Kelly et al. (1996) simulated a range of tillage and crop rotation options in the Upper Midwest and estimated their economic and environmental impacts over a 30-year period. They found that no-till rotations provided the greatest estimated net economic returns, followed by a conventional corn–soybean rotation; net returns on the two cover crop rotations were lowest, although they generated significant environmental benefits. Jatoe et al. (2008) simulated the impact of introducing environmentally beneficial crop rotations into potato production systems in Canada and found that the most environmentally protective rotations required substantial reduction in gross margins to producers.

Cover Cropping

As with many farming practices for improving sustainability, the economic performance of cover cropping is difficult to quantify. A holistic assessment of the economic performance of cover cropping would include estimates of direct and indirect costs, cost savings provided by the practice, and increased income as a result of improved yield. Snapp et al. (2005) reviewed economic costs of cover cropping internal to farms. A direct cost of cover cropping is establishment. The establishment cost is particularly high for leguminous cover crops—up to 10 times higher than that for grasses—because of the high seed costs and the large amount of seed necessary for establishment. Indirect on-farm costs of cover crops include hindering establishment of the succeeding cash crop as a result of slow warming of soil or delayed nutrient mineralization and unanticipated cover crop management problems that reduce the expected benefits. Another cost of cover cropping is the opportunity cost of income foregone from cash crops (Snapp et al., 2005).

Despite the aforementioned direct, indirect, and opportunity costs, cover cropping can provide many benefits that often lead to cost savings and improved productivity, including weed suppression and improved pest control. Indirect cost savings—as a result of improved soil fertility and overall improved health of the cropping system—accumulate over time and are difficult to quantify. Long-term economic analyses of the benefits and costs of cover cropping might provide valuable information to farmers and encourage adoption of such practices (Snapp et al., 2005).

Crop Nutrient Management Strategies

Nutrient input for crops usually accounts for 30 percent or more of total variable costs of production (Lu et al., 2000). Using on-farm nutrient sources such as green manure and animal manure could reduce input costs, but crop productivity might be compromised to some extent. Gareau (2004) conducted a meta-analysis of 120 studies to examine the economic profitability of using synthetic fertilizer versus cover crop-based or animal manure-based fertility treatments. The analysis suggests that conventional systems using commercial fertilizers had higher profits than organic systems using cover crop-based or animal manure-based nutrient management for most grain crops. Nonetheless, cover crop-based and animal manure-based nutrient management systems hold promise if they are used in an organic system, partly because of the price premium (Gareau, 2004).

Conservation Best Management Practices

A nutrient management plan is designed to balance plant nutrient requirements with purchased and on-farm nutrient inputs. A plan provides several benefits, including creating an optimum nutrient climate for plant growth, improving water quality, and improving farm profits by reducing inorganic fertilizer purchases (Maryland Cooperative Extension, 2009). Steinhilber (1996) noted that by giving nutrient credits to a preceding alfalfa crop, a farmer can save about $15–$30/acre in fertilizer costs from carryover nitrogen. A farmer can save $15–$30/acre in reduced inorganic fertilizer costs by giving credit to manure either applied in the previous year or in the current year.

However, not all farmers benefit equally from nutrient management planning. A crop farmer without any previous legume crops or application of organic sources of nutrients and who is already conducting soil testing and following accepted fertilizer recommendations might not save any fertilizer costs from a nutrient management plan (Steinhilber, 1996). A survey of 487 Maryland farmers showed that there can be biases in nutrient management planning (Lawley et al., 2009). Nutrient management plans were adopted more frequently by larger farm operations with grain and livestock and less on environmentally sensitive land. Biases also depended on who wrote the plan. Independent crop consultants tended to recommend increases in fertilizer uses. By comparison, farmers who were educated and certified to write their own plans recommended decreases in their own fertilizer use.

Poultry and dairy farmers in Virginia who are implementing nutrient management plans based on nitrogen and phosphorus can experience significant reductions in financial returns (Yang et al., 2000). Poultry litter (manure), in particular, is high in phosphorus. Repeated applications of poultry litter or dairy manure to the soil can raise the phosphorus levels of soil and cause runoff of phosphorus and reduction in water quality. A nutrient management plan based on nitrogen and phosphorus will limit the applications of manure or litter to the land because the phosphorus levels are too high. A farmer can incur significant costs in either reducing the herd or flock size, transporting excess manure or litter to different locations, or purchasing inorganic fertilizer to meet nitrogen requirements that could have been supplied by the manure that was shipped elsewhere.

The precision of fertilizer recommendations associated with nutrient management plans is also affected by weather and other variables during the growing season. Rajsic and Weersink (2008) compared ex ante recommendations of nutrient management plans with ex post analyses of optimal nitrogen application rates (based on actual field and weather conditions), and found that nutrient plans often recommend nitrogen application rates that are below optimum in Minnesota.

Precision Agriculture for Nutrient Management

Advances in technology have facilitated the development of farming equipment and management systems designed to apply agricultural inputs with greater precision, depending on site-specific soil and crop plant conditions (Zilberman et al., 2002). Most precision agriculture technology is based on Global Positioning Systems (GPS) that are used to map soil fertility levels, crop yields, and other indicators with a great deal of spatial accuracy (often within a few feet). That information can then be used to operate variable-rate application equipment that applies different amounts of agricultural inputs to specific parts of a crop field. Theoretically, precision agriculture systems can reduce use of unnecessary agricultural fertilizers, pesticides, water, or labor and thus minimize loss of nutrients and chemicals to the environment and improve farmers' net economic returns (Batte, 2000). A number of experimental studies have reported gains in productivity, input use efficiency, and economic returns from the use of precision agriculture systems across a range of production environments (Khosla et al., 2008). While enthusiasm about the promise of precision agriculture remains high, adoption by farmers has not met with initial expectations. Explanations for low adoption include farmer uncertainty about economic benefits, risk aversion (which contributes to continued overapplication of inputs as insurance against crop failure), and the fact that some of the social benefits of the technology (for example, reduced losses to the environment) do not accrue as economic gains for producers (Napier et al., 2000; Zilberman et al., 2002).

A growing number of experimental and long-term field studies suggest that impacts and economic benefits of precision farming practices can be variable across time and space (Koch et al., 2004; Rider et al., 2006; Tozer and Isbister, 2007; Bachmaier and Gandorfer, 2009; Biermacher et al., 2009). The net present value of nitrogen and phosphorus management can be affected by spatial and temporal variability in carryover nutrient levels from previous crops, water availability, weed or pest pressure, and weather conditions, all of which lead to volatility in crop yields and economic returns (Bullock and Lowenberg-DeBoer, 2007; Lambert et al., 2007). Economic returns tend to be greatest when variability in soil conditions at the subfield level are high and the relative gains from site-specific management is increased relative to uniform application of farm inputs (Swinton and Lowenberg-DeBoer, 1998; Isik and Khanna, 2002; Robertson et al., 2008). Precision agriculture approaches that require investments in expensive machinery and equipment are more profitable on larger farming operations that can spread the fixed costs of precision agriculture technology across more acres (Fernandez-Cornejo et al., 2001; Godwin et al., 2003). Economic returns are also sensitive to market conditions for farm inputs and commodities. In general, the higher the cost of farm inputs or farm commodity, the more likely precision agriculture will produce net economic gains (Khosla et al., 2008).

The profitability of many precision agriculture systems can be overestimated if uncertainties such as variable operating costs and other uncertainties inherent in cropping systems are not incorporated into economic models (Tozer, 2009). Several researchers have called for more long-term fundamental scientific research across a wider range of production environments to establish a more solid foundation for the design and management of precision agriculture systems (Isik and Khanna., 2003; Bullock and Lowenberg-DeBoer, 2007).

Integrated Pest Management

Field studies have shown that integrated pest management (IPM) can improve financial performance by reducing the cost of pesticide input, pest populations, and crop damage by pests. Trumble et al. (1997) compared two treatments of experimental celery plantings. The chemical standard treatment of nine applications of methomyl and permethrin were compared to an IPM program that included three or four applications of *Bacillus thuringiensis*, the need for which was determined by sampling insect populations for established thresholds. Both treatments resulted in less crop damage and better yields than the untreated control, but IPM had better economic returns than chemical treatment because of reduced input costs. Reitz et al. (1999) found similar results in field station trials and also conducted a commercial trial in collaboration with a celery producer in Ventura County, California. The IPM program that relied on biological control agents and rotations of selective, environmentally safe biorational insecticides (*Bacillus thuringiensis*, spinosad, tebufenozide)—applied only when pests exceeded threshold levels—resulted in 25 percent fewer pesticides used compared to the grower's program. The cost savings from reduced pesticide use were more than $250/hectare (Reitz et al., 1999).

Burkness and Hutchison (2008) compared the efficacy and economics of an IPM program that uses reduced-risk pesticides (that is, pesticides with minimal negative effects on beneficial insects) on basis of need determined by established threshold with a conventional grower-based program in cabbage production. They found that the IPM program was more effective in reducing pests and resulted in an average of 10.5 percent higher marketable yields than the conventional program. Although the IPM program did not reduce pesticide use in all years, the average pesticide use over four years was 24 percent lower in the IPM program than the conventional program. The lower pesticide expense and higher marketable yield on average resulted in higher average net returns (Burkness and Hutchison, 2008).

Few other articles examine the economics of pest management. A survey of articles on the topic from 1972 to 2008 shows that less than 1 percent of the articles include economic evaluations. Moreover, the economic analyses in at least 85 percent of the papers that included them were conducted by entomologists and not economists (Onstad and Knolhoff, 2009). Because economic performance can influence the rate of adoption of farming practices that improve sustainability, research on economic evaluations are important to the future of agricultural sustainability.

Business and Marketing Diversification Strategies

Diversification of crop and livestock enterprises represents an important component of many modern sustainable agricultural systems. However, there has been growing attention to efforts of some farmers to diversify their income by developing alternative agriculturally related enterprises and marketing strategies (Barbieri et al., 2008). Four types of farm business and marketing diversification strategies that pertain to sustainability are discussed in this section: value-trait or niche marketing, direct marketing, agritourism and recreation, and diversification of income through value-added processing and off-farm income. Research suggests that all those types of diversification (and others such as contract service work) are increasing in importance in the United States (Hinrichs and Lyson, 1995; Barbieri et al., 2008), Canada (Smithers et al., 2008), Europe (van der Ploeg et al., 2000; McNally, 2001; Ilbery and Maye, 2007), Australia, and New Zealand (Guthrie et al., 2006). Broader diversification strategies are typically motivated by dissatisfaction with economic pressures and returns from conventional markets (Renting et al., 2003). Generally, such strategies can offer economic benefits to the extent that they employ underutilized farm assets, are complementary to existing farming practices, increase the farmer share of income from consumer spending on retail food products, or reduce reliance on generic farm commodities as a source of farm business income (McInerney, 1991; McNally, 2001; Barbieri et al., 2008). The attractiveness of various business and marketing diversification strategies to farmers also depends on such nonfinancial reasons as impacts on leisure time, pleasurable work, compatibility with farm and nonfarm work commitments of household members (Anosike and Coughenour, 1990; Barlas et al., 2001), and farm type, size, and location.

Value-Trait Marketing

Consumer concerns about the safety and quality of modern farming and food systems have led to the rapid growth of markets for "value-trait" food products that offer particular traits that these consumers value. Most notable has been the increase in the organic market, which has grown at a rate of approximately 20 percent per year since 1990. Other value-traits that have established niche markets include sale of "local," "natural," and "fair-trade" foods, as well as "free-range," "pasture-raised," and "hormone-free" livestock products (Pollan, 2006). Factors that contribute to the growth of the organic market and other niche markets include consumers' preference and sustainability initiatives of large retailers. (See Chapter 6.)

Many farmers have recognized that emerging niche markets offer unique opportunities for diversifying farm business income and for differentiation in the market. The economic competitiveness of organic farming practices often depends on payment of price premiums by consumers seeking certified organic products. Farmers' ability to tap into those niche markets can be an obvious way to improve the economic sustainability of the farm enterprise.

While higher prices for products are possible, participation in niche and value-trait markets can generate new costs and challenges for the producer, including learning and adopting new production practices, as well as spending more management time to understand and establish new market channels and to interact with consumers, transport products to market outlets, and ensure consistency in the quality and supply of their value-trait farm products (Lyson et al., 2008).

In many situations, the development of successful value-trait food chains requires collective action by larger groups of producers and consumers (Conner, 2004), or development of institutional mechanisms to establish standards and certification systems or to maintain the integrity of product labels in the marketplace (Hatanaka et al., 2006), as discussed further in Chapter 5.

Relatively small niche markets also have challenges in balancing supply and demand. When price premiums are high and entry into the market is easy, markets are at risk of becoming oversaturated, and competition among producers can erode price premiums, which has happened for some organic products. Similarly, economic downturns can result in dramatic decreases in consumer demand for value-trait products if they are sensitive to price or income (Box 4-1). Moreover, producers who participate successfully in niche or value-trait markets generally need to be located relatively proximate to their consumer base, or at least close to a central processing or distribution facility that assists with marketing. Producers in more remote locations are likely to have fewer options to participate in value-trait food chains (Selfa and Qazi, 2005). In spite of the challenges, increasing numbers of producers are participating in the previously mentioned types of niche markets, in part because they prefer the market options over the intense competitive pressures and consolidation trends associated with mainstream market channels.

Direct Marketing

For various reasons, operators of small- and medium-sized farms have difficulty competing with large farms in the mainstream food marketing system. For example, they might be unable to provide sufficient quantities needed to fulfill the supply requirements of large corporate buyers, they might not be able to take advantage of economies of scale to reduce their production costs, and they are not likely to have the manpower or capital to meet criteria imposed by buyers to monitor compliance with increasingly complex food safety and quality standards (Hendrickson et al., 2001; Reardon et al., 2001). These difficulties motivate farmers to seek other venue for sales.

Many direct marketing approaches have been developed to meet the demand for value-trait products. Approaches to direct sales including the following:

- Farmers' markets.
- Farm stands or "U-Pick" operations.
- Community Supported Agriculture programs (CSAs).
- Sales to institutional food service, such as "farm-to-school" programs.
- Sales to local restaurants.
- Sales to local grocery or specialty stores.

For many reasons, direct marketing can be a viable strategy to increase the economic sustainability of a farming system (Hinrichs and Lyson, 1995; Feenstra, 2002). Initially,

---

BOX 4-1
Impacts of Economic Recession on Organic and Local Food Markets

Food products produced with organic practices or other farming practices for sale to niche markets have typically captured price premiums in the marketplace (Greene et al., 2009). Although those products are perceived by consumers to offer important traits, emerging niche markets might be particularly susceptible to changes in consumer disposable income associated with the spike in energy costs, credit crunch, and declining personal income levels associated with the economic recession that began in 2008 (Hills, 2008). Various sources provide mixed evidence about the impact of recent economic downturns on these markets. For example, analysis by the Nielsen Corporation in early 2009 suggested that growth in organic sales had stagnated (Nielsen News, 2009), and news reports suggested that organic dairy markets were negatively affected (Martin and Severson, 2008; Zezima, 2009). However, analysis by the Organic Trade Association suggests that organic sales continued to increase at double-digit rates in 2008 (OTA, 2009). Aside from the impacts of economic stress, the rise of private-label organic products provided by some grocery chains has reduced demand for some branded-label products, leading to lower market prices and reduced total dollar value of organic sales (Hills, 2008).

direct marketing allows a farmer to build social ties with the people who consume their food (Hinrichs, 2000; Lamine, 2005). The farmer-consumer relationship is built on trust and mutual exchange that can be more secure and long-lasting than anonymous market transactions in the mainstream food system (Granovetter, 1985; Kirwan, 2004). As such, producers might gain more control over the prices they receive for their products and reduce annual price volatility. By cutting out the role of food processors and retailers, direct marketing allows farmers to capture a larger share of the end consumer's food dollar. Farmers engaged in direct marketing also report satisfaction in knowing the people who consume their food and feel that they are contributing to the well-being of their local community (Hinrichs, 2000; Smithers et al., 2008). In turn, consumers might benefit by knowing more about where their food comes from, might have access to food that is perceived as fresher and more healthful, and are able to better appreciate the contributions of farming to their local landscape and community (Sharp and Smith, 2003; Smithers et al., 2008).

Two of the most common forms of direct marketing used by U.S. farmers are sales to local consumers through farmers markets and CSA arrangements. The majority of farmers' market vendors sell their products as organic, natural, or other value-trait products (Gillespie et al., 2007).

Although farmers' markets and CSAs have grown dramatically in number and size over the past 10–20 years, those market approaches cannot always provide a sustained income to participating farm households (Feenstra et al., 2003; Varner and Otto, 2008). Many surveys consistently find that the vast majority of producers at farmers' markets are relatively small-scale businesses that do not rely principally on farmers' markets income to support their household, either because they rely on off-farm income or because they have other commercial farming ventures that generate more net income (Brown and Miller, 2008). Although the scale of economic opportunities for farmers might be limited at farmers' markets, they have been an important opportunity for producers to develop business and marketing skills, and they play a major role in the creation of more localized food systems (Gillespie et al., 2007).

According to the U.S. Department of Agriculture (USDA-NAL, 2009), Community Supported Agriculture consists of:

> [a] community of individuals who pledge support to a farm operation so that the farmland becomes . . . the community's farm, with the growers and consumers providing mutual support and sharing the risks and benefits of food production. Typically, members or "share-holders" of the farm or garden pledge in advance to cover the anticipated costs of the farm operation and farmer's salary. In return, they receive shares in the farm's produce throughout the growing season, as well as satisfaction gained from reconnecting to the land and participating directly in food production. Members also share in the risks of farming, including poor harvests due to unfavorable weather or pests.

The CSA concept was brought to the United States by Jan VanderTuin from Switzerland in 1984. CSA projects in Europe date to the 1960s, when women's neighborhood groups approached farmers to develop direct, cooperative relationships between producers and consumers (Allen et al., 2006a). Two distinct types of CSAs have developed: (1) farmer-managed, subscription-based operations, which constitute 75 percent of all CSAs and (2) shareholder CSAs organized by a group of consumers, sometimes organized as not-for-profit organizations, who "hire" a farmer. The success of any CSA depends heavily on highly developed organizational and communication skills (Brown and Miller, 2008). Money raised by the sale of CSA shares is used as operating capital to finance farm production activities, and consumers typically receive weekly deliveries of fresh produce (and occasionally meat and eggs) from the farmers. CSAs allow producers to lock in their prices and receive their income up front, and consumers share in the risks of variability in output due to weather or pest conditions.

CSA customers report numerous social, economic, and nutritional benefits from participation in the arrangement (Farnsworth et al., 1996; Ostrom, 2007). In several studies reported by Brown and Miller (2008), most CSA farmers mainly depend on income from their CSA shares and reported gross farm incomes that ranged from $15,000–$35,000 per year. However, financial analyses have found that CSA farmers often fail to cover their full economic costs and suggest that typical share prices would need to double or triple to be competitive with market rates of return (Sabih and Baker, 2000; Oberholtzer, 2004; Lizio and Lass, 2005). This result is supported by surveys in which the majority of CSA producers were not satisfied with their ability to cover their operating costs or provide sufficient compensation for their work on the farm, although most were still very satisfied overall with their decision to have CSAs (Lass et al., 2003; Tegtmeier and Duffy, 2005; Ostrom, 2007).

For many farmers, participation in farmers' markets and CSAs is not well suited to their commodity mix or location, and there has been growing attention to intermediate-scale marketing mechanisms that allow farmers to sell their products directly to institutional consumers in local and regional markets (Lyson et al., 2008). Common approaches include direct sales to restaurants, schools, hospitals, and universities (Center for Integrated Agricultural Systems, 2001; Beery and Markley, 2007). Case study reports suggest that direct sales can be a significant source of income for small numbers of local farms, but that many logistical barriers must be overcome to expand the markets to larger groups of local farmers and institutions (Lawless et al., 1999; Kloppenburg and Wubben, 2001; Gregoire et al., 2005).

Some producers who are successfully engaged in market alternatives choose to diversify their marketing strategies, just as they often diversify their production strategies. Diversity in marketing enables them to take advantage of various market opportunities and avoid the risk of relying on one strategy alone—which can contribute to longer-term stability.

Agritourism and Fee Hunting

Agritourism is another strategy to diversify farm income (Nickerson et al., 2001; McGehee and Kim, 2004) and reduce economic risks (Che et al., 2005). Agritourism provides tourists with "genuine" rural products and experiences from farms while also supporting the agricultural enterprise. Weaver and Fennell (1997, p. 357) defined agritourism as "rural enterprises, which incorporate both a working farm environment and a commercial tourism component." Data collected by USDA suggest that 2–3 percent of U.S. farms reported direct farm income from agritourism activities in 2002 (Brown and Reeder, 2007). Agritourism is more popular in Europe, where a third or more of farms engage in this type of activity (Evans and Ilbery, 1992; Bernardo et al., 2004) than in the United States. Most income from agritourism in the United States is generated from on-farm sales and activities (NEASS, 2002; Allen et al., 2006).

The potential benefits of agritourism include increasing farm profitability, keeping farmers on the land, enhancing environmental conservation and management, promoting rural artisanal products, supporting rural traditions and cultural initiatives, and enhancing rural and urban relations (Sonnino, 2004). It can potentially promote socioeconomic development of rural areas and illustrate the multifunctionality of farming. Agritourism has proven to be a viable income-generating strategy when the appropriate investments are made and networks are formed, particularly when it is organized on a regional level (Whatmore and Thorne, 1998; Bender and Davis, 2000; Sonnino, 2004).

Fee hunting offers another potential source of farm income. In Texas, about 28 percent of farmland is leased for recreational hunting (USDA-ERS, 2005). Fee hunting could be an economic incentive for landowners to improve habitats for wildlife species, but not all landowners who offer fee hunting are actively managing their lands to provide habitats for wildlife (Ribaudo et al., 2008).

Off-Farm Income

Although most farms in the United States (and around the world) are essentially family businesses that rely mainly on farm family members for their labor force (Gasson and Errington, 1993; Hoppe et al., 2007), the majority of farm families also gain income from off-farm work. Nonfarm work or transfer payments are commonly used to supplement income from the farm business. USDA estimates that on 71 percent of U.S. farms, either the farm operator or his or her spouse works at an off-farm job. More than 90 percent of aggregate farm household income is from nonfarm sources, including off-farm wages and salaries, as well as transfer payments and investment income (Hoppe et al., 2007).

Some farm household members might enter into off-farm employment reluctantly, but positive benefits can be associated with the diversification of farm household income from nonfarm sources. For many farms, off-farm work is the only way to access affordable health care and retirement benefit plans (Bharadwaj and Findeis, 2007; Kennedy, 2009). Similarly, off-farm work experiences have been shown to increase exposure to new ideas and receptivity to innovative opportunities for emerging production and marketing practices in agriculture. Overall, having off-farm income sources can contribute significantly to the long-term stability of many farming households.

Quality of Life and Sustainable Farming System

Studies of the adoption of many farming practices and systems aimed to improve sustainability often emphasize the importance of qualitative impacts on the farm labor process and the farm family. Some farming practices and systems can provide opportunities to reduce unpleasant farm chores, and allow farmers to engage in diverse and varied types of work. For example, no-till farming practices are often preferred by farmers as much because of reductions in labor time and fuel use as for their conservation or economic benefits (Lithourgidis et al., 2005).

Although some practices or systems for improving sustainability require more time than conventional agriculture, it is not clear that the time investment in increasing labor and management decreases quality of life for farm households. Boerngen and Bullock (2004) found conventional farmers reported spending just over 3 hours/week "keeping up" with information about their production practices, while reduced-chemical and organic farmers reported a time investment of nearly 4 hours/week. The difference was found to be statistically significant, suggesting that chemical inputs and human capital might be economic substitutes. Farmers who adopted reduced-chemical practices reported a transition period of 1±2 years; during the transition period, they spent around 3 hours/week learning about reduced-chemical technology. Adopters of organic practices also reported a transition period of 1–2 years; during that period, they invested 5 hours/week learning about organic technology.

A pilot study in the United Kingdom also suggests that work on organic farms requires higher levels of human energy (or caloric) and effort expenditure—which might result in physical stress in workers—than work on conventional farms (Loake, 2001). In a study that compares the self-reported well-being and welfare of migrant farm workers in the United Kingdom, the workers of organic horticultural farms reported the same level of health as their counterparts of conventional farm. Organic farm workers, however, were reported to be happier than conventional farm workers. Statistical analyses suggested that the workers' happiness is correlated with the number and range of tasks performed each day (Cross et al., 2008). Practices that increase and contribute to wildlife habitat conservation—for example, buffer strips—have also been shown to increase quality of life of people who are working and living on or near farms, partly related to improved aesthetic attributes (Meares, 1997; Chiappe and Flora, 1998).

SOCIOECONOMIC ASPECTS OF SUSTAINABILITY AT THE COMMUNITY LEVEL

Farm Labor Conditions and Security

Labor conditions on the farm have rarely been the focus of analysis or attention even though social equity has always been considered one of the goals of sustainability. Few analysts have done systematic research or surveys to document labor benefits, practices, and trends of many farms in this sector. Despite the fact that many consumers are concerned about social justice in the agrifood system, such concrete social justice goals as the welfare of the farm labor force have rarely been considered and included (Allen, 2008). Among the farmers who expressed a deep desire to improve workers' labor conditions, many of them expressed that they do not know how and cannot afford to do so (Strochlic and Hamerschlag, 2006). Many employers of farm workers perceive that they cannot afford to provide living wages, health insurance, and other benefits to employees (Shreck et al., 2006; Kandel, 2008).

The California Institute of Rural Studies conducted a study of 12 farms in California that are reputed for offering good labor conditions to create a road map on best practices on labor management (Strochlic and Hamerschlag, 2006). The study reported a range of no-cost or low-cost practices that can improve farm labor conditions. Other practices such as year-round employment, compensation, and fringe benefits have monetary costs to the farmers, but also yield benefits to them. Some of the best practices in the study are highlighted below. They are divided into practices that incur no or little costs and practices with not-so-trivial costs.

Practices that incur no or little costs:

- Respectful treatment. Respectful treatment includes a broad range of issues such as a humane pace of work, respectful communication styles, direct communications between growers and workers, a healthy work environment, and a decision-making structure that recognizes the contribution of the workers.
- Nontraditional benefits. Nontraditional benefits include personal loans and access to food on the farm.
- Labor relations, communications, and decision making. Practices that foster good communications, such as employee orientation, regular meetings, and informal solicitation of advice, provide ways for workers' representation and ways for them to participate in decision making. Good communications can also improve personal relations between growers and workers.

- Health and safety issues. A healthy and safe work environment is among the factors that farm workers most appreciate. Farmers could limit the time spent on repetitive work that can incur physical ailment (for example, hand weeding).
- Diversity of tasks. Workers appreciate the ability to switch between tasks several times a day. Diversity of tasks not only relieves workers from the boredom of doing the same task all day, but also reduces potential health problems associated with stoop labor or repetitive stress.

Practices that have nontrivial costs:

- Compensation. Compensation includes, but is not limited, to wages. Other forms of compensation include pay increases, profit-sharing, over-time pay, and bonuses.
- Year-round work. Because farm work is highly seasonal, there is a high rate of seasonal unemployment. Farm workers appreciate year-round employment that provides job security and a steady income.
- Traditional benefits. Examples of traditional benefits include health insurance, holiday pay, vacation pay, and retirement plans.

As the researchers noted, "workers who are treated well and made to feel an integral part of the farm operation are more satisfied, more motivated and ultimately, more productive" (Strochlic and Hamerschlag, 2006, p. 2). Satisfied workers are more likely to stay on the job than dissatisfied ones, so that farmers who provide socially just labor conditions are likely to have high labor retention, reduced recruitment, and hence reduced training and supervision costs and high-quality work from farm workers. The committee is not aware of any study on whether some types of farms are more likely to adopt the best labor management practices than other types. Nonetheless, many of the best labor management practices listed above are not directly linked to any farming practices or approaches (with few exceptions such as year-round work, as some farming practices spread the work year-round so that the farmer is more likely to hire year-round labor than seasonal labor). In Schreck et al.'s (2006) survey, some organic farmers argue that fair and healthy labor conditions should be required by state or federal law for all farms.

Community Economic Security

One of the standards for evaluating the sustainability of a farming system is to examine the positive and negative impacts of farms on the economic security of their local community. This section summarizes scientific evidence for the community economic linkages associated with different types of farming systems.

Farming Practices for Improving Sustainability and Community Economic Security

Few empirical studies have been conducted on the social and economic impacts of improving farms' sustainability on their local communities. In much of the public discourse on agricultural sustainability, farms that have improving sustainability as one of their specific goals have been assumed to be predominantly smaller (or family-labor scale) operations with a strong ethic of responsibility toward the local community and greater commitment to purchasing inputs locally (Lasley et al., 1993; Horne and McDermott, 2001; Earles and Williams, 2005; CAFF, 2009; Sustainable Table, 2009). As such, a shift away from conventional farming to farming systems that improve sustainability is assumed by some analysts to reverse some of the local community economic declines linked to the growth of larger farms and more industrialized agriculture (Strange, 1988; Campbell, 1997). The counterargument is that reducing the use of commercial farm inputs and lower levels of output or productivity will create reduced economic spinoffs or net activity in the local or regional agribusiness economy.

The few empirical studies that consider whether farmers aiming to improve sustainability typically have different purchasing behaviors than conventional farmers have shown mixed results. Lockeretz (1989) found that lower-input systems contributed less per acre to the local economy than higher-input systems. Dobbs and Cole (1992) compared hypothetical net farm income and the effect on the local economy (including such backward linkage as economic impact on input supply firms and such forward linkages as economic impact on transportation, processing, and marketing firms) of five conventional farms in South Dakota if they were to convert to farms with improved sustainability (that is lowinput or organic). They found that total net farm income would be higher on three out of five sustainable farms, but dropped to one out of five if organic commodity premiums were ignored. Meanwhile, sustainable farms generated notably smaller backward and forward linkages.

In their summary of a four-state study, Goreham et al. (1995) found that farmers who are committed to using natural fertilizers and cropping systems and no herbicide or commercial fertilizers were less likely to obtain goods and services locally. They also traveled farther and to larger communities to obtain goods and services for their farms. The low use of local goods and services was attributed to the fact that local businesses were less likely to be able to provide the inputs or markets for their particular production practices or commodities. However, they also found that those same farmers were more likely to purchase locally produced farm products and generate more total spending per acre on local farm products. They noted that the absence of a critical mass of producers that use similar farming practices to improve sustainability might prevent the development of viable local input providers or markets to service their farms.

A different assessment was reported by Ikerd et al. (1996), who simulated the effects of shifting Missouri farmland from the Conservation Reserve Program (CRP) to either conventional farming or a farming system that included crop rotations, intensive management of inputs, reduced tillage, and intensively managed, pasture-based beef production. They estimated that the latter farm generated more direct purchases of farm inputs and services, local farm businesses retained a higher share of farm economic activity, and the increased demand for farm labor and management associated with the practices listed above generated more net farm income and household consumption when compared to the conventional farming system. That study was conducted in the U.S. North Central and Great Plains regions, and the particular commodity mixes and trade patterns in those regions might affect the conclusions.

Potential linkages have been identified between the diversity and resilience of some farming practices or farming systems type and enhanced community economic security. For example, organic farms might be more resilient in the face of poor weather (Pimentel et al., 2005), leading to more consistent yields and farm income in periods of adverse climactic conditions than conventional farms. As discussed above, farm diversification (both in terms of crop rotations and integration of crop and livestock production) can spread out the risks of climate, pest, disease, or market condition changes associated with any particular commodity, and can thus potentially increase the stability of economic returns at the community level.

Civic Agriculture, Local Foods, and Community Economic Security

The research summarized above suggests that there are no simple or consistent relationships between the size, structure, or production practices of local farms and an area's community economic vitality. Although farm production practices (in particular, input expenses) can be important sources of income for local businesses, the manner in which farm products are marketed can also have an impact on local community well-being and economic development. Specifically, the rise of local food marketing outlets such as farmers' markets, CSAs, or direct sales to local restaurants has been linked to social and economic vitality in local communities (Kloppenburg et al., 1996; Feenstra, 1997; Hinrichs et al., 2004).

Lyson (2004) argues in favor of a "civic" agriculture, in which direct social and economic ties between local farmers, local businesses, and local consumers become the organizing principle of a local agrifood system. He asserted that "civic agricultural enterprises have a much higher local economic multiplier than farms or processors that are producing for the global mass market. Dollars spent for locally produced food and agricultural products circulate several times more through the local community than money spent for products manufactured by multinational corporations and sold in national supermarket chains" (Lyson, 2004, p. 62). A counterargument is that the money spent on locally produced food probably offsets money spent locally on other nonfood products; hence, the net effect of civic agriculture is uncertain.

Few empirical studies examine the local or community-level economic impacts of "civic" agricultural activities. Direct marketing between farmers and consumers is likely to increase the share of food dollars captured by farmers and minimize the leakage of local agrifood dollars to the mainstream (and highly vertically integrated) food processing and retailing industries (Brown and Miller, 2008). Several studies have estimated that farmers' markets and CSAs can generate state-level economic impacts on the order of tens of millions of dollars and hundreds of jobs (Otto and Varner, 2005; Henneberry et al., 2008; Hughes et al., 2008). Others have documented increases in a wide range of ancillary local consumer spending activity among people who shop at downtown farmers' markets that could multiply the local economic development impacts (Lev et al., 2003, 2007; Oberholtzer and Grow, 2003). However, it is not clear what proportion of local direct farm marketing activity is replacing income or employment opportunities from conventional retail food outlets (Brown and Miller, 2008). There is evidence that a more civic-oriented approach to marketing is more common among farmers who use farming practices for improving sustainability than among farmers who use conventional practices (MacRae et al., 2007).

Although significant in absolute terms, direct marketing represents a small share of total U.S. farm sales (or food purchases) and may not be large enough to generate major community- or regional-scale economic growth impacts (Gale, 1999). Results of the U.S. Census of Agriculture (USDA-NASS, 2009) suggest the total amount of direct farm marketing tripled between 1992 and 2007, to a total of more than $1.2 billion annually (Figure 4-2). However, the total represented just 0.4 percent of total U.S. farm sales, and the number of farmers reporting direct sales increased much more slowly over the same period of time (to roughly 6 percent of all U.S. farms). Even at the national level, few counties report direct sales as a significant fraction of their local agricultural activity. Census data indicate that only 24 U.S. counties (out of more than 3,000) reported direct sales in 2002 that exceeded 10 percent of total county gross farm sales, and in only 1 county did direct sales exceed 20 percent of total farm sales.

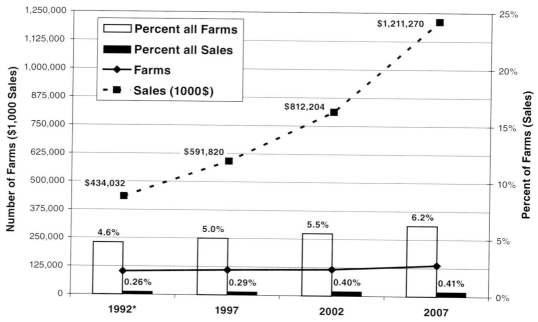

FIGURE 4-2 Importance of direct marketing of agricultural products in the United States, 1992–2007.
SOURCE: U.S. Census of Agriculture, 2007 (USDA-NASS, 2009).

Given the relatively small portion of local income and employment from agriculture in most U.S. counties, the direct effects of local food production currently are unlikely to serve as a major basis for local economic development, although economic multipliers might increase the total impact (Otto and Varner, 2005; Gillespie et al., 2007; Sonntag, 2008). The U.S. experience might not be representative of the potential of direct sales. Analysis of data from Europe (Renting et al., 2003) suggests that in some European Union countries (particularly France, Germany, and Italy), alternative food networks have become important components of rural development schemes. As localized food markets mature, however, there is an increasing concern that some of their distinctive benefits are being undermined by a gradual appropriation by mainstream food processors and retailers.

Community Well-Being

Various practices aimed at improving environmental sustainability can provide amenities and services that are seen as more attractive and desirable for well-being and quality of life for communities in general (Flora, 1995). Studies show that more diverse farm systems and diversified landscapes (for example, inclusion of noncrop vegetation) increase aesthetic attraction, provide more recreational opportunities for residents and tourists, and can help increase economic welfare. Practices that reduce surface runoff and improve surface water quality also increase aesthetic attraction. Deller et al. (2001) found predictable relationships between amenities, quality of life, and local economic performance, suggesting that diverse and integrated farming systems that contribute to natural amenities can increase quality of life for rural communities.

Direct marketing strategies link farmers with their local communities. Moreover, programs, such as farm-to-school programs, contribute to improving student nutrition, as noted by the National Farm to School Network (2009):

> These programs connect schools with local farms with the objectives of serving healthy meals in school cafeterias, improving student nutrition, providing health and nutrition education opportunities that will last a lifetime, and supporting local small farmers. Schools buy farm fresh foods such as fruits and vegetables, eggs, honey, meat, and beans on their menus; incorporate nutrition-based curriculum; and provide students experiential learning opportunities through farm visits, gardening and recycling programs. Farmers have access to a new market through schools and connect to their community through participation in programs designed to educate kids about local food and sustainable agriculture.

FOOD SECURITY, SAFETY, QUALITY, AND OTHER SOCIOECONOMIC DIMENSIONS

This section summarizes important issues related to agricultural sustainability and food systems at a broader regional and global scale, concerning food security, food quality, and ecosystem services.

Satisfying Human Food, Feed, and Fiber Needs

As discussed in Chapter 1, satisfying human food, feed, and fiber needs is one of the sustainability goals in agriculture. Although practices for improving sustainability require taking some land out of production (for example, maintaining wetlands and riparian buffer strips), many farming practices for improving environmental sustainability do not compromise productivity and might even enhance yield (for example, cover cropping, crop rotations, and integrated pest management), as reported in Chapter 3. The determination of the production potential associated with various farming practices or systems at a regional or global level is actually a complex result of several interacting factors: production potentials (typical per acre crop yields or indicators of livestock feed efficiency and growth rates), land and input requirements, and biophysical resource qualities (Smil, 2000). Many studies have shown that with the right conditions and management, low-input and organic systems can have yields, productivity, and economic returns that are comparable to conventional systems (Liebman et al., 2008; Posner et al., 2008).

Sustainable Agriculture and Food Access

The sustainable agriculture and food system movement in the United States and abroad has long embraced the goals of sustaining the economic viability of farm producers, while also seeking to ensure that low-income and underserved populations can access affordable and quality foods (Allen, 1999). Access to sufficient food depends on the affordability and availability of food at the local retail level, which in turn relates to marketing and sales well beyond the farm gate. Because the committee focused on production, marketing, and sales at the farm, the discussion on improving food access is limited to practices in which farmers can participate. The most common efforts of improving food access have focused on direct marketing channels, such as farmers' markets and CSAs. Strategies such as ensuring access to government nutritional programs by low-income customers (for example, the Supplemental Nutrition Assistance Program [SNAP, formerly the Food Stamp program], the Special Nutritional Supplemental Program for Women, Infants, and Children [WIC], and food banks), locating food outlets in low-income neighborhoods, encouraging growers to donate excess food to needy families, and having higher-income customers subsidize lower-income households (Donald and Blay-Palmer, 2006; Guthman et al., 2006) have been used. Other innovative initiatives such as farm-to-school projects and the rapidly growing development of community urban gardens are helping to improve sustainable agriculture's ability to deliver food to low-income populations in the United States.

Despite those initiatives, most direct marketing and value-trait food chains to date have predominantly benefited middle-income and upper-income households (Cone and Myhre, 2000; Kaltoft, 2001; Hinrichs and Kremer, 2002; Allen, 2004; Guthman, 2008). Although more than 20 percent of U.S. SNAP recipients have purchased food at a farmers market, those food purchases constituted only 0.02 percent of their total expenditures in 1997 (Ohls et al., 1999; Kantor, 2001). In the United States, access to food is primarily limited by insufficient financial resources (Nord et al., 2008); geographic distribution also limits the access of fresh produce in some cases (IOM and NRC, 2009).

Food Safety

Food safety concerns stem from the potential of contamination by pathogenic microorganisms and by agrichemicals. Although some practices that improve environmental sustainability could improve food safety (for example, reduced use of agrichemicals), others could increase the risk of microbial contamination (for example, the use of animal manure as fertilizer for crops).

Bacterial Pathogens in Natural Fertilizers and Irrigation Water

Use of animal manure as fertilizer recycles nutrients and can improve soil quality. However, if animal wastes are used in agricultural fields, the level of pathogens in the waste has to be controlled to reduce pathogen contamination of soil, surrounding water, and produce grown in the surrounding areas. Such food-borne pathogens as *Escherichia coli* O157:H7, *Salmonella* spp., and *Campylobacter* spp. might be present in livestock manure. Because animals could serve as a host reservoir for those pathogens, it is important to prevent contamination of animal products with fecal material and to reduce pathogen load (treat manure) prior to land application. Food crops consumed fresh or raw (such as fruits, vegetables, and nuts) might be susceptible to pathogen contamination. The California Leafy Green Marketing Agreement has identified best practices for soil amendments; raw manure or soil amendment that contain uncomposted, incompletely composted, or nonthermally treated animal manure are not to be applied to fields used for lettuce and leafy green production. If untreated manures were applied to fields intended for lettuce and leafy greens production, production of those crops would have to be delayed one year after the manure application (California Leafy Green Handler Marketing Board, 2008). Other key areas for potential contamination in the supply chain include transportation, processing, storage, or preparation, but they are not covered in this report.

Studies have shown that *E. coli* can persist in bovine feces (Wang et al., 1996; Elder et al., 2000) so that contamination of food products by bovine feces could be a vehicle for transmitting food-borne pathogens. The untreated or inadequately treated fecal matter could contaminate the soil, and the runoff water from the field could contaminate the irrigation water. Although competition with soil microorganisms and adverse environmental conditions can reduce the number of pathogens, the pathogens can survive in soil that is directly contaminated by fecal matter or indirectly contaminated by irrigation water for an extended period of time (Islam et al., 2004a,b). A study showed that *E. coli* O157:H7 can survive in manure-amended soil for an extended period of time, even in conditions as dry as 1 percent moisture content (Jiang et al., 2002). That study, which examined the survival of pathogens at various manure-to-soil ratios and soil temperatures, provides useful information on manure-handling practices to reduce the risk of *E. coli* O157:H7 contamination in fresh produce grown in fields with manure-amended soil. Improper aging of untreated manure can significantly increase the risk of *E. coli* contamination in preharvest produce (Mukhejee et al., 2007).

Brinton et al. (2009) surveyed the occurrence and levels of fecal pathogens in organic matter compost that is ready to be sold in the market. They quantified several pathogens in market-ready compost from 93 nonsludge processing facilities and found that only 1 compost contained *Salmonella*. However, 28 percent of the compost had levels of fecal coliforms that exceeded sludge hygiene limits set by the U.S. Environmental Protection Agency. Statistical analyses suggest that the size of the composting facility is correlated to levels of pathogens and that large pile size and immaturity of compost might contribute to high levels of pathogens. Nonetheless, the study shows that organic compost that is hygienic by common standards can be produced (Brinton et al., 2009). As with manure application, compost application would not compromise food safety if best management practices are used. Composting manure would kill such pathogens as *Salmonella* and *E. coli* if done properly (Edrington et al., 2009).

The risk of pathogen contamination from manure or manure-based compost could be reduced substantially with proper aging of manure and careful processing of compost. Manure and compost applications can be used as nutrient management strategies without compromising food safety.

Irrigation water also can be a source of bacterial contamination, particularly if the irrigation well is exposed to farm animal or wildlife feces (Doyle and Erickson, 2008) or contaminated by runoff from manure storage. Spray irrigation has been shown to spread pathogens from contaminated water to lettuce more effectively than drip irrigation (Solomon et al., 2002). Periodic testing of irrigation for pathogens would help reduce the incidence of microbial contamination.

Fungal Pathogens

Some fungi that grow on plants produce mycotoxins that have known toxic carcinogenic effects on humans (Magkos et al., 2003). Concerns have been raised that reduced pesticide use could result in higher incidence of fungal infections and hence higher levels of mycotoxins in food products (Doyle, 2006). Doyle (2006) reviewed a number of studies and found that mycotoxin levels do not differ significantly between grain or grain products that were grown organically (hence, no synthetic agrichemical application) and conventionally. The level of mycotoxins showed significant year-to-year variations and depend largely on climatic variations, rather than level of pesticide use.

Pesticide Residue

Fruits and vegetables that are grown with reduced or no synthetic pesticides are expected to have little pesticide residue. Small amounts of pesticide residue, even in produce that were grown without pesticide, are unavoidable because farmers cannot control all external sources of contamination (for example, spray drift) (Magkos et al., 2003; Doyle, 2006). Integrating data from the Pesticide Data Program of USDA, the Marketplace Surveillance Program of the California Department of Pesticide Regulation, and private tests by the Consumers Union, Baker et al. (2002) compared pesticide residue from foods in three market categories: organic, integrated pest management, and conventional. They found that produce from the conventionally grown category had the highest amount of pesticide residue. Organic produce had the lowest amount (about one-third that of conventionally produced fruits and vegetables) of pesticide residues and are less likely to contain multiple pesticide residues.

Food Quality and Nutritional Completeness

Producing quality food in terms of nutritional value and flavors is one of the objectives of satisfying human food needs. Along with food safety and price, nutrition and taste are among the values that consumers reported as most important to them (Lusk and Briggeman, 2009), even though taste and flavor attributes are partly subjective and difficult to measure and quantify. There are, however, studies that compare the nutritional quality of foods produced using different farming practices and systems. For example, Venneria et al. (2008) compared the nutritional characteristics, including fatty acids content, unsaponifiable fraction of antioxidants, total phenols, polyphenols, carotenoids, vitamin C, total antioxidant activity, and mineral composition, among genetically modified wheat, corn, and tomato crops and their nonmodified counterparts. Their study supported that genetically modified wheat, corn, and tomato crops are nutritionally similar to their nonmodified counterparts. Abouziena et al. (2008) compared the total soluble solids of fruits and vitamin C content of fruits from mandarin trees grown under different weed suppression treatments. They found no significant difference in total soluble solids of fruits among treatments, and vitamin C content was only significantly lower in the unweeded control. Hargreaves et al. (2008) examined antioxidant and vitamin C content in raspberries grown with two different organic composts (ruminant and municipal solid waste compost and compost teas) and did not observe any significant differences. In general, nutritional characteristics of crops are influenced by a multitude of factors including climatic variations, geographic locations, soil quality, cultivar, farming practices, and time of harvest. Many studies showed large year-to-year variations in the nutritional content of crops (Hargreaves et al., 2008; Koudela and Petkikova, 2008). Therefore, the effect of farming practices on nutritional characteristics, if any, is likely masked by the larger variability as a result of the other factors. The food quality and nutritional completeness of organic crops are discussed in Chapter 5.

Next Generation of Farmers

Farmers are the key to the vitality and sustainability of agriculture. As of 2008, about 40 percent of U.S. farmers are 55 years old or older (USDA-ERS, 2009), and one-fourth are at least 65 years old. Older farmers and landowners who control more than one-third of all U.S. farm assets are staying in farming longer than previous generations. Improved health and technological advances in farming equipment allow farmers to work in older age than farmers of previous generations. Farming is becoming popular as a part-time retirement activity (Gale, 2002). Although the turnover of farm assets will be gradual, many U.S. farmers will retire over the next decade. The graying of the farm population has led to concerns about what might happen to the large amount of farmland owned and managed by older farmers when they retire.

Efforts have been initiated to support beginning and entering farmers as a strategy to ensure a diverse and viable farm sector. Beginning farmers are also valued because they bring skill sets that complement traditional management and production technologies and can be a source of innovation and entrepreneurial activity (Ahearn and Newton, 2009). Programs that target beginner farmers include Future Farmers of America (FFA), which has more than 506,000 members across 50 states; 4-H, which has more than 6 million members in 50 states and 80 countries; the American Farm Bureau Federation Young Farmers and Ranchers Program; National Young Farmer Educational Association; International Farm Transition Network; American Farmland Trust; and Land Trust Alliance.

USDA provides financial assistance to beginning farmers and ranchers under its Direct Farm Ownership Down Payment Loan Program. The program provides retiring farmers the opportunity to transfer their land to future generations of farmers and ranchers. An individual requesting direct farm ownership assistance has to have participated in the business operations of a farm or ranch for at least three years, irrespective of whether the individual was the primary operator of the farm or ranch. Applicants are required to provide a down payment of at least 10 percent of the purchase price and meet all other direct farm ownership eligibility requirements to qualify for the Direct Farm Ownership Down Payment Loan Program. Critics of this program state that direct loan limits have not changed in years and have not kept pace with inflationary changes. More funding and better rates and terms are needed to encourage entry into farming (USDA, 2010).

Even with these programs, startup costs for farming is high and unaffordable for some. In addition, small-sized tracts of land that beginner farmers could afford are becoming increasingly rare. Beginner farmers who start out by renting land sometimes never have the opportunity to purchase farmland of their own because high land rental costs lower their profit margins. Contract farming requires large startup capital, and contract terms offer little long-term financial return or opportunities for young farmers to control and manage their own operation (Ahearn and Newton, 2009).

Some states have programs to link up retiring farmers with young aspiring farmers to meet their mutual needs and to preserve family farms. FarmLink and other similar programs maintain databases of retiring farmers and potential young farmers looking for an opportunity to gradually purchase or run a successful farming operation. Some states have created linking programs, but greater effort is needed at the federal and state level, as well as with farm associations and Cooperative Extension to train and support the next generation of farmers and provide access to farmland (DiGiacomo, 1996).

SUMMARY

The use of certain farming practices or systems is partly dependent on whether they provide reasonable economic returns. Yet, research on economic sustainability of farming practices and systems is sparse compared to research on environmental sustainability and productivity. Chapter 3 listed approximately 30 practices that can improve environmental sustainability, but the committee found economic studies on only a handful of those practices. Likewise, studies on social justice and community well-being related to farming practices and systems are lacking.

Conducting research on the social and economic performance of farming practices and systems is complicated by the fact that their economic "viability" is always influenced by the specific development and constellation of market and policy conditions. Similarly, social impacts or social "acceptability" of individual farms can be influenced as much by the behavior of key actors and the values of community members as by inherent qualities of specific production practices or farming systems. These complexities do not make research on social or economic sustainability impossible, but require a more extensive base of research findings and more complex research designs to draw strong conclusions. Given those limitations, review of the scientific literature by this committee suggests several important conclusions:

- The economic benefits of some farming practices accumulate over time as the farming system becomes more resilient. Long-term economic assessment of farming approaches would provide valuable information on economic sustainability of different practices.
- Although such strategies as direct marketing, CSA, and agritourism help to promote farm products and diversify farm income, financial security at the farm level remains a concern because many farms in the United States rely heavily on non-farm sources of income.
- Some practices for improving environmental sustainability also contribute to improving community well-being because they enhance the aesthetics of the community (for example, maintaining buffer strips).
- Other social facets, such as farm labor conditions, can be improved irrespective of farming practices or systems used for production. Social sustainability can be improved by limiting the number of hours on repetitive tasks and allowing workers to switch between several tasks in a day.
- Although some farmers reported that providing equitable wages and benefits to workers could be a financial constraint to their farms, some research and case studies have demonstrated the feasibility of designing production systems that are environmentally, economically, and socially sustainable. Hence, additional and sustained economic and socioeconomic research is necessary to complement the research on productivity and environmental sustainability and provide farmers with knowledge to design their systems to achieve the different sustainability goals simultaneously.

REFERENCES

Abouziena, H.F., O.M. Hafez, I.M. El-Metwally, S.D. Sharma, and M. Singh. 2008. Comparison of weed suppression and mandarin fruit yield and quality obtained with organic mulches, synthetic mulches, cultivation, and glyphosate. *HortScience* 43(3):795–799.
Ahearn, M., and D. Newton. 2009. Beginner Farmers and Ranchers. Washington, D.C.: U.S. Department of Agriculture Economic Research Service.
Allen, P. 1999. Reweaving the food security safety net: mediating entitlement and entrepreneurship. *Agriculture and Human Values* 16:117–129.
———. 2004. Together at the Table: Sustainability and Sustenance in the American Agrifood System. University Park: Pennsylvania State University Press.
———. 2008. Mining for justice in the food system: perceptions, practices, and possibilities. *Agriculture and Human Values* 25(2):157–161.
Allen, T., T. Gabe, and J. McConnon. 2006. The Economic Contribution of Agri-Tourism to the Maine Economy. Orono: University of Maine Department of Resource Economics and Policy.
Anosike, N., and C.M. Coughenour. 1990. The socioeconomic basis of farm enterprise diversification decisions. *Rural Sociology* 55(1):1–24.
Bachmaier, M., and M. Gandorfer. 2009. A conceptual framework for judging the precision agriculture hypothesis with regard to site-specific nitrogen application. *Precision Agriculture* 10(2):95–110.
Baker, B.P., C.M. Benbrook, E. Groth, and K.L. Benbrook. 2002. Pesticide residues in conventional, integrated pest management (IPM)-grown and organic foods: insights from three US data sets. *Food Additives & Contaminants* 19:427–446.
Barbieri, C., E. Mahoney, and L. Butler. 2008. Understanding the nature and extent of farm and ranch diversification in North America. *Rural Sociology* 73(2):205–229.

Barlas, Y., D. Damianos, E. Dimara, C. Kasimis, and D. Skuras. 2001. Factors influencing the integration of alternative farm enterprises into the agro-food system. *Rural Sociology* 66(3):342–358.

Barlett, P. 1993. American Dreams, Rural Realities: Family Farms in Crisis. Chapel Hill: University of North Carolina Press.

Batte, M.T. 2000. Factors influencing the profitability of precision farming systems. *Journal of Soil and Water Conservation* 55(1):12–18.

Beery, M., and K. Markley. 2007. Farm to hospital: supporting local agriculture and improving health care. Available at http://www.foodsecurity.org/F2H_Brochure.pdf. Accessed on September 10, 2009.

Bender, N.K., and N. Davis. 2000. Designing agricultural and nature-based tourism in Eastern Connecticut. In Small Town and Rural Economic Development: A Case Study Approach, P.V. Schaeffer and S. Loveridge, eds. Westport, Conn.: Praeger.

Bennett, J.W. 1982. Of Time and the Enterprise: North American Family Farm Management in a Context of Resource Marginality. Minneapolis: University of Minnesota Press.

Bernardo, D., L. Valentin, and J. Leatherman. 2004. Agritourism: if we build it, will they come? Paper read at 2004 Risk and Profit Conference, August 19–20, 2004, Manhattan, Kans.

Bharadwaj, L., and J.L. Findeis. 2007. Intrinsic and extrinsic motivations and participation in off-farm work among U.S. farm women Available at http://ageconsearch.umn.edu/bitstream/9850/1/sp07bh03.pdf. Accessed on September 15, 2009.

Biermacher, J.T., F.M. Epplin, B.W. Brorsen, J.B. Solie, and W.R. Raun. 2009. Economic feasibility of site-specific optical sensing for managing nitrogen fertilizer for growing wheat. *Precision Agriculture* 10(3):213–230.

Boerngen, M.A., and D.S. Bullock. 2004. Farmers' time investment in human capital: a comparison between conventional and reduced-chemical growers. *Renewable Agriculture and Food Systems* 19(2):100–109.

Brinton, W.F., P. Storms, and T.C. Blewett. 2009. Occurrence and levels of fecal indicators and pathogenic bacteria in market-ready recycled organic matter composts. *Journal of Food Protection* 72(2):332–339.

Brown, C., and S. Miller. 2008. Community supported agriculture. *American Journal of Agricultural Economics* 90(5):1296–1302.

Brown, D.M., and R.J. Reeder. 2007. Farm-Based Recreation: A Statistical Profile. Washington, D.C.: USDA Economic Research Service.

Bullock, D.S., and J. Lowenberg-DeBoer. 2007. Using spatial analysis to study the values of variable rate technology and information. *Journal of Agricultural Economics* 58(3):517–535.

Burkness, E.C., and W.D. Hutchison. 2008. Implementing reduced-risk integrated pest management in freshmarket cabbage: Improved net returns via scouting and timing of effective control. *Journal of Economic Entomology* 101(2):461–471.

CAFF (Community Alliance with Family Farmers). 2009. The Community Alliance with Family Farmers advocates for California family farmers and sustainable agriculture. Available at http://www.caff.org/. Accessed on September 16, 2009.

California Leafy Green Handler Marketing Board. 2008. Commodity specific food safety guidelines for the production and harvest of lettuce and leafy greens. Available at http://www.caleafygreens.ca.gov/trade/documents/LGMAAcceptedGAPs06.13.08.pdf. Accessed on September 9, 2009.

Campbell, D. 1997. Community-controlled economic development as a strategic vision for the sustainable agriculture movement. *American Journal of Alternative Agriculture* 12(1):37–44.

Center for Integrated Agricultural Systems. 2001. Dishing up local food on Wisconsin campuses. Available at http://www.cias.wisc.edu/farm-to-fork/dishing-up-local-food-on-wisconsin-campuses/. Accessed on September 10, 2009.

Chase, C.A., and O.S. Mbuya. 2008. Greater interference from living mulches than weeds in organic broccoli production. *Weed Technology* 22:280–285.

Che, D., A. Veeck, and G. Veeck. 2005. Sustaining production and strengthening the agritourism product: linkages among Michigan agritourism destinations. *Agriculture and Human Values* 22(2):225–234.

Chiappe, M.B., and C.B. Flora. 1998. Gendered elements of the sustainable agriculture paradigm. *Rural Sociology* 63(3):372–393.

Cone, C.A., and A. Myhre. 2000. Community-supported agriculture: a sustainable alternative to industrial agriculture? *Human Organization* 59(2):187–197.

Conner, D.S. 2004. Expressing values in agricultural markets: an economic policy perspective. *Agriculture and Human Values* 21(1):27–35.

Cross, P., R.T. Edwards, B. Hounsome, and G. Edwards-Jones. 2008. Comparative assessment of migrant farm worker health in conventional and organic horticultural systems in the United Kingdom. *Science of the Total Environment* 391(1):55–65.

Deller, S.C., T. Tsung-Hsiu, D.W. Marcouiller, and D.B.K. English. 2001. The role of amenities and quality of life in rural economic growth. *American Journal of Agricultural Economics* 83(2):352–365.

DiGiacomo, G. May 1996. Investing in the future—retiring farmers pass on their land, stewardship. *Farm Aid News and Views* 4(5).

Dobbs, T.L., and J.D. Cole. 1992. Potential effects on rural economies of conversion to sustainable farming systems. *American Journal of Alternative Agriculture* 7(1/2):70–80.

Donald, B., and A. Blay-Palmer. 2006. The urban creative-food economy: producing food for the urban elite or social inclusion opportunity? *Environment and Planning A* 38(10):1901–1920.

Doyle, M.E. 2006. Natural and Organic Foods: Safety Considerations. A Brief Review of the Literature. Madison: University of Wisconsin, Food Research Institute.

Doyle, M.P., and M.C. Erickson. 2008. Summer meeting 2007. The problems with fresh produce—an overview. *Journal of Applied Microbiology* 105(2):317–330.

Earles, R., and P. Williams. 2005. Sustainable agriculture: an introduction. Available at http://attra.ncat.org/attra-pub/PDF/sustagintro.pdf. Accessed on September 16, 2009.

Edrington, T.S., W.E. Fox, T.R. Callaway, R.C. Anderson, D.W. Hoffman, and D.J. Nisbet. 2009. Pathogen prevalence and influence of composted dairy manure application on antimicrobial resistance profiles of commensal soil bacteria. *Foodborne Pathogens and Disease* 6(2):217–224.

Elder, R.O., J.E. Keen, G.R. Siragusa, G.A. Barkocy-Gallagher, M. Koohmaraie, and W.W. Laegreid. 2000. Correlation of enterohemorrhagic *Escherichia coli* O157 prevalence in feces, hides, and carcasses of beef cattle during processing. *Proceedings of the National Academy of Sciences of the United States of America* 97(7):2999–3003.

Evans, N.J., and B.W. Ilbery. 1992. Farm-based accommodation and the restructuring of agriculture: evidence from three English counties. *Journal of Regional Studies* 8(1):85–96.

FAO (Food and Agriculture Organization). 2008. Soaring Food Prices: Facts, Perspectives, Impacts, and Actions Required. Rome: Author.

Farnsworth, R.L., S.R. Thompson, K.A. Drury, and R.E. Warner. 1996. Community supported agriculture: filling a market niche. *Journal of Food Distribution Research* 27(1):90–98.

Feenstra, G. 2002. Creating space for sustainable food systems: lessons from the field. *Agriculture and Human Values* 19(2):99–106.

Feenstra, G.W. 1997. Local food systems and sustainable communities. *American Journal of Alternative Agriculture* 12(1):28–36.

Feenstra, G.W., C. Lewis, C.C. Hinrichs, G.W. Gillespie, and D.L. Hilchey. 2003. Entrepreneurial outcomes and enterprise size in U.S. retail farmers' markets. *American Journal of Alternative Agriculture* 18:46–55.

Fernandez-Cornejo, J., S. Daberkow, and W.D. McBride. 2001. Decomposing the size effect on the adoption of innovations: agrobiotechnology and precision agriculture. *AgBioForum* 4(2):124–136.

Flora, C.B. 1995. Social capital and sustainability: agriculture and communities in the Great Plains and the Corn Belt. *Research in Rural Sociology and Development: A Research Annual* 6:227–246.

Friedmann, H. 1978. World market, state, and family farm: social bases of household production in an era of wage labor. *Comparative Studies in Society and History* 20:545–586.

Gale, F. 1999. Direct farm marketing as a rural development tool. *Rural Development Perspectives* 12(2):19–25.

———. 2002. The graying farm sector. Legacy of off-farm migration. *Rural America* 17:28–31.

Gareau, S.E. 2004. Analysis of plant nutrient management strategies: conventional and alternative approaches. *Agriculture and Human Values* 21(4):347–353.

Gasson, R., and A. Errington. 1993. The Farm Family Business. Wallingford, UK: CAB International.

Gasson, R., G. Crow, A. Errington, J. Hutson, T. Marsden, and D.M. Winter. 1988. The farm as a family business: a review. *Journal of Agricultural Economics* 39:1–41.

Gillespie, G., D. Hilchey, C. Hinrichs, and G.W. Feenstra. 2007. Farmers' markets as keystones in rebuilding local and regional food systems. In Remaking the North American Food System, C. Hinrichs and T. Lyson, eds. Lincoln: University of Nebraska Press.

Godwin, R.J., T.E. Richards, G.A. Wood, J.P. Welsh, and S.M. Knight. 2003. An economic analysis of the potential for precision farming in UK cereal production. *Biosystems Engineering* 84(4):533–545.

Goreham, G.A., G.A. Youngs, Jr., C. Hassebrook, and D.L. Watt. 1995. Community trade patterns of conventional and sustainable farmers. In Planting the Future: Developing an Agriculture that Sustains Land and Community, E.A. Bird, G.L. Bultena, and J.C. Gardner, eds. Ames: Iowa State University Press.

Granovetter, M. 1985. Economic-action and social-structure: the problem with embeddedness. *American Journal of Sociology* 91(3):481–510.

Greene, C., C. Dimitri, B.-H. Lin, W. McBride, L. Oberholtzer, and T. Smith. 2009. Emerging Issues in the U.S. Organic Industry. Washington, D.C.: U.S. Department of Agriculture Economic Research Service.

Gregoire, M.B., S.W. Arendt, and C.H. Strohbehn. 2005. Iowa producers perceived benefits and obstacles in mar-

keting to local restaurants and institutional foodservice operations. *Journal of Extension* 43(1): Article No. 1RIB1.
Guthman, J. 2008. Bringing good food to others: Investigating the subjects of alternative food practice. *Cultural Geographies* 15:431.
Guthman, J., A.W. Morris, and P. Allen. 2006. Squaring farm security and food security in two types of alternative food institutions. *Rural Sociology* 71(4):662–684.
Guthrie, J., A. Guthrie, R. Lawson, and A. Cameron. 2006. Farmers' markets: the small business counter-revolution in food production and retailing. *British Food Journal* 108(7):560–573.
Hargreaves, J., M.S. Adl, P.R. Warman, and H.P.V. Rupasinghe. 2008. The effects of organic amendments on mineral element uptake and fruit quality of raspberries. *Plant and Soil* 308(1–2):213–226.
Hatanaka, M., C. Bain, and L. Busch. 2006. Differentiated standardization, standardized differentiation: the complexity of the global agrifood system. In Between the Local and the Global, Research in Rural Sociology and Development, T. Marsden and J. Murdoch, eds. San Diego, Calif.: Elsevier Press.
Hendrickson, M., W.D. Heffernan, P.H. Howard and J.B. Heffernan. 2001. Consolidation in food retailing and dairy: implications for farmers and consumers in a global food system. Available at http://www.nfu.org/wp-content/uploads/2006/03/2001%20heffernan.pdf. Accessed on March 23, 2010.
Henneberry, S.R., H.N. Agustini, M. Taylor, J.E. Mutondo, B. Whitacre, and B.W. Roberts. 2008. The economic impacts of direct produce marketing: a case study of Oklahoma's farmers' markets. Available at http://ageconsearch.umn.edu/bitstream/6785/2/sp08he08.pdf. Accessed on September 16, 2009.
Hills, S. 2008. Organic market shows signs of slowdown. Available at http://www.foodnavigator-usa.com/Financial-Industry/Organic-market-shows-signs-of-a-slowdown. Accessed on March 23, 2010.
Hinrichs, C., G. Gillespie, and G.W. Feenstra. 2004. Social learning and innovation at retail farmers' markets. *Rural Sociology* 69:31–58.
Hinrichs, C.C. 2000. Embeddedness and local food systems: notes on two types of direct agricultural market. *Journal of Rural Studies* 16:295–303.
Hinrichs, C.C., and T.A. Lyson. 1995. Revisiting the role of state sector in the dual economy paradigm—assessing the effects of multiple work structures on earning. *Social Science Quarterly* 76(4):763–779.
Hinrichs, C.C., and K.S. Kremer. 2002. Social inclusion in a Midwest local food system project. *Journal of Poverty* 6:65–90.
Hoppe, R.A., P. Korb, E.J. O'Donoghue, and D.E. Banker. 2007. Structure and Finances of U.S. Farms: Family Farm Report, 2007 Edition. Washington, D.C.: U.S. Department of Agriculture.
Horne, J.E., and M. McDermott. 2001. The Next Green Revolution. Binghamton, N.Y.: Haworth Press.
Huggins, D.R., and J.P. Reganold. 2008. No-till: the quiet revolution. *Scientific American* 299:70–77.
Hughes, D.W., C. Brown, S. Miller, and T. McConnell. 2008. Evaluating the economic impact of farmers' markets in an opportunity cost framework: a West Virginia example. *Journal of Agricultural and Applied Economics* 40(1):253–265.
Ikerd, J., G. Devino, and S. Traiyongwanich. 1996. Evaluating the sustainability of alternative farming systems: a case study. *American Journal of Alternative Agriculture* 11(1):25–29.
Ilbery, B., and D. Maye. 2007. Marketing sustainable food production in Europe: case study evidence from two Dutch labelling schemes. *Tijdschrift Voor Economische En Sociale Geografie* 98(4):507–518.
IOM (Institute of Medicine) and NRC (National Research Council). 2009. The Public Health Effects of Food Deserts. Workshop summary. Washington, D.C.: National Academies Press.
Isik, M., and M. Khanna. 2002. Variable-rate nitrogen application under uncertainty: implications for profitability and nitrogen use. *Journal of Agricultural and Resource Economics* 27(1):61–76.
Isik, M., and M. Khanna. 2003. Stochastic technology, risk preferences, and adoption of site-specific technologies. *American Journal of Agricultural Economics* 85(2):305–317.
Islam, M., M.P. Doyle, S.C. Phatak, P. Millner, and X.P. Jiang. 2004a. Persistence of enterohemorrhagic *Escherichia coli* O157:H7 in soil and on leaf lettuce and parsley grown in fields treated with contaminated manure composts or irrigation water. *Journal of Food Protection* 67(7):1365–1370.
Islam, M., J. Morgan, M.P. Doyle, S.C. Phatak, P. Millner, and X.P. Jiang. 2004b. Fate of *Salmonella enterica* serovar Typhimurium on carrots and radishes grown in fields treated with contaminated manure composts or irrigation water. *Applied and Environmental Microbiology* 70(4):2497–2502.
Jatoe, J.B.D., E.K. Yiridoe, A. Weersink, and J.S. Clark. 2008. Economic and environmental impacts of introducing land use policies and rotations on Prince Edward Island potato farms. *Land Use Policy* 25(3):309–319.
Jiang, X.P., J. Morgan, and M.P. Doyle. 2002. Fate of *Escherichia coli* O157:H7 in manure-amended soil. *Applied and Environmental Microbiology* 68(5):2605–2609.
Kaltoft, P. 2001. Organic farming in late modernity: at the frontier of modernity or opposing modernity? *Sociologia Ruralis* 41(1):146–158.

Kandel, W. 2008. Profile of Hired Farmworkers, a 2008 Update. Washington, D.C.: U.S. Department of Agriculture Economic Research Service.

Kantor, L.S. 2001. Community food security programs improve food access. *Food Review* 24(1):20–26.

Katsvairo, T.W., and W.J. Cox. 2000. Tillage × rotation × management interactions in corn. *Agronomy Journal* 92(3):493–500.

Kelly, R.H., I.C. Burke, and W.K. Lauenroth. 1996. Soil organic matter and nutrient availability responses to reduced plant inputs in shortgrass steppe. *Ecology* 77(8):2516–2527.

Kennedy, W. 2009. Missouri farmers face tough choices getting health insurance. Available at http://www.access project.org/adobe/press/July11Joplin.pdf. Accessed on September 15, 2009.

Khosla, R., D. Inman, D.G. Westfall, R.M. Reich, M. Frasier, M. Mzuku, B. Koch, and A. Hornung. 2008. A synthesis of multi-disciplinary research in precision agriculture: site-specific management zones in the semi-arid western Great Plains of the USA. *Precision Agriculture* 9:85–100.

Kirwan, J. 2004. Alternative strategies in the UK agro-food system: interrogating the alterity of farmers' markets. *Sociologia Ruralis* 44(4):395–415.

Kloppenburg, J., and D. Wubben. 2001. Farm-to-school program provides learning experience. Available at http://www.cias.wisc.edu/wp-content/uploads/2008/07/rb74.pdf. Accessed on September 15, 2009.

Kloppenburg, J., Jr., J. Hendrickson, and G.W. Stevenson. 1996. Coming into the foodshed. *Agriculture and Human Values* 13(3):33–42.

Koch, B., R. Khosla, M. Frasier, D.G. Westfall, and D. Inman. 2004. Economic feasibility of variable rate nitrogen application utilizing site-specific management zones. *Agronomy Journal* 96:1572–1580.

Koudela, M., and K. Petkikova. 2008. Nutritional compositions and yield of sweet fennel cultivars—*Foeniculum vulgare* Mill. ssp *vulgare* var. *azoricum* (Mill.) Thell. *Horticultural Science* 35(1):1–6.

Lambert, D.M., J. Lowenberg-DeBoer, and G. Malzer. 2007. Managing phosphorus soil dynamics over space and time. *Agricultural Economics* 37:42–53.

Lamine, C. 2005. Settling shared uncertainties: local partnerships between producers and consumers. *Sociologia Ruralis* 45(4):324–345.

Lankoski, J., M. Ollikainen, and P. Uusitalo. 2006. No-till technology: benefits to farmers and the environment? Theoretical analysis and application to Finnish agriculture. *European Review of Agricultural Economics* 33(2):193–221.

Lasley, P., E. Hoiberg, and G. Bultena. 1993. Is sustainable agriculture an elixir for rural communities? Community implications of sustainable agriculture. *American Journal of Alternative Agriculture* 9(3):133–139.

Lass, D., A. Brevis, G.W. Stevenson, J. Hendrickson, and K. Ruhf. 2003. Community supported agriculture entering the 21st century: results from the 2001 National Survey. Available at http://www.cias.wisc.edu/wp-content/uploads/2008/07/csa_survey_01.pdf. Accessed on September 10, 2009.

Law, D.M., V. Bhavsar, J. Snyder, M.D. Mullen, and M. Williams. 2008. Evaluating solarization and cultivated fallow for Johnsongrass (*Sorghum halepense*) control and nitrogen cycling on an organic farm. *Biological Agriculture & Horticulture* 26(2):175–191.

Lawless, G., G.W. Stevenson, J. Hendrickson, and R. Cropp. 1999. The farmer-food buyer dialogue project. Available at http://www.uwcc.wisc.edu/info/ffbuyer/toc.html. Accessed on September 15, 2009.

Lawley, C., E. Lichtenberg, and D. Parker. 2009. Biases in nutrient management planning. *Land Economics* 85(1):186–200.

Lev, L., L. Brewer, and G. Stephenson. 2003. Research Brief: How Do Farmers Markets Affect Neighboring Businesses? Corvallis: Oregon State University.

Lev, L., G. Stephenson, and L. Brewer. 2007. Practical research methods to enhance farmers' markets. In Remaking the North American Food System, C. Hinrichs and T. Lyson, eds. Lincoln: University of Nebraska Press.

Liebman, M., L.R. Gibson, D.N. Sundberg, A.H. Heggenstaller, P.R. Westerman, C.A. Chase, R.G. Hartzler, F.D. Menalled, A.S. Davis, and P.M. Dixon. 2008. Agronomic and economic performance characteristics of conventional and low-external-input cropping systems in the central corn belt. *Agronomy Journal* 100(3):600–610.

Lithourgidis, A.S., C.A. Tsatsarelis, and K.V. Dhima. 2005. Tillage effects on corn emergence, silage yield, and labor and fuel inputs in double cropping with wheat. *Crop Science* 45(6):2523–2528.

Lizio, W., and D.A. Lass. 2005. CSA 2001: an evolving platform for ecological and economical agricultural marketing and production. Available at http://www.nesawg.org/pdf/CSA_2001_report.pdf. Accessed on September 10, 2009.

Loake, C. 2001. Energy accounting and well-being—examining UK organic and conventional farming systems through a human energy perspective. *Agricultural Systems* 70(1):275–294.

Lockeretz, W. 1989. Comparative local economic benefits of conventional and alternative cropping systems. *American Journal of Alternative Agriculture* 4:75–83.

Lu, Y.C., K.B. Watkins, J.R. Teasdale, and A.A. Abdul-Baki. 2000. Cover crops in sustainable food production. *Food Reviews International* 16(2):121–157.

Lusk, J.L., and B.C. Briggeman. 2009. Food values. *American Journal of Agricultural Economics* 91(1):184–196.

Lyson, T.A. 2004. Civic Agriculture: Reconnecting Farm, Food, and Community. Medford, Mass.: Tufts University Press.

Lyson, T.A., G.W. Stevenson, and R. Welsh. 2008. Food and the Mid-Level Farm: Renewing an Agriculture of the Middle. Cambridge, Mass.: MIT Press.

MacRae, R.J., B. Frick, and R.C. Martin. 2007. Economic and social impacts of organic production systems. *Canadian Journal of Plant Science* 87(5):1037–1044.

Magkos, F., F. Arvaniti, and A. Zampelas. 2003. Organic food: nutritious food or food for thought? A review of the evidence. *International Journal of Food Sciences and Nutrition* 54(5):357–371.

Martin, A., and K. Severson. 2008. Sticker shock in the organic aisles. Available at http://www.nytimes.com/2008/04/18/business/18organic.html. Accessed on March 23, 2010.

Martin, S.W., and J. Hanks. 2009. Economic analysis of no tillage and minimum tillage cotton-corn rotations in the Mississippi Delta. *Soil & Tillage Research* 102(1):135–137.

Maryland Cooperative Extension. 2009. What Is a Nutrient Management Plan? College Park: Author.

McGehee, N.G., and K. Kim. 2004. Motivation for agri-tourism entrepreneurship. *Journal of Travel Research* 43:161–170.

McInerney, J. 1991. A socioeconomic perspective on animal welfare. *Outlook on Agriculture* 20(1):51–56.

McNally, S. 2001. Farm diversification in England and Wales—what can we learn from the farm business survey? *Journal of Rural Studies* 17(2):247–257.

Meares, A. 1997. Making the transition from conventional to sustainable agriculture: gender, social movement participation, and quality of life on the family farm. *Rural Sociology* 62:21–47.

Mishra, A., H. El-Osta, M. Morehart, J. Johnson, and J. Hopkins. 2002. Income, Wealth and the Economic Well-Being of Farm Households. Washington, D.C.: U.S. Department of Agriculture Economic Research Service.

Mohr, R.M., D.A. Derksen, C.A. Grant, D.L. McLaren, M.A. Monreal, A.M. Moulin, M. Khakbazan, and R.B. Irvine. 2007. Effect of nitrogen fertilizer rate, herbicide rate, and soil disturbance at seeding on the productivity of a wheat-pea rotation. *Canadian Journal of Plant Science* 87(2):241–253.

Mukhejee, A., D. Speh, and F. Diez-Gonzalez. 2007. Association of farm management practices with risk of *Escherichia coli* contamination in pre-harvest produce grown in Minnesota and Wisconsin. *International Journal of Food Microbiology* 120(3):296–302.

Napier, T.L., J. Robinson, and M. Tucker. 2000. Adoption of precision farming within three Midwestern watersheds. *Journal of Soil and Water Conservation* 55:135–141.

National Farm to School Network. 2009. Farm to school. Available online at http://www.farmtoschool.org. Accessed on November 14, 2009.

NEASS (New England Agricultural Statistics Service). 2002. Vermont agritourism 2002. Available at http://www.nass.usda.gov/Statistics_by_State/New_England_includes/Publications/agtour04.pdf. Accessed on January 14, 2009.

Nickerson, N.P., R.J. Black, and S.F. McCool. 2001. Agritourism: motivations behind farm/ranch business diversification. *Journal of Travel Research* 40:19–26.

Nielsen News. 2009. Growth of organic sales slows with recession. Available at http://blog.nielsen.com/nielsenwire/consumer/sales-of-organic-products-dive-with-recession/. Accessed on August 26, 2009.

Nord, M., M. Andrews, and S. Carlson. 2008. Household Food Security in the United States, 2007. Washington, D.C.: U.S. Department of Agriculture Economic Research Service.

NRC (National Research Council). 1989. Alternative Agriculture. Washington, D.C.: National Academy Press.

Oberholtzer, L. 2004. Community Supported Agriculture in the Mid-Atlantic Region: Results of a Shareholder Survey and Farmer Interviews. Stevensville, Md.: Small Farm Success Project.

Oberholtzer, L., and S. Grow. 2003. Producer-Only Farmers Markets in the Mid-Atlantic Region: A Survey of Market Managers. Arlington, Va.: Winrock International.

Ohls, J.C., M. Ponza, L. Moreno, A. Zambrowski, and R. Cohen. 1999. Food stamp participants' access to food retailers. Available at http://www.mypyramidforkids.gov/ora/menu/Published/SNAP/FILES/Program Operations/retailer.pdf. Accessed on March 23, 2010.

Olmstead, J., and E.C. Brummer. 2008. Benefits and barriers to perennial forage crops in Iowa corn and soybean rotations. *Renewable Agriculture and Food Systems* 23(2):97–107.

Onstad, D.W., and L.M. Knolhoff. 2009. Finding the economics in economic entomology. *Journal of Economic Entomology* 102(1):1–7.

Ostrom, M. 2007. Community supported agriculture: is it working? In Remaking the North American Food System, C. Hinrichs and T. Lyson, eds. Lincoln: University of Nebraska Press.

OTA (Organic Trade Association). 2009. U.S. organic sales grow by a whopping 17.1 percent in 2008. Available at http://www.organicnewsroom.com/2009/05/us_organic_sales_grow_by_a_who.html. Accessed on March 23, 2010.

Otto, D., and T. Varner. 2005. Consumers, vendors, and the economic importance of Iowa farmers' markets: an economic impact survey analysis. Available at http://www.leopold.iastate.edu/research/marketing_files/markets_rfswg.pdf. Accessed on September 16, 2009.

Pimentel, D., P. Hepperly, J. Hanson, D. Douds, and R. Seidel. 2005. Environmental, energetic, and economic comparisons of organic and conventional farming systems. *BioScience* 55:573–582.

Pollan, M. 2006. One thing to do about food—a forum. *Nation* 283(7):16–17.

Posner, J.L., J.O. Baldock, and J.L. Hedtcke. 2008. Organic and conventional production systems in the Wisconsin Integrated Cropping Systems Trials: I. Productivity 1990–2002. *Agronomy Journal* 100(2):253–260.

Rajsic, P., and A. Weersink. 2008. Do farmers waste fertilizer? A comparison of ex post optimal nitrogen rates and ex ante recommendations by model, site and year. *Agricultural Systems* 97(1–2):56–67.

Reardon T., J. Codron, L. Busch, J. Bingen, and C. Harris. 2001. Global change in agrifood grades and standards: agribusiness strategic responses in developing countries. *International Food and Agribusiness Management Review* 2(3/4):421–435.

Reinhardt, N., and P. Barlett. 1989. The persistence of family farms in United States agriculture. *Sociologia Ruralis* 24:203–225.

Reitz, S.R., G.S. Kund, W.G. Carson, P.A. Phillips, and J.T. Trumble. 1999. Economics of reducing insecticide use on celery through low-input pest management strategies. *Agriculture Ecosystems & Environment* 73(3):185–197.

Renting, H., T.K. Marsden, and J. Banks. 2003. Understanding alternative food networks: exploring the role of short food supply chains in rural development. *Environment and Planning A* 35(3):393–411.

Ribaudo, M., L. Hansen, D. Hellerstein, and C. Greene. 2008. The Use of Markets to Increase Private Investment in Environmental Stewardship. Washington, D.C.: U.S. Department of Agriculture Economic Research Service.

Rider, T.W., J.W. Vogel, J.A. Dille, K.C. Dhuyvetter, and T.L. Kastens. 2006. An economic evaluation of site-specific herbicide application. *Precision Agriculture* 7:379–392.

Robertson, M.J., G. Lyle, and J.W. Bowden. 2008. Within-field variability of wheat yield and economic implications for spatially variable nutrient management. *Field Crops Research* 105(3):211–220.

Sabih, S.F., and L.B.B. Baker. 2000. Alternative financing in agriculture: a case for the CSA method. *Acta Horticulture* 524:141–148.

Schillinger, W.F., A.C. Kennedy, and D.L. Young. 2007. Eight years of annual no-till cropping in Washington's winter wheat-summer fallow region. *Agriculture Ecosystems & Environment* 120(2–4):345–358.

Selfa, T., and J. Qazi. 2005. Place, taste, or face-to-face? Understanding producer-consumer networks in "local" food systems in Washington State. *Agriculture and Human Values* 22(4):451–464.

Sharp, J.S., and M.B. Smith. 2003. Social capital and farming at the rural-urban interface: the importance of non-farmer and farmer relations. *Agricultural Systems* 76(3):913–927.

Shreck, A., C. Getz, and G. Feenstra. 2006. Social sustainability, farm labor, and organic agriculture: findings from an exploratory analysis. *Agriculture and Human Values* 23(4):439–449.

Smil, V. 2000. Feeding the World: A Challenge for the Twenty-first Century. Cambridge, Mass.: MIT Press.

Smithers, J., J. Lamarche, and A.E. Joseph. 2008. Unpacking the terms of engagement with local food at the farmers' market: insights from Ontario. *Journal of Rural Studies* 24(3):337–350.

Snapp, S.S., S.M. Swinton, R. Labarta, D. Mutch, J.R. Black, R. Leep, J. Nyiraneza, and K. O'Neil. 2005. Evaluating cover crops for benefits, costs and performance within cropping system niches. *Agronomy Journal* 97(1):322–332.

Solomon, E.B., C.J. Potenski, and K.R. Matthews. 2002. Effect of irrigation method on transmission to and persistence of *Escherichia coli* O157:H7 on lettuce. *Journal of Food Protection* 65(4):673–676.

Sonnino, R. 2004. For a "piece of bread"? Interpreting sustainable development through agritourism in southern Tuscany. *Sociologia Ruralis* 44(3):285–300.

Sonntag, V. 2008. Why local linkages matter: findings from the local food economy study. Available at http://www.sustainableseattle.org/Programs/localfoodeconomy. Accessed on May 20, 2010.

Steinhilber, P. 1996. Environmental protection can save the farm money. Paper read at Proceedings from the Animal Agriculture and the Environment, North American Conference, at Cornell University, N.Y.

Strange, M. 1988. Family Farming: A New Economic Vision. Lincoln: University of Nebraska Press.

Strochlic, R., and K. Hamerschlag. 2006. Best Farm Labor Practices on Twelve Farms: Toward a More Sustainable Food System. Davis: California Institute for Rural Studies.

Sustainable Table. 2009. Introduction to sustainability. Sustainable vs. industrial: a comparison. Available at http://www.sustainabletable.org/intro/comparison/. Accessed on September 16, 2009.

Swinton, S.M., and J. Lowenberg-DeBoer. 1998. Evaluating the profitability of site-specific farming. *Journal of Production Agriculture* 11(4):439–446.

Tegtmeier, E., and M. Duffy. 2005. Community Supported Agriculture in the Midwest United States: a regional characterization. Ames, Iowa: Leopold Center for Sustainable Agriculture.

Tozer, P.R. 2009. Uncertainty and investment in precision agriculture—is it worth the money? *Agricultural Systems* 100:80–87.

Tozer, P.R., and B.J. Isbister. 2007. Is it economically feasible to harvest by management zone? *Precision Agriculture* 8(3):151–159.

Triplett, G.B., and W.A. Dick. 2008. No-tillage crop production: a revolution in agriculture! *Agronomy Journal* 100(3):S153–S165.

Trumble, J.T., W.G. Carson, and G.S. Kund. 1997. Economics and environmental impact of a sustainable integrated pest management program in celery. *Journal of Economic Entomology* 90(1):139–146.

USDA. 2010. Farm loan programs. Available at http://www.usda.gov/documents/FARM_LOAN_PROGRAMS.pdf. Accessed on March 9, 2010.

USDA-ERS (U.S. Department of Agriculture Economic Research Service). 2005. Land use, value, and management: agricultural land management. Available at http://www.ers.usda.gov/briefing/LandUse/landmgmtchapter.htm. Accessed on February 10, 2010.

———. 2009. Farm household economics and well-being: beginning farmers, demographics, and labor allocations. Available at http://ers.usda.gov/Briefing/WellBeing/demographics.htm. Accessed on March 9, 2010.

USDA-NAL (U.S. Department of Agriculture National Agricultural Library). 2009. Community Supported Agriculture. Available at http://www.nal.usda.gov/afsic/pubs/csa/csa.shtml. Accessed on December 12, 2009.

USDA-NASS (U.S. Department of Agriculture National Agricultural Statistics Service). 2009. 2007 Census of Agriculture. Washington, D.C.: U.S. Department of Agriculture National Agricultural Statistics Service.

van der Ploeg, J.D., H. Renting, G. Brunori, K. Knickel, J. Mannion, T. Marsden, K. de Roest, E. Sevilla-Guzman, and F. Ventura. 2000. Rural development: from practices and policies towards theory. *Sociologia Ruralis* 40(4):391–408.

Varner, T., and D. Otto. 2008. Factors affecting sales at farmers' markets: an Iowa study. *Review of Agricultural Economics* 30(1):176–189.

Venneria, E., S. Fanasca, G. Monastra, E. Finotti, R. Ambra, E. Azzini, A. Durazzo, M.S. Foddai, and G. Maiani. 2008. Assessment of the nutritional values of genetically modified wheat, corn, and tomato crops. *Journal of Agricultural and Food Chemistry* 56(19):9206–9214.

Wang, G.D., T. Zhao, and M.P. Doyle. 1996. Fate of enterohemorrhagic *Escherichia coli* O157:H7 in bovine feces. *Applied and Environmental Microbiology* 62(7):2567–2570.

Weaver, D.B., and D.A. Fennell. 1997. The vacation farm sector in Saskatchewan: a profile of operations. *Tourism Management* 18(6):357–365.

Whatmore, S., and L. Thorne. 1998. Nourishing networks. Alternative geographies of food. In Globalising Food. Agrarian Questions and Global Restructuring, D. Goodman and M.S. Watts, eds. London: Routledge.

Yang, X., D. Bosch, T. Nordberg, and M.L. Wolfe. 2000. Phosphorus-based nutrient management planning on dairy/poultry farms: implications for economic and environmental risks. Paper read at American Agricultural Economics Association Annual Meeting, July 30–August 2, Tampa, Florida.

Zentner, R.P., G.P. Lafond, D.A. Derksen, and C.A. Campbell. 2002. Tillage method and crop diversification: effect on economic returns and riskiness of cropping systems in a Thin Black Chernozem of the Canadian Prairies. *Soil & Tillage Research* 67(1):9–21.

Zezima, K. 2009. Organic dairies watch the good times turn bad. Available at http://www.nytimes.com/2009/05/29/us/29dairy.html. Accessed on March 23, 2010.

Zilberman, D., M. Khanna, and L. Lipper. 2002. Economics of new technologies for sustainable agriculture. *Australian Journal of Agricultural and Resource Economics* 41(1):63–80.

5

Examples of Farming System Types for Improving Sustainability

One of the underlying themes of this report is the tension between the rapid specialization of much of U.S. agriculture in the last few decades and its resulting high production of individual commodities (Chapter 2) with the requirements of robustness, resilience, and appropriate levels of environmental integration in sustainable production systems (as discussed in Chapter 1). That tension revolves around the balance between the "industrial philosophy" and "agrarian philosophies" (Box 1-7) and varies among different commodities and environments. This chapter illustrates a few system types that lie within the complex matrix of that balance. They represent modifications within industrial approaches, and, in some cases, a more aggressive departure toward an agrarian approach. Chapters 3 and 4 highlight advances in the scientific understanding of different management practices and approaches that can contribute to improving productivity and environmental, economic, and social sustainability. The practices are central to the examples below because they are components of a larger farming system.

"System" is interpreted in a broad sense, from the individual farm agroecosystem to the wider ecological system or biome. The systems approach recognizes the importance of interconnections and functional relationships between different components of the farming system (for example, plants, soils, insects, fungi, animals, and water). It also stresses the significance of the linkages between farming components and other aspects of the environment and economy. Understanding how the components function individually and the outcomes each produces becomes the foundation of systems agriculture research. The aggregate outcome of applying those practices in concert cannot be predicted from simply combining the anticipated outcome of each practice because they interact with one another. In some instances, the combination of practices has complementary or synergistic relationships; in other instances, combining two practices might have unintended negative consequences.

A systems approach to agriculture is generally guided by an understanding of agroecology, as a scientific basis, and agroecosystem interactions. Agroecology applies ecological concepts and principles to the design and management of agricultural systems to improve sustainability (Gliessman, 1998; Altieri, 2004; Wezel and Soldat, 2009). Agroecology provides a framework to integrate the biophysical sciences and ecology for management of agricultural systems. It emphasizes the interactions among all agroecosystem components (for example, biophysical, technical, and socioeconomic components of the farming system) and recognizes the complex dynamics of ecological processes (Vandermeer, 1995). The approach aims to maintain "a productive agriculture that sustains yields and optimizes the use of local resources while minimizing the negative environmental and socio-economic impacts of technologies" (Altieri, 2000).

When used in agriculture, agroecosystems have been defined as "communities of plants and animals interacting with their physical and chemical environments that have been modified by people to produce food, fiber, fuel, and other products for human consumption and processing" (Altieri, 1995). Agroecosystem design has been recognized as an important part of an agroecological approach, which is a more holistic concept of integrated resource management and understanding complex interactions than a reductionist approach (Swift et al., 1996).

This chapter uses a few farming system types to illustrate how they combine practices and to discuss the potential environmental, social, and economic outcomes. (See Box 2-1 for articulation of the distinction between "farming system"—the integrated system of a single farm management entity—and a "farming system type"—aggregations of farming systems defined by commonalities of commodity, management practices, or farming system approach.) Specifically, the organic, integrated crop–livestock, pasture-based livestock, low-confinement hogs, and perennial grains system types are used in this chapter to represent commonalities of commodity, of specific management approach to those commodities, or of a particular philosophical or scientific approach to farming system management. The integrative perspective of how the components interact with each other in a system and the study of the potential outcomes of those interactions provide valuable information for designing, implementing, and operating a farming system that achieves multiple sustainability goals. Beyond the boundary of a farm, many elements of sustainability, such as product and market diversity and resilience, water resource quality and use, elements of ecosystem health, and community well-being, are highly influenced at landscape, watershed, and regional scales. Sustainability, thus, suggests and requires in most instances an appropriate mix and location of farming system types. The last part of this chapter discusses agricultural sustainability at the landscape level.

ORGANIC CROPPING SYSTEMS

The organic approach to farming, and specifically to cropping systems, is of scientific interest as an alternative type of system to the conventional type for several reasons:

- The organic approach is driven by a philosophy of using biological processes to achieve high soil quality, control pests, and provide favorable growing environments for productive crops, and by the prohibition of use of most synthetically produced inputs. For farm products to meet organic standards, farmers either substitute "organic" inputs (which are usually expensive) or use "biological structuring" (illustrated by use of practices described below) to achieve a high level of internal ecosystem services in their farming systems to permit high efficiency and productivity. Most productive organic farms are highly integrated and use what is referred to as a holistic approach to manage agricultural operations and their processes and impacts (Vandermeer, 1995; Gliessman, 1998; Altieri, 2004). (See the et al., 2006). The lower leachable nitrates in organic systems could be because they operate at lower levels of nitrogen application, and because nitrogen in organic systems is bound to organic fertilizers, such as composts and manures, when added or incorporated in the soil. Organically managed soils have been shown to store nitrogen more efficiently than their conventional counterparts (Clark et al., 1998). Other organic practices that minimize nitrogen losses are wide crop rotations, cover crops, and intercrops (Kasperczyk and Knickel, 2006). Although data on phosphorus loss from organic systems are limited, Lotter (2003) found phosphorus loss from leaching, runoff, and erosion in organic farming systems to be lower than in comparable conventional systems in all studies found.

The small nutrient surpluses in organic farms reduce the risk of nutrient (especially nitrogen) pollution from agriculture to rivers, lakes, wetlands, and coastal oceans. Han et al. (2009) reported that if farmers choose organic practices and reduce fertilizer use, nitrogen pollution levels could decrease to below present-day levels. They used existing data on nitrogen levels in rivers across 18 watersheds in the Lake Michigan basin and from five time intervals between 1974 and 1992. The researchers projected future nitrogen fluxes under three land-use and two climate scenarios: 1) business as usual, 2) increased dependence on organic farming, 3) increased fertilizer use from corn-based ethanol production, 4) a 5 percent increase in rainfall, and 5) a 10 percent increase in rainfall. The study revealed that the combined effect of 10 percent more rainfall and more ethanol production would increase nitrogen levels in rivers by 24 percent. However, increased use of organic farming practices could reduce nitrogen levels in rivers by 7 percent, even if rainfall increased by 10 percent. In southern Michigan, organic rotations using compost leached an average of 35 kg/ha of nitrogen per year compared to 53 kg/ha of nitrogen per year for conventional systems (Sanchez et al., 2004), a 34 percent reduction.

Weeds

Weed control is one of the greatest challenges to yield productivity and economic profitability in organic systems. Seeding in organic grain systems is typically conducted later in the spring than in conventional systems to take advantage of the nitrogen in cover crops and to give weeds an opportunity to emerge. Soybean is particularly susceptible to weed competition. Cavigelli et al. (2008) showed that the yield difference between organic and conventional soybean in a Maryland experiment could be explained solely by the increased weed problem in the organic field. In a Wisconsin study, corn yields were 72 to 84 percent of conventional production in years with wet conditions (Posner et al., 2008). Soybean yields under the same conditions were 64 to 79 percent of yields for the conventional crops. However, in years where weather conditions were favorable and weed pressure was low, yields from organic and low-input systems were comparable (Porter et al., 2003; Posner et al., 2008).

Organic farms tend to rely on hand labor for weed control more heavily than do conventional farms. In a survey of 59 tomato farms in Indiana, Hillger et al. (2006) found that farmers generally reported more hours of hand weeding for fields under organic management than for those under nonorganic management. Swezey et al. (2007) found that production cost of cotton grown under organic management is higher than nonorganic management primarily because of the greater hand-weeding costs and lower productivity. Although improvements have been made in tillage machinery for controlling weeds in organic systems, results from research and experience suggest that additional research is needed in economical weed control for those systems. (See also Chapter 3.)

Greenhouse-Gas Emissions

Organic crop production could have lower greenhouse-gas emissions than conventional production because the former does not use synthetic fertilizers or pesticides that require fossil fuel to produce. Meisterling et al. (2009) conducted a lifecycle assessment to compare the global warming potential and primary energy of conventional and organic wheat production. Their model estimated that the global warming potential of producing 0.67 kg (for a 1 kg loaf of bread) of wheat is 190 g of carbon dioxide equivalent (CO_2eq) using conventional production and 160 g CO_2eq using organic production. Those modeled estimates, however, include high uncertainties associated with N_2O emitted from fields and soil carbon sequestration because excess nitrogen input can increase N_2O emission in either conventional or organic production. Nitrous oxide release is correlated more with overall soil nitrogen levels and mineralization amounts than with source of nitrogen input. Loss of soil carbon and N_2O emissions can be reduced by using best management practices in either conventional or organic production (Meisterling et al., 2009). In a long-term ecological research experiment in Michigan, organic treatments were found to have nitrous oxide (N_2O) greenhouse warming similar to conventional no-till, low-input rotation with legumes and perennial alfalfa in spite of having no fertilizer N input (Robertson et al., 2000). (See also Table 3-1 in Chapter 3.) Net greenhouse warming potential for the organic system was less than half that of standard conventional with full tillage, but higher than for no-till due to the higher soil carbon gains from no-till. Systematic assessment of greenhouse-gas emissions of different cropping systems or system types over the lifecycle of crop production is sparse.

Economic Impact

The economics of organic cropping systems has considerable variation by regions of the United States and by different crops. Organic crop yields per acre are generally lower and labor requirements are often higher than in conventional agriculture systems. However, purchased input costs are less than conventional agriculture so that profits per acre are typically only slightly lower than conventional agriculture. Most organic farmers gain price premiums that range from 5 percent to more than 70 percent of the market price obtained by conventional products (Greene et al., 2009; USDA-ERS, 2009b). Fruits and vegetables account for more than 37 percent of organic food sales, which include processed products. The profits per acre of organic farming can significantly exceed those of conventional agriculture.

The most accurate comparisons between organic and conventional agriculture are seen across crop rotations rather than between specific crops. Moreover, organic agriculture is often a favorable alternative in regions where farmers lack access to synthetic inputs because of the inability to purchase inputs or absolute lack of physical access to inputs, or in regions with a large labor supply (as in many developing countries).

In a long-term farming systems trial at the Rodale Institute in Pennsylvania, the net returns per acre for the conventional system were slightly higher than the net returns per acre for the organic system without premiums during the period of 1991 to 2001 (Pimental et al., 2005). Production costs per acre for the organic system were lower. Total labor for the organic system was higher, but because it was spread more equally through the growing season, the organic system had fewer off-farm hired workers. Organic corn production over the 10-year period was more profitable per acre than conventional corn, but organic corn was not grown as often in the rotation because of the need for soil-building crops. When all land, cover crops, and input costs were calculated, given the frequency of each crop in the rotation, production costs per unit of output were 10 percent higher for organic corn, soybean, wheat, and hay. Delate et al. (2003), however, found net returns for corn within the organic corn–soybean–oat and corn–soybean–oat–alfalfa rotations were significantly greater than conventional corn–soybean rotation returns on the basis of the market prices for the year of study.

Lotter (2003) reviewed numerous comparisons of organic versus conventional agriculture in the United States and worldwide. He concluded that despite the lower yields of organic crops compared to conventional crops, organic systems can still be more profitable than conventional systems because of lower input costs and organic price premiums. When organic premiums were not included, conventional systems were generally more profitable. However, Welsh (1999) noted that the differences within a given system (for example, organic versus organic, conventional versus conventional) were often greater than the differences between the two systems and that the local environment had a greater effect on their relative performance. More specifically, Mahoney et al. (2004) found that the direct production costs for corn in a conventional two-year rotation were $60 per hectare more than corn produced in a two-year or four-year low-input rotation and $96 per hectare more than that of a four-year organic rotation. In soybean, the organic or low-input systems had a slight advantage of $13–$18 per hectare in savings over conventional production. The use of petroleum-based chemicals make nonorganic agriculture more vulnerable to the volatility of crude oil prices compared to organic agriculture (Scialabba, 2007).

Organic practices tend to be more labor intensive (Klepper et al., 1977; Pimental et al., 2005) and often need more intensive management time (Porter et al., 2003) than conventional agriculture. In general, unpaid family members provide a larger proportion of the overall farm labor (Tegegne et al., 2001; Macombe, 2007; MacRae et al., 2007). As a result, the economic performance of organic farming systems can depend heavily on the input costs attributed to unpaid family labor (Hanson et al., 1997; Brumfield et al., 2000). For example, a comparison of wheat farmers in the Mid-Atlantic found that organic farms were more efficient than conventional farms by $34/ha in terms of cash operating expenses. However, when opportunity costs, including unpaid family labor, were incorporated, the fortunes were reversed—organic costs exceeded those of conventional by almost $100/ha (Berardi, 1978). Organic farmers in this study also averaged four more hours of labor per hectare than their conventional colleagues.

In fruit and vegetable farms, an organic system with 50 percent organic premiums was more profitable than the conventional or integrated apple production systems (Reganold et al., 2001). For all three systems to break even (when cumulative net returns equal cumulative costs) at the same time, price premiums of 12 percent for the organic system and 2 percent for the integrated system would be necessary to match the conventional system. Walsh et al. (2008) noted that for organic apple production in the humid Mid-Atlantic, the organic price premium required to break even with the conventional production system was greater than the premiums currently offered by the market. Brumfield et al. (2000) reported that organic sweet corn was 2 percent more profitable than conventionally grown sweet corn in New Jersey. Economic analyses of organic production of California specialty crops also have shown higher profitability than conventional counterparts (Klonsky and Tourte, 1998).

The rapid rise in consumer demand for organic products and the concomitant growth of the organic market have brought important economic opportunities and benefits to producers, as discussed in Chapter 6. However, the ability of farmers to gain access to and advantages from the growing organic market depends partly on their marketing strategies and their location because of considerable regional variations.

Several economic analysts have also addressed questions about the scale of organic production. It is often argued that organic production is more conducive and successful for small- or medium-scale operations because organic farming usually requires more intensive management and labor requirements per unit of land, and because of biophysical aspects, such as difficulties in maintaining high levels of biodiversity at larger scales (Hall and Mogyorody, 2001). However, recent studies and prominent organic farming businesses, including several case studies in this report, show that large-scale organic farming systems can also be economically profitable and successful (for example, the Lundberg Family Farms and Stahlbush Island Farms described in Chapter 7). Indeed, by 2007 the average gross sales on U.S. organic farms (and degree of market concentration) were similar to farm sales by size category among conventional farms (USDA-NASS, 2009). Those sales data demonstrate clearly that most organic systems with their high levels of biological structuring through crop rotation, use of cover crops, IPM, and other commonly used organic practices can be applied across the full spectrum of scales if farmer monitoring and management systems are adequate.

Social Impact

Labor Practices

Most published literature and policy discussions about the treatment of farm labor in sustainable farming systems have focused on the example of organic farming. Formal standards for organic food production, however, do not typically include detailed requirements for treatment or compensation of the farm labor force (IFOAM, 2002; Guthman, 2004; USDA-AMS, 2009).

Some explanations for why organic farms might have progressive farm labor practices and workplace conditions (Duram, 2005) include: organic farmers typically use fewer risky agrichemicals, are more likely to use diversified livestock and cropping systems that are better able to employ labor throughout the year, and might be more likely to share an ideological commitment to environmental and social justice issues (Pretty, 1995; Guthman, 2004; Glenna and Jussaume, 2007). Nevertheless, organic farming systems in the United States have been criticized for relying heavily on mundane hand labor and for exploiting the labor of idealistic, young farm interns seeking to learn about farming by working on organic farms for a summer. In addition, the organic and sustainable farming social "movements" have spent much more time advocating for environmental issues than for the well-being and fair treatment of farmers or farm workers (Allen et al., 1991; Allen and Sachs, 1993).

Detailed empirical studies of the labor practices on organic and sustainable farms have only recently been conducted. In general, organic production entails greater use of labor per unit output, although there is a greater share of overall farm labor obtained from unpaid family members (Tegegne et al., 2001; Macombe, 2007; MacRae et al., 2007).

Although the labor required to produce individual crops using organic techniques might be high, the diverse cropping patterns (and the reintegration of livestock into traditional cropping systems) often associated with organic farming can spread labor demands evenly throughout the year (Nguyen and Haynes, 1995). In some cases, the distribution of labor-input needs over time reduces the need for hired workers or could provide greater opportunities for full-time permanent employment for farm workers.

Perceived high labor requirements are often cited as a critical barrier to adopting organic methods by conventional farmers (Schneeberger et al., 2002). But, at the same time, the increased labor associated with alternative farming practices has not diminished the work satisfaction of farmers or the likelihood of farm succession among farmers in France (Macombe, 2007). To some extent, machinery or management techniques can be developed or adapted to reduce labor needs in organic systems to levels similar to conventional practices (Peruzzi et al., 2007); a small fraction of public and private sector agricultural research and development has been conducted with that goal in mind (Dabbert et al., 2004).

Because of the large scale and heavy use of labor in Californian agriculture, several recent studies report data on the treatment of hired workers among organic farms in that state. Initially, Guthman (2004) reported that exclusively organic farms tended to pay higher wages to farm workers than farms that maintained both organic and nonorganic operations. However, larger farms of both types tended to pay higher wages and were more likely to offer benefits than small operations. Whether larger farms of either type tend to offer higher wages than their smaller counterparts was unclear.

An exploratory survey (Shreck et al., 2006) found that two-thirds of organic farmers in the survey hired workers (other than family members) for at least part of the year, but that just one-third of organic farms provided at least one basic health benefit to their workers. The provision of health insurance benefits was positively correlated with the overall scale of the farming operation. In addition, another study that compared wage and benefit practices of organic and conventional farms in California found that organic farms paid better wages and were more likely to offer profit-sharing (or produce-sharing) arrangements with their workers (Strochlic et al., 2008). However, conventional farms were more likely to offer their workers health insurance, paid time off, retirement plans, and employee manuals. Fair labor practices are not necessarily a result of organic farming. Whether farmers provide fair wages and good working conditions depends on their commitment to social justice, their perceived financial impacts on the farm as a result of such provision, and other conditions.

Food Adequacy

As discussed in Chapter 4, food security depends on multiple factors, including policies, prices, and access to food, but the first step is to ensure adequate production. Badgley et al. (2007) compiled data from multiple studies and estimated the global organic food supply by multiplying the amount of food in the 2001 food supply by a ratio comparing average organic to nonorganic yields. The authors suggested that organic farming could produce enough food on a global per capita basis to sustain the current human population, and potentially an even larger population, without increasing the agricultural land base. Their findings were based on a global dataset of 293 yield ratios for plant and animal production taken from previous studies that compare organic and nonorganic production systems (Badgley et al., 2007) and have been criticized by Cassman (2007). Although 74 percent of the studies used in the Badgley et al. dataset were from peer-reviewed journals (Badgley and Perfecto, 2007), Cassman (2007) stated that many studies "seem to be demonstrations and informal trials" and fail to meet reliable scientific standards. Another criticism is that a portion of their dataset was from Pretty and Hine's (2000) survey data from 52 developing countries, where many farms included as "organic" were only "close to organic." Nevertheless, their results, along with the Stanhill study (1990) mentioned earlier, suggest that organic methods of food production can contribute to feeding the current and future human population on the current agricultural land base.

Crop yields in organic and nonorganic systems were also discussed earlier in the context of farm economics. This committee did not consider whether a certain system type could feed the world because how each system type is managed can affect the farm's sustainability performance. Doran et al. (2007) argued that "[a] focus on existing conventional and emerging organic systems limits the possibilities." They suggested that "the emphasis should be on developing cropping systems that best contribute to a set of well-defined performance parameters that ensure adequate food supply and farm family income, treats farm labor well and farm animals humanely, and protects environmental quality and natural resources" (Doran et al., 2007, p. 78).

Food Quality and Nutritional Completeness

Although consumers often perceive organic fruits and vegetables as more nutritious than their conventional counterparts, the nutritional superiority of organic crops has not been unequivocally demonstrated. Such comparisons are often complicated by the interactive effects on nutritional quality of farming practices, soil quality, climate, plant genetics, and the time of harvest (Benbrook, 2005; Benbrook et al., 2008), which account for the inconsistent differences reported in more than 150 studies that compare nutritional content of organic and conventional crops (Woese et al., 1997; Benbrook, 2005).

Benbrook et al. (2008) identified peer-reviewed studies that compare nutrient levels in organic and conventional foods published in the scientific literature from 1980 to 2007. Mindful of the confounding factors discussed above, they reviewed the articles to identify scientifically valid "matched pairs" of measurements that include an organic and a conventional sample of a given food. For each matched pair, they also made sure that the same cultivars were planted in both the organic and conventional fields, and the differences in soil types and topography were minimized. They took into consideration the focus and location of the study to only include pairs that use analytical methods for nutrient analyses that they considered reliable. They identified 236 matched pairs across 11 nutrients. The organic product had higher nutrient content than the conventional in 61 percent of the cases; the opposite was true in 37 percent of the cases. No significant differences in nutrient content were observed in 2 percent of the cases. They concluded that organic plant-based food, on average, is more nutritious than nonorganic food (Benbrook et al., 2008).

A controlled, replicated plot study conducted on a 1.7-hectare plot within a 20-hectare commercial orchard in Washington compared the productivity and fruit quality of apples under organic and conventional production (Peck et al., 2006). That study found that organic apples had a higher level of total antioxidant activity than similar-sized conventional apples. The researchers of the study asked panels of consumers to do taste-testing, and the panels tended to rate organic apples to have equal or better overall acceptability, firmness, and texture than conventional apples.

Another study compared the influence of organic and conventional crop management practices on the flavonoid content in a tomato cultivar (*Lycopersicon esculentum* L. cv. Halley 3155) over 10 years (Mitchell et al., 2007). That study observed higher levels of three flavonoids in tomatoes grown in the organic system than in the conventional system. Chassy et al. (2006) did a similar comparison of flavonoids and ascorbic acid in organic and conventionally managed tomatoes and bell peppers over a three-year period. They used two varieties of tomatoes (*Lycopersicon esculentum* L. cv. Ropreco and Burbank) and two varieties of bell peppers (*Capsicum annum* L. cv. California Wonder and Excalibur). They found that, unlike in tomatoes, flavonoid and ascorbic acid contents in bell peppers were not much affected by cropping systems. They suggested that different crops respond differently to agronomic and environmental pressures, so statements about organic crops having greater nutritional content than conventional crops are overgeneralized. Pieper and Barrett (2009) confirmed Chassy et al.'s suggestion when they compared the quality and nutritional content of one variety of processing tomatoes (*Lycopersicon esculentum* var. AB2) produced under organic and conventional production systems on a commercial scale. Their study included data from three different growers for two production years. They found that nutritional quality of the one variety of processing tomato varied by growers and production systems. However, organically grown tomatoes in their study had significantly higher average soluble solids content and consistency than conventionally grown tomatoes.

Community Well-Being

As discussed in Chapter 4, one of the standards for evaluating the sustainability of a farming system type is whether the farms of that type have positive effects on the economic security of their local communities. Until the late 1990s organic agriculture was primarily oriented toward local and regional (mostly direct) sales and therefore contributed to the economic security of local communities. In 2004, 24 percent of organic sales were made locally and another 30 percent were made regionally (USDA-ERS, 2009a). Many Community Supported Agriculture (CSA) operations, for instance, are organic. Local, direct marketing to provide fresh produce to community markets is still a hallmark of a large segment of organic producers and remains one of the points of controversy among organic producers. With the enactment by the U.S. Department of Agriculture (USDA) of organic certification standards, large-scale production operations have became more common and marketing channels lengthened. The 2007 U.S. Census of Agriculture (USDA-NASS, 2009) lists 2673 crop farms (of the total 18,211 organic farms) that have sales of $50,000 or more per year, accounting for $1.02 billion in sales (of the total organic sales of $1.7 billion). Some see the trend towards larger organic farms as "an industrialization of organic production," but that trend is observed across agriculture as a whole.

Increased species richness and abundance and continuous blocks of woodland are thought to improve aesthetics of the community. It has been inferred that organic farming enhances biodiversity because it prohibits the use of synthetic agrichemicals. Several studies in Europe attempted to compare biodiversity in conventional and organic farms. One study relied on meta-analysis (Bengtsson et al., 2005) and found that organic farming seems to have positive effects on species richness and abundance. Its effects, however, vary between organism groups and across landscapes. Gibson et al. (2007) found organic farms had greater total areas of semi-natural habitat (woodland, field margins, and hedgerows combined) than conventional farms in the southwest of England. The organic farms they studied had more continuous blocks of woodland (with simpler perimeters than similarly sized patches on conventional farms), whereas woodland on conventional farms often consisted of more linear patches. Although a larger percentage of semi-natural habitat appears to occur in organic rather than conventional farms, the study did not explore the cause of that association.

ALTERNATIVE LIVESTOCK PRODUCTION SYSTEMS

Over the past 50 years, the most striking changes in the U.S. livestock sector reflect the increasing use of production systems in which animals are kept in full confinement and are fed fewer traditional forage crops and higher proportions of corn, soybean, and food processing byproducts (MacDonald and McBride, 2009). Nevertheless, the last 30 years have also witnessed growing interest in a number of alternative livestock production systems. The alternative systems include efforts to expand the integration of crop and livestock enterprises, intensive grazing management systems on dairy farms, and low-confinement integrated hog production practices. All three alternative systems take advantage of opportunities for greater on-farm cycling of nutrients, seek to mimic natural patterns of animal behavior, and respond to dissatisfaction by farmers and consumers with aspects of confinement livestock production systems.

Integrated Crop–Livestock Systems

Conventional economic wisdom suggests that specialized production systems have strong economic rewards. Specifically, scale economies (driven by new technologies, capital, and labor efficiencies) and commodity support policies have been linked to increasing farm specialization and a dramatic reduction in the average number of crops on typical farms, as well as the farm-level and regional-level separation of crop and livestock production enterprises (Hallam, 1993; Gardner, 2005; MacDonald and McBride, 2009). Economic challenges associated with diversification include higher management, labor, capital, and machinery requirements (Hendrickson et al., 2008; Wilkins, 2008).

Large specialized livestock facilities focus more on producing animals and purchase more of their livestock feed from off the farm than farming systems with both livestock and crops. That trend has led to a decline in available land for recycling livestock waste through cropping enterprises. Gollehon et al. (2001) reported a 40 percent decrease in available farmland per animal unit on U.S. farms from 1982 to 1997. The tendency for specialized livestock operations to purchase a higher percentage of their livestock feed requirements has led to growing imbalances in the supply of nutrients in livestock manure relative to the crop nutrient requirements in fields surrounding livestock operations at the farm, watershed, and regional levels (Kellogg et al., 2000; Ribaudo et al., 2003). At the same time, most U.S. cropland is managed as farms that do not use manures as an important source of nutrients. The 2007 Census of Agriculture reported that just 22 million acres of U.S. farmland received manure in 2007, less than 10 percent of the acreage that received chemical fertilizer treatments (USDA-NASS, 2009). That situation has led to serious waste disposal and water pollution issues around intensive livestock production, high use of fertilizer to replace the lost nutrients in land where animal feed crops are produced, and a 50 percent increase in global reactive nitrogen between 1890 and 1990 (Galloway and Cowling, 2002).

Evidence is increasing that integration of livestock into diverse cropping systems can produce important benefits (Sulc and Tracy, 2007). In particular, the ability to feed crops to livestock enables producers to capture and potentially recycle nutrients back to farm fields, which reduces the need for purchased fertilizers and enhances such desirable soil attributes as organic matter, water-holding capacity, and soil structure (Schiere et al., 2002; Entz et al., 2005; Hendrickson et al., 2007). Moreover, the ability of livestock to take advantage of underutilized resources (for example, less productive croplands that can be converted to pasture, periods of slack family labor demand, or unused crop residues) can improve the overall efficiency of the farm operation and capture new sources of income (Smil, 1999; Russelle et al., 2007). Livestock are often used to convert relatively low-value crops to high-value protein, which can potentially increase total farm returns on integrated crop–livestock farms (Anderson and Schatz, 2003).

Numerous studies have documented the economic benefits of integrated crop–livestock systems. Sulc and Tracy (2007) reviewed recent scientific studies of integrated crop–livestock farms in the U.S. Corn Belt, including the use of alfalfa in crop rotations, the use of annual or short-season pastures in rotation with grains, and the strategic grazing of crop residues. They reported that many of those systems have been shown to be economically competitive and offer environmental benefits when compared to specialized production systems typical of that region. Marois et al. (2002) found that adding cattle and forage rotations into traditional cotton–peanut production systems in the Southeastern United States produced increased whole-farm returns. A similar study contrasting cotton–forage–beef systems with traditional High Plains cotton monoculture systems found that the integrated system reduced irrigation water needs by 23 percent, reduced nitrogen fertilizer applications by 40 percent, and increased net farm profitability by up to 90 percent on a per-acre basis (Allen et al., 2005, 2008). Other studies revealed significant economic advantages of integrated beef–crop operations (Anderson and Schatz, 2003; Gamble et al., 2005; Franzluebbers and Stuedemann, 2008).

Integrated crop–livestock systems have been found to be particularly beneficial when conservation tillage practices are used (Franzluebbers and Stuedemann, 2008). The use of short-term and long-term pasture crops in rotations and the strategic placement of well-adapted forage crops on the landscape can provide particular environmental and economic benefits (Entz et al., 2002; Rotz et al., 2005; Russelle et al., 2007). At the same time, most evidence for successful crop and livestock integration has been linked to the use of ruminant livestock (beef, dairy, sheep, or goats) that can eat forages and crop residues; different challenges exist for finding productive synergies for monogastric livestock species such as poultry and hogs.

Management-Intensive Rotational Grazing Systems

Grazing systems encompass a diverse set of management strategies. Extensive low intensity pastoralist grazing systems have been prominent features of human society for millennia, and the bulk of the U.S. beef cow and sheep flock inventory continues to spend a considerable amount of their lives grazing on rangelands, pastures, and the residues of harvested crop fields. More recently, interest has surged in more intensive grazing management systems, particularly so-called "management-intensive rotational grazing" (MIRG). A key feature of most MIRG systems is the use of short-duration grazing episodes on relatively small paddocks, with longer rest periods that allow plants to recover and regrow before another grazing episode.

MIRG approaches have quickly emerged as a major alternative production system among dairy farms in the Upper Midwest and Northeast, the nation's "traditional dairy belt," characterized by humid temperate climates and the persistence of mixed crop–livestock farming operations. Many farms in those regions use hybrid systems that combine MIRG during grazing months and conventional confinement production in the winter (Kleinman and Soder, 2008). Surveys suggest that MIRG operations constitute more than 20 percent of dairy farms and produce more than 10 percent of milk in major dairy states such as Wisconsin, Pennsylvania, New York, and Vermont (USDA-REEIS, 2003; Winsten and Petrucci, 2003; Taylor and Foltz, 2006). The use of MIRG-like systems is also becoming more common among beef producers in the Great Plains and Southeast. Studies have documented social and economic benefits to farmers from the use of MIRG dairy production systems, including comparable or greater profitability per cow or unit milk output, and higher quality of life and greater levels of satisfaction for farm operators. (See Mariola et al., 2005, and Taylor and Foltz, 2006, for recent reviews.)

Environmental Impact of MIRG Systems

Early reports and farmer testimonials suggested the potential for the adoption of MIRG systems to improve environmental sustainability, including improved soil quality, reduced soil erosion, decreased input use, improved wildlife habitat, and potential for better sequestration of atmospheric carbon. Over the past 15 years, a considerable scientific literature has emerged to closely examine those claims (Mariola et al., 2005; McDowell, 2008; Taylor and Neary, 2008).

Scientific studies of conventional, extensive, or traditional pastoral grazing systems vastly outnumber studies of the more intensive forms of short-duration rotational grazing. Furthermore, many more studies focus on grazing in arid grassland or rangeland regions than on the temperate regions dominated by more productive cool-season grasses and forages. In both instances, it has become clear that livestock left in single grazed pastures for weeks or months at a time (continuous grazing) can generate overgrazed, sparse pastures with low persistence, diminished soil quality, and greater risk of soil erosion (Brummer and Moore, 2000; Teague and Dowhower, 2003). However, most MIRG systems are carefully monitored to manage intensity and timing of grazing to ensure continual ground cover and high-quality, high-yielding forage for livestock (Kanneganti and Kaffka, 1995; Paine et al., 1999; Hensler et al., 2007) without significantly diminishing ecosystem qualities.

The sections below provide scientific evidence regarding the impacts of grazing systems on soil quality, soil erosion, nutrient dynamics, greenhouse-gas emissions, biodiversity, and human health and nutrition. The committee found from its review of the literature that simple conclusions regarding an overall assessment of the environmental impacts of such systems cannot be drawn because environmental impacts depend heavily on at least three major factors: 1) local biophysical conditions, including climate, topography, and soil types; 2) the specific management practices used, including stocking rates, duration of grazing and rest periods, use of purchased fertilizers, and access to riparian areas; and 3) the types of "alternative" land uses against which the performance of grazing systems is compared.

Soil Quality and Soil Erosion

In general, when compared to more intensively cropped fields, soils under pasture management tend to accumulate soil organic matter (SOM), which favors the development of good soil structure (Soane, 1990; Tisdall, 1994; Kemp and Michalkand, 2005). In a series of paired comparisons, rotationally grazed pastures have been shown to have significantly more SOM in the top 12 inches of soil than conventional row crop fields (Dorsey, 1998) or extensively grazed or hayed pastures (Conant et al., 2003). In well-managed pastures, high SOM was associated with higher rates of soil biological activity than equivalent arable fields (Cuttle, 2008). Earthworm populations were 1.3 to 3.0 times higher in MIRG fields than cropped fields (Dorsey, 1998; Mele and Carter, 1999). Improved soil structure in MIRG pastures has been associated with reduced soil erosion and nutrient runoff compared to tilled corn fields (DeVore, 2001; Haan et al., 2006).

However, poor management of grazing fields, particularly in wet conditions and under high stocking rates, can lead to soil compaction and hoof print indentations or pocketing in the top 12 cm of soil, which can diminish soil quality, decrease water infiltration, and increase the potential for soil erosion and runoff of sediment, nutrients, and fecal matter (Evans, 1997; Greenwood and McKenzie, 2001; Cuttle, 2008). Compared to natural or forested landscapes, most grazing systems have greater potential for runoff of nutrients, agrichemicals, and fecal microbes, and for deterioration of aquatic stream ecosystems. Overgrazing, in particular, can lead to defoliation, exposure of the soil surface to direct rainfall impacts, reduced root density, and shifts in plant communities that diminish soil quality and increase soil erosion (Schacht and Reece, 2008). Because they often allow for increased livestock numbers per area of land, MIRG systems require higher levels of management to avoid deleterious impacts on soil compaction and to maintain sufficient vegetative cover.

Management of riparian areas is critical to controlling impacts of grazing on aquatic ecosystems (Wilcock et al., 2008). Allowing access by livestock to streams or stream banks can cause decreased riparian area vegetative cover (particularly tall trees), increase stream bank erosion, decrease channel stability and channel width, and increase stream water temperatures. Morphological changes in stream conditions linked to runoff from grazing landscapes and degraded riparian areas typically reduce the ability of aquatic systems to support healthy fish and macroinvertebrate populations (Allan and Johnson, 1997; Allan et al., 1997; Rutherford et al., 1999; Wilcock et al., 1999; Wilcock and Nagels, 2001).

Carbon, Greenhouse Gas, and Nutrient Dynamics

MIRG systems have been touted as environmentally friendly because a greater percentage of the farm's land use is comprised of untilled permanent or semi-permanent pastures and hayfields, because they typically use much lower levels of artificial fertilizers and agrichemicals (Pain and Jarvis, 1999; Kriegl and McNair, 2005), use less fossil fuels and equipment, and offer direct opportunities to recycle nutrients between livestock and farm fields (Taylor and Neary, 2008). In particular, when compared to row crop farming and extensive grazing systems, there is evidence that well-managed intensive grazing systems can sequester more atmospheric carbon and minimize losses of agricultural nutrients to surface and ground waters (Cuttle, 2008).

Scientific studies of rising global concentrations of greenhouse gases have identified grassland ecosystems as potentially important sinks for sequestering atmospheric carbon (Kucharik et al., 2003; Lal, 2006; Allard et al., 2007). Soils store a large proportion of the world's carbon (Amundson, 2001) such that small changes in soil carbon content can have a large effect on global carbon cycling. Studies of the conversion of tilled soils into native perennial grasses under the Conservation Reserve Program (CRP) suggest net increases in soil carbon (Reeder et al., 1998; Potter et al., 1999; Baer et al., 2000). In the Southeastern United States, pastures under MIRG management sequestered more soil carbon than continuously grazed or hayed fields (Conant et al., 2003). Net gains in soil carbon are highest in the first years of conversion from arable to untilled grasslands (Tyson et al., 1990). At a global scale, however, increased soil respiration as a result of global warming suggests that the world's grasslands could be experiencing net losses of carbon (Bellamy et al., 2005; Schipper et al., 2007).

On the other hand, grazing livestock have also been identified as a significant potential source of greenhouse-gas emissions (de Klein and Eckard, 2008). Worldwide, pastoral grazing sources are estimated to contribute roughly 8 percent of methane (CH_4) and 15 to 30 percent of total N_2O emissions (Clark et al., 2005). Methane emissions are primarily a function of the fermentation of feed in the rumens of grazing animals, mostly lost through the lungs, not flatulence (Torrent and Johnson, 1994). By contrast, CH_4 losses from animal excreta are trivial sources of net emissions.

Methane emissions are affected by feed and forage type and by the intensity of grazing management. One study shows that grain-finished cattle that spend some time in feedlots produce more CH_4 emissions from enteric fermentation per animal than grass-finished cattle. However, because of their efficient weight gain, grain-finished cattle produce 38 percent less CH_4 emission per unit beef produced than grass-fed cattle. Higher-quality forages, including legumes, also tend to yield less CH_4 in the rumen (Peters et al., 2010). Intensive grazing can decrease CH_4 per unit weight gain, but greater rates of forage production and consumption could increase total CH_4 emissions per hectare. Nitrous oxide emissions in grazing systems are primarily a byproduct of the denitrification process in soils. Important sources of nitrogen deposition in pastures are livestock urine, commercial fertilizers, and legume crops. Denitrification is accelerated under wet or anaerobic conditions, which can be aggravated by soil compaction and poaching in pastures (de Klein et al., 2001; Bolan et al., 2004). Emissions of both of types of greenhouse gases tend to increase with more productive pastures or intensive pasture management systems, because of higher soil nitrogen levels, rates of plant growth, and stocking rates.

MIRG farms experience different nutrient-cycling dynamics than traditional row crop and confinement agricultural systems. Although overall applications of nutrients to fields tend to be low in MIRG farms, a greater percentage of nutrients come from direct deposition by grazing animals (and less from commercial fertilizers) in intensive grazing systems. In general, lower levels of input use on MIRG farms provide fewer available nutrients (and much lower levels of pesticides and herbicides) than comparable row crop operations, therefore reducing the risks of losses to the environment. At the same time, deposition of manure and urine by grazing animals can be uneven, and areas of animal congregation (such as at watering troughs, feeding stations, under shade trees, and in overwintering fields) can become potential sites of nutrient build up. Field studies have reported mixed results on the impact of MIRG on nutrient cycling. Dorsey (1998) found that deep soil nitrate concentrations were significantly lower on MIRG fields than on low-intensity grazing or cropped fields. However, Stout et al. (1997) observed high nitrate losses underneath urine patches, which could contribute to ground water contamination at a field scale. Moreover, nitrate losses from grazing animals can be highly variable depending on rainfall patterns and levels of supplemental nitrogen fertilization (Stout et al., 2000).

Some have argued that MIRG systems offer environmental advantages over modern confinement livestock production systems that rely on harvested forage and grains as feed inputs and might have a sufficient land base on which to distribute livestock manure nutrients. Comparisons of intensive grazing and confinement livestock systems have produced mixed results. The mixed results reflect different assumptions about stocking rates, grazing practices, manure nutrient handling, and crop fertility management practices. Recent models for whole-farm nutrient budgets and a full accounting of farm-level nitrogen balances reveal little systematic difference between grazing and confined dairy operations (Rotz et al., 2002; Watson et al., 2002; Kleinman and Soder, 2008). Similarly, phosphorus-accounting models suggest that grazing operations are faced with similar challenges to effectively use phosphorus from animal manures on their fields as conventional confinement farms are (Sharpley, 1985; McDowell et al., 2007; Sharpley and West, 2008), partly because most confinement dairies in the United States still maintain active cropping operations. Whole-farming-system analyses of the risks of soil erosion, nutrient losses, and atmospheric greenhouse-gas emissions would account for losses within the livestock operations itself and in cropping farms and feed processing facilities that produce substantial portions of the feed inputs on many confinement farms.

There can be a tradeoff between managing farming systems to minimize nitrate (NO_3^-) losses to ground water resources or to reduce the loss of N_2O to the atmosphere. Many management practices designed to maximize denitrification efficiencies can reduce the threat of nitrate leaching (which benefits water quality), but can increase nitrous oxide emissions (which are potent atmospheric greenhouse gases). Using an intensive grazing system (particularly if it replaces reliance on traditional crop production for livestock feeds) could affect the balance between nitrate leaching into water and N_2O release into the atmosphere. For example, evidence suggests that denitrification efficiency under MIRG systems is higher than under a corn crop (70 to 90 percent versus 10 to 15 percent), in part due to subsurface soil environments that are richer in plant, microbial, and macrobiotic activities (Browne and Turyk, 2007). As a result, grazing farms contribute comparatively fewer nitrates to ground water, but might convert a higher percentage of nitrates and nitrites into N_2O and N_2 gases.

Ultimately, patterns of nutrient flows in pastures are affected by the impacts of grazing on soil structure, water infiltration, and soil microbial activity, as summarized above, and those interactions make it difficult to draw sweeping conclusions about the net environmental costs and benefits of grazing systems.

Biodiversity

Compared to most arable farming systems, pastures and grazed rangelands tend to have more diverse plant, insect, and animal populations. Research on MIRG farms has focused on the impacts of greater use of managed pastures on wildlife and bird species. Higher proportions of the land base on MIRG farms offer potentially suitable habitat for grassland bird species (Paine et al., 1995; Temple et al., 1999). However, bird counts suggest that grazed pastures have similar levels of overall bird species richness, dominance, and density compared to crop fields, but higher levels of rare and unusual species in the Upper Midwest (Renfrew and Ribic, 2001). The role of management appears to be critical—efforts to exclude livestock from some pasture areas during nesting season are important to generating wildlife benefits (Holechek et al., 1982; Koper and Schimiegelow, 2006). Impacts of grazing systems on aquatic ecosystems was discussed elsewhere, but studies show that well-managed MIRG systems (particularly control over livestock access to riparian areas) can generate favorable conditions for fish populations (Mosely et al., 1998; Lyons et al., 2000).

At the landscape scale, grazing systems can provide important habitat diversity to watersheds dominated by traditional row crop and hay production (Nassauer, 2008). However, different grazing management strategies are likely to produce distinctive impacts on the composition of plant communities at the field and landscape scale (Schacht and Reece, 2008). Well-managed rangeland grazing systems have been associated with greater spatial and temporal variability in species richness (Bakker, 1994; Patten and Ellis, 1995; Fuhlendorf and Smeins, 1999). Continuous grazing and overstocking on rangelands or pastures can result in the elimination of plant species that are preferred by grazing livestock and an evolution toward lower-quality and less palatable species. Conversely, intensive short-duration grazing systems (like MIRG) force livestock to eat a diversity of plants, although the systems are still likely to select for species that can survive under this form of grazing pressure. Impacts of intensive grazing on biodiversity are also likely to differ depending on rainfall conditions, which affect the ability of plants to recover from grazing episodes.

Economic Performance of MIRG Systems

Kriegl and Frank (2004) compared MIRG with traditional confinement (TC) systems (50 to 75 cows) in a stanchion barn with stored feed and family labor and with large modern confinement (LMC) systems that have more than 250 cows, milk cows in parlors and house cows in free stalls, and rely on hired help and stored feed. Their analysis was based on eight years of data. Table 5-1 summarizes 2002, the most recent year in their analysis. MIRG produced less milk per cow. However, MIRG's expenses were lower so that its cost per hundred-weight of milk produced was lower. Net farm income per hundred-weight of milk produced was higher for dairy farmers using MIRG and continued to be even when all labor charges (which were higher for LMC) were omitted. However, income per farm was lower for MIRG, compared to LMC, because MIRG systems had fewer cows and lower milk yields per cow. Similar patterns have been found in other studies (Rust et al., 1995; Hanson et al., 1998; Dartt et al., 1999; Conneman et al., 2000; Winsten et al., 2000; Gloy et al., 2002; Kriegl and McNair, 2005).

The scale of operation associated with confinement and MIRG dairy systems appears to be different. In general, confinement systems—particularly modern parlor or freestall

TABLE 5-1 Economic Indicators Comparing Three Systems of Dairy Production in Wisconsin, 2002

	MIRG[a]	TC[b]	LMC[c]
Pounds of milk/cow	15,644	19,490	22,403
Basic cost ($)/CWT[d] equivalent	$7.48	$7.69	$8.18
Net farm income from operations/CWT equivalent	$2.53	$0.91	$0.47
Net farm income from operations if all paid labor were omitted/CWT equivalent	$3.14	$2.12	$2.34
Net farm income from operations per farm	$31,928	$15,564	59,616

[a]Management-intensive rotational grazing.
[b]Traditional confinement.
[c]Large modern confinement.
[d]CWT—hundred weight.

SOURCE: Kriegl and Frank (2004).

confinement systems—have larger herds than rotational grazing operations. While their economic costs of production per unit output might not be notably lower, larger herd sizes enable confinement systems to generate greater gross income than intensive grazing systems as implemented in different regions of the country (Winsten et al., 2000; Kriegl and McNair, 2005). Smaller dairy operations might be better adapted for rotational grazing compared to larger operations, either because of management and logistical complexity, or because of the life style and income preferences of typical MIRG dairy farmers. Similarly, farmers who rely on traditional extensive pasture grazing practices have been disinclined to shift to more intensive rotational grazing techniques primarily because of perceived increases in labor and management required (Gillespie et al., 2008).

Social Performance of MIRG Systems

Labor Practices

Dairy farmers who use management-intensive rotational grazing emphasize that the approach allows them to spread their labor more evenly throughout the day and the growing season, enables their young children to participate in more farming activities, and gives them a better appreciation for nature and the environment (Ostrom and Jackson-Smith, 2000; Brock and Barham, 2009). However, not all sustainable farming practices necessarily confer improvements in the quality of the labor experience. Beef ranchers, for example, are more likely to prefer extensive grazing approaches than a MIRG system because of the higher total labor (and labor per cow) associated with a more intensive management regime (Gillespie et al., 2008). Many MIRG approaches to sustainable farming require significant investments (Nichols and Knoblauch, 1996) in time and learning by the farm operator.

Impact on Human Nutrition and Health

Although most operators of grazing livestock farms and ranches are drawn to grass-based systems for personal, social and economic reasons (Nichols and Knoblauch, 1996), a number of consumers are attracted to grass-raised meat products for perceived health benefits. In an extensive review of the scientific literature, Clancy (2006) found that meat products from pasture-raised cattle are associated with lower levels of total fat than meat from conventionally grain-finished animals. Similarly, meat and milk from pasture-raised animals has been shown to contain higher levels of particular kinds of fats (Martz et al.,

1997; Dhiman et al., 1999; Beaulieu, 2000; White et al., 2001). Specifically, ground beef from grass-fed cattle contains higher levels of conjugated linoleic acid (CLA), and milk from pasture-raised cows contains higher levels of both CLA and alpha-linolenic acid (ALA, an omega-3 fatty acid). CLA has been linked to reduced risk of heart disease and heart attacks, and omega-3 acids have been linked to the same benefits, plus potential reduced risks of cancer and immune system diseases.

Low-Confinement Integrated Hog-Producing Systems

Forces of Change in the Hog Sector

Hog production (in the United States and globally) has undergone sweeping changes in the last 30 years. The number of U.S. farms with hogs declined from 667,000 in 1980 to 75,442 in 2007, but the total animal inventory increased slightly from 62.3 to 67.8 million (Informa Economics, 2004; USDA-ERS, 2008). Of the total farms with hogs in 2007, 84 percent had less than 1,000 animals each and accounted for 2.3 percent of animals sold. The 2,850 operations with over 5,000 animals made up 3 percent of hog operations and accounted for 87 percent of hogs sold in 2007. Along with the dramatic change in size has come a significant shift in production practices and in facilities. The most important distinctions in type of operation are between confinement operations that are not integrated (except for manure disposal) with crop operations and operations that are highly integrated with crop production. Operations of each type can be classified by the U.S. Environmental Protection Agency as concentrated animal feeding operations (CAFOs), depending on size and methods of manure disposal. The distinction among alternatives, and the wide network of support farm-integrated systems, is clearly described by Gegner (2004).

The largest operations are driven by capital investments, often from meat-processing and marketing firms that are highly vertically integrated (MacDonald and McBride, 2009). The production facilities use full animal confinement, so that animals do not have access to the outdoors or to farm fields. Few of the large operations raise feed crops; most, if not all, purchase their feed inputs. Many of those operations typically specialize in a single stage of production such as animal finishing. Manure handling and facility cleaning is almost exclusively handled by liquid systems, with contracts to landowners who use the manure to produce a wide range of crops (MacDonald et al., 2009). Public controversy over many facets of animal raising and confinement has been escalating. Issues of concerns include animal welfare, widespread use of antibiotics for animal health, farm worker safety, safety of meat products, and environmental impacts on soil and water. These impacts range from nutrient and antibiotic loading on the land and in the waters surrounding the operations, and reduced air quality from volatile organics and other emissions from the large facilities. The debate is especially intense over the large-scale CAFOs for hogs, often referred to as "factory farms" given the scale of operations, their location, and the numbers of public issues surrounding them (Gurian-Sherman, 2008; Pew Commission on Industrial Farm Animal Production, 2008). This report does not assess the sustainability of CAFOs, but outlines in a section below the research needed for holistic evaluation and comparison of system types for each of the major areas of concern (system drivers).

Many small animal-producing farms (from a few hundred up to 3,000 acres) are typically structured for crop-livestock integration, producing feed crops and crop residues for bedding, and they often have at least a portion of the production cycle on rotated fields for farrowing or pasture. They typically use a dry manure-handling system and are almost exclusively owner operated. That type of farm ownership and structure is often a product of traditional farm ownership and operation coming from a crop base, a pasture-based livestock operation, or a mixed crop–livestock operation where farmers are intensifying their operation and often purchasing additional feed. They are responding to a range of economic factors, including opportunities for local and direct marketing, branding, and specialty products for niche markets. A field-crop–hog integrated system, often referred to as "extensive" agriculture, is defined as farming in which large areas of land are used with low to modest outlays in capital expenditure (Honeyman, 2005). Outdoor swine production is a system that allows the pigs outside access including contact with the soil and growing plants (Honeyman et al., 2001b). Most low-confinement alternative systems share many commonalities.

Low-confinement hog systems have followed a trajectory of development that differs from the more common CAFO operations. They resemble the more traditional systems of field farrowing from which they have evolved more closely than they do a CAFO. Most low-confinement hog systems are medium- to small-sized farms and have hundreds rather than tens of thousands of animals (Honeyman, 2005). Farmer groups, cooperatives, and individual farmers use a wide range of reduced-confinement swine systems, with adaptations to many farm environments and types. A subset of low-confinement, extensively raised hog farmers follow guidelines established for U.S. organic systems by the National Organic Standards Board (NOSB). International organic guidelines are similar (Padel et al., 2004). Other farmers use some variation of those practices to raise animals "sustainably," but most often they use chemicals for control of internal parasites or use alternative forms of nose rings for management of animal rooting and pasture or ground cover disruption. Three of the case-study farms in Chapter 7—the Rossman, Mormon Trail, and Thompson farms—raise hogs using bedding systems. Each of those farms raises feeder cattle on pasture, but none has extensive pasture for hogs. Scientific and technical resources available in the United States for alternative, low-confinement, extensive systems are summarized by SARE (2003) and Gegner (2004).

Guiding Principles

The primary guiding philosophy of all swine producers (CAFO and low-confinement) leads to an ultimate goal of maintaining animal health through management and provision of appropriate nutrition to optimize rates of growth and produce meat and carcass quality targeted to their specific markets, while having minimal adverse impact on the environment. In the discussion that follows, the committee applies the more stringent guidelines and literature from the "sustainable" and organic sectors as an example of production alternatives. There is extensive literature on the vastly differing strategies for both high-confinement and low-confinement systems.

NOSB sets the minimum requirement for organic systems in the United States. Many hog trade brands based on "humane" and "sustainable" criteria have most of the same guidelines, but they might differ in their latitude for control of internal parasites, tail docking, or use of nose rings. The guidelines highly influence the feeding, care, and handling of pigs and have a major influence on how pigs are housed, have access to grazing areas, and are allowed to socialize, which, in turn, highly influences the structure of farming systems. The most complex set of guidelines is outlined in the writings of Temple Grandin (2007, 2010) as "core standards." Farmers who use those principles are guided by a philosophy that animals (regardless of species) be treated with respect and allowed to fulfill their instinctive natural behaviors without damaging their environment. Specific factors include:

- Animals must be given opportunity to care for, interact with, and nurture their young.
- They must be able to build nests during farrowing.
- They must have sufficient space to exercise and socialize with herd mates.
- A dry area where all animals can lay down at the same time without soiling their bellies must always be available.
- Air quality must be maintained for good health, including ammonia levels not to exceed 25 parts per million (ppm), with 10 ppm as the goal.
- Pasture or bedding are the preferred environments.
- Nontherapeutic use of antibiotics is prohibited. Animals that have been administered antibiotics must be segregated and not sold under the organic or (most) sustainable brands.

Following "instinctive natural behaviors" requires that animals have access to the outdoors and to pastures for much of the year. Pasture care requires plant cover, with active rooting of the plants to cover a minimum portion of the area, normally at least 70 percent. That requirement, by itself, requires animal-crop integration and a land base sufficient to support the swine herd. It also requires housing to be decentralized to provide shelter either within or in proximity to the pastures. Many farmers use existing farm buildings converted from former dairy or more intensive systems. Hoop houses are recommended for most new construction (Honeyman et al., 2001a). The outdoor and hoop structure research of the Allee Demonstration Farm in Newell, Iowa, is a key source of technology for many swine growers, regardless of region. Specific guidelines for such housing and management practices are available from many state extension agencies and from the National Sustainable Agriculture Information Service (National Sustainable Agriculture Information Service [ATTRA] of the National Center for Appropriate Technology [NCAT], 2009). The Sustainable Agriculture Research and Education program has many materials available (SARE, 2000).

Organic and "natural" pork farms in the Corn Belt fit well into the summary of such systems by Honeyman (2005, p. 15): "Most natural pork markets require outdoor bedding settings, no subtherapeutic antibiotics of growth promoters, no animal by-products in feed, and family farm production settings." Animal housing is a critical factor for all farms. Organic requirements for space, bedding, and access to free space and to pasture are specific and auditable for each stage of animal growth according to USDA organic guidelines.

In addition, "Hoop-fed pigs have fewer aberrant behaviors and handle easier than confinement pigs. Health is similar except for an increase in internal parasites in hoop-fed pigs. Pigs in hoops are in larger groups than in confinement. Biosecurity in hoops is more difficult due to incoming bedding and open access" (Honeyman, 2005, p. 15). Organic farmers have a policy for rescue using antibiotics on occasional animals with a sickness problem, and then culling that animal from the certified market channel. Parasites are controlled with both careful sanitation and use of parasite-control chemicals. Because of parasite control (Baumgartner et al., 2003) and the problems of accessing certified-organic feed, few, if any, "natural" swine farmers raise organic hogs. Requirements for either bedded structures or pasture (or a combination) during grow-out place additional requirements on broad systems integration. The production of small grains for both feed and for bedding material, rotation of pasture to maintain mandatory levels of ground cover, and field conditions for the recycling of bedding and manure make overall farm integration an economic and environmental necessity.

Farrowing operations for organic and "natural" systems differ markedly from those of conventional systems. Sow health is maintained through diet and access to pasture for exercise. Deep bedding systems are used for farrowing huts or pens, often referred to as the Swedish deep-bedding system. When sows are seasonally field farrowed, a limit of 10 or fewer sows per acre, depending on soil type, is used to maintain 80 percent or more pasture rooting and cover. Therefore, farm infrastructure varies, depending on farm history, climate, soil type, land slope, and hydrological characteristics. Some of that variability is documented in an Iowa State University study of four collaborating farms in southern Minnesota over two winters in 2003–2004 (Serfling et al., 2006). The greatest departure from industry practice in hog raising is in the farrowing requirements of most organic and other niche marketing swine operations. The prohibition of farrowing crates and the requirement for deep bedding and nesting capabilities for the sows influence many other system characteristics. Temperature regulations for sows and piglets are critical in winter, particularly in northern climates. The four farms studied used a wide range of farrowing and feeding structures. Results showed that the study farms had 11.0 pigs born live per litter and 8.8 pigs weaned per litter. Those numbers are comparable to Minnesota averages of 10.1 born and 8.7 weaned, and to national, industry-wide averages of 10.0 born and 8.6 weaned.

Herd genetics is extremely important to all niche-marketing swine operations. Most specialty farmers use crossbreeds of Berkshire, Chester White, and Duroc in their herds to achieve high meat quality, effective farrowing, and high growth rates. The herds' performance is related to meat quality, animal behavior, and economics. Most cooperative sustainable animal operations recommend particular producers of semen and specific lines within those sources for farms using artificial insemination. Most niche-market brands have a certification and audit program for their contracting farmers.

Thompson Farm in Boone, Iowa, described in Chapter 7, is an example of a diversified pig-raising farm and has been a model for diversification and low-confinement swine production for many years. The 300-acre farm includes corn, soybean, oats, and hay, with 75 head of beef cattle and 75 hogs in a farrow-finish operation. The farm supports two families without outside hired help. The Thompsons learn and teach low-confinement principles and practices to many thousands of visitors to their farm.

Environmental Impacts of Low-Confinement Hog Systems

Nutrient Cycling, Odor Control, and Greenhouse-Gas Emissions
Nutrient cycling, odor control, and greenhouse-gas emissions are inseparably linked in pig raising, and determined by the way in which the many kinds of structures are managed for manure collection and removal. All farmers with organic or sustainable certification use various forms of dry bedding for most animal shelter areas (Honeyman, 1996). Farrowing guidelines require dry bedding, while grow-out shelters, whether converted traditional barns or hoop structures, are often set up with the Swedish dry-bedding system. Those areas are kept dry with bedding and ventilation. Manure and urine is thus incorporated in high-carbon, dry crop, or sometimes woody residues. The animal wastes are collected and handled for eventual field application as dry material with reasonably high carbon-to-nitrogen ratio, thereby reducing both ammonia and volatile organic loss. A key factor is the type, amount, and frequency of application in the structures, with availability and cost of bedding material often the limiting factors. There is an extensive literature on bedding systems for farrowing, with the most comprehensive and often referenced done by Honeyman (2005). Research on the direct effects of the confinement portion of those systems is sparse. There is no effluent, either through soil below the structures or to surrounding areas from

with humane standards to minimize bruising and to reduce stress-induced meat quality changes. The meat from every farm is tested at intervals by a pork quality testing service of Iowa State University for moisture, acidity, marbling, color, and other characteristics to assure brand quality. Lactic acid content is measured as an indicator of stress during transport and handling. Farmers use the ratings to rank themselves in providing quality product, with annual incentives calculated on the basis of their scores. In general, alterative producers and many niche market brands produce meat with higher fat content, a darker red color, and a more distinctive taste, determined both by genetics of the herds for meat type and by access to outdoor and open spaces, which influence meat texture.

As mentioned in the guidelines, many niche-market brands do not allow antibiotic use in any of their marketed animals. Except for certified organic products, others use standard parasite control chemicals. The prevalence of methicillin-resistant staphylococcus (MRSA)[1] increasingly reported for confinement operations in Iowa and in several other U.S. locations and several developed countries (Smith et al., 2009) is not considered to be an issue for most alternative systems. No extensive MRSA surveys have been done for those operations, but as of yet, none of the niche market or organic pigs, or workers who volunteered to be tested by Iowa State University, has tested positive.

Public Reaction

An increasing number of scientists are testing attitudes among rural residents and communities toward intensive animal agriculture (NRC, 2003). Studies indicate that rural residents and activists, while understanding the economic constraints that swine producers are under, strongly feel that large-scale confinement operations are, at least temporarily, eroding farmers' traditional base of support (Pew Commission on Industrial Farm Animal Production, 2008). Residents tend to be more tolerant of swine facilities when farmers are long-time residents and are active in the community than of new, large-scale industrial facilities coming in under corporate ownership and management (Reisner and Taheripour, 2007).

Summary

Extensive, alternative hog production systems can be equally if not more productive than large-scale specialized operations, even if the significantly higher externalities of the large operations relying on liquid manure handling are disregarded. Small, integrated operations fit better into acceptable patterns of landscape use in and around rural communities. Many such operations exist in many parts of the United States, and they can provide ample examples and data points for comparisons with large-scale systems. Comparative studies of systems types using a holistic approach, looking at economic, environmental, and social factors embedded in a sustainability matrix for efficiency, resistance, and resilience across landscapes, are needed to identify how each system performs with respect to each of the four sustainability goals and to explore how synergies are achieved in each systems type.

PERENNIAL AGRICULTURE SYSTEMS

Perennial crops generally have advantages over annuals in maintaining important ecosystem functions, particularly on marginal landscapes or where available resources are limited. Perennial grain agriculture, sometimes called natural systems agriculture, is an ecology-based approach to agricultural production in which perennial grain-producing

---

[1] MRSA is an infection caused by a strain of bacteria, *Staphylococcus aureus*, that is resistant to the broad-spectrum antibiotics commonly used to treat it.

crops are grown alone or in mixtures. Growing perennial grains for food or perennial grasses for biofuel can potentially increase carbon sequestration in soil and mitigate nutrient runoff from agricultural fields. Perennial grain and perennial grass-based biofuel systems are being developed, and the following sections discuss their potential contributions to various sustainability goals.

Perennial Grain System

Today most of humanity's food comes directly or indirectly (as animal feed) from cereal grains, legumes, and oilseed crops, all of which are annual crops. Replacing some of the single-season crops with perennials would create large root systems capable of preserving the soil and would allow cultivation in areas currently considered marginal (Figure 5-1) (Cox et al., 2006; Glover et al., 2007). Perennial plants reduce erosion risks, sequester more carbon, and require less fuel, fertilizer, and pesticides to grow than their annual counterparts (Glover et al., 2007). Plant breeders see several opportunities for perennial plants to maintain their perennial characteristic and produce high seed yield for the following reasons:

- Perennials have greater access to resources over a longer growing season.
- Perennials have greater ability to maintain the health and fertility of a landscape over longer periods of time.

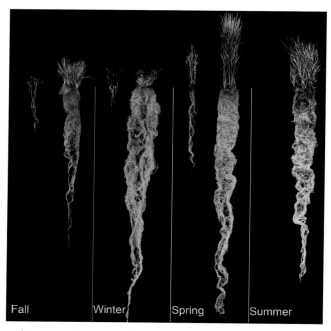

FIGURE 5-1 Root and top growth of annual wheat (at left in panels above) and its perennial relative, wheatgrass (at right in panels above), at four different times of year.
NOTES: Perennial crops have deeper root systems than annuals, providing access to more water and nutrients. Perennials also have a longer growing season, allowing more sunlight to be captured by the crop.
SOURCE: Glover (2010). Reprinted with permission from the American Association for the Advancement of Science.

- The unprecedented success of plant breeders in recent decades to select for negatively correlated characteristics in annual crops (such as seed yield and protein content) can be applied to perennial crop development. Recent advances in plant breeding, such as the use of marker-assisted breeding, genomic in situ hybridization, transgenic technologies, and embryo rescue, provide new opportunities for plant breeders to select for desired characteristics.

In the last seven years, plant breeders in the United States, Argentina, Australia, China, India, and Sweden have initiated plant genetic research and breeding programs to develop wheat, rice, corn, sorghum, sunflower, intermediate wheatgrass, and other species as perennial grain crops (Glover and Reganold, 2010). However, it could take 20 years to develop perennial wheat ready to be widely planted on farms. At present, it takes plant breeders more than a decade just to develop new varieties of annual wheat and ensure that they are ready to be widely grown for commercial use.

Impact

Comparisons of the effects of annual and perennial management systems on soil properties have shown that well-managed perennial systems compare more favorably than annual management systems. Robertson et al. (2000) found that perennial production systems of alfalfa, poplar trees, and perennial grass systems had higher levels of soil organic carbon and resulted in lower net greenhouse-gas emissions than annual cropping systems. Other researchers have similarly found positive effects of perennial vegetation on soil properties compared to annual cropping systems (Weil et al., 1993; Mummey et al., 1998; Karlen et al., 1999; Culman et al., 2010). Randall et al. (1997) also found perennial systems to be effective at reducing the potential for ground water contamination by nitrate leaching. Perennial grain agriculture is expected to provide similar benefits.

Perennial Grasses for Biofuels

Depending on landscape management, the use of cellulosic feedstock for biofuel production can avoid some of the social and environmental concerns associated with corn grain ethanol and soybean biodiesel. As noted by Robertson et al. (2008b, p. 49), "Biofuel sustainability has environmental, economic, and social facets that all interconnect. Tradeoffs among them vary widely by types of fuels, and where they are grown and, thus, need to be explicitly considered by using a framework that allows the outcomes of alternative systems to be consistently evaluated and compared. A cellulosic biofuels industry could have many positive social and environmental attributes, but it could also suffer from many of the sustainability issues that hobble grain-based biofuels, if not implemented the right way."

Impact on Food Security

Unlike corn grain ethanol and soybean biodiesel, the feedstock for cellulosic biofuels does not have to be grown on fertile cropland. Some dedicated fuel crops (for example, switchgrass and native grasses) can be grown on marginal lands that are not used for food and feed production (NAS-NAE-NRC, 2009b). Other lignocellulosic feedstocks include residual products from farming (for example, corn stover) or forestry operations (for example, residues from forest thinning). However, if dedicated fuel crops displace food crops, the social and environmental concerns pertaining to corn ethanol discussed in Box 2-2 would be relevant in the context of cellulosic biofuels.

Environmental Impact

Debate continues about the relative merits of crop mixtures, but agronomy and industry development have moved toward sole crops (for example, *Miscanthus* or switchgrass) suited to differing climates, soil types, and growing conditions. Perennial crops have many potential environmental benefits, including reduced soil erosion, greater efficiency of nutrient uptake, and greater attractiveness to wildlife. It is argued that the enhanced biodiversity of mixtures can add significantly to environmental benefits (Tilman et al., 2006), but the weight of evidence seems to indicate higher productivity and much greater ease of commercial production from monocultures.

Production research in the Corn Belt states includes model projections of potential of different land types (Nelson, R. for Kansas Biomass Committee, 2007) and a series of agronomic trials underway by USDA laboratories and by state universities. Research in Minnesota on the landscape positioning of perennial biomass crops using alfalfa, willow, poplar, cottonwood, false indigo, switchgrass, and a polyculture mix shows that different species can be used on lands with different characteristics (such as slopes, soil types, and water availability) to optimize biomass yields and improve environmental quality (Johnson et al., 2008). Existing precision geo-referenced yield monitoring of commercial grain crops can be a tool in that design. Benefits to such positioning not only include optimizing yield potential and field water availability, but also can greatly enhance nutrient recovery and cycling (Annex et al., 2007). Such energy crops, if properly placed in the landscape, have potential to increase productivity of land types, provide diversity of markets for farmers who produce food crops and animals, and could contribute to ecosystem services. They could serve as riparian buffers, filter strips, and nutrient traps and could stabilize fragile land on a gentle slope. They could, therefore, replace at least some of the noncommercial crops in set-aside and other programs that currently require government subsidy.

Lignocellulosic biofuels (including ethanol derived from biochemical conversion or gasoline and diesel derived from thermochemical conversion) have been estimated to have lifecycle greenhouse-gas emissions of close to zero (NAS-NAE-NRC, 2009b). If the lignocellulosic biomass is grown in an appropriate landscape, it can provide biofuel feedstock and enhance environmental quality and the quality of the resource base (NAS-NAE-NRC, 2009b). However, water use for converting biomass to liquid fuels could create competition for water with agricultural production. Biorefineries will likely be located close to where the biomass feedstock is produced. The amount of water required for processing biomass into ethanol is estimated to be 2–6 gallons per gallon of ethanol produced (Aden et al., 2002; Pate et al., 2007).

Economic Impact

Dedicated fuel crops (for example, *Miscanthus* and switchgrass) can be grown on lands that might not be suitable for other crops and can provide an additional income source for farmers, but they are a single-market commodity. If they are to be used for ethanol production, their demand will depend on oil price, the percentage of ethanol that can be blended in fuel, and the number of flex-fuel vehicles, as discussed in Box 2-2. They can be used to produce gasoline and diesel by thermochemical conversion. Thermochemical conversion technology is estimated to be ready for deployment by 2020, and the fuel products will be compatible with existing transportation-fuel infrastructure. In addition, dedicated fuel crops can be used for bioenergy production (NAS-NAE-NRC, 2009a).

GAPS IN EXISTING SCIENCE AT THE SYSTEMS LEVEL

In the preceding sections, four alternative system "types," including organic, integrated crop–livestock, management-intensive rotational grazing, and low-confinement hogs were illustrated. Each type was selected because of its overall enterprise mix and structure, with components and practices that ostensibly lead to complementarities and synergies of resource use and containment, positive impacts on their ecological and social environments, and resistance and resilience in each of its resource domains. Each has substantial claims being made for its contribution to sustainability. Each has a large and growing number of farms in many parts of the country that are, or could, serve as research cases. Organic farms certified in accordance with USDA standards, for example, are found for most commodities, in every part of the country, and across a range of farm sizes. The case-study farms of Chapter 7 represent organic and other system types that have departed from the traditional conventional farms and moved much further along the trajectory toward improved sustainability. The four types presented above are described by research data that focus on component pieces of the systems, much of which is summarized in Chapters 3 and 4. The paucity of reliable data on holistic descriptions of those operations in the United States is evident. Holistic comparisons between system types aimed at improving sustainability are not possible with present data, given the multidimensional nature of sustainability as defined in Chapter 2.

What are the resource constraints for a systems design environment, and which systems and sustainable management practices best fit within the biophysical conditions and meet the social and economic sustainability goals? A few of the constraints coming from data of the above alternative systems include:

- Land capability classification, sensitivity to runoff, soil erosion, or other loss.
- Biodiversity needs (diversity of plants, wildlife, soil organisms).
- Water availability, alternative demands, projections for future needs and sharing.
- Sensitivity to water quality degradation and downstream hypoxia.
- Probability of extreme climatic events (flooding, short-term changes in water availability).
- Population density and exposure to odors, noise, or other "nuisance" factors.
- Support for the business community, employment needs, and social viability of local communities.

Design Within Systems Types

The most common "systems approaches," and for the most part, the most useful, have been studies comparing integrative practices that are well-defined and are located within operating farm environments. Examples are tillage comparisons, cover crop integration into rotations, integrated pest management for particular crops such as tree fruit or certain field crops, and, in some instances, well-defined approaches such as organic and conventional approaches for a particular crop such as apples or cherries (Reganold et al., 2001; Peck et al., 2006). Such studies compare specific integrative practices within a whole-farm context, where appropriate interactions can be defined and measured (Drinkwater, 2002; Snapp and Pound, 2008). The on-farm systems studies are usually conducted under experimental management by farmers and last for 5–10 years to measure intermediate-term effects. Those studies can be significantly cost-effective, but require considerable farmer (or farmer–community-of-interest) interest and support (Carter et al., 2004).

Research personnel manage a number of well-known, long-term systems studies in the United States on experiment stations. The studies are designed to measure crop performance and often soil and environmental impact over 15–20 years or longer. A few studies have century-old duration (Paul et al., 1997). The more modern ones (designed over the last 30 or so years) are replicated and sometimes have large plots size with ancillary smaller-plot experiments (Temple et al., 1994; Robertson et al., 2008a). They use planned rotations and integrative management practices and are designed to study comparative agronomic, horticultural, and ecosystem processes (see Box 3-3 in Chapter 3). Those carefully designed "experiments" are more appropriately "research platforms" within which specific "factor studies," usually focused on specific biological process, can be conducted (Robertson et al., 2000; Hepperly et al., 2006; Cavigelli et al., 2008; Center for Integrated Agricultural Systems, 2009; University of California-Davis, 2009). Although those experiments measure specific farm enterprises (rotations and so on) within the context of local environments and ecologies, they are not whole-farm studies in that they do not represent true farm conditions of labor and equipment use, enterprise diversity, and a scale suitable to measure landscape-level impacts. Whole-farm studies are highly useful in quantifying biological processes and in calibrating models for use in widespread holistic systems comparisons. They are expensive to run, require stable and long-term institutional interest and funding, and have to be located within agricultural environments that represent large regional production zones for them to be both relevant and cost-effective.

A number of questions relating to environmental fragilities and constraints are important to consider. For example,

- Which systems of tillage and crop and animal diversity provide high nutrient flow rates (for high productivity) with farm- and landscape-level recycling and containment?
- What is the net nutrient flow into or out of production systems? For those with a large relative inflow (such as most large confinement operations), what is the "command" (distribution) area for nutrient dispersal for alternative scales of operations and what is the "life expectancy" of that land for phosphorus and other nutrient loading given alternatives for cropping or other use?
- Given the environmental conditions, which practices (for example, raising animals to optimize genetic immunity) or animal production systems would be least conducive to disease build up? What are the tradeoffs between those approaches and that of substitution of antibiotics?
- Could animal systems be designed to minimize the creation of odors and to optimize manure quality close to the source? If engineering solutions are to be depended upon to correct the problems after they occur, what are the tradeoffs?
- What are the economic, environmental, and social costs of the presence of antibiotic-resistant organisms within production facilities, in food products, or in the environment?
- What influence do subsidies to various sectors of agriculture have on the viability and productivity of alternative systems?
- What are the lifecycle energy costs for different systems types?

Holistic Comparisons Between Farming Systems Types

There are examples of whole-farm systems research in the United States (albeit with limited scope) that are highly productive and necessary, such as the examples cited above and in Chapter 3. "Whole-farm" studies, for the most part, have focused on economic performance or on efficiency of resource use such as energy, water, or output per unit of land. Those studies have been most useful where adequate numbers of farms have been selected to compare reasonably well-defined farming types. The organic versus "conventional" or nonorganic comparisons are a good example. Comparative studies work best where there are reasonably large numbers of comparable farms in a given geographical area (for instance, Lockeretz et al., 1984; Drinkwater et al., 1995).

Most "reasonable" models for sustainability of U.S. agriculture, if built on sound theoretical grounds, would include diversity in farm scale, structure, product output, and multifunctionality at each scale (Gibson et al., 2007). The objective of holistic comparisons, therefore, is not to identify the "best" system for each environment, but the relative strengths and tradeoffs for each alternative. It might well be that some conventional or alternative extremes of size or configuration are simply not worth trying to "fix." As an example, agriculture as it now exists in water overdraft areas is clearly not sustainable.

In conducting the comparisons, farms that represent working examples of each comparison type would be selected scientifically. Data would originate from and process models would be base-calibrated for the selected farms. For example, with each of the four alternative types described above, hundreds of farms located in differing environments could be used for case studies. In meeting the environmental needs and constraints and answering the research questions above, one approach is to set specific targets for product output for a given geographical area or environment (whether number of animals, or of crop product output), then calculate the impact of a set of farms or facilities of each alternative on the many parameters to be considered.

The lesson learned from systems research is that it is complex. For systems research to be successful and cost-effective, the research objectives would have to be clearly defined and hypothesis driven. The design and expected duration of the research would have to be consistent with resources and objectives. The required data would be not otherwise available from other, less complex experiments that provide the needed data and interactions. The research would be done with the least complicated design that meets the requirements of the research. In developed countries as in the United States, a proportion of the more complex experiments and research platforms have not met expectations for output in comparison to the resources invested.

BIOGEOPHYSICAL LANDSCAPE-LEVEL SUSTAINABILITY ANALYSIS AND PLANNING

Many qualities of a sustainable agriculture are both defined and managed at aggregate levels beyond field and farm boundaries at the community, watershed, and river basin scales (as discussed in Chapter 2). That generalization is especially true for many social and economic effects at aggregate levels. (See Chapter 4.) Most of the foregoing discussion in this report on practices and farming systems has focused on comparative function and impact as measured by productivity, efficiency of resource use, environmental impact, and ecosystem integration as implemented within farms and the immediate environments of those farms. In the coming decades, the environments of those farms will change, as discussed in Chapter 2, and market requirements and opportunities will evolve. The land form and soil types within a farm that have been major determinants of farm crop and animal selection and placement as guided by best management practices (USDA-NRCS, 2009) and evolving precision agriculture tools (Srinavasan, 2006) will be increasingly influenced by forces beyond the farm. Examples include regional water shortages, loading of major off-farm water bodies creating hypoxic zones, or new crop opportunities for energy production. The location and scale of animal enterprises will change, driven by such factors as regional phosphorus loading in the soil, water needs, or risk from flooding. Biological diversity and ecosystem health within landscapes are important factors across scales. Ecosystem health and "ecoagricultural" landscapes are of increasing sustainability concern and have a large and growing research literature (Scherr and McNeely, 2007; Van Bruggen, 2008). The diversity of crop, animal, and native vegetation areas within a landscape and the diversity of farm types that will provide them is thus important. The proportion of a landscape devoted to each farming system type, their positioning along gradients of land capability and to each other and to the practices employed all contribute to aggregate effects across larger scales. Those changes will be brought about by gradual changes within farms, as well as by multilevel policies to encourage structural changes across landscapes (Nassauer and Opdam, 2008).

Recent modeling application for two watersheds of the Upper Mississippi River basin illustrate the potential for shifting farming system enterprises, numbers, location, and balance to impact water quality (Burkart et al., 2005; Nassauer et al., 2007). Two watersheds in western Iowa counties having long-term river flow and water quality data, and a database of farming systems, crop acreage, and livestock census data, were modeled for sediment, nitrogen flow, water runoff, and other factors. Existing and three alternative crop, pasture, and animal component combinations were compared. Those studies then have formed the basis for subsequent landscape ecology studies, which made assumptions of altered percentages of the landscape occupied by the same types of farming systems found in the area (Nassauer and Opdam, 2008). Across many watersheds, the nitrate losses could be reduced by more than 40 percent by altering the percentages of the landscape occupied by the same farming system types. For specific watersheds such as Walnut Creek and Buck Creek, the losses could be reduced by up to 74 percent by changing the portion of the landscape devoted to cover crops and pasture, with actual increases in profitability. The most profitable alternative had a slight increase in the numbers of hogs, significant increase in cow-calf operations, more pasture and hay, and reduction in corn–soybean acreage. Societal costs and downstream impacts were estimated from fish kill and sediment-loading values. The potential benefits of introducing modern perennial energy crops into those systems are enormous. The availability of considerable stream flow and loading data and the availability of numerous models for different parameter flows are critical for that type of watershed analysis. Similar scenario studies have been conducted in Minnesota (Boody et al., 2005), which demonstrated that economic and environmental (water quality protection and conservation of biodiversity) benefits could be achieved through changes in agricultural land management without increasing public costs.

There are publicly available (digitized) sources of soil type, land classification, hydrology, climatology, and a host of demographic and other ecosystem parameters. Likewise, process-level models for most biogeophysical factors (such as nitrogen flows, carbon processes, and energy transformations and use) are under continual evaluation by scientists in the various agriculturally related societies and are in the public domain. Geographic information system (GIS) tools for agricultural use are evolving from precision agriculture research (Pierce and Clay, 2007), but modeling tools for use in the planning and assessment of agricultural landscapes are less well developed, and many are based on proprietary

Conant, R.T., J. Six, and K. Paustian. 2003. Land use effects on soil carbon fractions in the southeastern United States: I. Management-intensive versus extensive grazing. *Biology and Fertility of Soils* 38(6):386–392.

Conneman, G., J. Grace, J. Karszes, S. Marshman, E. Staehr, S. Schosek, L. Putman, B. Casey, and J. Degni. 2000. Intensive Grazing Farms New York. Ithaca, N.Y.: Cornell University.

Cox, T.S., J.D. Glover, D.L. Van Tassel, C.M. Cox, and L.R. DeHaan. 2006. Prospects for developing perennial-grain crops. *BioScience* 56(8):649–659.

Culman, S.W., S.T. DuPont, J.D. Glover, D.H. Buckley, G.W. Fick, H. Ferris, and T.E. Crews. 2010. Long-term impacts of high-input annual cropping and unfertilized perennial grass production on soil properties and belowground food webs in Kansas, USA. *Agriculture, Ecosystems & Environment* 137(1–2):13–24.

Cuttle, S.P. 2008. Impacts of pastoral grazing on soil quality. In Environmental Impacts of Pasture-Based Farming, R.W. McDowell, ed.Wallingford, UK: CAB International.

Dabbert, S., A.M. Haring, and R. Zanoli, eds. 2004. Organic Farming: Policies and Prospects. London: Zed Books.

Dartt, B.A., J.W. Lloyd, B.R. Radke, J.R. Black, and J.B. Kaneene. 1999. A comparison of profitability and economic efficiencies between management-intensive grazing and conventionally managed dairies in Michigan. *Journal of Dairy Science* 83:2412–2420.

de Klein, C.A.M., and R.J. Eckard. 2008. Targeted technologies for nitrous oxide abatement from animal agriculture. *Australian Journal of Experimental Agriculture* 48:14–20.

de Klein, C.A.M., R.R. Sherlock, K.C. Cameron, and T.J. van der Weerden. 2001. Nitrous oxide emissions from agricultural soils in New Zealand—a review of current knowledge and directions for future research. *Journal of the Royal Society of New Zealand* 31:543–574.

Delate, K., M. Duffy, C. Chase, A. Holste, H. Friedrich, and N. Wantate. 2003. An economic comparison of organic and conventional grain crops in a long-term agroecological research (LTAR) site in Iowa. *American Journal of Alternative Agriculture* 18(2):59–69.

DeVore, B. 2001. Same storm, different outcomes. *The Land Stewardship Letter* 19:2. Available at http://www.landstewardshipproject.org/lsl/lspv19n2.html#coverstory. Accessed on February 8, 2010.

Dhiman, T.R., G.R. Anand, L.D. Satter, and M.W. Pariza. 1999. Conjugated linoleic acid content of milk from cows fed different diets. *Journal of Dairy Science* 82:2146–2156.

Doran, J.W., F. Kirschenmann, and F. Magdoff. 2007. Balancing food, environmental and resource needs. *Renewable Agriculture and Food Systems* 22:77–79.

Dorsey, J.Z.J. 1998. Production performance of sheep under rotational (strip) grazing systems. Master's thesis. Alabama A&M University, Huntsville.

Drinkwater, L., D. Letourneau, F. Workneh, A.H.C. Van Bruggen, and C. Shennan. 1995. Fundamental differences between conventional and organic tomato agroecosystems in California. *Ecological Applications* 5(4):15.

Drinkwater, L.E. 2002. Cropping systems research: reconsidering agricultural experimental approaches. *HortTechnology* 12(3):355–361.

———. 2009. Ecological knowledge: Foundation for sustainable organic agriculture. Pp. 19–47 in Organic Farming: The Ecological System, C. Francis, ed. Madison, Wis.: American Society of Agronomy.

Duram, L.A. 2005. Good Growing: Why Organic Farming Works. Lincoln: University of Nebraska Press.

Entz, M.H., V.S. Baron, P.M. Carr, D.W. Meyer, S.R. Smith, and W.P. McCaughey. 2002. Potential of forages to diversify cropping systems in the northern Great Plains. *Agronomy Journal* 94(2):240–250.

Entz, M.H., W.D. Bellotti, J.M. Powell, S.V. Angadi, W. Chen, K.H. Ominski, and B. Boelt. 2005. Evolution of integrated crop-livestock production systems. In Grassland: A Global Resource, D.A. McGilloway, ed. Wageningen, the Netherlands: Wageningen Academic Publishing.

Evans, R. 1997. Soil erosion in the UK initiated by grazing animals: a need for a national survey. *Applied Geography* 17(2):127–141.

Flora, C., J.R. Black, T.J. Dalton, R. Kershegan, M. Liebman, S.N. Smith, S.S. Snapp, and G.K. White. 2004. Reintegrating Crop and Livestock Enterprises in Three Northern States. Ames: Iowa State University.

Franzluebbers, A.J., and J.A. Stuedemann. 2008. Soil physical responses to cattle grazing cover crops under conventional and no tillage in the Southern Piedmont USA. *Soil & Tillage Research* 100(1–2):141–153.

Fuhlendorf, S.D., and F.E. Smeins. 1999. Scaling effects of grazing in a semi-arid savanna. *Journal of Vegetation Science* 10:731–738.

Galloway, J.N., and E.B. Cowling. 2002. Reactive nitrogen and the world: 200 years of change. *Ambio* 31(2):64–71.

Gamble, B.E., G. Siri-Prieto, D.W. Reeves, and R.L. Raper. 2005. Forage and tillage systems for integrating winter-grazed stocker cattle in peanut production. Paper presented at the 27th Annual Southern Conservation Tillage Systems Conference, June 27–29, 2005, Clemson, S.C. Available at http://www.ars.usda.gov/SP2UserFiles/Place/66120900/IntegratedFarmingSystems/Gamble_05a.pdf. Accessed on February 8, 2010.

Gardner, B.L. 2005. The little guys are O.K. *New York Times*. March 7, 2005.

Gegner, L. 2004. Hog production alternatives. Available at http://attra.ncat.org/attra-pub/PDF/hog.pdf. Accessed on August 12, 2009.

Gibson, R.H., S. Pearce, R.J. Morris, W.O.C. Symondson, and J. Memmott. 2007. Plant diversity and land use under organic and conventional agriculture: a whole-farm approach. *Journal of Applied Ecology* 44:792–803.

Gillespie, J.M., W. Wyatt, B. Venuto, D. Blouin, and R. Boucher. 2008. The roles of labor and profitability in choosing a grazing strategy for beef production in the U.S. Gulf Coast Region. *Journal of Agricultural and Applied Economics* 40(1):301–313.

Glenna, L., and R. Jussaume. 2007. Organic and conventional farmers' opinions on GM crops and marketing strategies. *Renewable Agriculture and Food Systems* 22(2):118–124.

Gliessman, S.R. 1998. Agroecology: Ecological Processes in Sustainable Agriculture. Chelsea, Mich.: Ann Arbor Press.

Glover, J.D. 2010. Policy forum on increasing food and ecosystem security through perennial grain breeding. *Science* (In press).

Glover, J.D., C.M. Cox, and J.P. Reganold. 2007. Future farming: a return to roots? *Scientific American* 297:82–89.

Glover, J.D., and J.P. Reganold. 2010. Perennial grains: food security for the future. *Issues in Science and Technology* 26(Winter):41–47.

Gloy, B.A., L.W. Tauer, and W. Knoblauch. 2002. Profitability of grazing versus mechanical forage harvesting on New York dairy farms. *Journal of Dairy Science* 85(9):2215–2222.

Gollehon, N., M. Caswell, M. Ribaudo, R. Kellogg, C. Lander, and D. Letson. 2001. Confined Animal Production and Manure Nutrients. Washington D.C.: U.S. Department of Agriculture.

Golley, F.B. 1993. A History of the Ecosystem Concept in Ecology: More than the Sum of the Parts. New Haven, Conn.: Yale University Press.

Grandin, T. 2007. Behavioral principles of handling cattle and other grazing animals under extensive conditions. In Livestock Handling and Transport, T. Grandin, ed. Wallingford, UK: CAB International.

———. 2010. Improving Animal Welfare: A Practical Approach. Wallingford, UK: CAB International.

Greene, C., C. Dimitri, B.-H. Lin, W. McBride, L. Oberholtzer, and T. Smith. 2009. Emerging issues in the US organic industry. Available at http://www.ers.usda.gov/Publications/EIB55/EIB55.pdf. Accessed on February 8, 2010.

Greenwood, K.L., and B.M. McKenzie. 2001. Grazing effects on soil physical properties and the consequences for pastures: a review. *Australian Journal of Experimental Agriculture* 41(8):1231–1250.

Gurian-Sherman, G. 2008. CAFOs uncovered; the untold costs of confined animal feeding operations. Available at http://www.ucsusa.org. Accessed on August 12, 2009.

Guthman, J. 2004. Agrarian Dreams: The Paradox of Organic Farming in California. Berkeley: University of California Press.

Haan, M.M., J.R. Russell, W.J. Powers, J.L. Kovar, and J.L. Benning. 2006. Grazing management effects on sediment and phosphorus in surface runoff. *Rangeland Ecology and Management* 59(6):607–615.

Hall, A., and V. Mogyorody. 2001. Organic farmers in Ontario: an examination of the conventionalization argument. *Sociologia Ruralis* 41(4):399–422.

Hallam, A., ed. 1993. Size, Structure, and the Changing Face of American Agriculture. Boulder, Colo.: Westview Press.

Han, H., J.D. Allan, and D. Scavia. 2009. Influence of climate and human activities on the relationship between watershed nitrogen input and river export. *Environmental Science and Technology* 43(6):1916–1922.

Hanson, G.D., L.C. Cunningham, M.J. Morehart, and R.L. Parsons. 1998. Profitability of moderate intensive grazing of dairy cows in the Northeast. *Journal of Dairy Science* 81:821–829.

Hanson, J.C., E. Lichtenberg, and S.E. Peters. 1997. Organic versus conventional grain production in the mid-Atlantic: an economic and farming system overview. *American Journal of Alternative Agriculture* 12(1):2–9.

Harmon, J.D., D.S. Bundy, T. Richard, S.J. Hoff, and A. Beatty. 2002. Survey Monitoring of Environmental Factors from Bedded Swine Systems. Des Moines, Iowa: National Pork Board.

Harwood, R.R. 1990. A history of sustainable agriculture. In Sustainable Agricultural Systems, C.A. Edwards, R. Lal, P. Madden, R.H. Miller, and G. House, eds. Ankeny, Iowa: Soil and Water Conservation Society.

Hendrickson, J., G.F. Sassenrath, D. Archer, J. Hanson, and J. Halloran. 2008. Interactions in integrated US agricultural systems: the past, present and future. *Renewable Agriculture and Food Systems* 23(4):314–324.

Hendrickson, J.R., J.D. Hanson, D.L. Tanaka, and G. Sassenrath. 2007. Principles of integrated agricultural systems: introduction to processes and definition. *Renewable Agriculture and Food Systems* 23(4):265–271.

Hensler, A., D. Barker, M. Sulc, S. Loerch, and L.B. Owens. 2007. Management Intensive Grazing and Continuous Grazing of Hill Pasture by Beef Cattle. Coshocton, Ohio: U.S. Department of Agriculture Agricultural Research Service.

Hepperly, P.R., D.D. Douds, and R. Seidel. 2006. Long term analysis of organic and conventional maize and soybean cropping systems. In Long-Term Field Experiments. Pp. 15–31 in Organic Farming, J. Raupp, C. Pekrun, M. Oltmanns, and U. Köpke, eds. Bonn, Germany: International Society of Organic Agriculture Research.

Hillger, D.E., S.C. Weller, E. Maynard, and K.D. Gibson. 2006. Weed management systems in Indiana tomato production. *Weed Science* 54(3):516–520

Holechek, J.L., R. Valdez, S.D. Schemnitz, R.D. Pieper, and C.A. Davis. 1982. Manipulation of grazing to improve or maintain wildlife habitat. *Wildlife Society Bulletin* 10(3):204–210.

Honeyman, M.S. 1996. Sustainability issues of US swine production. *Journal of Animal Science* 74:1410–1417.

———. 2005. Extensive bedded indoor and outdoor pig production systems in USA: current trends and effects on animal care and product quality. *Livestock Production Science* 94:15–24.

Honeyman, M.S., and J.D. Harmon. 2003. Performance of finishing pigs in hoop structures and confinement during winter and summer. *Journal of Animal Science* 81(7):1663–1670.

Honeyman, M.S., J.D. Harmon, J.B. Kliebenstein, and T.L. Richard. 2001a. Feasibility of hoop structures for market swine in Iowa: pig performance, pig environment, and budget analysis. *Applied Engineering in Agriculture* 17(6):869–874.

Honeyman, M.S., J.J. McGlone, J.B. Kliebenstein, and B.E. Larson. 2001b. Outdoor pig production. In Pork Industry Handbook. West Lafayette, Ind.: Purdue University.

Honeyman, M.S., R.S. Pirog, G.H. Huber, P.J. Lammers, and J.R. Hermann. 2006. The United States pork niche market phenomenon. *Journal of Animal Science* 84(8):2269–2275.

Hueth, B., M. Ibarburu, and J. Kliebenstein. 2005. Business Organization and Coordination in Marketing Specialty Hogs: A Comparative Analysis of Two Firms in Iowa. Ames: Iowa State University Center for Agricultural and Rural Development.

IFOAM (International Federation of Organic Agriculture Movements). 2002. IFOAM basic standards for organic production and processing, chapter 8: social justice. Available at http://www.uni-kassel.de/fb11/fnt/download/frei/dII/IFOAM%20Standards.pdf. Accessed on February 8, 2010.

Informa Economics. 2004. Special report: the changing U.S. pork industry. Available at http://www.informaecon.com/LVNov1.pdf. Accessed on August 12, 2009.

Iowa State University Extension Service. 1996. Outdoor pig production: an approach that works. Available at http://www.thepigsite.com/articles/?Display=1119. Accessed on August 12, 2009.

Johnson, G., C. Sheaffer, H.J. Jung, U. Tschirmer, S. Banerjee, K. Petersen, and D.L. Wyse. 2008. Landscape and Species Diversity: Optimizing the Use of Land and Species for Biofuel Feedstock Production Systems. Houston, Tex.: American Society of Agronomy.

Kanneganti, V.R., and S.R. Kaffka. 1995. Forage availability from a temperate pasture managed with intensive rotational grazing. *Grass and Forage Science* 50(1):55–62.

Karlen, D.L., M.J. Rosek, and J.C. Gardner. 1999. Conservation reserve program effects on soil quality indicators. *Journal of Soil and Water Conservation* 54:439–444.

Kasperczyk, N., and K. Knickel. 2006. Environmental impacts of organic farming. In Organic Agriculture: A Global Perspective, P. Kristiansen, A. Taji, and J. Reganold, eds. Collingwood, Victoria, Australia: CSIRO Publishing.

Kellogg, R.L., C.H. Lander, D.C. Moffitt, and N. Gollehon. 2000. Manure Nutrients Relative to the Capacity of Cropland and Pastureland to Assimilate Nutrients: Spatial and Temporal Trends for the United States. Washington, D.C.: USDA Natural Resources Conservation Series and the Economic Research Service.

Kemp, D.R., and D.L. Michalkand. 2005. Grasslands for production and the environment. In Utilisation of Grazed Grass in Temperate Animal Systems, J.J. Murphy, ed. Wageningen, the Netherlands: Wageningen Academic Publishing.

Key, N., and M.J. Roberts. 2007. Commodity program payments and structural change in agriculture. *Journal of Agricultural and Resource Economics* 32(2):330–348.

Kleinman, P.J.A., and K. Soder. 2008. The impact of hybrid dairy systems on air, soil and water quality: focus on nitrogen and phosphorus cycling. In Environmental Impacts of Pasture-based Farming, R.W. McDowell, ed. Wallingford, UK: CAB International.

Klepper, R., W. Lockeretz, B. Commoner, M. Gertler, S. Fast, D. O'Leary, and R. Blobaum. 1977. Economic performance and energy intensiveness of organic and conventional farms in the corn belt: A preliminary comparison. *American Journal of Agricultural Economics* 59:1–12.

Klonsky, K., and L. Tourte. 1998. Organic agricultural production in the United States: debates and directions. *American Journal of Agricultural Economics* 80(5):1119–1124.

Koper, N., and F. Schimiegelow. 2006. Effects of habitat management for ducks on target and nontarget species. *Journal of Wildlife Management* 70(3):823–834.

Kramer, S.B., J.P. Reganold, J.D. Glover, B.J.M. Bohannan, and H.A. Mooney. 2006. Reduced nitrate leaching and enhanced denitrifier activity and efficiency in organically fertilized soils. *Proceedings of the National Academy of Sciences USA* 103:4522–4527.

Kriegl, T., and G. Frank. 2004. An Eight Year Economic Look at Wisconsin Dairy Systems. Madison: University of Wisconsin Center for Dairy Profitability.

Kriegl, T., and R. McNair. 2005. Pastures of Plenty: Financial Performance of Wisconsin Grazing Dairy Farms. Madison: University of Wisconsin Center for Integrated Agricultural Systems, Center for Dairy Profitability, and Program on Agricultural Technology Studies.

Kucharik, C.J., J.A. Roth, and R.T. Nabielski. 2003. Statistical assessment of a paired-site approach for verification of carbon and nitrogen sequestration on Wisconsin Conservation Reserve Program land. *Journal of Soil and Water Conservation* 58(1):58–67.

Kustermann, B., O. Christen, and K.J. Hulsbergen. 2010. Modelling nitrogen cycles of farming systems as basis of site- and farm-specific nitrogen management. *Agriculture Ecosystems & Environment* 135(1–2):70–80.

Lal, R. 2006. Carbon dynamics in agricultural soils. In Climate Change and Managed Ecosystems, J. Bhatti, R. Lal, M.J. Apps, M.A. Price, eds. Boca Raton, Fla.: CRC Press.

Lammers, P.J., D.R. Stender, and M.S. Honeyman. 2007. Niche Pork Production. Ames: Iowa State University, Iowa Pork Industry Center.

Lawrence, J.D., M.K. Muth, J. Taylor, and S.R. Koontz. 2007. Downstream meat marketing practices: lessons learned from the livestock and meat marketing study. Available at http://www.lmic.info/memberspublic/LMMA/LM-6_Fact%20Sheet_Downstream.pdf. Accessed on February 8, 2010.

Lockeretz, W., G. Shearer, D.H. Kohl, and R.W. Klepper. 1984. Comparison of organic and conventional farming in the Corn Belt. In Organic Farming: Current Technology and Its Role in a Sustainable Agriculture, D.F. Bezdicek and J.F. Power, eds. Madison, Wisc.: American Society of Agronomy, Crop Science Society of America, Soil Science Society of America.

Lotter, D.W. 2003. Organic agriculture. *Journal of Sustainable Agriculture* 21(4):59–128.

Lyons, J., S.W. Trimble, and L.K. Paine. 2000. Grass versus trees: managing riparian areas to benefit streams of central North America. *Journal of the American Water Resources Association* 36(4):919–930.

MacDonald, J.A., and W.D. McBride. 2009. The Transformation of U.S. Livestock Agriculture: Scale, Efficiency, and Risks. Washington, D.C.: U.S. Department of Agriculture Economic Research Service.

MacDonald, J.M., M.O. Ribaudo, M.J. Livingston, J. Beckman, and W. Huang. 2009. Manure Use for Fertilizer and for Energy: Report to Congress. Washington, D.C.: U.S. Department of Agriculture Economic Research Service.

Macombe, C. 2007. Work: a necessary sacrifice or a suffered chore? Labor and farm continuity in alternative agriculture in France. *Renewable Agriculture and Food Systems* 22(4):282–289.

MacRae, R.J., B. Frick, and R.C. Martin. 2007. Economic and social impacts of organic production systems. *Canadian Journal of Plant Science* 87(5):1037–1044.

Mader, P., A. Fliessbach, D. Dubois, L. Gunst, P. Fried, and U. Niggli. 2002. Soil fertility and biodiversity in organic farming. *Science* 296:1694–1697.

Mahoney, P.R., K.D. Olson, P.M. Porter, D.R. Huggins, C.A. Perillo, and R.K. Crookston. 2004. Profitability of organic cropping systems in southwestern Minnesota. *Renewable Agriculture and Food Systems* 19:35–46.

Mariola, M.J., K. Stiles, and S. Lloyd. 2005. The social implications of management intensive rotational grazing. An annotated bibliography. Available at http://www.cias.wisc.edu/wp-content/uploads/2008/07/grazebib.pdf. Accessed on September 16, 2009.

Marois, J.J., D.L. Wright, J.A. Baldwin, and D.L. Hartzog. 2002. A multi-state project to sustain peanut and cotton yields by incorporating cattle in a sod-based rotation. In 25th Annual Proceedings of the 25th Annual Southern Conservation Tillage Conference for Sustainable Agriculture, E.V. Santen, ed. Auburn: Alabama Agricultural Experiment Station and Auburn University.

Martz, F.A., H. Heymann, V. Tate, A. Clarke, and J. Gerrish. 1997. Quality of beef from pasture finished cattle. *Proceedings of the American Forage Grasslands Council* 6:218–222.

McDowell, R.W. 2008. Environmental Impacts of Pasture-Based Farming. Wallingford, UK: CAB International.

McDowell, R.W., D.M. Nash, and F. Robertson. 2007. Sources of phosphorus lost from a grazed pasture receiving simulated rainfall. *Journal of Environmental Quality* 36(5):1281–1288.

Meisterling, K., C. Samaras, and V. Schweizer. 2009. Decisions to reduce greenhouse gases from agriculture and product transport: LCA case study of organic and conventional wheat. *Journal of Cleaner Production* 17:222–230.

Mele, P.M., and M.R. Carter. 1999. Impact of crop management factors in conservation tillage farming on earthworm density, age structure and species abundance in south-eastern Australia. *Soil & Tillage Research* 50(1):1–10.

Mitchell, A., Y.J. Hong, E. Koh, D. Barrett, D.E. Bryant, R.F. Denison, and S. Kaffka. 2007. Ten-year comparison of the influence of organic and conventional crop management practices on the content of flavonoids in tomatoes. *Journal of Agricultural and Food Chemistry* 55(15):6154–6159.

Mosely, M., R.D. Harmel, R. Blackwell, and T. Bidwell. 1998. Grazing and riparian area management. In Riparian Area Management Handbook. Available at http://www.okstate.edu/OSU_Ag/e-952.pdf. Accessed on February 8, 2010.

Mummey, D.L., J.L. Smith, and G. Bluhm. 1998. Assessment of alternative soil management practices on N_2O emissions from US agriculture. *Agriculture, Ecosystems and Environment* 70:79–87.

NAS-NAE-NRC (National Academy of Sciences, National Academy of Engineering, and National Research Council). 2009a. Electricity from Renewable Resources: Status, Prospects and Impediments. Washington, D.C.: National Academies Press.

———. 2009b. Liquid Transportation Fuels from Coal and Biomass: Technological Status, Costs, and Environmental Impacts. Washington, D.C.: National Academies Press.

Nassauer, J.I. 2008. Different futures for farmland and the Gulf. Societal and environmental implications of alternative futures. Paper read at Clean Water Network Caucus, February 2008, Memphis, Tenn.

Nassauer, J.I., and P. Opdam. 2008. Design in science: extending the landscape ecology paradigm. *Landscape Ecology* 23:633–644.

Nassauer, J.I., M.V. Santleman, and D. Scavia. 2007. From the Corn Belt to the Gulf: Societal and Environmental Implications of Alternative Agriculture Futures. Washington, D.C.: RFF Press.

National Sustainable Agriculture Information Service (ATTRA) of the National Center for Appropriate Technology (NCAT). 2009. Livestock. Available at http://attra.ncat.org/livestock.html. Accessed on November 9, 2009.

Nelson, R. for Kansas Biomass Committee. 2007. Draft background paper. Available at http://kec.kansas.gov/biomass/Background_Nelson.ppt#256,1,DRAFT Background Paper Kansas Biomass Committee. Accessed on April 6, 2010.

Nguyen, M.L., and R.J. Haynes. 1995. Energy and labour efficiency for three pairs of conventional and alternative mixed cropping (pasture-arable) farms in Canterbury, New Zealand. *Agriculture Ecosystems and Environment* 52(2–3):163–172.

Nichols, M., and W.A. Knoblauch. 1996. What's motivating the grazers. *Hoard's Dairyman*, 404.

NRC (National Research Council). 2003. Air Emissions from Animal Feeding Operations: Current Knowledge, Future Needs. Washington, D.C.: National Academies Press.

Ostrom, M., and D.B. Jackson-Smith. 2000. The use and performance of intensive rotational grazing among Wisconsin dairy farms in the 1990s. Madison: University of Wisconsin–Madison, Program on Agricultural Technology Studies.

Padel, S., O. Schmid, and V. Lund. 2004. Organic livestock standards. In Animal Health and Welfare in Organic Agriculture, M. Vaarst, S. Roderick, V. Lund, and W. Lockeretz, eds. Wallingford, UK: CAB International.

Pain, B.F., and S. Jarvis. 1999. Ammonia emissions from agriculture: IGER Innovations. Available at http://www.aber.ac.uk/en/media/99ch8.pdf. Accessed on February 8, 2010.

Paine, L.K., G.A. Bartelt, D.J. Undersander, and S.A. Temple. 1995. Agricultural practices for the birds. *The Passenger Pigeon* 57(2):77–87.

Paine, L.K., D. Undersander, and M.D. Casler. 1999. Pasture growth, production, and quality under rotational and continuous grazing management. *Journal of Production Agriculture* 12(4):569–577.

Pate, R., M. Hightower, C. Cameron, and W. Einfield. 2007. Overview of Energy-Water Interdependencies and the Emerging Energy Demands on Water Resources. Los Alamos: Sandia National Laboratory.

Patten, R.S., and J.E. Ellis. 1995. Patterns of species and community distributions related to environmental gradients in an arid tropical ecosystem. *Journal of Vegetation Science* 117:69–79.

Paul, E.A., K. Paustian, E.T. Elliot, and C.V. Cole. 1997. Soil Organic Matter in Temperate Agroecosystems: Long-Term Experiments in North America. Boca Raton: Fla.: CRC Press.

Peck, G.M., P.K. Andrews, J.P. Reganold, and J.K. Fellman. 2006. Apple orchard productivity and fruit quality under organic, conventional, and integrated management. *HortScience* 41(1):99–107.

Peruzzi, A., M. Ginanni, M. Fontanelli, M. Raffaelli, and P. Barberi. 2007. Innovative strategies for on-farm weed management in organic carrot. *Renewable Agriculture and Food Systems* 22(4):246–259.

Peters, G.M., H.V. Rowley, S. Wiedemann, R. Tucker, M.D. Short, and M. Schulz. 2010. Red meat production in Australia: life cycle assessment and comparison with overseas studies. *Environmental Science & Technology* 44(4):1327–1332.

Pew Commission on Industrial Farm Animal Production. 2008. Economics of Industrial Farm Animal Production. Washington, D.C.: Pew Charitable Trust.

Pieper, J.R., and D.M. Barrett. 2009. Effects of organic and conventional production systems on quality and nutritional parameters of processing tomatoes. *Journal of the Science of Food and Agriculture* 89(2):177–194

Pierce, F.J., and D. Clay. 2007. GIS Applications in Agriculture. Boca Raton, Fla.: CRC Press.

Pimental, D., P. Hepperly, J. Hanson, D. Douds, and R. Seidel. 2005. Environmental, energetic, and economic comparisons of organic and conventional farming systems. BioScience 55:573–582.

Porter, P.M., D.R. Huggins, C.A. Perillo, S.R. Quiring, and R.K. Crookston. 2003. Organic and other management strategies with two- and four-year crop rotations in Minnesota. Agronomy Journal 95(2):233–244.

Posner, J.L., J.O. Baldock, and J.L. Hedtcke. 2008. Organic and conventional production systems in the Wisconsin integrated cropping systems trials: I. Productivity 1990–2002. Agronomy Journal 100(2):253–260.

Potter, K.N., H.A. Torbert, H.B. Johnson, and C.R. Tischler. 1999. Carbon storage after long-term grass establishment on degraded soils. Soil Science 164:718–725.

Pretty, J. 1995. Regenerating Agriculture: Policies and Practice for Sustainability and Self-reliance. Washington, D.C.: National Academy Press.

Pretty, J., and R.E. Hine. 2000. Reducing food poverty with sustainable agriculture: a summary of new evidence. Available at http://www.essex.ac.uk/ces/esu/occasionalpapers/SAFE%20FINAL%20-%20Pages1–22.pdf.

Randall, G.W., D.R. Huggins, M.P. Russelle, D.J. Fuchs, W.W. Nelson, and J.L. Anderson. 1997. Nitrate losses through subsurface tile drainage in Conservation Reserve Program, alfalfa, and row crop systems. Journal of Environmental Quality 26:1240–1247.

Reeder, J.D., G.E. Schuman, and R.A. Bowman. 1998. Soil C and N changes on conservation reserve program lands in the Central Great Plains. Soil & Tillage Research 47:339–349.

Reganold, J., A. Palmer, J. Lockhart, and N. Macgredor. 1993. Soil quality and financial performance of biodynamic farms in New Zealand. Science 206(5106):334–349.

Reganold, J.P., J.D. Glover, P. Andrews, and H.R. Hinman. 2001. Sustainability of three apple production systems. Nature 410:926–930.

Reisner, A.E., and F. Taheripour. 2007. Reaction of the local public to large-scale swine facilities. Journal of Animal Science 85:1587–1595.

Renfrew, R.B., and C.A. Ribic. 2001. Grassland birds associated with agricultural riparian practices in southwestern Wisconsin. Journal of Range Management 54:546–552.

Ribaudo, M., N. Gollehon, M. Aillery, J. Kaplan, R. Johansson, J. Agapoff, L. Christensen, V. Breneman, and M. Peters. 2003. Manure Management for Water Quality: Costs to Animal Feeding Operations of Applying Manure Nutrients to Land. Washington, D.C.: U.S. Department of Agriculture.

Robertson, G.P., E.A. Paul, and R.R. Harwood. 2000. Greenhouse gases in intensive agriculture: contributions of individual gases to the radiative forcing of the atmosphere. Science 1922–1925.

Robertson, G., V.G. Allen, G. Boody, E.R. Boose, N.G. Creamer, L.E. Drinkwater, J.R. Gosz, L. Lynch, J.L. Havlin, L.E. Jackson, S.T.A. Picket, L. Pitelka, A. Randall, A.S. Reed, T.R. Seastedt, R.B. Waide, and D.H. Wall. 2008a. Long-term agricultural research: a research, education and extension imperative. BioScience 58(7):640–645.

Robertson, G., V.H. Dale, O.C. Doering, S.P. Hamburg, J.M. Melillo, M.M. Wander, W.J. Parton, P.R. Adler, J.N. Barney, R.M. Cruse, C.S. Duke, P.M. Fearnside, R.F. Follett, H.K. Gibbs, J. Goldemberg, D.J. Mladenoff, D. Ojima, M.W. Palmer, A. Sharpley, L. Wallace, K.C. Weathers, J.A. Wiens, and W.W. Wilhelm. 2008b. Sustainable biofuels redux. Science 322:49–50.

Rotz, C.A., A.N. Sharpley, L.D. Satter, W.J. Gburek, and M.A. Sanderson. 2002. Production and feeding strategies for phosphorus management on dairy farms. Journal of Dairy Science 85:3142–3153.

Rotz, C.A., F. Taube, M.P. Russelle, J. Oenema, M.A. Sanderson, and M. Wachendorf. 2005. Whole-farm perspectives of nutrient flows in grassland agriculture. Crop Science 45(6):2139–2159.

Russelle, M.P., M.H. Entz, and A.J. Franzluebbers. 2007. Reconsidering integrated crop-livestock systems in North America. Agronomy Journal 99(2):325–334.

Rust, J.W., C.C. Sheaffer, V.R. Eidman, R.D. Moon, and R.D. Mathison. 1995. Intensive rotational grazing for dairy cattle feeding. American Journal of Alterative Agriculture 10:147–151.

Rutherford, I., B. Abernethy, and I. Prosser. 1999. Riparian land management technical guidelines. In Principles of Sound Management, S. Lovett and P. Price, eds. Canberra, Australia: Land and Water Resources Research Development Corporation.

Sanchez, J.E., R.R. Harwood, T.C. Willson, K. Kizilkaya, J. Smeenk, E. Parker, E.A. Paul, B.D. Knezek, and G.P. Robertson. 2004. Managing soil carbon and nitrogen for productivity and environmental quality. Agronomy Journal 96:769–775.

SARE (Sustainable Agriculture Research and Education). 2000. Animal production. Available at http://www.sare.org/coreinfo/animals.htm. Accessed on October 12, 2009.

———. 2003. Profitable pork: alternative strategies for hog producers. Available at http://www.sare.org/publications/hogs/profpork.pdf. Accessed on February 8, 2010.

Schacht, W.H., and P.E. Reece. 2008. Impact of livestock grazing on extensively-managed grazing lands. In Impacts of Pastoral Grazing on the Environment, R. McDowell, ed. Wallingford, UK: CAB International.

Scherr, S.J., and J.A. McNeely. 2007. Biodiversity conservation and agricultural sustainability: towards a new paradigm of "ecoagriculture" landscapes. *Philosophical Transactions of the Royal Society B: Biological Sciences* 363:477–494.

Schiere, J.B., M.N.M. Ibrahim, and H.v. Keulen. 2002. The role of livestock for sustainability in mixed farming: criteria and scenario studies under varying resource allocation. *Agriculture, Ecosystems & Environment* 90:139–153.

Schipper, L.A., W.T. Baisden, R.L. Parfitt, C. Ross, J.J. Claydon, and G. Arnold. 2007. Large losses of soil C and N from soil profiles under pasture in New Zealand during the past 20 years. *Global Change Biology* 13:1138–1144.

Schneeberger, W., I. Darnhofer, and M. Eder. 2002. Barriers to the adoption of organic farming by cash-crop producers in Austria. *American Journal of Alterative Agriculture* 17(1):24–31.

Scialabba, N.E.H. 2007. Organic farming. *Appropriate Technology* 34(4):31.

Serfling, D., M. Honeyman, and J. Harmon. 2006. Comparison of alternative winter farrowing techniques on four Niman Ranch cooperating farms in southern Minnesota. Iowa State University Animal Industry Report.

Sharpley, A.N. 1985. Depth of surface soil-runoff interaction as affected by rainfall, soil slope and management. *Soil Science Society American Journal* 49:1010–1015.

Sharpley, A.N., and C. West. 2008. Pressures on beef grazing in mixed production farming. In Environmental Impacts of Pasture-based Farming, R.W. McDowell, ed. Wallingford, UK: CAB International.

Shepherd, M., B. Pearce, B. Cormack, L. Philipps, S. Cuttle, A. Bhogal, P. Costigan, and R. Unwin. 2003. An Assessment of the Environmental Impacts of Organic Farming. Berkshire, UK: Elm Farm Research Centre.

Shreck, A., C. Getz, and G. Feenstra. 2006. Social sustainability, farm labor, and organic agriculture: findings from an exploratory analysis. *Agriculture and Human Values* 23(4):439–449.

Smil, V. 1999. Crop residues: agriculture's largest harvest. *BioScience* 49:299–308.

Smith, T.C., M.J. Male, A.L. Harper, J.S. Kroeger, G.P. Tinkler, E.D. Moritz, A.W. Capuano, L.A. Herwaldt, and D.J. Diekema. 2009. Methicillin-Resistant Staphylococcus aureus (MRSA) Strain ST398 is present in Midwestern U.S. swine and swine workers. *PLoS ONE* 4(1):e4258.

Snapp, S.S., and B. Pound, eds. 2008. Agricultural Systems: Agroecology & Rural Innovation for Development. Burlington, Mass.: Academic Press.

Soane, B.D. 1990. The role of organic matter in soil compactability: a review of some practical aspects. *Soil & Tillage Research* 16:179–201.

Srinavasan, A., ed. 2006. Handbook of Precision Agriculture: Principles and Application. Binghamton, N.Y.: Haworth Press.

Stanhill, G. 1990. The comparative productivity of organic agriculture. *Agriculture, Ecosystems and Environment* 30:1–26.

Stolze, M., A. Piorr, A. Häring, and S. Dabbert. 2000. The Environmental Impacts of Organic Farming in Europe. Stuttgart, Germany: University of Hohenheim Publishers.

Stout, W.L., S.L. Fales, L.D. Muller, R.R. Schnabel, and S.R. Weaver. 2000. Water quality implications of nitrate leaching from intensively grazed pasture swards in the northeast US. *Agriculture, Ecosystems & Environment* 77(3):203–210.

Stout, W.L., R.R. Schnabel, W.E. Priddy, and G.F. Elwinger. 1997. Nitrate leaching from cattle urine and feces in Northeast USA. *Soil Science Society of America* 61:1787–1794.

Strochlic, R., C. Wirth, A. Fernandez Besada, and C. Getz. 2008. Farm Labor Conditions on Organic Farms in California. Davis: California Institute for Rural Studies.

Sulc, R.M., and B.F. Tracy. 2007. Integrated crop-livestock systems in the U.S. corn belt. *Agronomy Journal* 99(2):335–345.

Swezey, S.L., P. Goldman, J. Bryer, and D. Nieto. 2007. Six-year comparison between organic, IPM and conventional cotton production systems in the Northern San Joaquin Valley, California. *Renewable Agriculture and Food Systems* 22(1):30–40.

Swift, M.S., J. Vandermeer, P.S. Ramakrishnan, J.M. Anderson, C.K. Ong, and B.A. Hawkins. 1996. Biodiversity and agroecosystem function. In Functional Roles of Biodiversity: A Global Perspective, H.A. Mooney, J.H. Cushman, E. Medina, O.E. Sala, and E.D. Schulze, eds. New York: J. Wiley and Sons.

Taylor, J., and J. Foltz. 2006. Grazing in the Dairy State: Pasture Use in the Wisconsin Dairy Industry. Madison: University of Wisconsin–Madison Center for Agricultural Systems (CIAS) and Program on Agricultural Technology Studies (PATS).

Taylor, J., and S. Neary. 2008a. How does managed grazing affect Wisconsin's environment? Madison: University of Wisconsin–Madison College of Agricultural and Life Sciences.

Teague, W.R., and S.L. Dowhower. 2003. Patch dynamics under rotational and continuous grazing management in large, heterogeneous paddocks. *Journal of Arid Environments* 53(2):211–229.

Tegegne, F., S.P. Singh, E. Ekanem, and S. Muhammad. 2001. Labor use by small-scale conventional and sustainable farmers in Tennessee. *Southern Rural Sociology* 17(17):66–80.

Temple, S. R., D. B. Friedman, O. Somasco, H. Ferris, K. Scow, and K. Klonsky. 1994. An interdisciplinary, experiment station-based participatory comparison of alternative crop management systems for California's Sacramento Valley. *American Journal of Alternative Agriculture* 9(1–2):64–71.

Temple, S.A., B.M. Fevol, L.K. Paine, D.J. Undersander, and D.W. Sample. 1999. Nesting birds and grazing cattle: accommodating both on Midwestern pastures. *Studies in Avian Biology* 19:196–202.

Thrupp, L.A. 2003. Growing Organic Wine Grapes Successfully. Hopland, CA: Fetzer Vineyards.

Tilman, D., J. Hill, and C. Lehman. 2006. Carbon-negative biofuels from low-input high-diversity grassland biomass. *Science* 314:1598–1600.

Tisdall, J.M. 1994. Possible role of soil-microorganisms in aggregation in soils. *Plant and Soil* 159:115–121.

Torrent, J., and D.E. Johnson. 1994. Methane Production in the Large Intestine of Sheep. Granada, Spain: CSIC Publishing Service.

Tynon, R., J. Mintert, M. Tokach, T. Schroeder, and M.R. Langemeier. 1994. Group marketing of hogs: organization, successes and guidelines. In Swine Center packet: Cooperative Extension Service, Kansas State University, Lawrence.

Tyson, K.C., D.H. Roberts, C.R. Clement, and E.A. Garwood. 1990. Comparison of crop yields and soil conditions during 30 years under annual tillage on grazed pasture. *Journal of Agricultural Science* 115:29–41.

University of California-Davis. 2009. Russell Ranch sustainable agriculture facility. Available at http://ltras.ucdavis.edu/. Accessed on October 10, 2009.

USDA-AMS (U.S. Department of Agriculture Agricultural Marketing Service). 2009. Home page for National Organic Program. Available at http://www.ams.usda.gov/AMSv1.0/ams.fetchTemplateData.do?template=TemplateA&navID=NationalOrganicProgram&page=NOPNationalOrganicProgramHome&resultType=&topNav=&leftNav=NationalOrganicProgram&acct=nop. Accessed on January 11, 2009.

USDA-ERS (U.S. Department of Agriculture Economic Research Service). 2008. Livestock, dairy, and poultry outlook: tables. Available at http://www.ers.usda.gov/publications/ldp/LDPTables.htm. Accessed on January 7, 2009.

———. 2009a. Agricultural Resource Management Survey (ARMS): resource regions. Available at http://www.ers.usda.gov/Briefing/arms/resourceregions/resourceregions.htm. Accessed on August 4, 2009.

———. 2009b. Organic production. Available at http://www.ers.usda.gov/Data/Organic/. Accessed on September 17, 2009.

USDA-NASS (U.S. Department of Agriculture National Agricultural Statistics Service). 2009. 2007 Census of Agriculture. Washington, D.C.: Author.

USDA-NRCS (U.S. Department of Agriculture National Resource Conservation Service). 2009. National Conservation Practice Standards. Available at http://www.nrcs.usda.gov/technical/standards/nhcp.html. Accessed on September 10, 2009.

USDA-REEIS (U.S. Department of Agriculture Research, Education, and Economics Information System). 2003. Organic Dairy Farming in Vermont: Profitability Analysis and Educational Programs. Burlington: University of Vermont.

USDA-SARE (U.S. Department of Agriculture Sustainable Agriculture Research and Extension). 2009. Transitioning to organic production. Organic farming systems overview. Available at http://www.sare.org/publications/organic/organic03.htm. Accessed on December 6, 2009.

Van Bruggen, A.H.C. 2008. Ecological principles underlying the functioning of sustainable farming systems and the general principles for the design of such systems. Presentation to the committee in Washington, D.C. on August 5, 2008.

Vandermeer, J. 1995. The ecological basis of alternative agriculture. *Annual Review of Ecology and Systematics* 26:201–224.

Walsh, C.S., A.R. Ottesen, J. Hanson, E. Leone, and M. Newell. 2008. The production of organic apples and asian pears to promote the sustainability of existing orchards and as alternative enterprises for new growers. Available at http://agroecol.umd.edu/files/Walsh%20Final%20Report%20HCAE%20Pub%202008-02.pdf. Accessed on April 6, 2010.

Watson, C.A., D. Atkinson, P. Gosling, L.R. Jackson, and F.W. Rayns. 2002. Managing soil fertility in organic farming systems. *Soil Use and Management* 18:239–247.

Weil, R.R., K.A. Lowell, and H.M. Shade. 1993. Effects of intensity of agronomic practices on a soil ecosystem. *American Journal of Alternative Agriculture* 8:5–14.

Welsh, R. 1999. The Economics of Organic Grain and Soybean Production in the Midwestern United States. Greenbelt, Md.: H.A. Wallace Institute for Alternative Agriculture.

Wezel, A., and V. Soldat. 2009. A quantitative and qualitative historical analysis of the scientific discipline of agroecology. *International Journal of Agricultural Sustainability* 7(1):3–18.

White, S.L., J.A. Bertrand, M.R. Wade, S.P. Washburn, J.T. Green, and T.C. Jenkins. 2001. Comparison of fatty acid content of milk from Jersey and Holstein cows consuming pasture or a total mixed ration. *Journal of Dairy Science* 84:2295–2301.

Wilcock, B., S. Elliott, N. Hudson, S. Parkyn, and J. Quinn. 2008. Climate Change Mitigation Measures: Water Quality Benefits and Costs. Wellington, New Zealand: Ministry for the Environment.

Wilcock, R.J., and J.W. Nagels. 2001. Effects of aquatic macrophytes on physico-chemical conditions of three contrasting lowland streams: a consequence of diffuse pollution from agriculture? *Water Science and Technology* 43:163–168.

Wilcock, R.J., J.W. Nagels, H.J.E. Rodda, M.B. O'Connor, B.S. Thorrold, and J.W. Barnett. 1999. Water quality of a lowland stream in a New Zealand dairy farming catchment. *New Zealand Journal of Marine and Freshwater Research* 33:683–696.

Wilkins, R.J. 2008. Eco-efficient approaches to land management: a case for increased integration of crop and animal production systems. *Philosophical Transactions of the Royal Society B-Biological Sciences* 363(1491):517–525.

Willer, H., M. Yussefi-Menzler, and N. Sorensen. 2008. The world of organic agriculture—statistics and emerging trends 2008. Available at http://orgprints.org/13123/2/willer-yussefi-sorensen-2008-final-tables.pdf. Accessed on February 8, 2010.

Winsten, J.R., R.L. Parsons, and G.D. Hanson. 2000. Differentiated dairy grazing intensity in the Northeast. *Journal of Dairy Science* 83(4):836–842.

Winsten, J. R., and B. T. Petrucci. 2003. Seasonal Dairy Grazing—A Viable Alternative for the 21st Century. DeKalb, Ill.: American Farmland Trust, Farms Division.

Woese, K., D. Lange, and P.O. Sjoden. 1997. A comparison of organically and conventionally grown foods—results of a review of the relevant literature. *Journal of the Science of Food and Agriculture* 74:281–293.

6

Drivers and Constraints Affecting the Transition to Sustainable Farming Practices

All individual farms, whether large or small, can contribute both positively and negatively to sustainability goals in various degrees. As discussed in Chapter 1, the committee views sustainability as a *process* that moves farming systems along a trajectory toward meeting societal-defined goals, as opposed to any particular end state. The determination of which sustainability goals are worth pursuing or the appropriate balance between gains along different dimensions of sustainability (for example, economic viability, ecosystem functioning, social responsibility, and food characteristics) is quintessentially a social choice. Depending on what is viewed as an adequate or optimal outcome, the necessary changes might range from incremental adjustments to existing farming practices to fundamental changes in the underlying structure, organization, and management of farming enterprises.

Earlier chapters discussed a wide range of farming practices and systems and reviewed scientific evidence for how they affect various indicators of sustainability. The earlier chapters, however, did not discuss the factors that influence farmer adoption of any of those practices or systems. Addressing the challenges outlined in Chapter 2 and meeting societal expectations for greater sustainability will depend on the ability and willingness of American farmers to adopt appropriate farming practices and systems.

This chapter analyzes some of the factors that influence farmers' ability and willingness to change production practices or to convert to new farming systems that move their farms along the sustainability trajectory. All farmers make decisions in a complex environment in which broad contextual factors, such as markets, public policies (including regulation), and social institutions, create opportunities and barriers to change. The first part of this chapter explores how trends in the ownership and diversity of markets increase or decrease the latitude with which farmers can make production decisions that allow them to move further along a sustainability trajectory.

For some farmers, production decisions are further shaped by incentives inherent in federal and state policies, including trade policies, federal Farm Bill programs, national energy policy, and regulations that address animal welfare or environmental impacts of farming. Many public policies influence farmers' management choices, and the influences they have on farmers' choices frequently depend on the farm type. For example, federal farm commodity, domestic food aid, and nutrition programs (which make up the bulk of national government expenditures under the Farm Bill) have considerable influence on industrial, larger commercial farms, while agrienvironmental, niche-market development, and land use policies (often implemented at the local level) tend to be more important to the viability of smaller farms (such as those with annual sales below $250,000). Policy influences are summarized in the second section of this chapter.

The third section provides an overview of public and private knowledge institutions that play a role in fostering a change in farmer behavior and decision making by generating new knowledge and innovations and disseminating them among farmers. Those institutions include not only national and state research and extension services, but also private institutions ranging from large agribusiness research and development divisions to farmer-based learning and networking groups.

All three major types of institutional contexts—markets, policies, and knowledge institutions—are shaped by larger societal forces that have particular goals and objectives. In agriculture, different stakeholder groups and social movement organizations are constantly working to shape the structure and behavior of public and private institutions. The fourth section of this chapter uses some of the key groups and organizations to illustrate how they work to shape the viability and movement of farms along a sustainability trajectory.

Two farmers facing similar contextual factors do not always respond in the same way to the incentives and disincentives created by markets, policies, and new knowledge. Decisions by individual farmers to pursue (or not to pursue) different farming practices and systems depend also on what sort of land or other resource endowments they have available; their existing farming approaches, knowledge, and skills; and their goals and motivations, including personal ethics, religious beliefs, or world view. The last part of this chapter examines evidence linking different characteristics of farmers with a willingness or likelihood to change production systems.

In the face of many different drivers and constraints, the large corn producer in the Midwest, the Southwestern cotton grower, the rancher in the Mountain states, the Southern part-time vegetable grower, the Northeastern dairy farmer, and the peri-urban hobby farmer face distinctly different challenges in their efforts to operate viable enterprises that preserve the natural resource base and produce all the additional benefits desired by society. Figure 6-1 presents a simplified illustration of the various types of influences on farmers, including broad contextual factors surrounding the agricultural system, the mediating role of local assets and farmer values, and the emerging trends that could facilitate movement along the sustainability trajectory or make it more challenging in the future than at present.

AGRICULTURAL MARKETS AS CONTEXTUAL FACTORS

Concentration in the Agrifood System

Farmers participate in agricultural markets as buyers and sellers. As buyers, they seek high-performing, competitively priced production inputs (for example, seeds, livestock, fertilizer, pesticides, fuel, and machinery). As sellers, some—those with less differentiated products—tend to compete mainly by lowering the cost of their production; thus, they are participants in a low-cost supply chain. Others—those with more differentiated products—tend to compete mainly by producing the most consumer-valued attributes of their product

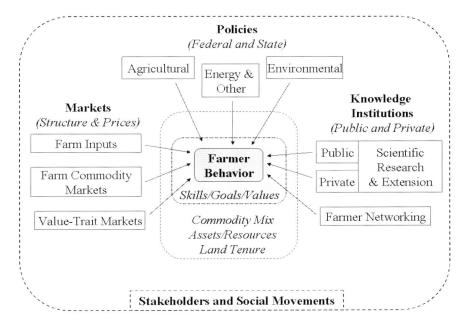

FIGURE 6-1 Drivers and constraints that could affect farmers' behaviors.

per dollar cost of production; thus, they are participants in value supply chains. Some sellers seek out markets that provide the best prices for their production, some ascertain what type of crop will be the most valuable for the given year, some pursue marketing strategies that reduce their marketing risks, and some produce crops supported by the Farm Bill. The pricing, ownership structure, and direction of markets for farm inputs and products influence farmer production decisions.

Markets for Farm Inputs

Over the last few decades, the degree of concentration of ownership and control among the major firms that supply farm inputs to U.S. farmers has steadily increased (Heffernan, 1999; Hendrickson and Heffernan, 2007). An example of that trend is the consolidation of the U.S. seed industry during the 1990s, when many small independent seed companies were acquired by or entered into joint ventures with major international corporations, including pharmaceutical and chemical firms like Monsanto, Dow Chemical, Dupont, and Aventis (Fernandez-Cornejo, 2004). As a result of the mergers, the top four U.S. commercial seed companies supply two-thirds of the corn seeds, half of soybean seeds, and almost 90 percent of cotton seeds in the United States. Consolidation among major chemical companies, farm cooperatives, and seed companies has produced similar concentration of market power among sellers of chemicals, fertilizers, and other key farm inputs (Heffernan et al., 1999). Consolidation of ownership in input markets could result in increased prices paid by farmers for inputs and a reduction in the variety and sources of available inputs. For example, the market concentration of genetically engineered (GE) hybrid corn seed has led to significant increase in seed price (Shi et al., 2008). Triple-stacked varieties with genes for corn borer and rootworm resistance and for herbicide tolerance added $39.50 (about 40 percent) to the price per bag of seeds on average compared to non-GE varieties seed in 2007. The consolidation of market power among the suppliers of farm inputs and the purchasers and processors of raw agricultural commodities has been linked to a diminishing farmer-share of the consumer food dollar (Gardner, 2002). Farmers might decide to increase their reliance on on-farm inputs to insulate their farms from rising input costs, but that relationship is difficult to document. Some farms discussed in Chapter 7 mention rising input costs as a contributing factor to their decisions to change their farming practices (for example, Brookview Farm and Thompson Farm).

Markets for Products

Rapid consolidation is apparent among firms that purchase and process major agricultural commodities. In the grain sector, for example, the top three or four companies control 60 percent of terminal grain handling facilities, over 80 percent of all corn exports, and two-thirds of soybean exports (Hendrickson et al., 2001). In the meat-packing sector, the largest four buyers control 84 percent of the beef market, 66 percent of the pork market, and 59 percent of the poultry market (Hendrickson and Heffernan, 2007).

Concurrently, market power in the food manufacturing, wholesaling, and retailing sectors has consolidated. For instance, the largest 10 U.S.-based food manufacturers control over half of the sales of food and beverages in the nation (Lyson and Raymer, 2000). The largest 50 food distributors (for example, Sysco, US Foodservice, SUPERVALU, and McLane Company) control more than half of the total food distribution market (Hoovers, 2008). Sysco Corporation alone controls 28 percent of the broadline (or nonspecialized) food service market (which represents half of U.S. food service distribution sales). The top four food service companies accounted for 27 percent of all wholesale food sales in 2001 (Harris et al., 2002). Although those firms used to be major suppliers of food to independent grocery stores, the consolidation in the retail grocery sector has led to the development of integrated internal sourcing and distribution networks controlled by grocery chains.

Recent mergers and acquisitions in the food manufacturing and distribution industries are thought to reflect a response to pressure from an increasingly consolidated food-retailing sector. Recent data suggest that the share of total retail-food sales in the largest five U.S. grocery store chains increased from 24 percent in 1997 to almost 50 percent in 2006 (Hendrickson and Heffernan, 2007). In the 100 largest U.S. cities, four firms controlled an average of more than 72 percent of the local grocery store market (Kaufman, 2000). As recently as 1998, independent retailers and smaller grocery store chains accounted for just 16 percent of the U.S. food retail market (Stanton, 1999).

A consequence of the trends summarized above is that most farmers are facing increasingly consolidated and vertically integrated output markets for their agricultural commodities. Decisions to use farming practices that might promote various aspects of sustainability are likely to be conditioned by the unique market opportunities and constraints presented by the increasingly large system.

Growth in scale and vertical integration could contribute to greater efficiencies, economies of scale, and lower transaction costs throughout the agrifood system, which could potentially benefit consumers, but the degree of concentration and consolidation among agribusinesses might create monopolistic or monopsonistic conditions and correspondingly anti-competitive behavior (Sexton, 2000; Barkema and Novack, 2001; Fulton and Giannakas, 2001). The shift in market power from the farm to output and food-retailing sectors is also likely to shift decision-making power and authority away from the farm operator. An example is the increased use of production and marketing contracts between farmers and commodity marketers and food processors in the United States.

Although more than half of U.S. sales of farm products still occur in open commodity markets, the share of total U.S. farm sales sold through contracts has risen from 11 percent in 1969 to 41 percent in 2005 (MacDonald and Korb, 2008). Contracting of farm products in 2005 was particularly prevalent in certain commodity sectors, such as poultry (94 percent of production), hogs (76 percent), dairy (59 percent), vegetables (54 percent), and fruits and nuts (64 percent).

Contracts provide farmers with some level of certainty in the price they will receive in the market, quantities to be sold, or attributes of the product that are most valuable to the end consumer, and, in some cases, remove some of the marketing risks (USDA-ERS, 1996; Kunkel et al., 2009). Two common types of products contracts are marketing contracts and production contracts. Marketing contracts are written agreements between farmers and contractors that specify the price and outlet for the commodity before the commodity is produced. Typically, production and management decisions are left to the farmer. However, some marketing contracts can specify quality requirements (USDA-ERS, 1996) that can create pressures on producers to deliver standardized products and varieties to meet specified standards. Those contractual terms can force farmers to use production practices to meet quality or cosmetic requirements that might not be suited to local ecological conditions. Hence, they might create disincentives for the use of some farming practices that could enhance sustainability (Busch and Bingen, 2005; Bingen and Busch, 2007).

Production contracts usually specify the production inputs to be supplied by contractors, the quantity and quality of products, and how the farmers are compensated (USDA-ERS, 1996; Kunkel et al., 2009). Because production contracts shift the locus of control from farmers to contractors, farmers who produce under contract often cannot adopt innovative practices that might promote various aspects of sustainability in addition to productivity in their farming systems unless the contracts specify the practices.

On the other hand, retailers are interested in meeting the demands of their consumers. As consumers become more demanding about how their food is produced, some retailers have begun to require different production practices from their suppliers. Such shifts in retailer demands can create conditions where contractors might require their producers to use particular types of practices that improve multiple aspects of sustainability. Consumer demand as a driver of improving agricultural sustainability is discussed in the next section.

Emerging Markets

Changes in Consumer Preferences

Studies and surveys suggest that consumers' preference for foods that are perceived to be grown using "sustainable practices" and that are considered to be natural or healthy is increasing, and the demand is having an impact on agricultural markets. U.S. food consumers are increasingly requesting foods that are pesticide free, hormone free, fair traded, eco-friendly, locally grown, cruelty free, and otherwise associated with "ethical" approaches to production (Bell, 2004; Pollan, 2006; Packaged Facts, 2007). A nationwide study published in 2007 found that consumer awareness and acceptance and practices that relate to sustainability has been shifting (The Hartman Group, 2008). The study estimated that U.S. retail sales of grocery products that include some form of ethical claim reached nearly $33 billion in 2006, an increase of more than 17 percent from 2005. That amounts to roughly 6 percent of the $550 billion spent by U.S. households in grocery stores annually (Food Marketing Institute, 2009).

Most consumers want foods that they perceive to be safe, nutritious, tasty, and environmentally friendly (Food Marketing Institute, 2008; Lusk and Briggeman, 2009). Some consumers believe that organic products taste better (The Hartman Group, 2008). Some are increasingly interested in foods that are grown locally, because freshness and support to local producers are important to them (Allen, 2004). However, what consumers think they are getting out of "natural" and "organic" foods might not always correspond to the reality of what they are consuming (Demeritt, 2006; see also discussion on nutritional quality in Chapter 4). Nonetheless, retail establishments have taken notice of consumers' interest, because those consumers are willing to pay a premium to get these types of products.

When directed toward issues of health, environmental quality, and food quality, the growing power of consumers within the U.S. food system can become a force for driving farming systems toward increasing sustainability (Allen, 2004). Farmers, particularly those who are engaged in direct sales or value supply chains, tend to adjust their production practices in response to consumers' demand. Moreover, the consolidation of the agrifood industry presents a situation where changes in the purchasing behavior of a few large institutional actors toward purchases of food with value traits can have major influences on farmers' production practices (Sligh and Christman, 2003; Dimitri and Oberholtzer, 2008).

Sustainability Initiatives

Driven by changes in consumer preference, the last decade has seen a great upsurge of interest in the idea of sustainability among numerous influential food-marketing companies, including large multinational retailers (such as Wal-Mart and Costco), large supermarkets (such as Safeway and Kroger), expansive restaurant chains (such as Starbucks and McDonald's), and very large food processors, distributors, and food service providers (such as Unilever, Nestle, Tyson, Sysco, and Sodexho). Sustainability initiatives are well developed in Europe (Fulponi, 2006). Box 6-1 describes a recent global food industry initiative to coordinate the efforts to improve sustainability. Several retailers have developed sustainability standards or "green" guidelines, and some retailers require their food suppliers to use the standards or guidelines to meet the sustainability goals that each retailer considers important. Several industry groups and trade associations also have established guidelines or standards to encourage the use of practices that can improve sustainability among their suppliers and vendors. Although some skeptics criticize those efforts and question the overall sustainability of such mega-corporations, the initiatives have an important impact in the food system because they drive changes toward increasing sustainability in the supply chain by affecting purchasing decisions, food processing and transport systems, and agricultural production practices at the farm level (Doane, 2005; Aragón-Correa and Rubio-López, 2007).

Organic Food Markets

Organic price premiums have been documented as early as the 1970s (Greene, 2001). Access to price premiums is important to the economic viability of organic production because production costs tend to be higher than in conventional production (McBride and Greene, 2007, 2008). Although organic food markets began as a niche market found mainly in health-food stores and local food cooperatives, organic foods and beverages have become increasingly mainstream. Consumer demand for organic products has grown rapidly during the 1990s and the first decade of the 21st century (Dimitri and Oberholtzer, 2008). Organic sales account for approximately 3 percent of total U.S. food sales (USDA-ERS, 2009b), and the market has maintained a growth rate of about 20 percent per year in retail sales since 1990, as shown in Figure 6-2 (The Hartman Group, 2008). By comparison,

BOX 6-1
The Sustainable Agriculture Initiative Platform

According to its website, "the Sustainable Agriculture Initiative (SAI) Platform is an organization created by the food industry to communicate and to actively support the development of sustainable agriculture. SAI Platform supports agricultural practices and agricultural production systems that preserve the future availability of current resources and enhance their efficiency" (SAI Platform, 2009). It attempts to address the three aspects of sustainability—economic, social, and environmental—and to involve all stakeholders of the food chain. Many global food industry companies, such as Coca-Cola, General Mills, Kellogg's, Fonterra, McDonald's, Nestle, Sara Lee, and Kraft Foods, are included in the initiative.

Among the industries' sustainability initiatives, Wal-Mart's is one of the most widely publicized (see http://walmartstores.com/sustainability/). With nearly 4,000 stores in the United States and more than 2,200 stores internationally, the company wields tremendous economic power in the retail system. In addition to improving energy efficiency, increasing energy conservation, and reducing wastes, the company has been expanding its purchases of organic products. Wal-Mart has become the largest seller of organic milk and the largest buyer of organic cotton in the world (Gunther, 2006). Wal-Mart also started the Heritage Agriculture Program to encourage farms within a day's drive of one of Wal-Mart's warehouses to grow crops and supply them to its local stores (Kummer, 2010).

Analysts have criticized the company by pointing out its continued weaknesses in employee policies, contribution to structural inequities, price reductions that are potentially unfair to organic producers, and other inadequacies (Tocco and Anderson, 2007). However, it is working with nonprofit environmental groups and other advisors and suppliers to establish metrics and standards for sustainability attributes, and to encourage changes broadly in the supply chain. Because of the scale of the company, Wal-Mart's initiative has great influence on hundreds of food suppliers, among other types of suppliers, who are being asked to use production and processing processes that meet the company's sustainability goals (Vandenbergh, 2007).

---

the conventional food market has grown at a rate of only 4 to 5 percent annually over the same period. The organic sector certified by the U.S. Department of Agriculture (USDA) is the fastest-growing segment of food sales in North America. As USDA-certified organic foods become popular, an increasing number of mainstream retail establishments carry them (Sligh and Christman, 2003). In 2008, the U.S. organic food industry was estimated to have generated almost $21 billion in consumer sales, and more than two-thirds of U.S.

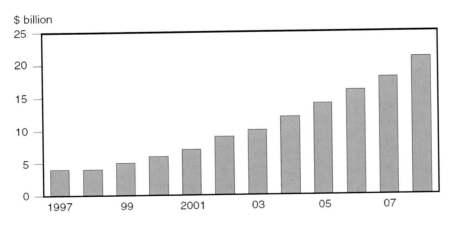

FIGURE 6-2 Growth in U.S. retail sales of certified organic food products. Reprinted with permission from Nutrition Business Journal.

consumers bought organic products at least occasionally (Dimitri and Oberholtzer, 2009; Greene et al., 2009; The Hartman Group, 2008). The demand in organic products and the maintenance of a price premium contribute to motivating some farmers to transition to organic production.

Although the acreage for organic production doubled in the United States between 1997 and 2005, consumer demand outpaces supply. Consumer demand seems to play a small role in driving an increase in organic production as the overall adoption rate of organic agriculture is still low—about 0.5 percent of U.S. cropland and pastureland was certified organic in 2005 (Greene et al., 2009). Other factors affect farmers' decision to transition to organic production. As mentioned earlier, organic production costs more than conventional production, and farmers are likely to have lower economic returns in the first few years of transition as a result of lower yields and inability to access organic premiums until the transition is completed. Although organic food markets are frequently supply constrained, few organic food handlers have worked to assist farmers to make the transition toward organic production (Dimitri and Oberholtzer, 2008).

Direct-Sales Markets

Direct-sales markets have the potential to enhance the feasibility of using farming approaches for improving sustainability and motivating farmers to use them. The various direct-sales markets provide important new opportunities for farms that use practices and farming systems that can meet the demands of those new consumers. As discussed in Chapter 4, direct marketing allows farmers to capture a larger share of the end consumer's food dollar. Farmers, in particular small or mid-sized farms that make most of their sales as direct sales, can afford to use practices that enhance environmental and social sustainability that are not necessarily the lowest cost and still make a profit. However, in many cases, direct-sales farmers have to take on responsibilities, such as doing their own marketing, in addition to producing the products.

Many farms have been able to tap into increasing consumer interest in local sources of foods and public perceptions that food transported long distances not only adds to the atmospheric carbon burden, but also tastes less fresh than local foods. In addition, an increasing number of consumers seem to want to develop a closer connection to the farms that produce their food. That interest provides impetus to the development of localized food markets that allow farmers to bypass mainstream distribution channels and market their products directly to consumers at the local and regional levels (Allen, 2004; Hinrichs and Lyson, 2007). Some farmers who are unable to compete in, or are locked out of, distant markets have been able to build a thriving local business (Allen and Hinrichs, 2007). Direct sales are proportionally more common in small and mid-sized operations (firms with sales under $250,000, which generated 57 percent of all U.S. direct sales according to the 2007 Census of Agriculture) than on larger farms (USDA-NASS, 2009).

Farm production sold directly to consumers has grown rapidly since 1980 and tripled between 1992 and 2007. In 2007, about 6 percent of all U.S. farms engaged in direct marketing to consumers, generating more than $1.2 billion in gross receipts (USDA-ERS, 2009a). Direct sales occur primarily through farmers' markets, followed in importance by Community Supported Agriculture (CSA). Growth in traditional direct marketing in roadside stands and in U-pick operations has been more modest than farmers' markets and CSAs. The amount of locally produced foods sold directly to consumers is expected to grow over the next decade and a half, but at a slower rate than the growth rates from 2005 to 2008. Meanwhile, local sourcing of food supplies by institutions such as restaurants, schools,

DRIVERS AND CONSTRAINTS

universities, and especially grocery stores and chains is likely to be an important source of growth for direct sales in the next 15 years.

Farmers' Markets and Farm Stands

Between 1980 and 2007, the number of farmers' markets nationally nearly quadrupled from an estimated 1,200 to 4,385 (USDA-AMS, 2008a), and they generated total vendor sales of more than $1 billion in 2005 (Ragland and Tropp, 2009). Sales at farmers' markets grew between 2.5 to 5 percent annually from 1996 to 2006 (Cobb, 2008; Ragland and Tropp, 2009). Farmers attribute the sudden increase in demand to families concerned about food safety of distantly produced or imported foods, their need for a greater sense of "community," and their desire to talk to a person growing their food (Hinrichs, 2000; Lamine, 2005; Smithers et al., 2008). The direct connection between farmers and consumers allows farmers to adjust their production practices in response to consumer demand. Although the rate at which new farmers' markets are formed is expected to slow, the number in America could reach 6,000 by 2015, with some 65,000–180,000 small farmers and vendors generating gross revenues approaching $1.5 billion. Although the projected revenue represents only a small percentage (1.5 to 2 percent) of the gross revenue of the U.S. food system from retail and from hotels, restaurants, and institutions, farmers' markets and farm stands represent a marketing channel that can support the use of practices for improving sustainability in many small and mid-sized farms. Farmers' markets will continue to be community based, run by farmer or community volunteers, and open seasonally in public spaces.

Community Supported Agriculture

Two CSA projects in the United States emerged in New England in 1986. By 2009, an estimated 2,877 CSAs operated in all 50 states (Table 6-1; Local Harvest, 2009). Data from the 2007 Census of Agriculture included a question on CSAs for the first time, and the results suggest that there could be as many as 12,000 farms that describe themselves as participating in a CSA (USDA-NASS, 2009). Assuming an average of 50 subscribers each (Lass et al., 2003), the CSAs listed on the Local Harvest website (Table 6-1) are estimated to supply almost 150,000 U.S. households with various vegetables and other produce during the growing season.

The original idea of CSA was to reestablish a sense of connection to the land for urban dwellers and to foster a strong sense of community with a social justice goal to provide

TABLE 6-1 Estimated Number of CSA Farms by State as Measured by USDA and Local Harvest (LH) Website

	USDA	LH		USDA	LH		USDA	LH		USDA	LH		USDA	LH
AK	6	9	HI	3	13	ME	32	65	NJ	16	46	SD	8	44
AL	7	20	IA	39	70	MI	40	141	NM	16	20	TN	56	17
AR	4	18	ID	16	41	MN	35	99	NV	1	15	TX	75	24
AZ	9	24	IL	20	93	MO	18	62	NY	101	209	UT	15	2
CA	81	179	IN	12	53	MS	2	4	OH	31	110	VA	86	32
CO	27	71	KS	8	34	MT	3	17	OK	4	15	VT	89	36
CT	22	44	KY	15	50	NC	26	95	OR	45	120	WA	152	60
DE	4	7	LA	3	9	ND	2	8	PA	69	161	WI	148	71
FL	15	42	MA	60	113	NE	5	15	RI	10	16	WV	16	9
GA	5	57	MD	36	67	NH	21	54	SC	4	24	WY	11	4

SOURCE: Local Harvest (2009); USDA-NASS (2009).

food security for disadvantaged groups (Allen et al., 2006). Some CSAs also are involved in farmland preservation, and they offer insurance against unexpected disruptions of the food supply line. Another aim of CSA was to enlist support from urban consumers for local agriculture that emphasizes various aspects of sustainability. A key concept of early CSA organizers was to assert local control over a food system that was growing increasingly consolidated and to offer small farmers a fair return for their products. Although CSAs currently only serve a small proportion of the U.S. consumer food market, the CSA model offers an alternative approach to mainstream marketing channels for producers and consumers in some regions.

Farm to Institutions

One of the most rapidly growing forms of direct marketing consists of direct sales from farmers to institutions, particularly schools, hospitals, and government agencies. Farm-to-school programs are emerging all over the United States. As of October 2009, farm-to-school programs had been established in a total of 42 states and are estimated to serve approximately 8,943 schools in 2,065 school districts (Occidental College, 2009).

The other types of direct marketing institutional arrangements for food service—for hospitals and other institutions—are more difficult to monitor and measure than farm-to-school programs, partly because the other types have been established at such a rapid pace in recent years. Several hospitals throughout the country are developing "farm-to-hospital" linkages that bring fresh, healthful food to medical facilities and offer new markets for local farmers. For example, in Billings, Montana, the Community Food Campaign urged a local hospital, Billings Deaconess Clinic, to procure food locally. The clinic amended the contract with its food provider and now procures locally raised turkey. Sutter Maternity and Surgery Center in Santa Cruz, California, buys almost 20 percent of its produce from trainee farmers working on the Agriculture and Land-Based Training Association (ALBA) farm in nearby Salinas. Duke University in Durham, North Carolina, holds a weekly farmers' market in an area between its clinic and hospital. The establishment of farmers' markets at several Kaiser Permanente Hospitals in California has sparked discussions about the need for a company-wide food policy to bring fresh food to patients, visitors, and surrounding communities. These few examples show how health care facilities and hospitals around the United States are creating new opportunities for food procurement and provision that can potentially improve environmental sustainability (by decreasing the distance of food delivery), economic sustainability of farms (via new market opportunity), and social sustainability (by providing access to fresh food). Linking local farms and hospitals has the potential to improve the freshness, quality, and nutritional value of hospital food while opening new institutional markets to small farmers (Beery and Markley, 2007).

Several government agencies at federal, state, and local levels have established new marketing arrangements that enable direct marketing with farmers. Employees have initiated many of those initiatives (for example, USDA cafeterias in Washington, D.C., and the California Environmental Protection Agency's food service in Sacramento).

Grades, Standards, and Certification Labels

Marketing tools such as grades and standards, certifications, labels, and branding can create niches of profitability for farmers whose products meet specific requirements related to the nature of their products or the way in which they were produced. In many instances, standards reflect characteristics that make farm commodities and food products easier to handle, process, and transport. In other cases, standards are designed to maintain a consistent level of quality, the cosmetic appearance of fresh food products, or the safety of food products at the retail level.

Grades and Standards

Standards are the measures by which products, processes, and producers are judged, whereas grades are the categories used to implement the standards (Busch and Bingen, 2006). Grades and standards can define what is to be traded on the market, establish agreed-upon production processes, fix levels of consistent product quality, and make possible the location of exportable production around the world by ensuring compatible products and processes (Busch and Bingen, 2006). For example, U.S. federal law allows growers and handlers of many fruit, vegetable, and specialty crop products to develop formal marketing orders as standards and grades to coordinate production, processing, and marketing of specific commodities. Marketing orders are often used to raise money for production research, marketing research and development, and advertising. In addition, they can be used by industry actors to create binding rules (enforced by USDA) regarding allowable production, packaging, and handling practices designed to ensure consistency, quality, safety, and cosmetic appearance of food products (USDA-AMS, 2008b). USDA also provides standardized grading, certification, and inspection services as a service to commodity sectors that voluntarily want those services as part of their marketing strategies. However, the need for global harmonization of standards and local adaptability of farming systems to meet such standards remains unresolved (Vogl et al., 2005).

Although marketing grades and standards theoretically create similar expectations for all producers and can communicate to consumers about the attributes of the products, they can affect the ability of producers to use certain production practices. For many fruits and vegetables, grades and standards might inhibit the use of practices that can improve environmental sustainability if they affect the ability of the product to meet the grade and standard that leads to the highest price. For example, increasingly strict defect action levels (DALs), established by the U.S. Food and Drug Administration partly to enhance safety from microbial contamination (sometimes by mandating the production practices to be followed), have been linked to increased pesticide use in many commodity sectors (Hart and Pimentel, 2002), although Lichtenberg (1997) found an opposite effect on apple production. Mandatory behaviors and financial assessments under marketing order rules could disadvantage some producers and handlers of organic products because they might not benefit from generic commodity research, supply control, and marketing efforts supported by those orders (Carman et al., 2004). Similarly, standards established for farm worker protection might encourage the replacement of hand harvesting with mechanical harvesting and reduce employment opportunities for farm workers (Friedland et al., 1981).

Aside from governmental rules, privately developed systems of grades and standards designed to ensure safety, quality, and appearance have increasing influence on the way that food is produced in the United States (Henson and Reardon, 2005; Hatanaka et al., 2006). Private standards systems and third-party certification are increasingly replacing "hard regulation" (that is, traditional regulatory) approaches to governing international trade (Hatanaka et al., 2005; Ponte and Gibbon, 2005; Higgins and Hallström, 2007). Global coalitions of private firms that set harmonized standards for food have been emerging. They facilitate coordination of production and distribution (Nadvi and Waltring, 2003) and protect the firms' reputation for consistent quality and safety (Fulponi, 2006). Harmonized standards also can secure competitive advantages for the coalition of firms and exclude access to firms and producers not in the coalition, and thereby serve as a private governance tool in the food system (Fulponi, 2006).

Sustainable Agriculture Standards, Certification, and Eco-label Programs

A market trend that favors improving agricultural sustainability has emerged, as discussed in the earlier section on emerging markets. Standards and certification programs specifically designed for marketing "sustainably produced" foods in the United States and internationally are intended to establish measurable criteria or guidelines for food producers and distributors, provide verification to the public, and support claims about sustainability and environmental sensitivity. Many different types of organizations, including nongovernmental organizations, trade associations, food industry groups, cooperatives, regional organizations, and some university departments, are involved in establishing and administering these kinds of certification and standard-setting programs. Through such programs, the groups are attempting to verify producers' and distributors' efforts to reduce environmental impacts, while also gaining market opportunities to meet rapidly growing demand for "green" products. Although certification and eco-label programs are considered voluntary, an increasing number of food processors, retailers, and distributors require producers who sell to them meet these "sustainability standards." The programs are therefore becoming important forces that drive change in practices by producers, in many cases making it possible for large-scale production systems to provide the now standardized products (Vogl et al., 2005). Certification also offers producers additional benefits, such as greater marketplace recognition, and might facilitate greater information exchange among participating farmers (Klonsky and Tourte, 1998).

Organic agriculture certification is probably the most well-known and well-established food standard related to environmental concerns in the United States and around the world. As of 2002, organic production standards in the United States have been regulated by federal law, and they are administered and enforced by the USDA National Organic Program—NOP (7 C.F.R. Part 205). By law, any product labeled "organic" is required to be produced and certified according to certain standards, including eliminating synthetic pesticides and fertilizers (USDA-AMS, 2008b). The creation of USDA Organic Certification codified a set of practices that emerged from an agroecological approach to production that emphasizes the use of naturally occurring tools for controlling pest, pathogens, and weeds, and the elimination of synthetic inputs. The certification program was created to regulate competition and provide uniform information to consumers.

USDA organic certification standards provide consistency, but the extent to which they should include criteria beyond environmental or health goals, and the specific practices allowed on organically certified farms, have been the subject of much debate and controversy. Although the original Organic Foods Production Act was passed by Congress in 1990, it took 12 years to develop and promulgate formal rules to implement the USDA organic certification label. The resulting program focuses on health and environmental issues, but generally does not address labor, social, economic, or community welfare goals that are important objectives of many proponents of organic farming systems (Guthman, 2004; Fürst et al., 2005; Bittman, 2009).

Partly in reaction to limitations in the formal USDA organic certification standards, programs that have developed alternative or more broadly construed "sustainability" standards for agricultural practices have grown rapidly. Some of these programs include traceability and tracking of the origin of products, or require the analysis of the complete lifecycle environmental impact in the supply chain of each product. Many have evolved into developing "eco-labels," which are logos or seals that identify products or companies that meet certain environmentally preferable standards or criteria. Eco-labels are generally intended to enable people to identify, buy, and sell products and services that are considered environmentally sensitive (Big Room Inc., 2009). More than 100 different eco-label programs have been identified for food or agriculture products worldwide, excluding private labels developed by individual companies (Big Room Inc., 2009). Other groups have developed standards that address nonenvironmental goals, such as fair trade, fair labor, and livestock production practices perceived as more humane (Brown and Getz, 2008; Food Alliance, 2009). Table 6-2 illustrates some of the eco-labels and sustainability certification programs in the food and agriculture sector.

The labeling programs have different levels of quality control and different processes for verification. Typically, companies apply to an organization for the right to use its eco-label on their products. The applicant pays an initial fee and undergoes some kind of inspection or audit. If it successfully meets the standard, it pays a fee to use the label, and in most cases, it is required to have a regular audit to ensure continual adherence to the standards. Although the eco-label would allow producers to access a niche-market and could enhance marketplace recognition of their products, the fees for certification and the paperwork could deter producers from using the label.

The increase in the use of various eco-labels and environmental certification programs in the food sector has created some confusion among consumers and has created challenges for producers who are often being asked to fulfill several distinct standards by different buyers. As a result, eco-labels and certification programs might not be as effective in motivating producers to adopt certain practices for improving sustainability to seek certification as they could be. Some consumer advocacy organizations and government agencies are also concerned about the lack of oversight, consistency, and quality control of such programs. Consequently, some organizations (including Scientific Certification Systems and ANSI, Keystone Center, and USDA) have initiated efforts to develop national sustainable agriculture standards, which are intended to be similar to the Forest Stewardship Council (FSC) certification program in the forest sector, to create score cards, or to monitor different certification and eco-labeling programs.

Marketing Institutions for Mid-Sized Commercial Farmers: Branding

Branding is a method for defining a product as unique and building customer loyalty. By establishing a proprietary brand that consumers associate with desirable qualities, producers can "create" a market for themselves. Marketing studies find that the brand represents a set of values; therefore, the brand gives the consumer confidence in the product (see also the section on low-confinement hog systems in Chapter 5). Branding can provide opportunities to reward farmers for using certain socially desired production practices. However, how branded products actually differ from nonbranded products is not always clear, particularly when without governmental or private sector standards against which brands can be held accountable. Efforts to develop individual farm brands—such as grass-fed or natural beef—can be time consuming and difficult for producers, and rewards depend on the ability to develop regional or national markets for their branded products (Gwin, 2009). Branding can also be undermined by the rise of private-label products in many grocery chains. Efforts to develop private-label organic products, for example, are thought to be reducing demand for some branded organic companies' products (Hills, 2008).

TABLE 6-2 Examples of Sustainable Agriculture Standards and Eco-Label Programs

Program/Initiative	Commodity or Food Sector	Sustainability Scope	Geographic Coverage	Participants
Better Sugarcane Initiative	Sugar	Broad	Global	Companies, nongovernmental organizations (NGOs), unions
California Sustainable Winegrowing Program	Winegrapes	Broad	California	Winegrowers
Common Code for the Coffee Community	Coffee	Broad	Global	Companies, NGOs, unions
Demeter—Biodynamic	All	Ecosystems, biodiversity, spiritual	Global	Not available
Eurep GAP Protocols	Horticultural crops, coffee newly added	Mainly environmental, some social	Global	Companies
Food Alliance	All	Mainly environmental, some social	United States	Companies, university, stakeholders
Fish Friendly Farming	Winegrapes and other crops	Environmental, focused on water	California, Oregon	NGOs, companies, government
Global Environmental Management Initiative	Many food products (and other commodities)	Environmental	Global	Companies (mainly multinational corporations)
Good Agricultural Practices Standard (GAP)	All	Safe handling, environmental, some social	United States	Companies, government
Humane Dairy Checklist	Dairy animals	All	Global	Companies
ICCO Sustainable Cocoa Program	Cocoa	Mainly social	Global	Companies, NGOs, government
IISD/UNCTAD Sustainable Commodities Initiative	Selected commodities, including coffee	All	Global	
International Federation of Organic Agriculture Movements (IFOAM)	Organic certification for all agriculture products	Mainly environmental	Global	Companies, NGOs, unions, farmers
Keystone Center	All	Broad (still in process in 2009)	United States and beyond	Multiple stakeholder, mainly companies
Marine Stewardship Council	Wild and caught seafood/fish	Mainly environmental	Mostly temperate waters	Companies, NGOs
Protected Harvest (Healthy Grown)	Potatoes, winegrapes, and developing others	Mainly environmental	North America	Companies, NGOs
Rainforest Alliance Standards for Sustainable Agriculture	Coffee, bananas, and others	Broad	Global	Companies, NGOs

TABLE 6-2 Continued

Program/Initiative	Commodity or Food Sector	Sustainability Scope	Geographic Coverage	Participants
SAI 8000	All	Social	Global	Companies
SAI Platform	Selected crops, including peas, potatoes, and coffee	Broad	Global	Companies
Salmon Safe	Several crops	Environmental, focused on water quality and fish protection	Western United States, mainly Oregon	Companies, NGOs
Bird-Friendly Certification	Coffee	Environmental	Global	Farmers, NGOs, for example, Smithsonian, Rainforest Alliance, others
Social Accountability in Sustainable Agriculture (SASA)	All	Broad	Global	Research program comparing standards of 4 schemes of the International Social and Environmental Accreditation and Labelling Alliance

Emerging Markets for Ecosystem Services

Some agricultural practices can provide beneficial ecosystem services (see detailed discussion in Chapter 3)—for example, regulation functions such as water quality and nutrient cycling, pollination, and habitat for wildlife and beneficial species; supporting services, such as soil fertility, soil structure and carbon, and carbon sequestration; and aesthetic and cultural services, such as open space and cultural heritage (Heal and Small, 2002; Swinton et al., 2007).

When managed sustainably, agricultural ecosystems have the potential to deliver diverse ecosystem services and to mitigate past ecosystem disservices (Swinton et al., 2007). For example, Boody et al. (2005) examined the impacts of a range of practices, such as cover crops, riparian buffer strips, conservation tillage, crop rotation, and other increases of vegetative cover, and they noted that "environmental and economic benefits can be attained through [these] changes in agricultural land management without increasing public costs. The magnitude of these benefits depends on the magnitude of changes in agricultural practice. Environmental benefits include improved water quality, healthier fish, increased carbon sequestration, and decreased greenhouse gas emissions, while economic benefits include social capital formation, greater farm profitability, and avoided costs" (Boody et al., 2005, p. 27). The quantity and quality of services that are produced by agriculture in a given location generally depend on the joint actions of many farmers and other resource users.

Many ecosystem services are appreciated by society and can be interpreted as capital assets (Heal and Small, 2002). However, ecosystem services can be indirect, underappreciated, and, in general, undervalued. Because of the lack of markets for ecosystem services, farmers cannot add to their revenues when supplying these services (Swinton et al., 2007). For example, pollination services, which have recently become threatened by honeybee colony collapse disorder, contribute to fruit, nut, and vegetable production worth $75 billion in 2007 (USDA, 2007). Since farm producers typically receive no economic payments for possible ecosystem services (for example, pollination) and other social benefits provided by their farming operations, they have little incentive to adopt management practices that provide ecosystem services if the practices incur additional costs.

Because of the missing markets for valued ecosystem services, there has been an argument for and experimentation with public provision of such market signals by providing ecosystem service markets. The underlying assumption for providing ecosystem markets is that paying for ecosystem results will encourage innovation and achieve cost-efficiencies in providing services (Shabman and Stephenson, 2007). However, others argue that such a provision is sometimes tantamount to paying farmers not to pollute or harm public resources (Box 6-2).

Payment for Environmental Services: Beneficiary Pays

Although practical applications with ecosystem markets are limited, emerging experiences demonstrate promise and value to farmers who are using conservation practices. For example, to compensate for a lack of markets for ecosystem services, some have proposed using a payment for environmental services (PES) approach in which the beneficiaries of an ecosystem service pay for the provision of that service. The PES approach can be contrasted with the so-called "green payments," where the general taxpayer subsidizes farmers for desired outcomes or for adoption of desired agricultural practices. The green payments are usually discussed as replacing or augmenting Farm Bill commodity subsidy payments, which are discussed later in this chapter. In PES approaches, the beneficiaries might be a private industry (for example, a private water treatment facility), but the beneficiaries are often represented by some agency or institution. For example, Bohlen and his colleagues (2009) reported on developing a PES pilot program in the northern Everglades of Florida, where beneficiaries (in this case, Florida state agencies) compensate ranchers for providing

---

BOX 6-2
Public Attitudes Toward Private Land Management

Public attitudes toward private land management have undergone a long evolution in U.S. history. For much of U.S. history, the role for public programs in agriculture was to enhance the productivity of croplands. Thus, taxpayers subsidized public projects to provide the agriculture sector with low-cost energy, irrigation water, and transportation and to enhance farmers' incomes. That "progressive conservation" attitude has been challenged by a rise in environmental attitudes that has meant the redesign of some public programs. For example, farmers might have received taxpayer funds and technical assistance to drain and fill wetlands at one time; now they might receive funds to protect wetlands. Most environmentally focused programs for agriculture operate on the premise that farmers do not want to harm public resources but need information, financial support, and technical assistance if they are to avoid doing so (Batie et al., 1986).

Because of that history, disputes about whether farmers should be paid for positive environmental outcomes or penalized for environmental damages occur frequently. The ultimate choice is a political one, but with few exceptions, such as the Endangered Species Act, policy makers have favored voluntary, incentive-based approaches to obtain public benefits from agriculture. Thus, payment for the provision of positive ecosystem services is generally viewed as the appropriate public response—a viewpoint that reflects a belief that farmers have the property rights to pollute or harm ecosystem services, but will be stewards if provided compensation. A contrasting viewpoint is one in which farmers should be regulated and penalized if they harm public resources.

objectives, are important in the cost-effective design of the carbon sequestration trading and offset markets.

Nevertheless, the carbon-trading programs are garnering interest. Several proposed federal climate bills include cap-and-trade or offset approaches that will limit carbon emissions and increase the value of carbon credits. Passage of those bills could lead to the development of additional market or trading opportunities for farmers who are using "carbon-friendly" practices (AFT, 2009). The increasing attention to carbon sequestration might encourage management practices that increase carbon sequestration, reduce greenhouse-gas emissions, and have other benefits for producers and have positive social impacts. For example, conservation tillage, cover crops, habitat buffers, or other vegetative cover can be profitable for farmers and are useful to increase carbon storage and other ecosystem services (Kolk, 2008).

Role of Valuation of Ecosystem Services

In part because of the increasing interest in providing "markets" for ecosystem services, interest in providing monetary values of ecosystem services such as carbon sequestration, water quality, soil carbon, and wildlife habitat is growing. Valuations can emerge from market trading when willing buyers and sellers of offset credits come to agreements on prices. Valuations can also be estimated using nonmarket valuation techniques, such as contingent valuation, hedonic valuation, and cost-based or factor-income approaches (Swinton et al., 2007). Nonmarket valuation techniques require distinctive accounting methods and policies and integration of ecology into agricultural economics (Heal and Small, 2002). Although valuation of ecosystem services is a relatively new field of economics, methods and knowledge for estimating the value of ecosystem services and for identifying cost-effective policy designs to create incentives for agriculture producers to provide them are developing rapidly (Casey et al., 1999).

PUBLIC POLICY AS A CONTEXTUAL FACTOR

A large number of public policies and programs influence farmer decision making: production credits, environmental regulations, liability rules, tax incentives, transportation policies, antitrust legislation, credit availability, intellectual property rights, disaster payments, education and research, crop insurance policies, international and domestic trading rules, and private contracts. They include incentive programs to address environmental goals (such as improving air, water, or grazing lands quality), they might involve public investment in infrastructure (for example, road, sewer, or fiber optic locations), or they might be macroeconomic policies (for example, tax or labor policies) (Batie, 2001).

So extensive and complicated are the numbers of programs and policies that affect farming that it is nearly impossible for policy analysts, economists, and others to predict all their direct and indirect impacts. In addition, only after some period of time will sufficient data and information become available to know how the policies have affected farmer behavior and the collective result of that behavior. This section focuses on some policies and programs with particular relevance to sustainable farming practices and systems.

The Food, Conservation and Energy Act of 2008

The Farm Bill, established in 1936 and reenacted every four to five years by the U.S. Congress, provides the legal framework for taxpayer support of the agricultural and rural economy, including the promotion of conservation and provision of nutritional assistance to needy families. Each Farm Bill has its own name; the 2008 bill is called The Food, Conservation and Energy Act. The Farm Bill is a major influence on many, but not all, producers' choice of crops and land management decisions. About 26 percent of farms, representing 53 percent of all farmland (Figure 6-3), receive Farm Bill commodity payments; an additional 10 percent of all farms and 8 percent of farmland also receive conservation payments without participating in commodity payments (Claassen et al., 2007a). Major provisions of the Farm Bill are the commodity support programs, the crop insurance and disaster programs, conservation programs, and nutritional assistance programs.

Commodity Support Programs

Some provisions of the Farm Bill subsidize the cost of producing commodity crops such as cotton, rice, corn, wheat, and soybean and the cost of producing dairy products, peanuts, and sugar. These Farm Bill crops and products can be grown using a variety of practices and systems, including organic production practices. None of the Farm Bill commodity payments go to nondairy livestock, fruits, and vegetables. Current commodity support programs include direct payments to farmers based on the historical number of acres planted of a particular commodity, and others are tied to target prices that trigger payments. In 2003–2006, the Farm Bill provided more than $63.4 billion in farm subsidies to recipients (not all of whom were commodity producers) (Environmental Working Group, 2009). Commodity payments provide strong incentives to plant program crops or produce dairy products and to maximize yields.

Partly because of a perception that such subsidies might put nonsubsidized fruits and vegetables at a price disadvantage relative to Farm Bill crops, some argue that a portion of Farm Bill payments should be directed toward fruits and vegetable production. In other words, they suggest that "full planting flexibility" be given to existing Farm Bill participants so that these farmers could plant fruits and vegetables instead of traditional Farm Bill crops. However, there is little evidence that "full planting flexibility" would result in sig-

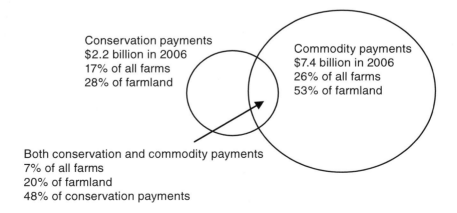

FIGURE 6-3 Commodity payments are large relative to conservation payments; overlap is modest. DATA SOURCE: 2001–2003 Agricultural Resource Management Survey (ARMS).

nificant increases in the production of fruits and vegetables (Fumasi et al., 2006; Thornsbury et al., 2007). Direct subsidies to existing fruit and vegetable growers, however, would likely result in an expansion of such production and a lowering of market prices.

Much of the value of commodity payments tends to be capitalized into the price of fixed assets such as land, so that if program payments were to decline, so would farmland prices. High farmland prices can raise the cost of producing crops and dairy products, but farmland can also serve as collateral for loans if a farmer owns the land. However, higher-priced farmland can serve as a barrier to entry for new farmers. Even if new farmers are able to purchase land, they might find the high costs of land make it harder for them to be the low-cost producer.

The commodity support programs in Farm Bills have been criticized for encouraging farming practices that move away from sustainability. Unless effectively inhibited by environmental regulation or other requirements, the commodity programs encourage farmers to convert lands—sometimes marginal lands—to agricultural uses and to use intensive production practices to produce commodity crops. Intensive production practices could lead to farmers using excess inputs (for example, pesticides and fertilizers), neglecting conservation practices, and modifying the hydrology of the landscape extensively (for example, via destruction of wetlands, drainage, or irrigation). The result can be loss of wildlife habitat and biodiversity and substantial leaking of dissolved salts, pathogens, and various soil and farm nutrients from the farm that impair the functional integrity of ecosystems (NRC, 1993a; Claassen et al., 2001; Lubowski et al., 2006).

Other criticisms are that the commodity support programs of the Farm Bill accelerate the loss of mid-sized farms and the consolidation of agricultural lands under fewer operators and thereby can undermine the economic viability of rural communities. Because the commodity support programs have resulted in more lands dedicated to feed grains such as corn and soybean than would otherwise be the case, the Farm Bill has also been accused of being a major force in producing a food supply inconsistent with USDA's dietary health guidelines (Hamm and Bellows, 2003). Abundant and low-cost feed grains have reduced the costs of confined animal feeding, which some critics claim has led to factory-like practices for the production of animal products (Pew Charitable Trusts and Johns Hopkins Bloomberg School of Public Health, 2008).

On the other hand, it is difficult to know how much of the land use changes and management would have been the same without the commodity support programs. The Farm Bill might have only accelerated underlying forces toward consolidation, certain crops, and neglect of off-farm costs such as water pollution (Dimitri et al., 2005). Outcomes associated with farm policy deregulation in other countries provide some insights into the U.S. Farm Bill's influences on sustainability outcomes, although each country's situation has its own contextual factors that can influence the outcomes. Box 6-4 summarizes some findings of deregulation studies.

The 2008 Farm Bill deviates from earlier Farm Bills in that it provides financial support to farmers to convert to organic production. The 2008 Farm Bill mandates a five-fold increase in research funds for organic production with two research priorities: assessing conservation and environmental outcomes of organic practices and developing new and improved seed varieties for use in organic production systems. It also includes other provisions that might encourage farmers to adopt organic production or other conservation practices, including technical assistance on organic conservation practices and priority given to qualified beginning and socially disadvantaged producers, owners, or tenants who use the loans to convert to organic or other production systems that could improve sustainability (Green, 2009).

> **BOX 6-4**
> **Other Countries' Experiences with Farm Policy Deregulation**
>
> Declining Canadian subsidies of agriculture did not result in diversification of wheat crops in Saskatchewan, Canada. Rather, wheat plantings and a continued reliance on specialized production seem to have expanded (Bradshaw, 2004). In contrast, in southwestern British Columbia, there was a loss of the local fruit and processing industry, forcing diversification away from the production of processed fruits and vegetables, but not in fresh crops (Fraser, 2006). Those studies suggest that the impact of farm policies on crop and livestock specialization cannot be generalized, because responses to deregulation appeared to have been context and locality specific.
>
> Farm policy deregulation in New Zealand and Australia has been followed by fewer but larger farms, increased productivity from adoption of new technologies and improved management practices, and diversification of farm businesses and increased off-farm income (Harris and Rae, 2004; Anderson et al., 2007). Real prices of farmland in New Zealand have recovered from the deregulation shock (Lattimore, 2006; Anderson et al., 2007). Overall, the economic and environmental effects of the New Zealand deregulation appear to be positive (Vitalis, 2007).
>
> Elimination of subsidies in Australia and New Zealand included nonagricultural subsidies, and agricultural subsidies in those countries were smaller than agricultural subsidies in the United States or the European Union. In New Zealand, after deregulation, agriculture's share of Gross Domestic Product (GDP) increased from 7 percent in the late 1980s to 9 percent in the early 2000s (Anderson et al., 2007). The increase in share might be an explanation for why farmland prices rebounded after substantial declines in the mid-1980s.

Crop Insurance and Disaster Payments

Federal crop insurance programs subsidize the cost of private insurance purchased by farmers in the event of crop failure and enable them to be eligible for disaster assistance. Many banks require farmers to buy crop insurance as a precondition for a loan. Crop insurance subsidies are calculated as a percentage of the total premium (Westcott and Young, 2000). The total premiums depend on the risks of loss associated with each crop on each acre of land, and therefore vary across crops and farms. As a result, the premium subsidies are higher for coverage for production of riskier crops and for production on riskier lands. The premium structure potentially encourages farmers to engage in risky practices and to keep some low-productivity land and some environmentally sensitive lands, such as those with highly erodible soils, in production (Westcott and Young, 2000). A higher proportion of low-quality land than the national average for cultivated cropland is subject to such subsidized payments (Lubowski et al., 2006). Models of crop insurance effects on production decisions suggest that resulting land use effects (such as shifting extensive land from hay and pasture to corn production) increase total chemical use (Wu, 1999).

Crop insurance can be a disincentive for farmers to diversify their production enterprises. One of the reasons that some farmers diversify their production enterprises is to reduce income variation (Donoghue et al., 2009). Lowering the risk of income variation in specialized production systems by crop insurance coverage might motivate farmers to use those systems because of their ease of management and potentially lower production costs relative to diversified systems. The Federal Crop Insurance Reform Act of 1994 raised insurance subsidies for farmers substantially. An analysis of data from the 1992 to 1997 Census of Agriculture and the records of crop insurance adoption from USDA's Risk Management Agency suggest that farmers increased specialization in response to increases in subsidized crop insurance coverage (Donoghue et al., 2009). The increase in crop insurance purpose is to progressively eliminate tariff and quota barriers to agricultural trade, facilitate science-based sanitary and phytosanitary standards, and increase cross-border cooperation on issues associated with agricultural labor and the environment among the three member nations (USDA-ERS, 2009e).

Because global and regional trade agreements often limit the independence of individual countries' farm and environmental programs, there are concerns that they could hamper an individual country's ability to support agricultural sustainability attributes (Tothova, 2009). Trade agreements often include some trade-legal mechanisms for countries to protect desired landscapes, reduce emissions of greenhouses gases, dictate certain production practices (such as the use of pesticides, growth hormones, genetically engineered crops, or cruelty-free animal care), or demand certain labor working conditions (such as fair trade). However, because many disputes involve disagreements about acceptable risks for food safety or appropriate ethics for fair and humane production practices, no clear line of demarcation ascertains when such trade protection is a legitimate tool to accomplish a country's desired sustainability outcomes or desire for self-sufficiency in times of food shortages versus when it is a means of protecting domestic producers from international trade competition. If it is the latter, then such protection could violate international trade agreements. Some recent WTO trade disputes between the United States and the European Union about hormones in meat or the planting of genetically engineered crops indicate the difficulty of determining the line of demarcation between achieving domestic policy objectives and protecting domestic producers in such complex issues that involve the attributes of goods or production methods. As a result, legitimate concerns have arisen that sustainability attributes might be omitted or impaired in the pursuit of more liberalized trade (Batie and Schwiekhardt, 2009; Tothova, 2009) or that the imposition of sustainability requirements in traded products (for example, nongenetically engineered) attributes could limit U.S. farmers' ability to profitably supply commodities that have those attributes. Indeed, the failure of the 2009 Doha WTO trade rounds to reach a resolution as to how to manage countries' domestic sustainability objectives in the context of liberalized trade emphasizes the complexity of these issues. WTO might have to reconsider its policies to provide a balance between the objectives of each country's domestic sustainability and trade liberalization (Batie and Schweikhardt, 2009).

Energy Policy

The Energy Independence and Security Act (EISA) of 2007 was enacted to "to move the United States toward greater energy independence and security, to increase the production of clean renewable fuels, to protect consumers, to increase the efficiency of products, buildings, and vehicles, to promote research on and deploy greenhouse gas capture and storage options, and to improve the energy performance of the Federal Government, and for other purposes." EISA was signed into law on December 19, 2007, and took effect in January 1, 2009. A subtitle within the bill, the Renewable Fuel Standard (RFS), is designed to promote the production and consumption of biofuels by requiring minimum annual levels of renewable fuels in U.S. transportation fuels. Previously, the Energy Policy Act of 2005 had established a national renewable fuel standard that mandated an increased use of renewable fuels from 4 billion gallons per year in 2006 to 7.5 billion gallons per year in 2012. EISA amended that standard to set forth a phase-in for renewable fuel volumes beginning with 9 billion gallons in 2008 and ending at 36 billion gallons in 2022. The 2007 EISA requires 10.5 billion gallons of conventional ethanol to be used in 2009, 12 billion gallons in 2010, and continual increases until a maximum of 15 billion gallons in 2015 through 2022.

The bill also requires cellulosic biofuels to be used starting in 2010, rising dramatically from 100 million gallons in 2010 to 16 billion gallons in 2022 (Table 6-3).

To facilitate market penetration of biofuels, the federal government has adopted several policies that subsidize biofuels production. Specifically, the USDA Commodity Credit Corporation's (CCC) Bioenergy Program, announced in 2000, has made cash payments to commercial bioenergy (ethanol and biodiesel) producers in the United States who increase their bioenergy production from eligible commodities. The program was aimed to expand markets for agricultural commodities and to promote the use of biofuels (USDA-FSA, 2000; USDA, 2000). The American Jobs Creation Act of 2004 provided a $0.51 tax credit per gallon of ethanol blended to companies that blend ethanol and a $0.50–$1.00 per gallon tax credit, depending on the feedstock, to biodiesel producers.

The impact of those policies on American agricultural practices has been profound. As discussed in Chapter 2, the corn ethanol and soybean biodiesel industries emerged rapidly and contributed to shifts in the acreages planted to those crops. Whether and how cellulosic biofuels consumption mandates will influence farmers' choice of crops or management systems are yet to be seen. In each case, new biofuel policies and market opportunities will

TABLE 6-3 Mandated Consumption Targets for Various Biofuels Under the 2007 Energy Independence and Security Act (EISA)

Year	(In billions of gallons)				
	Conventional Biofuel[a]	Advanced Biofuel[b]	Cellulosic Biofuel[c]	Biomass-based Diesel[d]	Total RFS
2008	9.0	—	—	—	9.000
2009	10.5	0.600	—	0.500	11.100
2010	12.0	0.950	0.100	0.650	12.950
2011	12.6	1.350	0.250	0.800	13.950
2012	13.2	2.000	0.500	1.000	15.200
2013	13.8	2.750	1.000	[e]	16.550
2014	14.4	3.750	1.750	[e]	18.150
2015	15.0	5.500	3.000	[e]	20.500
2016	15.0	7.250	4.250	[e]	22.250
2017	15.0	9.000	5.500	[e]	24.000
2018	15.0	11.000	7.000	[e]	26.000
2019	15.0	13.000	8.500	[e]	28.000
2020	15.0	15.000	10.500	[e]	30.000
2021	15.0	18.000	13.500	[e]	33.000
2022	15.0	21.000	16.000	[e]	36.000

[a]Conventional biofuel means renewable fuel that is ethanol derived from corn starch.

[b]Advanced biofuel means renewable fuel, other than ethanol derived from corn starch, that has lifecycle greenhouse-gas emissions, as determined by the administrator, after notice and opportunity for comment, that are at least 50 percent less than baseline lifecycle greenhouse-gas emissions.

[c]Cellulosic biofuel means renewable fuel derived from any cellulose, hemicellulose, or lignin that is derived from renewable biomass and that has lifecycle greenhouse-gas emissions, as determined by the administrator, that are at least 60 percent less than the baseline lifecycle greenhouse-gas emissions.

[d]Biomass-based diesel means renewable fuel that is biodiesel and that has lifecycle greenhouse-gas emissions, as determined by the administrator, after notice and opportunity for comment, that are at least 50 percent less than the baseline lifecycle greenhouse-gas emissions.

[e]At least 1.000 (specific amount to be determined by the administrator).

SOURCE: EPA (2009a).

provide complex new incentives and disincentives for producers to use farming practices described in Chapter 3 (for example, crop rotations, cover crops, reduced tillage, and integrated pest management), which will affect the sustainability performance of the U.S. farm sector. Science-informed policy is important if the positive social and environmental attributes of biofuel production are to be achieved (Robertson et al., 2008). Research is needed to design standardized metrics and approaches and to develop decision-support tools to identify and quantify environmental, food versus fuel, water use, or other potential sustainability tradeoffs for biofuel production (Williams et al., 2009). Otherwise, the effect of biofuel production on resilience of the food production system and environmental quality is uncertain (Naylor, 2008).

Environmental Regulation

The Environmental Protection Agency (EPA) has a number of laws and programs that could affect agricultural producers. (Similarly, many states have their own regulations.) A description of the major federal laws and programs that could affect farmers and their requirements are listed on EPA's website (EPA, 2007). This section uses a few of those regulations to illustrate their potential influence on farmers' adoption of various farming practices and systems and their effect on improving sustainability in U.S. agriculture. Other federal agencies also administer regulations that affect farmers' decisions. The Department of the Interior, for example, administers the Endangered Species Act, and the Food and Drug Administration provides food safety guidelines and administers food safety requirements.

Clean Air Act

The Federal Clean Air Act established the National Ambient Air Quality Standards (NAAQS) for criteria pollutants: carbon monoxide (CO), particulate matter (PM), lead (Pb), nitrogen dioxide (NO_2), sulfur dioxide (SO_2), and ozone (O_3). It also regulates 188 hazardous air pollutants. The law aims to control "major sources" of emissions that exceed the specified thresholds. Although agricultural production is not exempt from the statute, most farms do not exceed the specified thresholds (Copeland, 2009). Among the six criteria air pollutants, two (PM and NO_2) are emitted by animal feeding operations (AFOs). Volatile organic compounds are also emitted by livestock production facilities. Large AFOs that exceed the emissions thresholds are regulated by the Clean Air Act and are required to apply for permits. In addition to the Clean Air Act, some livestock operations are subject to reporting requirements under the Comprehensive Environmental Response, Compensation, and Liability Act (CERCLA, the Superfund law) and the Emergency Planning and Community Right-to-Know Act (EPCRA) if large quantities of certain hazardous substances are released into the environment, including ambient air. The Clean Air Act, CERCLA, and EPCRA might motivate owners and operators of large operations to adopt technologies or practices that reduce the emissions of criteria pollutants and hazardous pollutants so that they do not exceed the thresholds of emissions to be considered "major sources," but the acts only apply to a small percentage of farm operations.

Clean Water Act

The objective of the Clean Water Act is to "restore and maintain the chemical, physical, and biological integrity of the nation's waters by addressing point and nonpoint pollution sources, providing assistance to publicly owned treatment works to improve wastewater treatment, and maintaining the integrity of wetlands" (EPA, 2007). The statute distinguishes between point source and nonpoint source pollution. CAFOs are considered a point source of pollution. Under EPA's revised regulations in 2003, AFOs that meet a specific regulatory threshold number of animals (see EPA, 2002, for the threshold number of animals) are defined as CAFOs for purposes of permit requirements and discharge allowances. CAFOs are required to obtain a National Pollutant Discharge Elimination System (NPDES) permit and are subject to effluent limitations on pollutant discharges. An unpermitted CAFO has an option to certify to the permitting authority that the CAFO does not discharge or propose to discharge and thus avoid the necessity of a permit. Other animal feeding operations are not required by the Clean Water Act to obtain a NPDES permit; however, they might be regulated by state programs. As a result, only about 15,500 of the largest animal feeding operations that confine cattle, dairy cows, swine, sheep, chickens, laying hens, and turkeys, or about 6.5 percent of all animal confinement facilities in the United States, are required to obtain NPDES permits under the Clean Water Act (Copeland, 2009). The Clean Water Act contributes to preventing those 6.5 percent of CAFOs from unpermitted discharges into water bodies, but does not necessarily encourage them to improve in environmental sustainability. Studies suggest that the Clean Water Act and Clean Air Act regulations, along with state regulations, encouraged a shift in production of hogs from the Southeast to the Midwest during the late 1990s to the mid 2000s. The regional shifts were in response to regulations that require reducing waste and odor from large manure lagoons and reducing land applications of manure (Key et al., 2009).

Most other agricultural activities are considered nonpoint sources of pollution, which are generally governed by state water quality planning provisions of the act. Section 319 of the Clean Water Act was enacted to guide states in conducting nonpoint source assessments and in developing and implementing nonpoint source management programs, but there are no federal regulatory requirements. The states have a wide array of enforceable mechanisms for controlling nonpoint source pollution, some of which pertain to agriculture. Some states prohibit the discharge of pollutants or waste without a permit and have enforceable laws for the control of erosion of sediments, but agricultural enterprises are often exempt from those laws (Environmental Law Institute, 1997). Most state plans rely primarily on voluntary programs to promote the adoption of practices for reducing nonpoint source pollution (Feather and Cooper, 1995). Such voluntary programs as education and technical assistance seem to be most effective in encouraging adoption if the practices involve only small, inexpensive changes in the farming operation and are profitable to the farmer. Incentive payments or cost-sharing can encourage adoption of best management practices, but the adoption rates vary across practices and geographic areas (Feather and Cooper, 1995).

In addition to regulating discharges, the Clean Water Act provides funding to states, territories, and Indian tribes to support a wide variety of activities aimed at reducing nonpoint pollution under Section 319(h). For example, Indiana uses some of its Section 319(h) funds in an agricultural cost-share program that encourages the implementation of BMPs.

Food Quality Protection Act

Many regulations are designed to ensure food safety, most of which affect processes beyond the farm gate. One exception is the Food Quality Protection Act (FQPA) of 1996, which can affect on-farm practices. Signed into law on August 3, 1996, FQPA amended the two major pesticides law at that time—the Federal Insecticide, Fungicide, and Rodenticide

Conjunctive Use

Conjunctive use of ground and surface water acknowledges the inherent hydrological interconnections between these apparently different sources of supply. At its simplest, conjunctive use entails the reliance on surface water during times of average or above average precipitation and runoff. During drought periods or other times when surface water availability is constrained, use shifts to ground water, which tends to be buffered to some extent from the variability to which surface water is subject. One suggestion is to manage ground water as a reservoir for use during periods of surface water shortfall and recharge it during periods of normal or above normal availability of surface water. Sophisticated schemes of conjunctive use employ managed recharge whereby excess surface waters are captured and transformed into ground water. Managed recharge can be accomplished simply through the use of surface spreading or through direct injection, which generally requires significant investment in facilities (Jury and Vaux, 2007). Economic-engineering models have been developed to optimize the California water supply system, and these models suggest that water transfers and exchanges and conjunctive use could contribute to improved water use performance (Draper et al., 2003; Jenkins et al., 2004).

Animal Welfare Regulations

Several states have passed legislation, termed animal welfare legislation, which mandates certain practices to follow for livestock production. Maine and California, for example, have statutes that require any enclosures or tethers confining specified farm animals allow the animals to fully extend their limbs or wings, lie down, stand up, and turn around for the majority of the day. Specified animals include calves raised for veal, egg-laying hens, and pregnant pigs. Exceptions are made for transportation, fairs, rodeos, research, and veterinary purposes (Swanson, 2009). The new statutes reflect a growing public interest in the treatment of animals, and many producer associations are responding with their own codes of ethical treatment (Swanson, 2009). Because those statutes, which include fines and imprisonment penalties for noncompliance, are now law, other states will likely add similar legislation; some producers will have to change management of their livestock and poultry.

KNOWLEDGE INSTITUTIONS AS CONTEXTUAL FACTORS

Research and experimentation, whether by scientists or farmers, contribute to the development of specific agricultural tools and approaches that help farmers address various sustainability objectives. Previous chapters provided numerous examples of farming practices designed to improve sustainability, most of which have been shaped by public and private sector scientific research and development activities. Moreover, understanding the effects of different types of farming practices or systems on productivity, the environment, and economic and social outcomes is informed by both formal scientific studies and years of farmers' experiences with those approaches. Options that farmers have and their decisions to use a particular practice or type of production system are thus shaped by the availability of appropriate technologies or techniques, and by their understanding of the impacts of the practices on their farm, the environment, their community, and the nation as a whole. Importantly, the breadth of possible agricultural sustainability research topics greatly exceeds the time and resources available to scientists and farmers. As a result, choices about the organization and focus of agricultural research activities can drive and constrain movement along the sustainability trajectory.

Publicly Funded Agricultural Research and Extension

Many of the changes over the past century in agricultural sector productivity, farmer quality of life, livestock animal health, and environmental performance associated with U.S. agriculture have been shaped by public sector investments in agricultural research, education, and extension programs at the state and federal level (NRC, 2002). Numerous studies have shown that investments in public agricultural research have demonstrated consistent economic rates of return of $10 in social benefits for every $1 invested (Fuglie and Heisey, 2007).

The large public investment in agricultural research and extension has long been justified based on the social benefits—in terms of improved factor productivity and a sustained supply of inexpensive and high-quality food—associated with technical gains in farming practices and growing labor efficiency in the farm sector. Examples of public welfare benefits include the often-cited high rates of return on public research investments (Huffman and Evenson, 2006). Publicly funded scientific research has been viewed as necessary because of inherent problems in the economic incentives of private sector inventors, who might find it impossible to capture the benefits of new innovations through market mechanisms because of insecure intellectual property rights (Fuglie et al., 2000).

Land-Grant Universities

Most publicly funded research and extension takes place at the state level, where a network of Land-Grant Universities (LGUs) and State Agricultural Experiment Stations (SAES) spends over $3.3 billion annually on agricultural research, education, and extension programs (Holt, 2007). The federal vision for the LGUs began with the 1862 Morrill Act, which gave states parcels of public land ("land grants") to sell to raise revenues to create public universities in every state in the Union. The Morrill Act gave LGUs their teaching responsibilities and was expanded with the 1887 Hatch Act to create SAES, which use a mix of state and federal funds to support the research activities of LGU scientists working on rural, food, and agricultural research topics. In 1914, the Smith-Lever Act provided additional funds to support the development of agricultural extension systems designed to extend new scientific knowledge, technologies, and management ideas to farmers and others in rural America. The tripartite LGU mission of teaching, research, and extension is a uniquely American invention designed to democratize institutions of higher learning, apply the principles of science to solve applied problems in agriculture and industry, and ensure that the benefits of research are made widely available to people throughout society (McDowell, 2001).

U.S. Department of Agriculture

In addition to the SAES system, USDA conducts a large proportion of publicly funded agricultural research and extension. Specifically, the USDA Research, Education and Economics (REE) mission area includes three major agencies that carry out important agricultural scientific research at the federal level. The largest is the Agricultural Research Service (ARS), which maintains a network of regional research stations organized into eight geographic areas. Research within the ARS system currently occurs within 19 national

DRIVERS AND CONSTRAINTS

program areas within 4 broader areas: animal production and protection; crop production and protection; nutrition and food safety and quality; and natural resources and sustainable agriculture (Knipling and C.E. Rexroad, 2007). The other major REE agencies are the Economic Research Service (ERS), the National Agricultural Statistics Service (NASS), and the National Institute of Food and Agriculture (NIFA; formerly known as Cooperative State Research, Education, and Extension Service).

Distribution of Federal Funds for Agricultural Research

In discussions about improving the sustainability of U.S. agriculture, one of the most commonly suggested policy levers is the use of publicly funded research programs to bolster innovation and overcome obstacles to innovative agricultural practices and farming systems. The distribution of public agricultural research and development (R&D) funding from different sources is highlighted in Figure 6-5, which illustrates the importance of public agricultural research funding in three different ways. At the top of the figure, the entire federal research and development budget is disaggregated to highlight the importance of expenditures on research channeled through USDA. In 2005, USDA managed roughly 2 percent of total federal R&D. The middle of Figure 6-5 disaggregates the total amount of public spending on agricultural research from all sources in 2005. USDA research spending

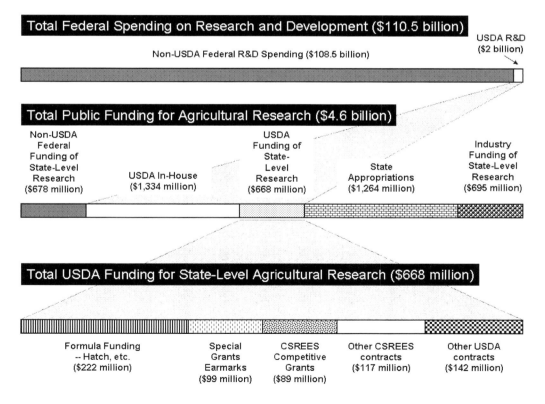

FIGURE 6-5 Estimated importance of publicly funded agricultural research in the United States, by major funding source, 2005.
DATA SOURCE: Pollak (2009), and Schimmelpfennig and Heisey (2009).

makes up just over 40 percent of the total public agricultural science portfolio, and roughly two-thirds of the USDA budget is committed to in-house spending through the ARS, ERS, and NASS, with an additional $668 million sent to state-level agricultural researchers (primarily at LGUs) through various programs administered by USDA-NIFA. Other important public sources of agricultural research funds include state appropriations ($1.3 billion, or roughly 27 percent of the national total), funding of state-level research by other non-USDA federal agencies (15 percent), and private industry funding of state-level research (15 percent). For example, California's SAES spent more than $276 million in 2005 alone, an amount that exceeds the entire USDA competitive research grants budget (Holt, 2007).

The bottom of Figure 6-5 illustrates the distribution of USDA funds that are provided to state-level agricultural researchers and institutions. Although the original Hatch Act established a formula for the distribution of federal funds to each LGU based on rural populations and farm numbers, "formula funds" are currently one-third of the total USDA support of the SAES system. The most important source of funding for the SAES are appropriations from state legislatures, some portion of which is required to match any federal formula funds received. State appropriations contribute roughly 40 percent of the total SAES expenditures (Holt, 2007).

A second major mechanism for distributing USDA funds is through national peer-reviewed competitive research grant programs, renamed Agriculture and Food Research Institute (AFRI; formerly the National Research Initiative) in 2008. NIFA administers this program (Jacobs-Young et al., 2007). The USDA Competitive Grants Program began in response to a National Research Council report (NRC, 1972) that criticized the quality of agricultural research in the United States and the traditional formula funding approach to allocating resources. The original USDA Competitive Grants Program was created in 1977 and received $15 million in congressional appropriations in its first year. In 1989, another NRC report (NRC, 1989b) and a report by the U.S. Congress' Office of Technology Assessment (U.S. Congress Office of Technology Assessment, 1990) recommended expanding and refocusing the USDA Competitive Grants Program with a funding level of at least $500 million per year. Congress created the new National Research Initiative (NRI) in 1990, but has never appropriated more than $190 million for the program, which is well below the maximum level authorized in the founding legislation.

In 2008, USDA reorganized its research programs under the umbrella of AFRI, modeled after the successful National Science Foundation (NSF) and National Institutes of Health (NIH). Expenditures by the AFRI program areas in FY 2009 are listed in Table 6-4. Overall, AFRI allocated 57 percent of grant funds to plant and animal productivity research; 17.6 percent to food safety; 17.2 percent to renewable energy, natural resources, and the environment; 2.8 percent to agricultural systems; and 5 percent to agricultural economics and rural communities. Much of AFRI's research focuses on process-level science and component interactions in applied agricultural systems, so it is positioned between the basic research of NSF programs and the applied research done through farmer collaboration. Most AFRI research is integrated among biological and natural resource-based disciplines, but a small part has significant economic or social science integration at the systems level. In some program areas, research is integrated through the mandatory inclusion of educational and extension outreach activities to disseminate results of applied research projects.

Other sources of USDA funding for the SAES system include special grants (typically legislative earmarks for specific research programs in particular states) and cooperative research agreements between CSREES (or other USDA agencies) and state research institutions. In most years, those projects have received federal research allocations that have exceeded the size of competitive granting programs.

TABLE 6-4 USDA-AFRI Competitive Grants Funding by Program, 2009 Request for Applications

Program Category/Name	Million $	Integrated[a]
ALL PROGRAMS COMBINED:	189,050	$64,200
Plant Health and Production and Plant Products	73,450	
Arthropod and Nematode Biology and Management	12,500	no
Plant Biology	12,250	no
Microbial Genomics	11,000	no
Applied Plant Genomics CAP[b]	10,000	yes
Microbial Biology: Microbial Associations with Plants	7,400	no
Plant Genome, Genetics, and Breeding	6,500	no
Plant Breeding and Education	6,500	yes
Plant Biosecurity	4,300	yes
Protection of Managed Bees CAP	3,000	yes
Animal Health and Production and Animal Products	35,000	
Animal Genome, Genetics, and Breeding	11,000	no
Animal Health and Well-Being	11,000	no
Animal Growth and Nutrient Utilization	4,500	no
Animal Reproduction	4,500	no
Integrated Solutions for Animal Agriculture	4,000	yes
Food Safety, Nutrition, and Health	33,300	
Food Safety and Epidemiology	11,200	mixed
Human Nutrition and Obesity	11,000	yes
Improving Food Quality and Value	6,500	no
Bioactive Food Components for Optimal Health	4,600	no
Renewable Energy, Natural Resources, and Environment	32,500	
Air Quality	5,000	yes
Biology of Weedy and Invasive Species in Agroecosystems	4,600	yes
Managed Ecosystems	4,500	yes
Global and Climate Change	4,500	no
Enhancing Ecosystem Services from Agricultural Lands	4,500	no
Water and Watersheds	4,300	no
Soil Processes	4,100	no
Sustainable Agroecosystems Science Long-Term Agroecosystem Program	1,000	yes
Agriculture Systems and Technology	5,400	
Biobased Products and Bioenergy Production Research	5,400	no
Agriculture Economics and Rural Communities	9,400	
Agricultural Prosperity for Small and Medium-Sized Farms	4,800	yes
Agribusiness Markets and Trade	4,600	no

[a] Integrated projects require major efforts in at least two of three areas: research, outreach, education.
[b] CAP = Coordinated Agricultural Project.

Broadening Review of Public Competitive Grant Programs

The issue of balancing competitive and non-peer-reviewed (formula funding) mechanisms for allocating federal research dollars to agriculture has been discussed in some research and policy literature (Huffman and Evenson, 2006; Schimmelpfennig and Heisey, 2009). Most observers argue that a competitive peer-reviewed allocation process would more closely approximate the process used in most other federal agencies (for example, NIH and NSF) and would be more likely to generate high-quality basic science research, to encourage scientists to address national priorities, and to create greater incentives for the publication and dissemination of scientific research findings (NRC, 2000c, 2003).The proportion of total SAES funding delivered through competitive grant processes has increased significantly in the past 25 years (Schimmelpfennig and Heisey, 2009). Supporters of the formula funding system highlight the need for states to be able to set their research agendas to respond to regional agricultural problems and sustain locally adapted crop and livestock management systems (Huffman and Evenson, 2006). If formula funds are eliminated or drastically reduced, LGU research and extension faculty would spend a greater proportion of their state-funded time writing proposals for federal grants (Huffman et al., 2006). As a result, they will likely spend an increased proportion of time conducting research funded by grants based on federal priorities and a decreased proportion of their time addressing state-level research needs. Some experimental stations could lose matching state funds that are tied to the amount of federal formula funds received. At present, formula and state funding allow scientists from different disciplines to undertake projects that require sustained multiyear efforts to reach research objectives. In some states, a significant reduction in formula funds might erode their overall capacity to undertake agricultural research and result in the closure of outlying research facilities and research farms (Huffman et al., 2006). However, programs supported by formula funds need to be reviewed periodically to ensure that they are productive and responsive to the state's agricultural research needs.

In one study, Rubenstein et al. (2003) confirmed that competitive grants are more likely to support basic research among SAES institutions, but also tend to award federal grants to a smaller number of high-status state research universities than formula funding. Overall, they did not observe much change in the substantive focus of federally funded agricultural research. Rather, they conclude, "competitive funding seems even more focused on production cost reduction [research objectives] than [is] formula funding" (Rubenstein et al., 2003; p. 359). Similarly, a recent ERS report suggested that shifts toward competitive funding did not have much effect on the overall amount of basic research conducted in the SAES and that most public agricultural research funds (both basic and applied) are still spent on farm commodity research, with relatively small fractions devoted to issues of natural resource and environmental topics, family and community research, and investigation of markets, economics, and policy topics (Schimmelpfennig and Heisey, 2009).

Private Sector Agricultural Research

Under pure free market conditions, economists would expect private sector research to underinvest in research because the private sector would be unable to capture the full economic returns from its investments. As a result, for much of its history, the U.S. public agricultural research and extension community has viewed its role as increasing the viability and competitiveness of farmers and private sector agricultural and food companies (Kloppenburg, 1988). Private sector spending on agricultural research, however, has increased substantially and now makes up more than half of total expenditures in the United States (Figure 6-6).

Several factors have encouraged the rapid rise of private agricultural research in recent decades (Caswell and Day-Rubenstein, 2006). First, a series of important legislative and legal changes in U.S. and international patent law in the 1970s allowed the patenting of biological inventions, including new agricultural crop varieties and genetically engineered crops and livestock. This expanded protection of intellectual property rights enabled pri-

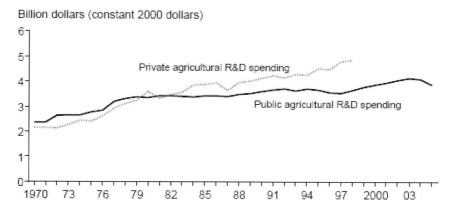

FIGURE 6-6 Real public and private agricultural research and development expenditures in the United States since 1970.
SOURCE: Schimmelpfennig and Heisey (2009).

vate firms to capture a greater share of the benefits of their research efforts than they used to. Second, the dramatic growth in molecular biology and genetics and genomics created a new set of tools well suited to developing proprietary agricultural products.

Private agricultural research can be expected to focus nearly exclusively on near-market R&D and on topics likely to create products that can be sold in the marketplace—for example, developing new chemical technologies, breeding and genetic engineering of certain crops, and developing agricultural equipment for large-scale farms (Reilly and Schimmelpfennig, 2000; Heisey et al., 2005). It is unreasonable to expect investor-owned firms to spend scarce research dollars unless they are able to control benefits from their innovations (Alston et al., 2000). Because of those potential incentives, private sector research focuses primarily on productivity and production efficiency. However, some of the private sector research certainly contributes to other sustainability goals. For example, irrigation technologies developed mostly by the private sector contributes to improving water use efficiency and reducing runoff from overirrigation or precision agriculture technologies contribute to improving efficiencies of water, nutrient, or pesticide use.

In a study of the history of plant breeding, Kloppenburg (1988) suggested that the development of hybrid corn seed was particularly well suited to private sector research because farmers who use hybrid seeds have to buy their seedstock from private seed companies every year (as opposed to open-pollinated varieties that could be saved for replanting). Although the commercialization of new agricultural technologies is essential to rewarding private sector investment, concern is growing about the increasing dominance of a few seed companies and their control over new genetic engineering (GE) crops and technologies. The concerns include seed company patents and control on genes, DNA fragments, and GE technologies. They also include seed companies' ability to block independent research on GE crop performance and environmental impacts, when such research has the potential to provide useful information on how to best grow those crops and give timely alerts to potential drawbacks. A similar pattern has been observed in the development of hybrid poultry varieties, which poultry processors and integrators almost universally now required (Fuglie et al., 2000). Green et al. (2007) suggested that the private sector has shown great capacity to sustain gains in livestock productivity, but that gains in production efficiencies have generated side effects on animal well-being and longevity in modern production environments.

Division of Labor Between Public and Private Agricultural Research

Incentives for public agencies and private firms to fund agricultural research are different (Spielman and von Grebmer, 2004). Public agencies typically are mandated to fund research topics of wide social significance with outcomes that have public goods characteristics (for example, nonexcludability and nonrivalry). The research is likely to require longer time horizons to yield results or to cater to end users with limited purchasing power or market access. Private firms are expected to conduct research that could increase a company's profits and hence the likelihood to engage in research that will potentially result in marketable products.

The differences in incentives for research between the two sectors and the rise of privately funded research are associated with an increasing division of labor between public and private sector agricultural institutions (Dahlberg, 1985). The public sector has been increasingly responsible for conducting basic research and is less engaged in applied research that can produce innovations or products that farmers can directly use (Reilly and Schimmelpfennig, 2000; Caswell and Day-Rubenstein, 2006). The growing division of labor has also led to increased calls for a redirection of public agricultural research away from productivity-increasing or other commodity-focused topics, toward areas of research that can generate public benefits but are unlikely to receive attention from the private sector (Huffman and Evenson, 2006). Examples of such research include environmental and natural resource conservation, food safety and nutrition, poverty reduction, research on public policy impacts, and research targeting small, disadvantaged, or underserved groups of farmers or consumers (Fuglie, 2000; NRC, 2003). Spielman and von Grebmer (2004) saw opportunities for increasing interactions between public and private sector agricultural institutions despite the difference in incentives and growing division of labor in research. For example, the large biotechnology companies have patented products and processes that can advance the public research agenda, and the public institutions have plant genetic resources (for example, germplasm collection) and access to local knowledge resources that could be useful to private firms. However, the authors concluded that "public-private partnerships are significantly constrained by insufficient accounting of the actual and hidden costs of partnership; persistent negative perceptions across sectors; undue competition over financial and intellectual resources; and a lack of working models from which to draw lessons and experiences" (Spielman and von Grebner, 2004, p. 38).

Expanding Beyond Productivity Research

Efforts to reform and refocus public agricultural research programs typically highlight research priorities that would enhance the sustainability performance of most U.S. farms by addressing sustainability goals in addition to productivity and production efficiency (see Box 6-6 and Dahlberg, 1985; Sauer, 1990; Duram and Larson, 2001). Some of the most intense efforts have been within the area of agricultural conservation and environmental issues. A growing number of integrated disciplinary and interdisciplinary research projects have been devoted to improving the understanding of the environmental impacts of typical farming practices and assessing the effectiveness of conservation practices designed to minimize these impacts. The proportion of research spending on productivity and traditional commodity-focused research relative to total USDA research spending declined

> **BOX 6-6**
> **National Research Council and Other Reports Call for Reform of the U.S. Public Agricultural Research, Education, and Extension System**
>
> The National Research Council (NRC) has facilitated a number of studies of the U.S. agricultural research system (see list below). Those reports have consistently called for increased funding for competitive agricultural research programs. They have also called for greater attention to emerging issues in food safety, health and nutrition, protection of environmental and natural resources, and rural community well-being (to complement conventional research goals designed to increase the productivity and competitiveness of the U.S. farm sector). Similar reports by agricultural foundations, policy researchers, and advocacy groups have echoed these concerns (Rockefeller Foundation, 1982). For example, in the book **Agricultural Research Policy**, Vernon Ruttan argued that:
>
> > society should insist that agricultural science embrace a broader agenda that includes a concern for the effects of agricultural technology on health and safety of agricultural producers, a concern for the nutrition and health of consumers, a concern for the impact of agricultural practices on the aesthetic qualities of both natural and artificial environments, and a concern for the quality of life in rural communities. (Ruttan, 1982, pp. 350–351)
>
> More than 10 years later, the Council for Agricultural Science and Technology published a report entitled **Challenges Confronting Agricultural Research at Land Grant Universities** (CAST, 1994). The report targets four areas as research priorities for the SAES: the environment, sustainable production systems, economies of rural communities, and consumer interests (for example, food safety and quality). Although many examples of tangible changes in the structure and administration of CSREES competitive grant programs exist, recent reports and criticisms of the public agricultural science system suggest that fundamental problems still remain (NRC, 2003; Huffman and Evenson, 2006; Robertson et al., 2008).
>
> Selected NRC reports on the U.S. agricultural research system, and their dates of publication beginning with the most recent, are as follows:
>
> - 2009: **Transforming Agricultural Education for a Changing World**
> - 2008: **Agriculture, Forestry, and Fishing Research at NIOSH**
> - 2003: **Frontiers in Agricultural Research: Food, Health, Environment and Communities**
> - 2002: **Publicly Funded Agricultural Research and the Changing Structure of U.S. Agriculture**
> - 2000: **NRI: A Vital Competitive Grants Program in Food, Fiber and Natural Resources Research**
> - 1999: **Sowing the Seeds of Change: Informing Public Policy in the Economic Research Service of USDA**
> - 1996: **Colleges of Agriculture at the Land Grant Universities: Public Service and Public Policy**
> - 1995: **Colleges of Agriculture at the Land Grant Universities: A Profile**
> - 1994: **Investing in the NRI: An Update of the Competitive Grants Program of the USDA**
> - 1989: **Investing in Research: A Proposal to Strengthen the Agricultural, Food and Environmental System**

from 1986 to 1997, whereas the relative proportion of spending on research on water, air, soil, forests, wildlife, sustainable resources management, disease control, and community impacts increased (NRC, 2002).

Nevertheless, a large proportion of public agricultural research (both within USDA and throughout the SAES system) remains focused on improving productivity and production efficiency systems (Caswell and Day-Rubenstein, 2006). An increasing number of specific federal and state agricultural research, education, and extension programs are explicitly designed to support agricultural systems that focus on more than one or two goals of sustainability. In this section, the committee uses some programs and institutions as examples to illustrate how they support agriculture that balances multiple sustainability goals.

Federal Sustainable Agriculture Research Programs

Several government programs have gained increases in funding to support innovative approaches in alternative or "systems" agriculture in the United States since 1989. The funding available through the federal government has been dedicated to research, education, extension, and technology transfer projects.

One growing category of public funding involves "integrated" or systems science funding within the major USDA competitive grant programs. For example, many NIFA research programs[2] listed in Table 6-4 include language encouraging an interdisciplinary or systems approach, and some are explicitly designed to improve understanding of agroecological processes at the landscape or watershed scale. In addition, other federal agencies are investing in agrienvironmental research—for example, the EPA water quality research program, Pesticide Environmental Stewardship Program, and Sustainable Agriculture Partnerships Grants (regional). Meanwhile, NSF has funded basic research into plant genomics on economically important crops and has supported development of a university and industry collaborative center to study integrated pest management, and a number of agriculture and food-oriented research projects.

However, observers of the projects funded through past grant cycles have commented that long-term, systems-oriented research is still largely lacking among the projects funded by USDA and other federal agencies (Robertson et al., 2008). Moreover, as noted above, competitive grant programs represent only a small percentage of the total public agricultural research portfolio. Unless research agendas within USDA-ARS, USDA-ERS, and the larger SAES system incorporate similar language and priorities, long-term, systems-oriented research will unlikely constitute a substantial portion of the public research portfolio.

The difficulty in reshaping conventional agricultural scientific institutions has led to support for the creation of new programs and funding streams specifically aimed at ecological and sustainable agriculture (Batie and Taylor, 1994). A program that addresses not only agricultural productivity but also other sustainability goals in the United States is the Sustainable Agriculture Research and Education (SARE) program within USDA-NIFA. Since its inception in 1988, SARE has provided grants for farmer research and education projects, supported farmer learning networks and alternative market development efforts, and sustained professional development programs designed to provide opportunities for training extension specialists and researchers within the SAES system. Most SARE programs require a university partner to strengthen research input and to include SARE experience and results in the classroom. The program increased in funding from $4 million in 1988 to almost $19 million in 2009, an increase of more than 160 percent even after controlling for inflation (Figure 6-7). An administrative council in each USDA region governs SARE. The council is coordinated through a central office and administered through a contract with one of the LGUs in the region. Each region has a technical committee that oversees the awarding of grants in each research category. The administrative councils and technical committees have representation from farmers, agri-industry, and educational institutions. The SARE program is often cited as an example of how federal funding can effectively and efficiently achieve results in the adoption of sustainable farming practices (Allen, 2004). It has been

---

[2]The five priority areas of NIFA in 2010 are global food security and hunger, climate change, sustainable energy, childhood obesity, and food safety.

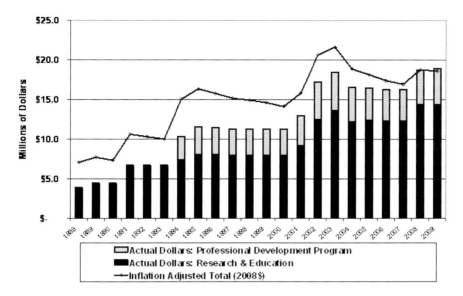

FIGURE 6-7 USDA funding for the Sustainable Agriculture Research and Education (SARE) program, 1988–2009.
DATA SOURCE: USDA-NIFA.

praised for its strong reliance on participatory approaches to agricultural research and training. Participatory approaches to research are discussed in more detail below.

State and Civil Society Support for Sustainable Agriculture

Every state supports its own agricultural experiment station and associated extension service. As mentioned above, the total for state support for experiment station maintenance and research was $1.3 billion in 2005. Funding normally goes through the state land-grant university and is matched by federal money in several ways. In Michigan, for instance, the budget from state appropriations has been around $80 million each year, which funds some 80 percent of the Michigan Agricultural Experiment Station annual base budget and supports 15 experiment stations and about 3,000 associated workers (Michigan State University Board of Trustees, 2009). The work encompasses activities of some 300 scientists in six colleges at Michigan State University: Agriculture and Natural Resources, Communication Arts and Sciences, Engineering, Natural Science, Social Science, and Veterinary Medicine. The $80 million is leveraged at an average rate of 2.3 from federal, foundation, industry, and other funding sources. The programs might be reduced because of the budgetary problems as a result of the economic downturn in 2008–2009.

In addition to a department of agriculture, most states have departments of natural resources and the environment that provide services, regulatory activity, and education to agriculture. The Michigan Department of Agriculture is the primary agency for interaction with the Great Lakes Commission (established in 1955) to protect and manage the Great Lakes watersheds. Nearly every Michigan county, for instance, falls within the mandate of the commission, which opens several avenues of funding for agricultural research and extension to protect and manage surface and ground water through agriculture. Most agencies that focus on agriculture in Michigan have influence through direct farmer support, policy, and regulatory activity. In Michigan, as in most states, many applied research programs are shared and supported among these agencies. Agricultural "service centers" throughout the state provide "one-stop shopping" of state and federal agencies programs for farmers and the public. On-farm research and other programs, such as SARE, are usually supported through state-funded infrastructure. Identifying and quantifying agricultural research that focuses on multiple sustainability goals in the complex mix of activities in the state-sponsored programs is difficult. Many states have specific programs for improving sustainability with identified budgets, but they capture only a small portion of the relevant work. Most agricultural research funded at the state and regional levels from state or regional collaborative publicly-funded programs is targeted toward applied research and technology development for production efficiency and for environmental protection or remediation. Coordinating all on-farm and farmer-involved agriculture research for improving sustainability within the states with the existing network of agencies and programs would be important to avoid redundancy and ensure efficiency.

University Sustainable Agriculture Programs

Many land-grant universities and other colleges and university departments have established programs to support agricultural research that focus on more than one goal of sustainability. An increasing amount (and percentage) of funding in university agriculture colleges has been dedicated to systems approaches to agriculture. USDA's Alternative Farming Systems Information Center of the National Agricultural Library compiles a list of educational and training opportunities in sustainable agriculture (Thompson, 2009). Most colleges and universities have some research programs on agricultural sustainability. A few of those programs are noted in Table 6-5 as examples.

Beyond designated "sustainable agriculture" programs, a much larger percentage of publicly funded research is directly or indirectly oriented to improve the sustainability of farming systems in the United States. A study of publicly funded agricultural research suggests that roughly 21 percent of funds were directed toward natural resources and environmental topics, up from 17 percent in 1998 (Caswell and Day-Rubenstein, 2007). Although comparable data for the 1980s are not readily available, it has been suggested that the state-funded and federally funded research portfolio at most LGUs has expanded from focusing on the sustainability goal of improving output and reducing production costs to encompassing a wider range of goals including environmental impacts, social and economic well-being, food safety, and animal welfare (NRC, 2002). Calls for increasing the "sustainability" share of the portfolio remain strong (Sustainable Agriculture Coalition, 2005; Robertson et al., 2008). At the same time, the broader shift toward privately funded research in the overall agricultural research system is less likely to embrace those topics (NRC, 2002).

Cooperative Extension

The Cooperative Extension System (CES) is a partnership of LGUs with federal, state, and local governments to enable the delivery of educational programs and information at the local level. As of 2009, there were about 2,900 extension offices in the United States. NIFA defines six national priorities for CES: 4-H youth development, agriculture, leadership development, natural resources, family and consumer sciences, and community and

TABLE 6-5 Examples of Agriculture Programs in Universities Aimed at Improving Sustainability

University Sustainable Agriculture Program/Dept	Year Formed	Budget ($) 1st Year	Budget ($) 2007	Staff (FTE)	Number of Affiliated Faculty
University of California Sustainable Agriculture Research and Extension Program (SAREP)	1987	775,000	574,000	6	10
North Carolina State University and North Carolina A&T State University Center for Environmental Farming Systems (CEFS)	1994	0	120,000[a]	21	37
Washington State University Center for Sustaining Agriculture and Natural Resources	1992[b]	40,000[c]	340,000	12	>100
University of Maine Sustainable Agriculture (SAG) Program	1988	N/A[d]	N/A[d]	4.5	8–10
Colorado State University Interdisciplinary Program in Organic Agriculture	2005	0	2,500	1.25	9
Clemson University	2000	10,000[e]	40,000	1	Unknown

[a]Amount listed does not include funding provided by the North Carolina Department of Agriculture and Consumer Services for infrastructure and operation of the field facility.
[b]The center was legislatively mandated in 1991 and became operational in 1992.
[c]Amount listed does not include salaries of part-time director and part-time administrative staff.
[d]Budget for sustainable agricultural program is not separated from the departmental budget.
[e]Amount listed does not include director's salary.

economic development (USDA-NIFA, 2009). In addition, CES is developing a nationally coordinated Internet information system called eXtension to provide up-to-date and specialized information and educational programs on a wide range of topics (eXtension, 2010).

In 1999, the Kellogg Commission on the Future of State and Land-Grant Universities (1999) published *Returning to Our Roots: The Engaged Institution* and urged land-grant universities to expand their mission beyond outreach and service to full engagement with their communities. In 2002, the Extension Committee on Organization and Policy (ECOP) of the Association of Public and Land-Grant Universities published the report *The Extension System: A Vision for the 21st Century* (2002). It outlined a vision for CES to address contemporary issues and to respond to changing societal needs. Later, ECOP (2010) published a list of strategic opportunities for extension in the report *Strategic Opportunities for Cooperative Extension*. The strategic opportunities listed include:

- Sustain profitable plant and animal production systems.
- Prepare youth, families, and individuals for success in the global workforce and all aspects of life.
- Create pathways to energy independence.
- Ensure an abundant and safe food supply for all.
- Assist in effective decision making regarding environmental stewardship.
- Assist communities in becoming sustainable and resilient to the uncertainties of economics, weather, health, and security.
- Help families, youth, and individuals to become physically, mentally and emotionally healthy.

This list broadly encompasses the four goals for sustainability described in Chapter 1 of this report.

Some observers charge that extension has experienced "mission creep" and should return to a focus on agriculture, while others argue that extension should serve a broad national purpose (Hefferan, 2004; McDowell, 2004). The traditional extension model is based on the "extension expert" who provides research and educational information to solve local problems. Although this model has clear benefits, it does not facilitate regionally based collaborative approaches to problem-solving or catalyze local stakeholders to develop their capacity to solve socially shared problems (Sandmann and Vandenberg, 1995; Pigg and Bradshaw, 2003).

A catalytic model shifts the role of extension from a leader expert to a coordinator and facilitator. The Food System Economic Partnership (FSEP) program, which combines effort from five county administrations in southeast Michigan, farm organization leaders, food industry heads, community groups, food system and economic development experts, and resource providers, is an example of the new model. FSEP's mission is to improve the viability of the agricultural sector in the region; provide economic revitalization opportunities for urban areas; improve consumer understanding of what is produced, processed and marketed in the region; and improve farmers' understanding of consumer needs. Lyson (2004) describes such initiatives as "civic agriculture," which can provide new market opportunities for producers and can enhance the social and human capital of the community in which it is embedded. When value chains are shortened to bring producers and consumers closer together, social and environmental values can be more fully articulated and monitored.

As CES expands beyond its traditional mandate, federal funding for extension through the Smith-Lever Act has decreased by $68 million over the past two decades (Association of Public and Land Grant Universities, 2009). Moreover, cost-sharing in the costs of extension programs by state and local governments has also declined in many cases. The funding decrease resulted in the elimination of some county and regional extension positions and reduced face-to-face services that extension programs had offered in the past. Some states are replacing county programs with in-state regional centers. Regionalization of extension programs could lower costs compared to individualized state programs (Laband and Lentz, 2004).

Some universities are shifting their extension resources toward sustainable agriculture and the green economy and away from traditional production agriculture to increase their competitiveness for funding. With half of all current farmers in the United States likely to retire in the next decade, extension programs will be critical in providing educational and networking opportunities for new farmers and for developing networks of information and resources needed to move U.S. agriculture toward greater levels of sustainability.

Farmer Participation and Innovation in Research and Development

Although an increasing proportion of public agricultural research focuses on enhancing the sustainability performance of modern framing practices, many cutting-edge or more broadly systemic alternative farming systems in the United States have been developed by farmers and continue to benefit from farmer innovation and experimentation (Kloppenburg, 1991; Hassanein, 1999). The heavy reliance on participatory research methods and farmer innovation is partly explained by the slow growth of publicly funded research on many emerging farming systems. It also demonstrates that research on complex systems dynamics can benefit enormously by taking place within the context of actual working farms. Recognition of the importance of farmer and knowledge innovation is reflected by the fact that many programs on agricultural sustainability include active participation of farmers and local communities in the R&D process of new practices (Wortmann et al., 2005). The USDA Alternative Farming Systems Information Center lists 91 U.S. organizations devoted to sustainable agriculture. Most are civil society organizations, including farmer-organized groups, regional and national civil society centers, and associations. Many organizations either support or conduct farm-based research. The Practical Farmers of Iowa, for instance, has been sponsoring and organizing on-farm research in Iowa for more than 20 years (Box 6-7).

Participatory approaches typically involve local farmers in agenda-setting, rely on farmers as critical players of innovation and development of new agricultural systems, and often include technical advisors and scientific researchers (Pretty et al., 1995; Farrington and Martin, 1998; International Institute for Environment and Development, 2005). When farmers are engaged as partners with scientists in innovation, development, extension, and outreach processes, technology adaptation and adoption have often been more effective and sustained over time than if farmers were not engaged. Such participatory farmer-centered approaches contrast with the "technology transfer" strategy, which tends to be "top-down" and "one-way" (from scientists or research centers to farmers) and is not always well suited

---

BOX 6-7
Examples of Organizations That Promote Farmer Participation in Research

Practical Farmers of Iowa

Practical Farmers of Iowa (PFI) is a nonprofit, educational organization that began in 1985 and now has more than 700 members in Iowa and neighboring states. PFI's mission is "to research, develop and promote profitable, ecologically sound and community-enhancing approaches to agriculture." PFI members are engaged in what they describe as a "movement to farm in ways that are both profitable and respectful of the natural environment." PFI supports research and information-sharing through its Farming Systems Program and On-Farm Research projects, and through resources available in a newsletter and website. Recognizing that farmers learn best from other farmers, PFI helps farmers connect with peers through Field Days, conferences, and a web-based listserve. PFI also works closely with researchers and extension professionals from Iowa State University to address a diversity of issues, trials, and practices that are generally tied to sustainable farming themes. The members have completed more than 600 replicated on-farm experiments since 1987. (See PFI's website at www.practicalfarmers.org for more information.)

California Sustainable Winegrowing Alliance

The California Sustainable Winegrowing Alliance (CSWA), launched in 2002, encourages practices that are sensitive to the environment, responsive to societal needs and interests, and economically feasible to implement. CSWA includes a sustainable wine program (SWP) that establishes voluntary standards of practices in all aspects of grape and wine production to promote sustainability. SWP facilitates peer-to-peer education about those practices through workshops, reporting, and other activities. In addition, it developed a comprehensive self-assessment workbook, the **California Code of Sustainable Winegrowing Practices Self-Assessment Workbook for the California Wine Community**, to guide growers and vintners in their efforts to improve sustainability of their systems. As of late 2008, CSWA involved more than 1500 growers and vintners who had completed a comprehensive self-assessment evaluation, and CSWA educational seminars have reached more than 5,000 growers, vintners, and vineyard or winery managers. The combination of SWP self-assessment and educational activities is designed to enable a cycle of continuous improvement for increasing sustainability, and to continually engage growers and vintners actively in the learning process. (See SWP's website at www.sustainablewinegrowing.org for more information.)

for problem-solving, particularly in complex and variable environments (Farrington and Martin, 1998).

Examples of successful participatory agriculture programs in the United States, in addition to the Practical Farmers of Iowa, include the Biologically Integrated Farming Systems program in California, the California Sustainable Winegrowing Alliance, and Wisconsin potato farmers (see Box 6-7). In *Agroecology in Action* (Warner, 2006), Warner documents 32 case studies that illustrate the effectiveness of farmer–researcher partnerships in catalyzing changes toward increasing sustainability in agriculture. In Europe, participatory learning programs and methodologies have been developed systematically in agriculture programs (Wageningen International, 2009).

Many organizations that have participatory agriculture programs provide a grassroots public education and awareness function. They often focus on upcoming public policy and program formulation. Few, if any, do direct lobbying, as most are nonprofit. Collectively these educational and service organizations provide a sizeable infrastructure and momentum to drive sustainable agriculture and to guide public policy. They have websites, and most provide electronic access to their research reports and to other sustainable agriculture literature and information. They form an important link to agricultural development and are a driver of agriculture toward sustainability.

Structuring Systems Research for Improving Agricultural Sustainability

Agricultural research is largely organized by discipline. The disciplinary research has raised awareness of the importance of environmental, economic, and social sustainability, in addition to increasing productivity, in agriculture. That research also led to incremental improvements in improving sustainability, particularly in environmental sustainability. Much more research on economic sustainability and community well-being is needed to complement the existing research on productivity and environmental sustainability. Yet, Chapters 3 to 5 in this report show the importance of interconnections and functional relationships between different components of the farming system.

The slow movement of public scientists toward more holistic and agroecological approaches to agricultural science and technology has been explored by several recent studies (Morgan and Murdoch, 2000; Vanloqueren and Baret, 2008). Aside from the influence of formal research program funding priorities, long-term cultural and cognitive routines of agricultural scientists generate assumptions about the current and future importance of different kinds of agricultural systems and influence their views of the viability of alternative approaches to scientific research (Blattel-Mink and Kaslenholz, 2005). Moreover, institutional and disciplinary reward mechanisms, publication opportunities, and increasingly specialized skill sets mitigate against the likelihood that young agricultural scientists will be successful pursuing careers using interdisciplinary, holistic, or alternative technological approaches (Lattuca, 2001; NRC, 2005). Efforts to overcome these institutional barriers include incentive grants, new interdisciplinary units, and new modes of faculty hiring and evaluation (Creso, 2008).

Several applied research programs in major land-grant and other universities and centers (such as the Leopold Center in Iowa, the Michael Fields Agricultural Institute in Wisconsin, and The North Central Regional Center for Rural Development at Iowa State University) have "research platforms" that provide infrastructure, partner research linkages, and access to databases that encourage and support interdisciplinary research beyond traditional biological integration to economics and social sciences. By facilitating coordination and providing access to databases, such platforms offer short- to intermediate-term options for scientists from a range of disciplines to add to ongoing systems research activities. To maintain and sustain those activities, the platforms (centers) require reasonably long-term funding for supporting systems research and for providing matching funds for certain grants. Those established institutions and the platforms they support have built linkages with civil society groups and programs, and they give rapid and easy access to the complex agricultural public and civil infrastructure necessary for systems-scale and landscape-scale studies.

STAKEHOLDERS AND SOCIAL MOVEMENTS

The development and evolution of markets, policies, and knowledge institutions in U.S. agriculture are shaped by a wide range of stakeholders (including civil society actors) and social movements. The diverse arrangement of civil society actors—including farmer organizations, farm commodity groups, food industry and environmental interest groups, global corporations, public health advocates, immigration and labor activists, civil rights groups, and others—also influence consumer, public, and farmer attitudes and perceptions of the sustainability of current food production practices, as well as the behaviors of key actors throughout the agrifood system. Collectively, these efforts appear to comprise a broad and important social movement that is influencing markets and farmers' choices in significant ways. The social movement includes a diverse and growing collection of people participating to achieve some desired social change and social, political, and economic agricultural reforms (Thompson, 2010).

In terms of public policy, sustainability continues to be a particularly contested concept, with dramatically different interpretations advocated by mainstream industrial farmers and their representatives and by alternative or ecological agricultural stakeholders. Different stakeholders have their preferences of which aspects of sustainability are the most important or of which indicators of system performance should guide public policies. For example, there is much dissension about which indicators are most appropriate for measuring farm viability, ecological impacts, and animal welfare (Thompson, 2010). Questions include: Do economic indicators need to include affordability for low-income consumers? Should public subsidies be included when evaluating economic sustainability? Do measures of the costs of production need to be adjusted for off-farm impacts associated with farming activities? Do changes in disease risk to animals that are outdoors need to be factored into animal welfare norms? Ultimately, the answers to those questions determine whose values are going to count in the development of indicators and norms. Different agricultural stakeholders lobby in favor of public policies that reflect their particular beliefs and interests, seek to create markets or influence the purchasing decisions of the general public, and support public investments in the R&D that support the types of agricultural production systems that they believe are most desirable.

A Brief History of Agricultural Stakeholders and Social Movements

For much of the middle 20th century, a small number of interest groups and stakeholders were deeply involved in public policy discussions about American agriculture. Since the establishment of the major U.S. farm commodity programs in the 1930s, the "conventional" stakeholders included farmers, commodity groups, and agribusinesses, and the focus of most debates on agricultural policy related to the levels and types of government subsidies and protections of different commodity sectors from market downturns (Browne, 1988). Until recently, many mainstream farm producers and producer organizations have opposed the movement for sustainability (Thompson, 2010). Although there are many exceptions to such opposition nowadays, the organizations and interest groups that participate in conventional agricultural politics remain strong and numerous. They focus on working with mainstream agrifood actors to maximize food output, increase the competitiveness of U.S. farm products in international trade, keep food prices down, and provide protections for the incomes of farmers, processors, and distributors. In other words, those groups tend to ascribe to and lobby for the more industrial philosophy of agriculture outlined in Box 1-7 (Chapter 1).

Since the 1970s, increasing awareness of unintended environmental, health, and food security problems have led some environmental and consumer groups to become engaged in farm and food policy debates. Their efforts have focused on creating new government programs (such as expanded food assistance and nutrition programs) or regulations (such as the Clean Water Act) to address specific problems associated with modern agricultural production systems. For the most part, their efforts led to parallel policies and programs that created new incentives (or disincentives) for farmers without confronting core elements of agricultural commodity or research policies and programs, or promoting dramatic changes in the organization or practices of farming enterprises.

More recently, a qualitative change has occurred in the size, sophistication, and organization of groups seeking to promote a fundamental transformation of the U.S. farm and food system. Some of the most effective efforts have been led by nonprofit organizations, including farmer-based groups and community organizations, and broader coalitions and national nonprofit organizations. (See the detailed discussion of the "Good Food Movement" in Box 6-8.) Hundreds of these organizations and groups have sprouted up in all parts of the United States and in other countries. Books and films on food and food production practices, many of which promote the agrarian philosophy discussed in Box 1-7 (Chapter 1), have gained broad public audiences. Producer associations and commodity groups are recognizing that they have to address environmental, community, and social and animal welfare issues that were not widely recognized concerns in the past. In general, these groups help to spread information, education, and technical support regarding practices that can achieve multiple sustainability goals, and some groups also promote policy and institutional changes to support improvement in agricultural sustainability. (See Box 6-7 regarding the Practical Farmers of Iowa and the California Sustainable Winegrowing Alliance, as examples.)

Many organizations have worked to create new marketing opportunities, such as farmers' market associations and nonprofit sustainable certification programs, or to promote changes in public food-buying and eating habits. There are more than 200 "Slow Food" local chapters in the United States. Chefs and educators also have formed influential organizations, such as the Chefs Collaborative, an organization that has contributed to increased demand for foods produced in ways that balance various sustainability goals. In some situations, organic farmers have formed local or state organizations tied to their particular concerns or educational needs, such as the Hawaii Organic Farmers Association, Texas Organic Farmers and Gardeners Association, and Northeast Organic Farming Association. Some organizations have emerged to address special interests of minority or immigrant communities, such as the Hmong Farmers Association in Washington State, the Agriculture and Land-Based Training Association (ALBA) for Latino farm workers and farmers in central California (ALBA, 2009), and Growing Power, a nonprofit organization in Milwaukee that supports people from diverse backgrounds through community food systems, providing high-quality, safe, healthful, and affordable food to all in the community, and directly benefits inner city youth (Growing Power, Inc., 2009).

DRIVERS AND CONSTRAINTS

> **BOX 6-8**
> **The Good Food Movement**
>
> Recently, a coalition of diverse social movements has grown in size and significance around the so-called Good Food Movement (see Figure 6-8, Flora 2009). These groups are distinctive in that they are increasingly raising broad challenges to the dominant, conventional agrifood system in the United States. The new social movements that make up the Good Food Movement are often based on an identity that transcends economic interests. They are attempting to transition the conventional agriculture sector toward agrifood systems that have several key attributes: green, healthy, fair, affordable, and local (W.K. Kellogg Foundation, 2009). Principles of sustainability included in this movement reflect strong interests in community well-being, ecosystem health, and economic security. The diagram below shows the array of social movement organizations (SMOs) associated with the Good Food Movement and their areas of overlap. A helpful way to understand the diverse drivers that bring different groups to the Good Food Movement is the Multiple Capitals Framework, which is an analytical tool for organizing and explaining land-based social movements (Flora and Flora, 2008). The growth and development of the Good Food Movement is likely to shape future public discussions about agriculture and food markets, policies, and research agendas.
>
> The social movements seem to be gaining strength and influence as represented by the response of retailers and others to market opportunities that reflect the Good Food Movement's desired agricultural attributes. As such, social movements like this one could provide new market opportunities for farmers and influence farmers' production decisions.
>
>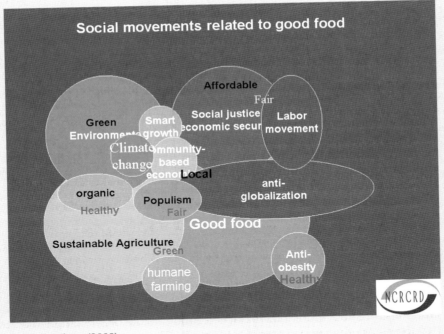
>
> SOURCE: C. Flora (2009).

The growth and connections among such organizations have increased through the use of the Internet and networking via web-based technology. As a result of Internet connections, producers, consumers, and others logically have far greater opportunities to acquire information about a range of practices and issues related to the food they produce and consume, and to exchange ideas and concerns. At the same time, other social networking has continued to be important in many areas, especially for producers who are attracted to workshops and meetings, farmers' markets, and other activities where they can talk and observe. For example, Communities of Practice that link local food movements provide ways for producers, consumers, and activists to come together to improve their good food practices. The Alternative Farming Systems Information Center lists many associations and nonprofit organizations in the United States (outside of universities and colleges) that have state-level training and education opportunities in sustainable agriculture (USDA-NAL, 2009).

Although the majority of those nonprofit groups are local or regionally oriented, some of them have formed into broader initiatives or coalitions at the state and national levels. They provide support and information to farmers and consumers. Some groups engage in advocacy related to sustainable agriculture. Some well-known organizations that have had influential roles at the national level in recent years include the American Farmland Trust, ATTRA—National Sustainable Agriculture Information Service, Organic Farming Research Foundation, Organic Center, Henry Wallace Center, and Heifer International (Morgan, 2010). The SARE program and its regional offices also play a major role by supporting the work of local farmers, groups, and organizations.

The National Sustainable Agriculture Coalition (NSAC), formed from a merger of two national policy groups linked to a network of local and regional groups, is an important national-level policy group. NSAC's vision of agriculture is one where "a safe, nutritious, ample, and affordable food supply is produced by a legion of family farmers who make a decent living pursuing their trade, while protecting the environment, and contributing to the strength and stability of their communities" (NSAC, 2009). Members include 73 environmental, rural development, faith-based, research (including university centers), and social justice organizations, and producer organizations. NSAC activities include gathering input from farmers and ranchers and from grassroots organizations that work directly with them, developing policy through participatory issue committees, providing direct representation in Washington, D.C., on behalf of its membership, and strengthening the capacity of member groups to promote citizen engagement in the policy process.

The combined efforts of these coalitions and other allies have focused on influencing the content of the Farm Bill, annual legislative budget allocations, and agency policy decisions. For example, NSCA successfully advocated to increase funds to the SARE program of USDA, and it proposed and defended provisions for funding farmland conservation programs and organic research in the recent Farm Bills. In addition, dozens of private foundations have increased support to sustainable and organic farming organizations in recent years. (See a partial list of resources at USDA-NAL, 2007).

DIVERSITY OF FARMER RESPONSES TO CONTEXTS

Market, policy, and institutional contexts are important drivers of the overall trajectory of U.S. agriculture, but how individual farmers respond to the incentives and disincentives created by those contexts differs widely. The diversity of farmers, farm types, and farming practices in the United States is testimony to the fact that past market, policy, and knowledge conditions have not dictated the detailed paths that farmers have pursued.

To encourage movement toward an increasingly sustainable farm sector, knowing the reasons why farmers have not universally adopted practices and systems that can maximize societal sustainability goals and objectives would be helpful (Rodriguez et al., 2009). Some farmers might be relatively satisfied with their current farming practices and do not think the reasons are strong enough to change their behaviors. The benefits of some practices might accrue off-farm, but the farmer must bear the costs. Some farmers might be reluctant to change their practices, even if they want to, because they lack the basic information to decide whether and how to change, or receive conflicting advice from government agencies, land-grant universities, agribusinesses, and financial lenders. Others might perceive changes in their farming practices to be too complex, the perceived additional operating costs might be viewed as too high, or the return on investment might seem either too risky or too lengthy for their planning horizon (NRC, 1993b). Some farmers might find that financial institutional managers, landlords, or business partners are resistant to their ideas for innovation or change (NRC, 1993b).

The sections below discuss research on factors that influence farmers' decisions to adopt new practices or systems designed to improve the sustainability impacts associated with their farming activities. Important influences on farmer behavior include the opportunities and challenges posed by variability in local biophysical conditions, local farm and non-farm economic conditions, and each individual farm's existing stock of physical assets. They also reflect differences in the skills, goals, and values of individual farmers (and their families). However, past research does not necessarily predict future outcomes. Ongoing public debates, emerging market opportunities, new programs and policies, and innovations in farming practices and systems are likely to continue to affect farmer perceptions, goals, and behaviors.

Local Conditions and Farm Sustainability

The performance of any new farming practice or system is mediated by the diverse biophysical resource conditions found in different regions of the United States. Important biophysical resources include soil quality, topography, climate, and water availability. Farms in areas with longer growing seasons and more moderate winters, for example, are able to cultivate particular kinds of crops that do not thrive in harsher climates, but also experience distinctive patterns of weeds, pests, and disease problems. Landscapes with greater variability in soils or topographic features might be particularly well suited to diversified farming systems that take advantage of diverse local biophysical niches. Flat homogenous production conditions might enable larger-scale specialized operations. Biophysical resource differences suggest that the specific tillage practices, crop rotations, and pest and disease management strategies appropriate for a particular region (for example, the Corn Belt) might differ from those for another region (for example, the humid Southeast or arid West). Thus, farming practices and system redesigns for improving sustainability often need to be tailored for micro climates and soil conditions; few practices and system redesigns will be appropriate for all situations.

Options for increasing the sustainability of U.S. farms are also shaped by the regional availability of a supportive agribusiness infrastructure and markets for particular farm products. Traditional vegetable production or dairy farming areas, for example, tend to have a critical mass of input suppliers, agribusiness professionals, and marketing and processing facilities that can easily handle those particular commodities. Similarly, farms located in remote areas are less likely to have opportunities to take advantage of emerging urban-oriented value-trait food markets. The lack of an appropriate agricultural infrastructure can constrain the ability of farmers to adopt new commodities or marketing approaches on local farms. Efforts to increase sustainable farming systems would need to focus as much on infrastructure and market development as on production techniques or practices.

At the farm level, historic investments in buildings and equipment suited to the production of particular commodities can present obstacles to the rapid retooling or restructuring of the farm production process (Barham et al., 1998; Clark, 2008). In particular, U.S.

commercial farms have become much larger in scale and increasingly specialized (Gardner, 2002; MacDonald, 2007). One of the most striking specialization trends has been the increasing separation of crop and livestock production (Powell and Unger, 1998; Russelle et al., 2007). Specialized crop farms have typically expanded in scale through the use of highly specialized machinery. Specialized livestock farms have usually made significant investments in buildings and equipment to support modern production processes (Boetel et al., 2007). Gollehon et al. (2001) documented a 40 percent decline in available land for recycling livestock waste through cropping enterprises on U.S. farms between 1982–1997, which can generate structural problems for the effective recycling of livestock nutrients through local crops (Kellogg et al., 2000; Naylor et al., 2005; Ribadau and Gallehon, 2006). Efforts to encourage greater diversification in crop enterprises, or to reintroduce integrated crop–livestock enterprises, could be complicated by commitments to modern specialized production systems and the current geographic separation of crops and livestock in many parts of the United States. Conversely, regions that have maintained highly diversified or integrated production systems might be well positioned to take advantage of many types of farming practices for improving sustainability (Singer et al., 2007; Prokopy et al., 2008).

The impact of government programs and policies on local farmer behavior can be expected to vary across regions. Farm commodity programs have the largest influence in communities where local farmers have the ability and propensity to cultivate program crops. The relative competitiveness of bids to enroll lands in CRP depends on the comparable erodibility of lands across regions and other environmental factors, and on the levels of productivity and opportunity costs associated with expected economic returns from locally dominant farming activities.

Farm and Farmer Characteristics and the Use of Sustainable Agricultural Practices

The characteristics of farms and farmers most likely to adopt new farming practices have been researched extensively (Rogers, 2003). A major subset of that research focuses specifically on the adoption of agricultural conservation practices (Prokopy et al., 2008) or the use of organic or other farming techniques to improve sustainability. The literature suggests that farmer demographic characteristics, knowledge and skills, and goals and values are weakly correlated with the likelihood of using agricultural practices to improve sustainability. Empirical studies have been able to explain only a small percentage of variation in adoption of those farming practices (Napier et al., 2000; Fuglie and Kascak, 2001). However, some patterns and lessons learned shed light on the ways that farm and farmer characteristics are likely to influence the ability or propensity of individual farms to increase their use of sustainable practices or approaches.

Farm Characteristics

Most researchers assume that economic factors—including costs, benefits, economies of scale, uncertainty, and policy incentives and disincentives—influence the attractiveness of different types of farming production practices and overall trajectories of technological and structural change in agriculture (Chavas, 1997, 2001; Halloran and Archer, 2008). That assumption holds true for most practices designed to enhance the sustainability performance of U.S. agriculture (Rodriguez et al., 2009). However, the relative economic costs and benefits of any given farming practice can differ across farm types and regions and are influenced by a wide range of farm characteristics, including scale, enterprise diversity, land tenure, indebtedness, and sunk costs associated with previous investments (Hall and Leveen, 1978; Hallam, 1993; Barham et al., 1998; Marra et al., 2004).

advantages available from organic markets (Best, 2005; Lockie and Halpin, 2005; Lamine and Bellon, 2009).

Farmer actions are also shaped by perceptions of the objective realities of their political, economic, and natural contexts, which can vary from individual to individual, even within the same location (Bieling and Plieninger, 2003; Burton, 2004; Ahnstrom et al., 2009). Heightened awareness about environmental impacts associated with farming practices, for example, can increase use of conservation BMPs (Prokopy et al., 2008).

The local sociocultural context of farming can influence the willingness of farmers to experiment and use nontraditional farming practices. For example, peer pressure to conform to dominant ideas of what "real farmers" would do can hinder the spread of farming practices that push the envelope of those definitions (Bell, 2004; Leeuwis, 2004; Rodriguez et al., 2009). However, Flora (1995) found that in communities accepting of innovation in general, farmers were more likely to adopt practices for improving sustainability.

Although many farmers share common goals, values, or motivations, most scholars recognize that distinct subgroups of farmers emphasize different priorities and thus pursue different technological or management strategies (Vanclay and Lawrence, 1994; van der Ploeg, 2003; Bell, 2004). For some farmers, differences in their choices of farming practices are related to their moral convictions about what constitutes "good farming"; for others, practical considerations dominate (Schoon and Grotenhuis, 2000). Differences in attitudes and values exist within subgroups of farmers (such as those using conservation practices or organic farming techniques). Padel (2001) emphasized that economic and noneconomic motivations are not mutually exclusive. Farmers are also not isolated individuals who make decisions entirely on the basis of innate preferences or characteristics. Their goals and motivations are best viewed as dynamic states shaped by their relationships to key actors across a range of social contexts (for example, interactions with other farmers, friends, neighbors, extension workers, and regulators) (Leeuwis, 2004). Complex combinations of different individual motivations are reflected in complex management styles that respond to similar pressures and incentives in unique ways (Nowak and Cabot, 2004).

SUMMARY

The decisions of farmers to use particular farming practices and their ability to move forward along the sustainability trajectory are influenced by many external forces, such as markets, public policies, available science, technology, knowledge and skills, and the farmers' own values, resources, and land tenure arrangements. The market, policy, and knowledge structure are in turn influenced by efforts of broad social movements and organized interest groups that have different perspectives about how agriculture should be organized and how food should be produced and distributed. A discussion of the sustainability of U.S. agricultural practices and farming systems is incomplete without discussion of some of the driving forces for improving sustainability and trends. Understanding the drivers and the trends can direct policy attention to where changes can be made to influence farmers' decisions to effectively address the challenges identified in Chapter 2. Key points from this chapter are summarized below.

Markets

- Increasing concentration of ownership and control in the U.S. agricultural sector can, through influences on costs, prices, and contractual arrangements, reduce farmers' flexibility in selecting farming practices. American farmers' decisions about planting,

input use, investments, and marketing are heavily influenced by the prevailing prices of and access to farm commodity markets.
- Increasing interest by consumers in various types of "value-trait" foods, including organic, natural, free-range, hormone-free, local, direct, or family-farm raised foods, has increased incentives for producers to use production practices that can serve those markets.
- Sustainability initiatives by major food retailers can contribute to moving more farmers toward practices and systems that can improve sustainability, as defined by the retailers. As the supply of some types of value-trait foods catches up to demand, there are concerns that price premiums received by producers would diminish, and therefore would reduce incentives to use associated farm practices.
- Access to local niche and direct-sales markets has allowed many small and medium-sized farm to find economically viable options to conventional commodity outlets and to use farming practices that could improve sustainability. Despite rapid growth, direct sales to consumers are less than 1 percent of total U.S. farm sales.
- Marketing tools such as certification, grades and standards, and branding can create niches of profitability for farmers whose produce meets specific requirements. Those tools have mixed effects on driving farmers to improve sustainability. Federal involvement in the creation and enforcement of standards for organic food labels has contributed to rapid growth in the sector, but the federal standard is narrowly focused on environmental and health issues. An increasing number of "sustainability" labels and certifications have broader definitions than the federal organic standard. They might identify food products produced in ways that address labor, community, or animal welfare concerns. However, the proliferation of sustainability labels and certifications has created some confusion among both producers and consumers and has led to calls for harmonization of standards across states and nations. Disputes over the stringency of certification or labeling requirements affect the types of practices that farmers will be allowed to use when producing for these markets.
- Interest has increased in public and private programs designed to create markets for ecosystem services to compensate farmers who use ecologically beneficial practices. Examples include payment for environmental services, cap-and-trade, and offset trading markets. To date, those programs have been experimental and have produced only modest success in changing producer behaviors.

Public Policies

- Major federal commodity and crop insurance programs have been linked to a decrease in the diversity of cropping systems, increases in the use of external farm inputs, and the extensive hydrologic modification of landscapes in the United States. Insurance programs encourage more acres to be planted to major commodity crops and on marginal or risky agricultural landscapes, and can disadvantage producers of commodities not covered by these programs.
- Although conservation programs have encouraged use of a range of best management practices, the voluntary nature of the programs and a lack of targeting in implementation have limited their impacts on air and water quality. Federal conservation programs have focused the use of cost-sharing to reduce the financial expenses associated with particular farm practices. However, minimal standards for qualifying for cost-sharing dollars and decisions about which types of practices are eligible for support have failed

to create effective incentives to use some important types of farming practices for improving sustainability, such as complex long-term crop rotations, information-intensive or management-intensive farming systems, and alternative livestock systems.
- One possible policy response to reach more sustainability objectives would be to redirect and redesign the Farm Bill programs so that they cost-effectively pursue such goals as controlling nonpoint pollution via landscape management, sequestering carbon on agricultural lands, providing habitats for wildlife, reducing pesticide use, and enhancing farmers' knowledge and skills on available practices and approaches to achieving various sustainability objectives.
- Traditionally, federal nutritional assistance—which represents a large proportion of the Farm Bill budget—had little impact on markets for foods produced in more sustainable ways. In the last few years, an increasing portion of the food assistance and school lunch program budget has been designed to encourage consumption of fresh fruits and vegetables and to be available for purchase of foods directly from local farmers. The changes could influence the adoption of farming practices for improving sustainability, but any effects have yet to be documented.
- To date, many environmental regulations have exempted agricultural operations, but recent changes to the Clean Air Act, the Clean Water Act, the Food Quality and Protection Act, the Endangered Species Act, and food safety guidelines have had important impacts on farmers' use of agrichemicals, conservation practices, and management of livestock wastes. Some of those impacts on farmers' behavior could lead to unintended consequences, such as accelerating structural change toward larger farms and tradeoffs among different environmental objectives (for example, landscape biodiversity, water quality, greenhouse-gas emissions, and food safety).
- U.S. agricultural markets are influenced by trade and trade policies. The most influential trade policy is that of WTO. Some recent WTO trade dispute resolutions have raised legitimate concerns as to whether certain sustainability attributes might be omitted or impaired by the pursuit of trade liberalization. Emerging debates over federal and state legislation addressing farm labor working conditions and farm animal welfare could become important drivers of farm practices and changes in the organization of farming systems in the United States.

Knowledge Institutions

- As of 2009, the U.S. public research system spends almost $5 billion per year on agricultural research and development. Although roughly a third of public research spending is devoted to examining environmental, natural resource, social, and economic aspects of farming practices, the other two-thirds is focused on improving the productivity and efficiency of conventional farming systems. A relatively small portion of public dollars has been devoted to the investigation of complex farming systems and long-term agricultural research.
- Cooperative extension has expanded its mandate beyond traditional agriculture to provide outreach and services to address a broad set of communities and their needs, despite declining funding from federal, state, and local governments. The regionalization of extension programs was proposed as one way to reduce costs. Extension would function as a catalytic coordinator and facilitator and work with a broad constituency to improve the viability of the agricultural sector in a region. A nationally coordinated eXtension Internet system is being developed to provide timely and specialized infor-

mation and educational programs to the public. The role and mission of cooperative extension to provide education and networking opportunities for farmers is critical to moving agriculture toward improved sustainability.
- Privately funded industry research in the United States has surpassed the public sector in spending to develop new farm practices and technologies. Research on the impact of farming practices and systems on various indicators of sustainability is much less common in the private sector, because such research is unlikely to lead to the development of marketable products or otherwise generate financial returns on private research and investments.
- A small but increasing amount of public research dollars is invested in programs devoted specifically to research on sustainable farming systems. The effects of those programs on the development and adoption of new farming practices or systems is not well documented.
- Much of the technical and managerial innovation in improving agricultural sustainability has occurred through farmer innovation and experimentation. Recognition of the wealth of farmer knowledge about farming practices and systems for improving sustainability has led to increased public support for farmer-to-farmer mentoring programs and farmer learning networks.
- Opportunities for collaboration between public agricultural scientists and farmer innovators are still relatively rare but offer a potentially important model for future research and development efforts.

Stakeholders and Social Movements

Agricultural markets, policies, and knowledge institutions are shaped by the relative influence and power of various stakeholders and interest groups in the United States. Public awareness of issues of sustainability will depend, in part, on the organization and efforts of individuals and groups that are dissatisfied with the current agrifood system. The number of groups and organizations that are working to support innovative farmers, develop new markets, and expand public programs designed to move agriculture toward sustainable agriculture has been increasing. However, their efforts and successes are constrained by the needs and goals of conventional agrifood system stakeholders, who have traditionally dominated public discussions of farming issues. Changes in markets, policies, and research institutions to encourage sustainable agriculture will depend on finding areas of common ground or negotiated compromises among various stakeholders.

Diversity of Farmer Responses

- Although market, policy, and institutional contexts are important drivers of the trajectory of U.S. agriculture, the response of individual farmers to the incentives and disincentives created by those contexts can be quite diverse. Local biophysical resources, proximity to markets, and existing investments in land, equipment, and buildings all affect the adoption of farming practices to improve the sustainability of U.S. agriculture.
- Farm business attributes (such as scale and land tenure arrangements), farm household characteristics (including age, formal education, and lifecycle stage), and farmer values and beliefs can affect the ability and desire of individual farmers to use many of the practices mentioned in Chapters 3, 4, and 5 of this report.

REFERENCES

Adler, R., and M. Straube. 2000. Watersheds and the integration of U.S. water law and policy: bridging the great divides. *William and Mary Environmental Law and Policy Review* 25(1):998–1067.
AFT (American Farmland Trust). 2009. AFT's $10,000 steward of the land award. Available at http://www.farmland.org/programs/award/default.asp. Accessed on August 17, 2009.
Ahnstrom, J., J. Hockert, H.L. Bergea, C.A. Francis, P. Skelton, and L. Hallgren. 2009. Farmers and nature conservation: what is known about attitudes, context factors and actions affecting conservation? *Renewable Agriculture and Food Systems* 24(1):38–47.
ALBA (Agriculture and Land-Based Training Association). 2009. Welcome. Available at http://www.albafarmers.org/index.html. Accessed on February 21, 2010.
Allen, P. 2004. Together at the Table: Sustainability and Sustenance in the American Agrifood System. University Park: Pennsylvania State University Press.
Allen, P., and C. Hinrichs. 2007. Buying into "buy local": agendas and assumptions of U.S. local food initiatives. In Constructing Alternative Food Geographies: Representation and Practice, L. Holloway, D. Maye, and M. Kneafsy, eds. Kidlington, UK: Elsevier Press.
Allen, P.A., J. Guthman, and A. Morris. 2006. Meeting Farm and Food Security Needs Through Community Supported Agriculture and Farmers' Markets in California. Santa Cruz, Calif.: Center for Agroecology and Sustainable Food Systems.
Alston, J.M., P.G. Pardey, and V.H. Smith. 2000. Financing agricultural R&D in rich countries: what's happening and why? In Public-Private Collaboration in Agricultural Research, K.O. Fuglie and D.E. Schimmelpfennig, eds. Ames: Iowa State University Press.
Anderson, K., R. Lattimore, P. Lloyd, and D. MacLaren. 2007. Distortions to agricultural incentives in Australia and New Zealand. World Bank Policy Research Paper 4471. Paper read at 51st Annual Conference of the Australian Agricultural and Resource Economics Society, February 13–16, at Queenstown, New Zealand.
Aragón-Correa, J.A., and E.A. Rubio-López. 2007. Proactive corporate environmental strategies: myths and misunderstandings. *Long Range Planning* 40:357–381.
Arha, K., T. Josling, D.A. Sumner, and B.H. Thompson, eds. 2007. U.S. Agricultural Policy and the 2007 Farm Bill. Stanford, Calif.: Woods Institute for the Environment.
Association of Public and Land Grant Universities. 2009. Greater funding needed for the Cooperative Extension System, to provide knowledge and information to citizens of the United States. Available at http://www.land-grant.org/docs/FY2010/KB/smith-lever.pdf. Accessed on February 26, 2010.
Atari, D.O.A., E.K. Yiridoe, S. Smale, and P.N. Duinker. 2009. What motivates farmers to participate in the Nova Scotia environmental farm plan program? Evidence and environmental policy implications. *Journal of Environmental Management* 90(2):1269–1279.
Atwood, J.D., and A. Hallam. 1993. Farm structure and stewardship of the environment. In Size, Structure, and the Changing Face of American Agriculture, A. Hallam, ed. Boulder, Colo.: Westview Press.
Barbarika, A. 2001. The Conservation Reserve Program Statistics, 2001. Washington, D.C.: U.S. Department of Agriculture Farm Service Agency.
Barham, B.L., J.P. Chavas, and O.T. Combes. 1998. Sunk costs and the natural resource extraction sector: analytical models and historical examples of hysteresis and strategic behavior in the Americas. *Land Economics* 74(4):429–448.
Barkema, A., and N. Novack. 2001. The new U.S. meat industry. Available at http://www.extension.iastate.edu/agdm/articles/others/BarJuly01.htm. Accessed on February 22, 2010.
Batie, S. 2001. Public programs and conservation on private lands. In Private Lands, Public Benefits: A Policy Summit on Working Lands Conservation. Washington, D.C.: National Governors' Association.
Batie, S.S. 2009. Green payments and the US Farm Bill: information and policy challenges. *Frontiers in Ecology and the Environment* 7(7):380–388.
Batie, S.S., and D.B. Taylor. 1994. Institutional issues and strategies for sustainable agriculture: view from within the land-grant university. *American Journal of Alternative Agriculture* 9(1&2):23–27.
Batie, S.S., and R. Horan. 2002. Green payments policy. In The 2002 Farm Bill: Policy Options and Consequences, J. Outlaw and E.G. Smith, eds. Oak Brook, Ill.: Farm Foundation.
Batie, S.S., and D.B. Schweikhardt. 2009. Societal concerns as wicked problems: the case of trade liberalization. Paper presented at the OECD Workshop on the Economic and Trade Implications of Policy Responses to Societal Concerns, November 2–3, 2009, Paris, France.
Batie, S.S., L.A. Shabman, and R. Kramer. 1986. U.S. agriculture and natural resources policy: past and future. In The Future of the North American Granary: Politics, Economics and Resource Constraints in North American Agriculture, C.F. Runge, ed. Ames: Iowa State University Press.

Beery, M., and K. Markley. 2007. Farm to hospital: supporting local agriculture and improving health care. Available at http://www.foodsecurity.org/F2H_Brochure.pdf. Accessed on September 10, 2009.

Bell, M.M. 2004. Farming for Us All: Practical Agriculture and the Cultivation of Sustainability. University Park: The Pennsylvania State University Press.

Beretti, M., and D. Stuart. 2008. Food safety and environmental quality impose conflicting demands on Central Coast growers. *California Agriculture* 62(2):68–73.

Best, H. 2005. Organic farmers in western Germany: is there a decline in environmental concern? Paper read at the XXI Congress of the European Society for Rural Sociology, August 21-26, Keszthely, Hungary.

Bieling, C., and T. Plieninger. 2003. "Stinking, disease-spreading brutes" or "four-legged landscape managers"—livestock, pastoralism and society in Germany and the USA. *Outlook on Agriculture* 32:7–12.

Big Room Inc. 2009. Ecolabelling.org. Who's deciding what's green? Available at http://www.ecolabelling.org/. Accessed on November 17, 2009.

Bingen, J., and L. Busch. 2007. Agricultural Standards: The Shape of the Global Food and Fiber System. Dordrecht, the Netherlands: Springer.

Bird, E.A.R., G.L. Bultena, and J.C. Gardner. 1995. Planting the Future: Developing an Agriculture That Sustains Land and Community. Ames: Iowa State University Press.

Bittman, M. 2009. Eating food that's better for you, organic or not. *New York Times*, March 21, 2009.

Blattel-Mink, B., and H. Kaslenholz. 2005. Transdisciplinarity in sustainability research: diffusion conditions of an institutional innovation. *International Journal of Sustainable Development and World Ecology* 12(1):1–12.

Boetel, B.L., R. Hoffman, and D.J. Liu. 2007. Estimating investment rigidity within a threshold regression framework: the case of U.S. hog production sector. *American Journal of Agricultural Economics* 89(1):36–51.

Bohlen, P.J., S. Lynch, L. Shabman, M. Clark, S. Shukla, and H. Swain. 2009. Paying for ecosystem services on agricultural lands: an example from the Northern Everglades. *Frontiers in Ecology and the Environment* 7(1):46–55.

Boody, G., B. Vondracek, D.A. Andow, M. Krinke, J. Westra, J. Zimmerman, and P. Welle. 2005. Multifunctional agriculture in the United States. *BioScience* 55(1):27–38.

Bradshaw, B. 2004. Plus c'est la meme chose? Questioning crop diversification as a response to agricultural deregulation in Saskatchewan Canada. *Journal of Rural Studies* 20:35–48.

Brown, S., and C. Getz. 2008. Privatizing farm worker justice: regulating labor through voluntary certification and labeling. *Geoforum* 39(3):1184–1196.

Browne, W.P. 1988. Private Interests, Public Policy and American Agriculture. Lawrence: University Press of Kansas.

Burfisher, M.E., and E.A. Jones. 1998. Regional trade agreements and U.S. agriculture. Washington, D.C.: U.S. Department of Agriculture Economic Research Service.

Burton, R.J.F. 2004. Reconceptualizing the "behavioral approach" in agricultural studies: a socio-psychological perspective. *Journal of Rural Studies* 20:359–371.

Busch, L., and J. Bingen. 2005. Standards and standardization. In Agricultural Governance, J. Beckert and M. Zafirovsky, eds. Boston: Routledge.

———. 2006. Introduction: a new world of standards. In Agricultural Standards: The Shape of the Global Food and Fiber System, J. Bingen and L. Busch, eds. Dordrecht, the Netherlands: Springer.

Butler, S.J., J.A. Vickery, and K. Norris. 2007. Farmland biodiversity and the footprint of agriculture. *Science* 315:381–384.

Buttel, F.H. 2006. Sustaining the unsustainable: agro-food systems and environment in the modern world. In Handbook of Rural Studies, P. Cloke, T. Marsden, and P. Mooney, eds. Thousand Oaks, Calif.: Sage.

California Leafy Green Handler Marketing Board. 2007. About LGMA. Available at http://www.caleafygreens.ca.gov/about/lgma.asp. Accessed on November 19, 2009.

Canada, C., and J. Zinn. 2005. Environmental Quality Incentives Program (EQIP): Status and Issues. Washington, D.C.: Congressional Research Service.

Capper, J.L., R.A. Cady, and D.E. Bauman. 2009. The environmental impact of dairy production: 1944 compared with 2007. *Journal of Animal Science* 87(6):2160–2167.

Carman, H.F., K. Klonsky, A. Beaujard, and A.M. Rodriguez. 2004. Marketing Order Impact on the Organic Sector: Almonds, Kiwifruit, and Winter Pears. Davis: Giannini Foundation of Agricultural Economics, University of California.

Casey, F., A. Schmitz, S. Swinton, and D. Zilberman. 1999. Flexible Incentives for the Adoption of Environmental Technologies in Agriculture. Boston: Kluwer Academic Publishers.

CAST (Council for Agricultural Science and Technology). 1994. Challenges confronting agricultural research at land grant universities. Available at http://www.cast-science.org/websiteUploads/publicationPDFs/chal_is.pdf. Accessed on February 22, 2010.

DRIVERS AND CONSTRAINTS

Caswell, M., and K. Day-Rubenstein. 2006. Agricultural Research and Development. Washington, D.C.: U.S. Department of Agriculture Economic Research Service.

———. 2007. Agricultural research and development. In Agricultural Resources and Environmental Indicators, K. Wiebe and N. Gollehon, eds. New York: Nova Science Publishers, Inc.

Cattaneo, A., D. Hellerstein, C. Nickerson, and C. Myers. 2006. Balancing the Multiple Objectives of Conservation Programs. Washington, D.C.: U.S. Department of Agriculture Economic Research Service.

CDFG (California Department of Fish and Game). 1934. Stream Survey: Scott River, Tributary to Klamath River. Sacramento: Author.

Chavas, J.P. 1997. An analysis of the source and nature of technical change: the case of U.S. agriculture. *Review of Economics and Statistics* 79:482–492.

———. 2001. Structural change in agricultural production: economics, technology and policy. In Vol. 1, Handbook in Agricultural Economics, B. Gardner and G. Rausser, eds. Amsterdam: Elsevier Science.

Claassen, R., and M. Ribaudo. 2006. Conservation Policy Overview. In Agricultural and Environmental Resources Indicators, 2006 edition, W. L. and N. Gollehon, eds. Washington, D.C.: U.S. Department of Agriculture Economic Research Service.

Claassen, R., L. Hansen, M. Peters, V. Breneman, M. Weinberg, A. Cattaneo, P. Feather, D. Gadsby, D. Hellerstein, J. Hopkins, P. Johnston, M. Morehart, and M. Smith. 2001. Agri-environmental Policy at the Crossroads: Guideposts on a Changing Landscape. Washington, D.C.: U.S. Department of Agriculture Economic Research Service.

Claassen, R., K. Wiebe, and L. Hansen. 2003. Farmers' choices and the role of environmental indicators in the development of soil conservation policy. Paper read at OECD Expert Meeting on Soil Erosion and Soil Biodiversity Indicators, March 25–26, Rome, Italy.

Claassen, R., V. Brenenman, S. Bucholitz, A. Cattaneo, R. Johansson, and M. Morehart. 2004. Agri-environmental Compliance in US Agriculture Policy: Past Performance and Future Potential. Washington, D.C.: U.S. Department of Agriculture Economic Research Service.

Claassen, R., M. Aillery, and C. Nickerson. 2007a. Integrating Commodity and Conservation Programs: Design Options and Outcomes. Washington, D.C.: U.S. Department of Agriculture Economic Research Service.

Claassen, R., A. Cattaneo, and R. Johansson. 2007b. Cost-effective design of agri-environmental payment programs: U.S. experience in theory and practice. *Ecological Economics* 65(4):737–752.

Clark, J. 2008. The relational geography in newly restructured, consumptive spaces: developing a model for agricultural geography of US peri-urban areas. Paper read at 2008 American Association of Geographers Annual Meeting, April 15–19, Boston, Mass.

Cobb, K. 2008. To market, to market to buy a fat tomato. *Ninth District Fedgazettte* 9–12.

Cooley, H., J.Christian-Smith, and P. Gleick. 2009. Sustaining California Agriculture in an Uncertain Future. Oakland, Calif.: Pacific Institute.

Copeland, C. 2009. Air Quality Issues and Animal Agriculture: A Primer. Washington, D.C.: Congressional Research Service.

Cox, C. 2007. US agriculture conservation policy and programs: history, trends and implications. In U.S. Agricultural Policy and the 2007 Farm Bill, K. Arha, T. Josling, D.A. Sumner, and B.H. Thompson, eds. Stanford, Calif.: Woods Institute for the Environment.

Creso, S.M. 2008. Interdisciplinary strategies' in U.S. research universities. *Higher Education* 55 (5):537–552.

Dahlberg, K., ed. 1985. New Directions for Agriculture and Agricultural Research. Totowa, N.J.: Rowman and Allenheld Press.

Demeritt, L. 2006. Presentation at the President's Organic Keynote Breakfast. CHFA Expo East, Toronto, ON, Hartman Group Report.

Dimitri, C., and L. Oberholtzer. 2008. The U.S. Organic Handling Sector in 2004. Washington, D.C.: U.S. Department of Agriculture Economic Research Service.

———. 2009. Marketing U.S. Organic Foods: Recent Trends from Farms to Consumers. Washington, D.C.: U.S. Department of Agriculture Economic Research Service.

Dimitri, C., A. Effland, and N. Conklin. 2005. The Twentieth Century Transformation of U.S. Agriculture and Farm Policy. Washington, D.C.: U.S. Department of Agriculture Economic Research Service.

Doane, D. 2005. The myth of CSR. Available at http://www.uwlax.edu/faculty/knowles/eco310/MythofCSR.pdf. Accessed on February 22, 2010.

Dobbs, T.L. 2008. Economic and policy conditions necessary to foster sustainable farming and food systems: U.S. policies and lessons from the European Union. Presentation to the Committee on March 27, 2008, Kansas City, Mo.

Doering, O. 2008. Biofuel implications for agriculture and the environment. Presentation to the Committee on March 27, 2008, in Kansas City, Mo.

Donoghue, E.J., M.J. Roberts, and N. Key. 2009. Did the Federal Crop Insurance Reform Act alter farm enterprise diversification. *Journal of Agricultural Economics* 60:80–104.

Draper, A.J., M.W. Jenkins, K.W. Kirby, J.R. Lund, and R.E. Howitt. 2003. Economic-engineering optimization for California water management. *Journal of Water Resources Planning and Management–ASCE* 129(3):155–164.

Duram, L.A., and K.L. Larson. 2001. Agricultural research and alternative farmers' information needs. *Professional Geographer* 51(1):84–96.

ECOP (Extension Committee on Organization and Policy). 2002. The Extension System: A Vision for the 21st Century. Washington, D.C.: National Association of State Universities and Land-Grant Colleges.

———. 2010. Strategic Opportunities for Cooperative Extension. Washington, D.C.: National Association of State Universities and Land-Grant Colleges.

Environmental Law Institute. 1997. Enforceable State Mechanisms for the Control of Nonpoint Source Water Pollution. Washington, D.C.: Author.

Environmental Working Group. 2009. EWG farm subsidy database. Available at http://farm.ewg.org/farm/regionsummary.php?fips=00000. Accessed on November 18, 2009.

EPA (U.S. Environmental Protection Agency). 2002. Concentrated animal feeding operations. Clean Water Act requirements. Available at http://www.epa.gov/npdes/pubs/cafo_brochure_regulated.pdf. Accessed on October 12, 2009.

———. 2007. Major existing EPA laws and programs that could affect agricultural producers. Available at http://www.epa.gov/agriculture/agmatrix.pdf. Accessed on October 12, 2009.

———. 2008. A National Ecosystem Services Research Partnership. Available at http://www.epa.gov/osp/ftta/ESRP_CRADA_Broch.pdf. Accessed on November 17, 2009.

———. 2009a. The new specific volume standards for cellulosic biofuel, biomass-based diesel, advanced biofuel, and total renewable fuel by year. Available at http://www.epa.gov/OMS/renewablefuels/rfs2-4standards.pdf. Accessed on November 18, 2009.

———. 2009b. Federal Insecticide, Fungicide, and Rodenticide Act (FIFRA). Available at http://www.epa.gov/agriculture/lfra.html. Accessed on December 13, 2009.

———. 2010. Highlights of the Food Quality Protection Act of 1996. Available at http://www.epa.gov/opp00001/regulating/laws/fqpa/fqpahigh.htm. Accessed on February 24, 2010.

Evergreen Funding Consultants. 2009. Washington conservation markets study—final report. Available at http://www.farmland.org/programs/states/wa/documents/WAConservationMarketsStudyReport_27Jan2009.pdf. Accessed on February 26, 2010.

eXtension. 2010. eXtension homepage. Available at http://www.extension.org/. Accessed on February 26, 2010.

FAO (Food and Agriculture Organization). 2009. Conservation agriculture. Available at http://www.fao.org/ag/ca. Accessed on June 8, 2009.

Farrington, J., and A.M. Martin. 1998. Farmer participatory research: a review of concepts and recent fieldwork. *Agricultural Administration and Extension* 29:247–264.

Faunt, C.C. 2009. Groundwater availability of the Central Valley Aquifer, California. *U.S. Geological Survey Professional Paper 1766*.

Feather, P.M., and J. Cooper. 1995. Voluntary Incentives for Reducing Agricultural Nonpoint Source Water Pollution. Washington, D.C.: U.S. Department of Agriculture Economic Research Service.

Feng, H., P. Gassman, L. Kurkalova, S. Secchi, and C. Kling. 2004. Targeting efficiency in the Conservation Security Program. Available at http://www.card.iastate.edu/iowa_ag_review/winter_04/article2.aspx. Accessed on November 18, 2009.

Fernandez-Cornejo, J. 2004. The Seed Industry in U.S. Agriculture: An Exploration of Data and Information on Crop Seed Markets, Regulation, Industry Structure, and Research and Development. Washington, D.C.: U.S. Department of Agriculture Economic Research Service.

Flora, C.B. 1995. Social capital and sustainability: agriculture and communities in the Great Plains and the Corn Belt. *Research in Rural Sociology and Development: A Research Annual* 6:227–246.

———. 2009. Good food as a social movement. Presentation at meeting of the Agriculture, Food, and Human Values Society, May 30, State College, Pa.

Flora, C.B., and J.L. Flora. 2008. Rural Communities: Legacy and Change. Boulder, Colo.: Westview Press.

Food Alliance. 2009. Welcome to Food Alliance. Available at http://www.foodalliance.org/. Accessed on November 17, 2009.

Food Marketing Institute. 2008. Natural and organic foods. Backgrounder report. Available online at http://www.fmi.org/docs/media/bg/natural_organic_foods.pdf. Accessed on January 9, 2010.

———. 2009. U.S. Grocery Shopper Trends 2009. Arlington, Va.: Author.

Fortrin, N.D. 2009. Food Regulations: Law, Science, Policy and Practice. Hoboken, N.J.: John Wiley and Sons.

Fraser, E.D.G. 2006. Crop diversification and trade liberalization: linking global trade and local management through a regional case study. *Agriculture and Human Values* 23(3):271–281.
Friedland, W.H., A.E. Barton, and R.J. Thomas. 1981. Manufacturing Green Gold: Capital, Labor and Technology in the Lettuce Industry. Cambridge, Mass.: Cambridge University Press.
Fuglie, K.O. 2000. Trends in agricultural research expenditures in the United States. In Public-Private Collaboration in Agricultural Research, K.O. Fuglie and D.E. Schimmelpfennig, eds. Ames: Iowa State University Press.
Fuglie, K.O., and C.A. Kascak. 2001. Adoption and diffusion of natural-resource-conserving agricultural technology. *Review of Agricultural Economics* 23(2):386–403.
Fuglie, K.O., and P.W. Heisey. 2007. Economic returns to public agricultural research. EIB-10. Washington, D.C.: U.S. Department of Agriculture Economic Research Service.
Fuglie, K.O., C.A. Narrod, and C. Neumeyer. 2000. Public and private investments in animal research. In Public-Private Collaboration in Agricultural Research, K.O. Fuglie and D.E. Schimmelpfennig, eds. Ames: Iowa State University Press.
Fulponi, L. 2006. Private voluntary standards in the food system: the perspective of major food retailers. *Food Policy* 31:1–13.
Fulton, M., and K. Giannakas. 2001. Organizational commitment in a mixed oligopoly: agricultural cooperatives and investor-owned firms. *American Journal of Agricultural Economics* 83(5):1258–1265.
Fumasi, R., J. Richardson, and J. Outlaw. 2006. Lifting the Fruit and Vegetable Restriction: Potential Impacts on Cropping Preference in the Lower Rio Grand Valley, Texas. Paper read at the Southern Agricultural Economics Annual Meetings, February 5–6, Orlando, Fla.
Fürst, M., J. Casale, and B. Wilhelm. 2005. Recommendations for Inspection of Social Standards. Tools and Methodologies for Implementation of Chapter 8 on Social Justice of the IFOAM Basic Standards. Bonn, Germany: International Federation of Organic Agricultural Movements.
Gagnon, S., J. Makuch, and T. Sherman. 2004. Environmental effect of USDA conservation programs. A conservation effects assessment bibliography. Available at http://www.nal.usda.gov/wqic/ceap/CEAP01complete.pdf. Accessed on February 22, 2010.
Gardner, B.L. 2002. American Agriculture in the Twentieth Century: How It Flourished and What It Cost. Cambridge, Mass.: Harvard University Press.
Gollehon, N., M. Caswell, M. Ribaudo, R. Kellogg, C. Lander, and D. Letson. 2001. Confined Animal Production and Manure Nutrients. Washington, D.C.: U.S. Department of Agriculture.
Green, R.D. 2009. ASAS Centennial Paper: future needs in animal breeding and genetics. *Journal of Animal Science* 87(2):793–800.
Green, R.D., M.A. Qureshi, J.A. Long, P.J. Burfening, and D.L. Hamernik. 2007. Identifying the future needs for long-term USDA efforts in agricultural animal genomics. *International Journal of Biological Sciences* 3(3): 185–191.
Greene, C. 2001. U.S. Organic Farming Emerges in the 1990s: Adoption of Certified Systems. Washington, D.C.: U.S. Department of Agriculture Economic Research Service.
Greene, C., C. Dimitri, B.-H. Lin, W. McBride, L. Oberholtzer, and T. Smith. 2009. Emerging Issues in the US Organic Industry. Washington, D.C.: U.S. Department of Agriculture Economic Research Service.
Greiner, R., L. Patterson, and O. Miller. 2009. Motivations, risk perceptions and adoption of conservation practices by farmers. *Agricultural Systems* 99:86–104.
Growing Power, Inc. 2009. Growing Power, Inc. Home. Available at http://www.growingpower.org. Accessed on February 21, 2010.
Gunther, M. 2006. The green machine. *Fortune*. Available at http://money.cnn.com/magazines/fortune/fortune_archive/2006/08/07/8382593/index.htm. Accessed on November 20, 2009.
Gurian-Sherman, D. 2008. CAFOs Uncovered. The Untold Costs of Confined Animal Feeding Operations. Cambridge, Mass.: Union of Concerned Scientists.
Guthman, J. 2004. Agrarian Dreams: The Paradox of Organic Farming in California. Berkeley: University of California Press.
Gwin, L. 2009. Scaling-up sustainable livestock production: innovation and challenges for grass-fed beef in the US. *Journal of Sustainable Agriculture* 33(2):189–209.
Hall, B.F., and E.P. Leveen. 1978. Farm size and economic efficiency: the case of California. *American Journal of Agricultural Economics* 60:589–600.
Hallam, A., ed. 1993. Size, Structure, and the Changing Face of American Agriculture. Boulder, Colo.: Westview Press.
Halloran, J.M., and D.W. Archer. 2008. External economic drivers and US agricultural production systems. *Renewable Agriculture and Food Systems* 23(4):296–303.

Hamm, M.W., and A.C. Bellows. 2003. Community food security and nutrition educators. *Journal of Nutrition Education and Behavior* 35(1):37–43.

Hanson, J., R. Dismukes, W. Chambers, C. Greene, and A. Kremen. 2004. Risk and risk management in organic agriculture: views of organic farmers. *Renewable Agriculture and Food Systems* 19(4):218–227.

Harris, D., and A. Rae. 2004. Agricultural policy reform and industry adjustment in Australia and New Zealand. Paper presented at International Agricultural Trade Research Consortium, June 6–7, Philadelphia, Pa.

Harris, J.M., P.R. Kaufman, S.W. Martinex, and C. Price. 2002. The U.S. Food Marketing System, 2002: Competition, Coordination, and Technological Innovations into the 21st Century. Washington, D.C.: U.S. Department of Agriculture Economic Research Service.

Hart, K.A., and D. Pimentel. 2002. Cosmetic standards (blemished food products and insects in food). In Encyclopedia of Pest Management, D. Pimentel, ed. New York: Marcel Dekker, Inc.

Hassanein, N. 1999. Changing the Way America Farms: Knowledge and Community in the Sustainable Agriculture Movement. Lincoln: University of Nebraska Press.

Hatanaka, M., C. Bain, and L. Busch. 2005. Third-party certification in the global agrifood system. *Food Policy* 30:354–369.

———. 2006. Differentiated standardization, standardized differentiation: the complexity of the global agrifood system. In Between the Local and the Global: Confronting Complexity in the Contemporary Food Sector, J. Murdoch and T. Marsden, eds. Amsterdam: Elsevier.

Heal, G.M., and A.A. Small. 2002. Agriculture and ecosystem services. In Handbook of Agricultural Economics, B.L. Gardner and G.C. Rausser, eds. Amsterdam: Elsevier.

Hefferan, C. 2004. Extension programming across the natural, socioeconomic, and political landscapes. Keynote address at the 4th Natural Resource Extension Professionals Conference, May 17, Wheeling, W. Va.

Heffernan, W., M. Hendrickson, and R. Gronski. 1999. Consolidation in the food and agriculture system. Available at http://www.nfu.org/wp-content/uploads/2006/03/1999.pdf. Accessed on February 22, 2010.

Heisey, P.W., J.L. King, and K.D. Rubenstein. 2005. Patterns of public-sector and private-sector patenting in agricultural biotechnology. *AgBioForum* 8(2&3):73–82.

Hendrickson, M., and W. Heffernan. 2007. Concentration of agricultural markets. Available at http://www.nfu.org/wp-content/2007-heffernanreport.pdf. Accessed on December 11, 2009.

Hendrickson, M., W. Heffernan, P. Howard, and J. Heffernan. 2001. Consolidation in food retailing and dairy: implications for farmers and consumers in a global food system. Available at http://www.libertyparkusafd.org/lp/NatureFirst%20USA/reports%5CConsolidation%20in%20Food%20Retailing%20and%20Dairy%20-%20Implications%20for%20Farmers%20and%20Consumers%20in%20a%20Global%20Food%20System.pdf. Accessed on February 22, 2010.

Henson, S., and T. Reardon. 2005. Private agri-food standards: implications for food policy and the agri-food systems. *Food Policy* 30:241–253.

Higgins, W., and K.T. Hallström. 2007. Standardization, globalization and rationalities of government. *Organization* 14:685–704.

Hills, S. 2008. Organic market shows signs of a slowdown. Available at http://www.foodnavigator-usa.com/Financial-Industry/Organic-market-shows-signs-of-a-slowdown. Accessed on April 6, 2010.

Hinrichs, C., and T.A. Lyson, eds. 2007. Remaking the North American Food System: Strategies for Sustainability. Lincoln: University of Nebraska Press.

Hinrichs, C.C. 2000. Embeddedness and local food systems: notes on two types of direct agricultural market. *Journal of Rural Studies* 16:295–303.

Holt, D.A. 2007. Agricultural research management in US land-grant universities: the state agricultural experiment station system. In Agricultural Research Management, G. Loebenstein and G. Thottappilly, eds. Dordrecht, the Netherlands: Springer.

Hoovers. 2008. Industry profile: food distributors. Available at http://premium.hoovers.com/subscribe/ind/fr/profile/basic.xhtml?ID=43. Accessed on October 13, 2009.

Horan, R., J. Shortle, and D.G. Ambler. 1999. Green payments for nonpoint pollution control. *American Journal of Agricultural Economics* 81(50):1210–1215.

Howitt, R., and K. Hansen. 2005. The evolving western water markets. *Choices* 20(1):59–64.

Huffman, W., and R.E. Evenson. 2006. Science for Agriculture: A Long-Term Perspective. 2nd edition. Ames, Iowa: Blackwell Publishing.

Huffman. W.E., G. Norton, G. Traxler, G. Frisvold, and J. Foltz, 2006. Winners and losers: formula versus competitive funding of agricultural research. *Choices* 21(4):269–274.

International Institute for Environment and Development. 2005. Participatory learning and action. Available at http://www.planotes.org/. Accessed on November 20, 2009.

Jacobs-Young, C., M.A. Mirando, and A. Palmisano. 2007. USDA-CSREES National Research Initiative support for agricultural research: the competitive grants program in the United States. In Agricultural Research Management, G. Loebenstein and G. Thottappilly, eds. Dordrecht, the Netherlands: Springer.

Jenkins, M.W., J.R. Lund, R.E. Howitt, A.J. Draper, S.M. Msangi, S.K. Tanaka, R.S. Ritzema, and G.F. Marques. 2004. Optimization of California's water supply system: results and insights. *Journal of Water Resources Planning and Management–ASCE* 130(4):271–280.

Jury, W.A., and H.J. Vaux, Jr. 2007. The emerging global water crisis: managing scarcity and conflict between water users. *Advances in Agronomy* 95:1–85.

Kaufman, P.R. 2000. Consolidation in Food Retailing: Prospects for Consumers and Grocery Suppliers. Washington, D.C.: U.S. Department of Agriculture Economic Research Service.

Kellogg Commission on the Future of State and Land-Grant Universities. 1999. Returning to our roots: the engaged institution. Available at http://www.cpn.org/topics/youth/highered/pdfs/Land_Grant_Engaged_Institution.pdf. Accessed on February 26, 2010.

Kellogg, R.L., C.H. Lander, D.C. Moffitt, and N. Gollehon. 2000. Manure Nutrients Relative to the Capacity of Cropland and Pastureland to Assimilate Nutrients: Spatial and Temporal Trends for the United States. Washington, D.C.: U.S. Department of Agriculture Natural Resources Conservation Series and the Economic Research Service.

Key, N., W.D. McBride, and M. Ribaudo. 2009. Changes in Manure Management in the Hog Sector: 1998–2004. Washington, D.C.: U.S. Department of Agriculture Economic Research Service.

Khanna, M., and A.W. Ando. 2009. Science, economics and the design of agricultural conservation programmes in the US. *Journal of Environmental Planning and Management* 52 (5):575–592.

Klonsky, K., and L. Tourte. 1998. Organic agricultural production in the United States: debates and directions. *American Journal of Agricultural Economics* 80(5):1119–1124.

Kloppenburg, J. 1991. Social theory and the de/reconstruction of agricultural science: local knowledge for an alternative agriculture. *Rural Sociology* 56(4):519–548.

Kloppenburg, J.R., Jr. 1988. First the Seed: The Political Economy of Plant Biotechnology, 1492–2000. New York: Cambridge University Press.

Knipling, E.B., and J.C.E. Rexroad. 2007. Linking priorities and performance: management of the USDA Agricultural Research Service research portfolio. In Agricultural Research Management, G. Loebenstein and G. Thottappilly, eds. Dordrecht, the Netherlands: Springer.

Kummer, C. 2010. The great grocery smackdown—will Walmart, not Whole Foods, save the small farm and make America healthy? Available at http://www.theatlantic.com/magazine/archive/2010/03/the-great-grocery-smackdown/7904/. Accessed on March 9, 2010.

Kolk, J. 2008. Carbon credits and agriculture. *Green Matters* 26(Winter):1.

Kunkel, P.L., J.A. Peterson, and J.A. Mitchell. 2009. Agricultural Production Contracts. Farm Legal Series, June 2009. St. Paul: University of Minnesota.

Laband, D.N., and B.F. Lentz. 2004. Which universities should provide extension services? *Journal of Extension* 42(2).

Lambert, D., P. Sullivan, R. Claassen, and L. Foreman. 2006. Conservation-Compatible Practices and Programs. Who Participates. Washington, D.C.: U.S. Department of Agriculture Economic Research Service.

Lambert, D.M., P. Sullivan, R. Claassen, and L. Foreman. 2007. Profiles of US farm households adopting conservation compatible practices. *Land Use Policy* 24:72–88.

Lamine, C. 2005. Settling shared uncertainties: local partnerships between producers and consumers. *Sociologia Ruralis* 45(4):324–345.

Lamine, C., and S. Bellon. 2009. Conversion to organic farming: a multidimensional research object at the crossroads of agricultural and social sciences—a review. *Agronomy for Sustainable Development* 29:97–112.

Lass, D., A. Brevis, G.W. Stevenson, J. Hendrickson, and K. Ruhf. 2003. Community supported agriculture entering the 21st century: results from the 2001 National Survey. Available at http://www.cias.wisc.edu/wp-content/uploads/2008/07/csa_survey_01.pdf. Accessed on September 10, 2009.

Lattimore, R. 2006. Farm subsidy reform dividends. Paper read at the North American Agrifood Market Integration Consortium Meeting, June 1–2, Calgary, Alberta.

Lattuca, L.R. 2001. Creating Interdisciplinarity: Interdisciplinary Research and Teaching among College and University Faculty. Nashville, Tenn.: Vanderbilt University Press.

Leeuwis, C. 2004. Communication for Rural Innovation: Rethinking Agricultural Extension. 3rd edition. Oxford, UK: Blackwell Publishing.

Leshy, J. 2009. Notes on a progressive national water policy. *Harvard Law and Policy Review* 3:133–159.

Lichtenberg, E. 1997. The economics of cosmetic pesticide use. *American Journal of Agricultural Economics* 79(1):39–46.

Lobley, M., and C. Potter. 1998. Environmental stewardship in UK agriculture: a comparison of the environmentally sensitive area programme and the countryside stewardship scheme in South East England. *Geoforum* 29(4):413–432.

Local Harvest. 2009. Local Harvest. Available at http://www.localharvest.org/. Accessed on February 22, 2010.

Lockie, S., and D. Halpin. 2005. The conventionalization thesis reconsidered: structural and ideological transformation of Australian organic agriculture. *Sociologia Ruralis* 45:284–307.

Loftus, T., and S.E. Kraft. 2003. Enrolling conservation buffers in the CRP. *Land Use Policy* 20:73–84.

Lubowski, R., S. Bucholtz, R. Claassen, M. Roberts, J. Cooper, A. Gueorguieva, and R. Johansson. 2006. Environmental Effects of Agricultural Land-Use Change: The Role of Economics and Policy. Washington, D.C.: U.S. Department of Agriculture Economic Research Service.

Lund, V., S. Hemlin, and W. Lockeretz. 2002. Organic livestock production as viewed by Swedish farmers and organic initiators. *Agriculture and Human Values* 19:255–268.

Lusk, J.L., and B.C. Briggeman. 2009. Food values. *American Journal of Agricultural Economics* 91(1):184–196.

Lyson, T., and A.L. Raymer. 2000. Stalking the wily multinational: power and control in the US food system. *Agriculture and Human Values* 17(2):199–208.

Lyson, T.A. 2002. Advanced agricultural biotechnologies and sustainable agriculture. *Trends in Biotechnology* 20(5):193–196.

Lyson, T.A. 2004. Civic Agriculture. Reconnecting Farm, Food, and Community. Medford, Mass.: Tufts University Press.

Lyson, T.A., G.W. Stevenson, and R. Welsh. 2008. Food and the Mid-level Farm: Renewing an Agriculture of the Middle. Cambridge, Mass.: MIT Press.

MacDonald, J., and P. Korb. 2008. Agricultural Contracting Update, 2005. Washington, D.C.: U.S. Department of Agriculture Economic Research Service.

MacDonald, J.M. 2007. Trends in the structure of agricultural markets. Presentation to the committee on December 11, 2007, in Washington, D.C.

Marra, M., D.J. Pannell, and A.A. Ghadim. 2004. The Economics of Risk, Uncertainty and Learning in the Adoption of New Agricultural Technologies: Where Are We on the Learning Curve? Raleigh: North Carolina State University.

McBride, W.D., and C. Greene. 2007. A comparison of conventional and organic milk production systems in the U.S. Paper read at American Agricultural Economics Association Annual Meeting, July 29–August 1, Portland, Ore.

———. 2008. The profitability of organic soybean production. Paper read at American Agricultural Economics Association Annual Meeting, July 27–29, Orlando, Fla.

McDowell, G., 2004. Is extension an idea whose time has come—and gone? *Journal of Extension* 42(6).

McDowell, G.R. 2001. Land Grant Universities and Extension into the 21st Century. Ames: Iowa State University Press.

Michigan State University Board of Trustees. 2009. About the MAES. Available at http://www.maes.msu.edu/about.htm. Accessed on November 19, 2009.

Morgan, D. 2010. The Farm Bill and Beyond. Economic Paper Series 2010. Washington, D.C.: German Marshall Fund of the United States.

Morgan, K., and J. Murdoch. 2000. Organic vs. conventional agriculture: knowledge, power and innovation in the food chain. *Geoforum* 31:159–173.

Nadvi, K., and F. Waltring. 2003. Making sense of global standards. In Local Enterprises in the Global Economy: Issues of Governance and Upgrading, H. Schmitz, ed. Cheltenham, UK: Edward Elgar.

Napier, T.L., J. Robinson, and M. Tucker. 2000. Adoption of precision farming within three Midwestern watersheds. *Journal of Soil and Water Conservation* 55:135–141.

Naylor, R., H. Steinfeld, W. Falcon, J. Galloway, V. Smil, E. Bradford, J. Alder, and H. Mooney. 2005. Losing the links between livestock and land. *Science* 310:1621–1622.

Naylor, R.L. 2008. Managing food production systems for resilience. In Principles of Natural Resource Stewardship: Resilience-Based Management in a Changing World, F.S. Chapin, G.P. Kofinas, and C. Folke, eds. New York: Springer.

NFU (National Farmers' Union). 2009. Carbon Credit Program. Available at http://nfu.org/issues/environment/carbon-credits. Accessed on October 8, 2009.

Nidever, S. 2009. Westside farmer sells water for $77 million. Available at http://hanfordsentinel.com/articles/2009/08/25/news/doc4a941aa622e70892149469.prt. Accessed on February 22, 2010.

Nowak, P. 1992. Why farmers adopt production technology? *Journal of Soil and Water Conservation* 47(1):14–16.

Nowak, P.J. 1987. The adoption of agricultural technologies: economic and diffusion explanations. *Rural Sociology* 52(2):208–220.

Nowak, P.J., and P. Cabot. 2004. The human dimension of resource management programs. *Journal of Soil and Water Conservation* 59(6):128A–135A.
NRC (National Research Council). 1972. Report of the Committee on Research Advisory to the U.S. Department of Agriculture. Washington, D.C.: National Academy Press.
———. 1989a. Alternative Agriculture. Washington, D.C.: National Academy Press.
———. 1989b. Investing in Research: A Proposal to Strengthen the Agricultural, Food and Environmental System. Washington, D.C.: National Academy Press.
———. 1993a. Soil and Water Quality. Washington, D.C.: National Academy Press.
———. 1993b. Sustainable Agriculture and the Environment in the Humid Tropics. Washington, D.C.: National Academy Press.
———. 1999. New Strategies for America's Watersheds. Washington, D.C.: National Academy Press.
———. 2000a. Investigating Groundwater Systems on Regional and National Scales. Washington, D.C.: National Academy Press.
———. 2000b. Watershed Management for Potable Water Supply: Assessing the New York City Strategy. Washington, D.C.: National Academy Press.
———. 2000c. A Vital Competitive Grants Program in Food, Fiber and Natural Resources Research. Washington, D.C.: National Academy Press.
———. 2002. Publicly Funded Agricultural Research and the Changing Structure of U.S. Agriculture. Washington, D.C.: National Academy Press.
———. 2003. Frontiers in Agricultural Research: Food, Health, Environment, and Communities. Washington, D.C.: National Academies Press.
———. 2005. Facilitating Interdisciplinary Research. Washington, D.C.: National Academies Press.
———. 2008. Hydrology, Ecology, and Fishes of the Klamath River Basin. Washington, D.C.: National Academies Press.
———. 2009. Science and Decisions: Advancing Risk Assessment. Washington, D.C.: National Academies Press.
NSAC (National Sustainable Agriculture Coalition). 2009. National Sustainable Agriculture Coalition. Available at http://sustainableagriculture.net/. Accessed on November 20, 2009.
NSTC (National Science and Technology Council) and BLM (Bureau of Land Management). 2001. California water rights fact sheet. Available at http://www.blm.gov/nstc/WaterLaws/california.html. Accessed on February 25, 2010.
Occidental College. 2009. Farm to school. Available at http://www.farmtoschool.org/. Accessed on October 30, 2009.
Oliveira, V. 2009. The Food Assistance Landscape: FY 2008 Annual Report. Washington, D.C.: U.S. Department of Agriculture Economic Research Service.
Packaged Facts. 2007. Ethical Consumers and Corporate Responsibility: The Market and Trends for Ethical Products in Food and Beverage, Personal Care and Household Items. Rockville, Md.: Author.
Padel, S. 2001. Conversion to organic farming: a typical example of the diffusion of an innovation. *Sociologia Ruralis* 41(1):40–61.
Peck, J.C. 2007. Groundwater management in the High Plains aquifer in the United States: legal problems and innovations. In The Agricultural Groundwater Revolution: Opportunities and Threats to Development, M. Giordano and K.G. Villholth, eds. Wallingford, UK: CAB International.
Perrot-Maitre, D. 2006. The Vittel Payments for Ecosystem Services: A "Perfect" PES Case? London: International Institute for Environment and Development.
Pew Charitable Trusts, and Johns Hopkins Bloomberg School of Public Health. 2008. Putting Meat on the Table: Industrial Farm Animal Production in America: A Report of the Pew Commission on Industrial Farm Animal Production. Washington, D.C.: Pew Charitable Trusts.
Pigg, K., and Bradshaw, T. 2003. Catalytic community development: a theory of practice for changing rural society. Pp. 385–296 in Challenges for Rural America in the Twenty-First Century. D.L. Brown and L.E. Swanson, eds. University Park: Pennsylvania State University Press.
Plummer, R., A. Spiers, R. Summer, and J. FitzGibbon. 2007. The contributions of stewardship to managing agroecosystem environments. *Journal of Sustainable Agriculture* 31(3):55–84.
Pollak, M.F. 2009. FY2008 data show downward trend in federal R&D funding. Science Resources Statisics Info-Brief NSF 09-309. Washington, D.C.: National Science Foundation.
Pollan, M. 2006. The Omnivore's Dilemma: A Natural History of Four Meals. New York: Penguin Books.
Ponte, S., and P. Gibbon. 2005. Quality standards, conventions and the governance of global value chains. *Economy and Society* 34:1–31.
Powell, J.M., and P.W. Unger. 1998. Alternatives to crop residues for sustaining agricultural productivity and natural resource conservation. *Journal of Sustainable Agriculture* 11(2–3):59–84.

Pratt, K.B. 1994. Water banking: a new tool for water management. *The Colorado Lawyer* 21(3):595–597.
Pretty, J.N., I. Guijt, J. Thompson, and I. Scoones. 1995. Participatory Learning and Action: A Trainer's Guide. London: International Institute for Environment and Development.
Prokopy, L.S., K. Floress, D. Klotthor-Weinkauf, and A. Baumgart-Getz. 2008. Determinants of agricultural best management practice adoption: evidence from the literature. *Journal of Soil and Water Conservation* 63(5): 300–311.
Ragland, E., and D. Tropp. 2009. USDA National Farmers Market Manager Survey 2006. Washington, D.C.: U.S. Department of Agriculture Agricultural Marketing Service.
Reilly, J.M., and D.E. Schimmelpfennig. 2000. Public-private collaboration in agricultural research: the future. In Public-Private Collaboration in Agricultural Research, K.O. Fuglie and D.E. Schimmelpfennig, eds. Ames: Iowa State University Press.
Ribaudo, M., and N. Gollehon. 2006. Animal Agriculture and the Environment. Washington, D.C.: U.S. Department of Agriculture Economic Research Service.
Robertson, G., V.G. Allen, G. Boody, E.R. Boose, N.G. Creamer, L.E. Drinkwater, J.R. Gosz, L. Lynch, J.L. Havlin, L.E. Jackson, S.T.A. Picket, L. Pitelka, A. Randall, A.S. Reed, T.R. Seastedt, R.B. Waide, and D.H. Wall. 2008. Long-term agricultural research: a research, education and extension imperative. *BioScience* 58(7):640–645.
Rockefeller Foundation. 1982. Science for Agriculture: Report of a Workshop on Critical Issues in American Agriculture. New York: Author.
Rodriguez, J.M., J.J. Molnar, R.A. Fazio, E. Snydor, and M.J. Lowe. 2009. Barriers to adoption of sustainable agricultural practices: change agent perspectives. *Renewable Agriculture and Food Systems* 24(1):60–71.
Rogers, E. 2003. Diffusion of Innovations. 5th edition. New York: Free Press.
Rubenstein, K.D., P.W. Heisey, C. Klotz-Ingram, and G.B. Frisvold. 2003. Competitive grants and the funding of agricultural research in the United States. *Review of Agricultural Economics* 25(2):352–368.
Russelle, M.P., M.H. Entz, and A.J. Franzluebbers. 2007. Reconsidering integrated crop-livestock systems in North America. *Agronomy Journal* 99(2):325–334.
Ruttan, V.W. 1982. Agriculture Research Policy. Minneapolis: University of Minnesota Press.
SAI Platform. 2009. SAI Platform. Available at http://www.saiplatform.org/home/home-page.html. Accessed on February 22, 2010.
Saltiel, J., J.W. Bauder, and S. Palakovich. 1994. Adoption of sustainable agricultural practices: diffusion, farm structure and profitability. *Rural Sociology* 59:333–349.
Sandmann L.R., and L.Vandenberg.1995. A framework for 21st century leadership. *Journal of Extension* 33(6).
Sauer, R. 1990. Meeting the challenges to agricultural research and extension. *American Journal of Alternative Agriculture* 5(4):184–187.
Schimmelpfennig, D., and P. Heisey. 2009. U.S. Public Agricultural Research: Changes in Funding Sources and Shifts in Emphasis, 1980–2005. Washington, D.C.: U.S. Department of Agriculture Economic Research Service.
Schnepf, M., and C. Cox. 2006. Environmental Benefits of Conservation on Cropland: The Status of Our Knowledge. Ankeny, Iowa: Soil and Water Conservation Society.
Schoon, B., and T. Grotenhuis. 2000. Values of farmers, sustainability and agricultural policy. *Journal of Agricultural and Environmental Ethics* 12:17–27.
Sexton, R.J. 2000. Industrialization and consolidation in the US food sector: implications for competition and welfare. *American Journal of Agricultural Economics* 82(5):1087–1104.
Shabman, L., and K. Stephenson. 2007. Achieving nutrient water quality goals: bringing market-like principles to water quality management. *Journal of the American Water Resources Association* 43(4):1076–1089.
Shabman, L., K. Stephenson, and W. Shobe. 2002. Trading programs for environmental management: reflections on the air and water experiences. *Environmental Practice* 4:153–162.
Shi, G., J.-P. Chavas, and K. Stiegert. 2008. An Analysis of Bundle Pricing: The Case of the Corn Seed Market. Madison: University of Wisconsin.
Singer, J.W., S.M. Nusser, and C.J. Alf. 2007. Are cover crops being used in the U.S. Corn Belt? *Journal of Soil and Water Conservation* 62(5):353–358.
Sligh, M., and C. Christman. 2003. Who owns organic? The global status, prospects and challenges of a changing organic market. Available at http://www.rafiusa.org/pubs/OrganicReport.pdf/. Accessed on February 22, 2010.
Smithers, J., J. Lamarche, and A.E. Joseph. 2008. Unpacking the terms of engagement with local food at the farmers' market: insights from Ontario. *Journal of Rural Studies* 24(3):337–350.
Spielman, D.J., and K. von Grebmer. 2004. Public-Private Partnerships in Agricultural Research: An Analysis of Challenges Facing Industry and the Consultative Group on International Agricultural Research. Washington, D.C.: International Food Policy Research Institute.

Stanton, J. 1999. Support the independent grocer—or else. *Food Processing* 60(2):36.
Stephenson, K., S. Aultman, T. Metcalfe, and A. Miller. 2009. An evaluation of nutrient trading options in Virginia: a role for agriculture? Paper prepared for the 2009 Annual Meeting of the Southern Agricultural Economics Association, January 31–February 3, Atlanta, Ga.
Strange, M. 1988. Family Farming: A New Economic Vision. Lincoln: University of Nebraska Press.
Sustainable Agriculture Coalition. 2005. Reinvigorating public plant and animal breeding for a sustainable future. Available at http://www.sacdev.org/wp-content/uploads/2008/10/seedsbreeds2005.pdf. Accessed on November 19, 2009.
Swanson, J.C. 2009. Animal welfare: stay informed, maintain best practices. *Michigan Dairy Review* 14(3):1–2.
Swift, M., A.M. Izac, and M. Noordwijk. 2004. Biodiversity and ecosystem services in agricultural landscapes—are we asking the right questions? *Agriculture Ecosystems and Environment* 104:113–134.
Swinton, S.M., and S.S. Batie. 2001. FQPA: pouring out (in?) the risk cup. *Choices* 16(1):14–17.
Swinton, S.M., F. Lupi, G.P. Robertson, and S.K. Hamilton. 2007. Ecosystem services and agriculture: cultivating agricultural ecosystems for diverse benefits. *Ecological Economics* 64(2):245–252.
Texas Center for Policy Studies. 1995. Texas Environmental Almanac. Available at http://www.texascenter.org/almanac/QUANTITYCH1P2.HTML. Accessed on November 19, 2009.
The Hartman Group. 2008. The Many Faces of Organic. Bellevue, Wash.: Author.
Thompson, B. 2009. Educational and Training Opportunities in Sustainable Agriculture. Washington, D.C.: U.S. Department of Agriculture Agricultural Research Service and National Agricultural Library.
Thompson, P. 2010. The Agrarian Vision: Sustainability and Environmental Ethics. Lexington: University Press of Kentucky.
Thronsbury, S., L. Martinez, and D. 2007. Michigan: a state at the intersection of the debate over full planting flexibility. Available at http://ddr.nal.usda.gov/bitstream/10113/33691/1/CAT31036247.pdf. Accessed on March 3, 2010.
Tocco, T., and S. Anderson. 2007. Wal-Mart's sustainability initiative: a civil society critique. Available at http://www.laborrights.org/sites/default/files/publications-and-resources/CounterSustainability.pdf. Accessed on February 22, 2010.
Tothova, M. 2009. The Trade and Trade Policy Implications of Different Policy Responses to Societal Concerns. Paris: OECD Publishing.
U.S. Congress Office of Technology Assessment. 1990. Agricultural Research and Technology Transfer Policies for the 1990s. A Special Report of OTA's Assessment on Emerging Technology: Issues for the 1990s. Washington, D.C.: Author.
U.S. Department of Interior. 2005. Reclamation—managing water in the West. Available at http://www.usbr.gov/uc/progact/salinity/pdfs/PR22.pdf. Accessed on April 6, 2010.
———. 2009. Colorado River Basin Salinity Control Program. Available at http://www.usbr.gov/uc/progact/salinity/. Accessed on October 8, 2009.
U.S. Fish and Wildlife Service. 2009. Endangered species habitat conservation planning. Available at http://www.fws.gov/endangered/hcp/index.html. Accessed on November 19, 2009.
USBR (U.S. Bureau of Reclamation). 2000. Klamath Project Historical Operation. Klamath Falls, Ore.: U.S. Bureau of Reclamation, Mid-Pacific Region, Klamath Basin Area Office.
USDA (U.S. Department of Agriculture). 2000. Glickman announces $300 million bioenergy program. Available at http://www.usda.gov/news/releases/2000/10/0379.htm. Accessed on August 4, 2009.
———. 2007. Secretary of USDA Mike Johanns addressed the issue of honey bee colony collapse disorder in USDA satellite feed on July 5. http//w3.usda.gov/agency/oc/bmtc/scripts.htm. Accessed on December 10, 2008.
USDA-AMS (U.S. Department of Agriculture Agricultural Marketing Service). 2008a. Fruit, vegetable and specialty crop marketing orders. Available at http://www.ams.usda.gov/AMSv1.0/ams.fetchTemplateData.do?more=A.OptionalText1&template=TemplateA&page=FVMarketingOrderLandingPage. Accessed on November 17, 2009.
———. 2008b. Organic labeling and marketing information. Available at http://www.ams.usda.gov/AMSv1.0/getfile?dDocName=STELDEV3004446. Accessed on October 8, 2009.
USDA-ERS (U.S. Department of Agriculture Economic Research Service). 1996. Farmers' Use of Marketing and Production Contracts. Washington, D.C.: Author.
———. 2009a. Agricultural Resource Management Survey (ARMS): resource regions. Available at http://www.ers.usda.gov/Briefing/arms/resourceregions/resourceregions.htm. Accessed on August 4, 2009.
———. 2009b. Organic agriculture: organic market overview. Available at http://www.ers.usda.gov/briefing/organic/demand.htm. Accessed on October 13, 2009.
———. 2009c. Organic agriculture: maps and images gallery: results from the 2007 Census of Agriculture. Available at http://www.ers.usda.gov/Briefing/Organic/Gallery/AgCensus2007.htm. Accessed on November 20, 2009.

———. 2009d. World Trade Organization (WTO): background. Available at http://www.ers.usda.gov/Briefing/wto/background.htm. Accessed on November 18, 2009.

———. 2009e. NAFTA, Canada, and Mexico. Available at http://www.ers.usda.gov./Briefing/NAFTA/. Accessed on November 18, 2009.

USDA-FNS (U.S. Department of Agriculture Food and Nutrition Service.) 2009. WIC farmers' nutrition program. Available at http://www.fns.usda.gov/wic/FMNP/FMNPfaqs.htm. Accessed on November 18, 2009.

USDA-FSA (U.S. Department of Agriculture Farm Service Agency). 2000. Commodity Credit Corporation (CCC) announces bioenergy program sign up. Available at http:// www.fsa.usda.gov/Internet/FSA_File/bioenergy_release_1654.pdf. Accessed on August 4, 2009.

———. 2009. Conservation Reserve Program. Available at http://www.fsa.usda.gov/FSA/webapp?area=home&subject=copr&topic=crp. Accessed on November 18, 2009.

USDA-NAL (U.S. Department of Agriculture National Agricultural Library) 2007. Sustainable agriculture research funding resources. Available at http://www.nal.usda.gov/afsic/pubs/agnic/susagfunding.shtml. Accessed on November 20, 2009.

———. 2009. Educational and training opportunities in sustainable agriculture. Available at http://www.nal.usda.gov/afsic/pubs/edtr/EDTR2009Orgs.shtml. Accessed on February 21, 2010.

USDA-NASS (U.S. Department of Agriculture National Agricultural Statistics Service). 2009. 2007 Census of Agriculture. Washington, D.C.: Author.

USDA-NIFA (U.S. Department of Agriculture National Institute of Food and Agriculture). 2009. Extension. Available at http://www.csrees.usda.gov/qlinks/extension.html. Accessed on February 26, 2010.

USDA-NRCS (U.S. Department of Agriculture Natural Resources Conservation Service). 2009a. Conservation Security Program. Available at http://www.nrcs.usda.gov/Programs/CSP/. Accessed on November 18, 2009.

———. 2009b. Environmental Quality Incentives Program. Available at http://www.nrcs.usda.gov/PROGRAMS/EQIP/. Accessed on November 18, 2009.

USDA-RMS (U.S. Department of Agriculture Risk Management Agency). 2008. Crop insurance works for organic producers. Available at http://farm-risk-plans.usda.gov/pdf/organic_producers.pdf. Accessed on November 17, 2009.

Utah Division of Water Rights. 2009. Frequently asked questions. Available at http://www.waterrights.utah.gov/wrinfo/faq.asp. Accessed on November 19, 2009.

van der Ploeg, J.D. 2003. The Virtual Farmer: Past, Present and Future of the Dutch Peasantry. Assen, the Netherlands: Royal Van Gorcum B.V.

Van Es, J.C., and P. Notier. 1988. No-till farming in the United States: research and policy environment in the development and utilization of an innovation. *Society and Natural Resources* 1(1):93–107.

Vanclay, F., and G. Lawrence. 1994. Farmer rationality and the adoption of environmentally sound practices: a critique of the assumptions of traditional agricultural extension. *European Journal of Agricultural Education and Extension* 1(1):59–90.

Vandenbergh, M.P. 2007. The new Wal-Mart effect: the role of private contracting in global governance. *UCLA Law Review* 54(4):913.

Vanloqueren, G., and P.V. Baret. 2008. Why are ecological, low-input, multi-resistant wheat cultivars slow to develop commercially? A Belgian agricultural 'lock-in' case study. *Ecological Economics* 66(2–3):436–446.

Vitalis, V. 2007 Agricultural subsidy reform and its implications for sustainable development—the New Zealand experience. *Journal of Integrative Environmental Sciences* 4(1):21–40.

Vogl, C.R., L. Kilcher, and H. Schmidt. 2005. Are standards and regulations of organic farming moving away from small farmers' knowledge? *Journal of Sustainable Agriculture* 26:5–26.

W.K. Kellogg Foundation. 2009. About Food and Society Program. Available at http://www.wkkf.org/Default.aspx?tabid=90&CID=19&ItemID=5000185&NID=5010185&LanguageID=0. Accessed on November 20, 2009.

Wageningen International. 2009. Participatory learning and actions. Available at http://portals.wi.wur.nl/ppme/?Participatory_Learning_and_Action. Accessed on November 20, 2009.

Walker, B., and D. Salt. 2006. Resilience Thinking. Washington, D.C.:Island Press.

Wall Street Journal, The. 2009. California's man-made drought—the green war against San Joaquin Valley farmers. Available at http://online.wsj.com/article/SB10001424052970204731804574384731898375624.html. Accessed on April 6, 2010.

Warner, K.D. 2006. Agroecology in Action. Cambridge, Mass.: The MIT Press.

Westcott, P.C., and E.C. Young. 2000. U.S. Farm Program benefits: links to planting decisions and agricultural markets. Available at http://www.ers.usda.gov/publications/agoutlook/oct2000/ao275e.pdf. Accessed on February 26, 2010.

Wheeler, W.B. 2002. Role of research and regulation in 50 years of pest management in agriculture—prepared for the 50th anniversary of the Journal of Agricultural and Food Chemistry. *Journal of Agricultural and Food Chemistry* 50(15):4151–4155.

Williams, P.D., D. Inman, A. Aden, and G. Heath. 2009. Environmental and sustainability factors associated with next-generation biofuels in the U.S.: What do we really know? *Environmental Science and Technology* 43(13):4763–4775.

Wortmann, C.S., A.P. Christiansen, K.L. Glewen, T.A. Hejny, J. Mulliken, J.M. Peterson, D.L. Varner, S. Wortmann, and G.L. Zoubek. 2005. Farmer research: conventional experiences and guidelines for alternative agriculture and multi-functional agro-ecosystems. *Renewable Agriculture and Food Systems* 20(4):243–251.

WTW (Water Transfer Workgroup). 2002. Water transfer issues in California. Available at http://www.waterboards.ca.gov/waterrights/water_issues/programs/water_transfers/docs/watertransferissues.pdf. Accessed on February 22, 2010.

Wu, J.J. 1999. Crop insurance, acreage decisions, and nonpoint-source pollution. *American Journal of Agricultural Economics* 81:305–320.

7

Illustrative Case Studies

The committee that authored the report *Alternative Agriculture* (NRC, 1989) commissioned case studies of 14 farms to illustrate the wide range of alternative farming systems that were operating in the United States in the 1980s. Those case studies provided important details about the complexity of individual farming practices and the diversity of approaches used to improve the sustainability of farming systems. The committee for this report commissioned two sets of case studies. Initially, the committee followed up with the operators or owners of the 14 case-study farms featured in *Alternative Agriculture* to see whether their approaches to farming have changed over time and how their approaches have affected the viability of their farms.

In addition to following up with those case-study farms, the committee selected nine new farms to serve as informative case studies for this report. The purpose of the new case studies was to illustrate the diverse production and management practices used to improve sustainability across different farm commodity types, to highlight how innovative producers address common challenges associated with moving toward greater sustainability, and to better understand the role of larger social, economic, and institutional contexts in the emergence and development of these farms. The assumption is that successful farmers, operating in real-world environments, are a key source of knowledge for innovative agricultural production systems design and management. The case-study farms were chosen to provide insight into how different segments of U.S. agriculture are implementing sustainability concepts in the 21st century. Each farm is located within biogeophysical, economic, and sociopolitical realms with scales varying from family to farm, local landscape, local community, regional, and global. Their stories are necessarily complex, and their success at improving different measures of sustainability illustrates the balancing of various goals and objectives faced by most real-world farmers. In most cases, their farming systems build on positive synergies and interactions among different aspects of the social and natural elements of their farms, which are manifested at various scales (for example, households, farms, watersheds, and niche markets). Those interactions are dynamic and are expected to change in response to or anticipation of opportunities and constraints. Many of the key technologies and interaction processes for sustainability have been well studied by scientists, and are summarized in the main body of this report. But the current state of farming systems science is far from adequate to effectively model, in a holistic way, the complexities of interaction illustrated by the case-study farms.

Follow-up of the Case Studies Featured in <u>Alternative Agriculture</u>

The report *Alternative Agriculture* (NRC, 1989) used 11 case studies that included 14 farms to provide in-depth examples of the wide range of "alternative" agricultural farming systems discussed in the report. The case studies were conducted in 1986 and included 5 integrated crop and livestock farms[1], 7 fruit and vegetable farms[2], 1 beef cattle ranch[3], and 1 rice farm[4]. The current committee attempted to contact these farms to find out how they have performed since 1986. This type of longitudinal study is a valuable way to identify the factors that influence long-term successes and challenges, and to highlight the organizational and management strategies associated with a farm's long-term viability.

STATUS OF THE FARMS

The committee was able to obtain current information and contacted the persons who owned or operated the farms for 10 out of the original 14 farms (Table 7-1). The committee was unable to confirm the operating status of 4 of the farms because it could not locate or reach the current owners, operators, or other people interviewed in 1986. Of the 10 farms contacted, 2 were no longer in business. One farm (Stephen Pavich and Sons) reported that they had ceased operation in 2001 as a result of several unfortunate events, including three insurance claims during the time that the farm was expanding. Although the case-study farm business no longer exists, Stephen Pavich Sr.'s children remain involved in agriculture as farmers or agricultural consultants. Another original case-study farmer (Ted Winsberg of Green Cay Produce) has retired, although other people are farming some of the land he used to farm.

Although Rex Spray of Spray Brothers' Farm and Mel Coleman were interested in providing an update of their farms, they were unable to participate in in-depth interviews because of time constraints. Both farms were still in operation in 2008. Spray Brothers' Farm was using mostly the same management practices and crop rotations reported in *Alternative Agriculture*. Rex said that the economic viability of the farm has not changed, but he indicated that weeds and changes in weather patterns were two of his biggest concerns.

When the committee reached Mel Coleman of Coleman Natural Foods (known as Coleman Natural Beef in NRC, 1989) in late 2007, he said that the family's cattle ranch had reduced the size of its herd considerably since 1989 because of drought. Coleman Natural Foods, however, was in a better financial position in 2007 than in the 1980s because of expanded product lines and a premium for its certified-organic and naturally raised livestock

---

[1]Spray Brothers Farm, BreDahl Farm, Sabot Hill Farm, Kutztown Farm, and Thompson Farm.
[2]Ferrari Farm, Hundley Farms, Winsberg Farm, Garguillo Farm, Barfield Farm, Stephen Pavich and Sons, and Kitamura Farm.
[3]Coleman Natural Beef.
[4]Lundberg Family Farms.

TABLE 7-1 Description and status of farms featured as case studies in the report *Alternative Agriculture* (NRC, 1989)

Farm Name	Location	Primary Products in the 1980s	Acres Operated in the 1980s	Status in 2008
Spray Brothers	Morgan, Ohio	Milk, beef, vegetables, small grains, soybean	720 acres	In operation. Declined to be interviewed.
Mormon Trail Farm (known as BreDahl Farm in NRC, 1989)	Adair County, Iowa	Lamb, beef, swine, vegetables, small grains, soybean	160 acres	In operation. Farmed acreage = 320 acres. Livestock = 80 cows, 55 ewes, a small flock of sheep, and broiler operation that was being scaled back.
Brookview Farm (known as Sabot Hill Farm in NRC, 1989)	Goochland County, Va.	Beef, forage, cash grain	3,530 acres of land, of which 815 acres were farmed 500 beef cattle	In operation. Farmed acreage = 980 acres. Livestock = 140 beef brood cows and 140 calves.
Kutztown Farm	Kutztown, Pa.	Beef, forage, small grains, soybeans	305 acres 250–290 beef cattle	In operation. Farmed acreage = 400 acres. Livestock = 117 cows.
Thompson Farm	Boone County, Iowa	Milk, swine, vegetables, soybeans, forage	282 acres 50 cows, 90 swine	In operation. Farmed acreage = 300 acres. Livestock = 50 beef cattle and 90 sows.
Ferrari Farm	Linden, Calif.	Vegetables, small fruit, nuts	223 acres	In operation. Farmed acreage = 450 acres.
Hundley Farms	Loxahatchee, Fla.	Vegetables, oranges, sugarcane, cattle	5,640 acres	Could not reach owner.
Ted Winsberg's Farm	Palm Beach, Fla.	Peppers	350 acres	Retired.
John Garguillo's Farm	Naples, Fla.	Tomatoes	1,300 acres	Could not find information on farm.
Fred Barfield's Farm	Immokalee, Fla.	Vegetables, oranges, beef, cattle	1,550 acres, 1,000 Beefmaster cattle, 1,200 cow mixed breed, commercial herd	Could not find information on farm.
Pavich Family Farms	Maricopa County, Ariz. Delano, Calif. Kern County, Calif.	Small fruit and grapes, vegetables	1,432 acres	Ceased operation.
Kitamura Farm	Colusa County, Calif.	Tomato, vine seeds, beans	305 acres	Could not find information on farm.
Coleman Natural Beef	Saguache, Colo.	Beef	21,500 acres owned, 13,000 leased, 250,000 available by grazing permits, 2,500 beef cattle	In operation. Declined to be interviewed.
Lundberg Family Farms	Richvale, Calif.	Rice	3,200 acres	In operation. Farmed acreage = 14,000 acres. Cash crop = rice

ILLUSTRATIVE CASE STUDIES

and products. The beef division of Coleman Natural Foods was sold to Meyer Natural Angus (another firm) in June 2008.

The committee obtained information on Kutztown Farm from Jeff Moyer of Rodale Institute. However, that case-study farm is not included because NRC could not reach the owner of the family farm to obtain informed consent for using it as a case study.

The committee commissioned consultant Susan Smalley (Michigan State University) to conduct follow-up interviews on 8 of the 14 farms featured in *Alternative Agriculture* (NRC, 1989) using a protocol designed by the committee (see Appendix D for protocol). The follow-up interviews included the 2 farms no longer in business and 5 farms still in operation—Mormon Trail Farm (known as BreDahl Farm in NRC, 1989), Brookview Farm (known as Sabot Hill Farm in NRC, 1989), Thompson Farm, Ferrari Farm, and Lundberg Family Farms.

COMMONALITIES AMONG THE FARMS

The follow-up interviews include examples of farms that used conventional and organic practices. Although organic farming has become much more common since the mid-1980s, two of the farms (Mormon Trail Farm and Thompson Farm) suggested that conversion did not seem suitable for their situation. The three originally organic farms (Brookview Farm, Ferrari Farm, and the Lundberg Family Farms), however, have increased the proportion of or shifted completely to organic production. Nonetheless, all of the 1989 case-study farms still in operation appear to exhibit qualities that are associated with movement toward greater sustainability (for example, robustness, resistance, and resilience). In the follow-up interviews, many farmers emphasized the importance of maintaining or building up their natural resource base and maximizing the use of internal resources as key parts of their farming strategies. Those farming philosophies are consistent with the committee's discussion on environmental sustainability and the importance of a closed nutrient cycle (Chapter 3).

Almost all the five restudied farms are using farming practices and management strategies similar to those described in the 1989 report. That observation reinforces that farming systems can maintain or improve natural resource quality and maintain economic sustainability over time. The crop farms emphasize the importance of careful soil management and use crop rotations and cover crops to reduce erosion and manage fertility. Crop diversity has also remained a key feature of these farms, some of which have increased crop diversity (for example, the Lundberg Family Farms). The farms with livestock each continue to pursue management practices that do not use hormones or antibiotics.

Despite strong continuity in their core farming practices, most of the 1989 case-study farms have adjusted and adapted their mix of crops and livestock, their scale of operations, and their marketing strategies in response to changes in environmental conditions, their family situations, customers' preferences, and market opportunities. Their ability to make changes in operations to adapt to new contexts reflects a form of resilience that was discussed in Chapter 2.

Most of the 1989 case-study farms participate in nontraditional commodity and direct-sales markets to some extent (for example, Ferrari Farm sells a small proportion of its fruit at a farmers' market and Brookview Farm sells most of its products via direct sales). They produce some, if not all, their products for value-trait markets—for example, organic crops and organic or naturally raised livestock.

Each of the five farms continues to rely heavily on family members for labor and management of farming operations. The Lundberg Family Farms also hires a number of nonfamily workers, through the use of "good" labor practices discussed in Chapter 4, and recently was named California Workplace of the Year by the Employer Resource Institute and the top Small Workplace by a nonprofit group, Winning Workplaces™.

Although all the 1989 case-study farms still in operation reported to be robust and successful, a few of the farms highlighted threats to their long-term viability. Those threats include high land-rental costs and rising land values associated with development pressure. Operators of two case-study farms in California (Bryce Lundberg and Wayne Ferrari) mentioned availability of water for farming as one of their concerns. Other challenges mentioned include spread of new weed species.

Mormon Trail Farm

Mormon Trail Farm (known as the BreDahl Farm in NRC, 1989) is located about 60 miles southwest of Des Moines, directly on the historic pioneer route that runs through southern Iowa. The home place, now owned by Clark BreDahl and his wife, Linda, has been in the BreDahl family since 1927. In 1974, Clark began operating the farm where he and three other siblings grew up. In the 1980s, he and Linda cash rented the farm from his mother. Since their interview for the original *Alternative Agriculture* study, they have been able to purchase the 160-acre home place along with an additional quarter section a few miles away. Although the farm consists of 320 acres, it is smaller than the Adair County average of 407 acres, according to the 2007 Census of Agriculture. The farming operation is similar in many ways to what it was 20 years ago, although several components have changed as conditions have dictated.

FARMING PHILOSOPHY

The BreDahls' farming philosophy has not changed over the years. They try to operate the farm using internal resources to the largest extent possible. Farm inputs are evaluated on the basis of how they will affect net income rather than gross production. Sometimes, maximum yields and maximum profits move in a direct correlation, but frequently they do not. The BreDahls stress that concerns for family and the environment often rank equally with economics when final decisions are made. Two examples illustrate this philosophy.

First, the BreDahls emerged from the farm crisis of the mid-1980s in stable financial condition. They could have attempted to expand the farm's land base more rapidly, but chose instead to pursue additional careers for which both had college training and previous job experience. It also fit with Clark's idea of diversifying the family's income as much as possible. Shortly after the original study, Linda quit her full-time teaching job to be a stay-at-home mother and farming partner while the children were young. After a 12-year hiatus, she returned to the classroom and, at that time, Clark also accepted a part-time job in the communications field. Thus, one of the biggest challenges with the farming operation in the years since has been tailoring the enterprise mix to fit the available labor supply.

Second, the BreDahls have countered state and national trends by steadily decreasing the amount of row crops grown on their farm, moving instead to more grass and livestock—a combination they feel is better suited to their highly erodible soils and the environment. They have also found success with several smaller niche enterprises that, cumulatively, have made a significant contribution to the farm's income.

MANAGEMENT FEATURES

Crop Rotations and the Soil

The overall mix of crops and livestock on the BreDahl farm has shifted steadily toward livestock over the past 20 years. In 2008, the BreDahls produced only 25 acres of field corn and 45 acres of soybean (the first they had grown in five years) on their 300-plus acres. Typically 40–50 acres of oats and alfalfa are grown, while the acreage of mixed grass-legume pasture has increased significantly. Intensive row-crop production is confined to the best upland soils (Sharpsburg, Macksburg, and Winterset) and river bottoms (mostly Colo-Ely) with high corn suitability ratings and little slope. On rotated acres, corn, soybean, oats, alfalfa or clover, and occasionally turnips or rye are grown. A small apple orchard and planting of fall-bearing raspberries are microenterprises.

Although crop rotations and manure applications remain essentially unchanged at the farm, planting practices have shifted heavily towards no-till. For the most part, contact herbicides are used that leave little or no long-term residue. In some cases, resistant corn varieties are used to eliminate insecticide applications. In rare instances where second-year corn has been grown, hybrids with genetic resistance to rootworms have performed as well or better than first-year corn. A heavy emphasis on grass, no-till planting practices, and row crops grown on only the best land have helped cut soil losses from the farm to near zero, a fact affirmed by soil tests for organic matter that show a steady rise.

Long-term yield trends on both the home farm where Clark's father was an early adopter of soil conservation practices more than 60 years ago and the neighboring farm which they purchased have been rising steadily, and year-to-year variations have become smaller. Soils in the area of southwest Iowa where Clark farms tend to have wet, clay outcroppings on hillsides that were once relatively unproductive. Combine yield monitors in recent years, however, have verified little or no difference in yields for many of these spots. Clark attributes the positive changes to better soil aeration resulting from installation of field drainage tile, manure applications to improve soil fertility, no-till planting practices that enhance soil structure, and the regular inclusion of deep-rooted legumes to penetrate any remaining hardpan soils. Crop varieties have also improved.

Livestock

Livestock numbers (and species) on the BreDahl Farm have also undergone changes. In the 1980s, the BreDahls had two flocks of sheep—one a flock of 40 registered Rambouillet ewes and another of approximately 150 commercial crossbred ewes bred annually to produce market lambs. They also maintained a small sow herd in a farrow-to-finish swine operation, and sometimes purchased lightweight beef calves that were fed to market on homegrown grains.

Two successful livestock enterprises that emerged on the BreDahl farm during the 1990s were the custom finishing of feeder lambs for other owners and the addition of broiler chickens sold direct from the farm to customers. Lambs were brought in from ranches in western range states, finished on grain, and marketed to Midwestern packing plants. The BreDahls were paid a monthly fee for the animals' care, plus the feed they ate. The arrangement worked well because it gave western ranchers a new marketing option for their lambs while adding a predictable source of income for the BreDahls' labor and a portion of their grain.

Broiler chickens emerged as an enterprise almost by accident. A few chickens raised strictly for family use while Linda was pregnant with their second child mushroomed by word of mouth into a business that eventually attracted customers from as far away as Des Moines and Omaha. An additional benefit was that chicken customers visiting the farm frequently asked about other meats, leading to the direct sale of beef, lamb, and, initially, pork. Broiler chickens were a labor-intensive project, but worked well for their young family. For 17 years, the BreDahls grew, processed, and direct-marketed up to 1,200 broilers per summer season with the help of their daughters. As their last daughter leaves home, Clark and Linda are in the process of phasing back broiler production, but direct sales of beef and lamb remain strong. Pork is no longer included in the direct-sale mix as outdated facilities, poor prices, and competing outlets for their labor led the BreDahls to exit pork production in the late 1990s.

Part of the product appeal of the BreDahls' poultry, beef, and lamb was that they were marketed as produced "naturally" without any introduced hormones or antibiotics. While the BreDahls experimented briefly with organic production, they learned that most of their meat customers were more interested in natural husbandry practices than organic certification. This was especially true as those customers discovered that organically produced meats came at a significantly higher price due to higher production costs.

Although direct sales represent only a small part of total beef and lamb production (about 10 percent), they have been an important source of additional revenue to the farm. Because the BreDahls have the ability to set their own prices on direct-sale items and those prices remain relatively stable, they provide critical income support, especially in years when commodity prices are low.

In 2008, the BreDahls still ran about 55 commercial crossbred ewes along with a small flock of registered Finnsheep. Lack of competitive markets for their commercially grown lambs has limited growth of that enterprise. Partially offsetting that obstacle, however, are lambing percentages well above industry averages with the inclusion of the prolific Finnsheep in their crossbreeding mix.

A big gain in livestock numbers has been growth of a commercial cow herd. In 2008, the herd numbered approximately 80 cows, with ownership of all calves retained through slaughter. Growth of the herd, started in 1988, was "haphazard" until Clark began identifying all newborn calves with ear tags and keeping detailed records on their growth characteristics and carcass merit. He also began purchasing bulls based on their expected progeny differences (EPDs). EPDs use measured data to scientifically rate an animal's ability to pass along key genetic traits to its offspring. Some of the traits evaluated include birth, weaning and yearling weights, calving ease, and factors related to carcass quality. The combination of using performance-tested bulls, maintaining detailed cow herd records, and basing cull decisions on hard data has led to steady gains in productivity. Feedlot and carcass data gathered on all slaughter cattle have charted similar improvements in rate of gain, percentage of carcasses grading choice or prime, and number of animals qualifying for the Certified Angus Beef® designation. Participation in a producer-verified program (PVP) that documents the origin, age, and history of each animal nets the BreDahls an additional premium for their commodity cattle that end up in lucrative foreign markets.

As pasture acres have increased, so has the intensity of production. The BreDahls use a system of planned grazing that limits the amount of time animals spend in each pasture. By grazing quickly, removing the animals, and giving the forage more time to rest and regrow, stands are improved and production increased. Electric fence, centralized watering sites, and careful attention to maintaining mixed stands of grasses and legumes have been key tools to making the system work.

Another change related to the livestock business has evolved out of necessity. In recent years, the farm has gone to composting nearly all the dead animals. They started by composting waste parts from broiler chickens that were processed on the farm. The practice worked so well that when commercial rendering companies quit accepting sheep and drastically increased their fees for removing cattle, the BreDahls began composting those species as well. Using mainly cornstalks, straw, and other bedding materials as their carbon source, they have found the practice to be clean, odor-free, and much less expensive than commercial disposal. The process also results in valuable recycled nutrients to return to their fields as fertilizer. All of these efforts represent examples of using internal resources—time and knowledge—to add value to the farm's production.

LEARNING NETWORKS

Clark has been involved with several farm organizations—Farm Bureau, Practical Farmers of Iowa, and numerous sheep and cattle groups, and he has served on the steering committees of two university research farms in his area. He likes attending field days and open houses to observe first-hand how practices work, often recalling his father's advice: "You won't live long enough to make all the mistakes yourself, so do your best to learn from others."

He has seen what he considers a positive shift in attitude in recent years by some of the larger farm organizations and the Iowa Department of Agriculture to provide more support for small, beginning, and niche farming operations. Overall, he thinks that type of assistance is more readily available today than it was 15 or 20 years ago. Countering that, in his opinion, are traditional farm commodity programs that reduce risk for established farmers, but make it harder for beginning farmers to compete.

USE OF GOVERNMENT PROGRAMS

As in the 1980s, the BreDahls today use government programs selectively. Twenty years ago, Clark felt that most family farms would be better off without commodity support programs. He also concluded that those programs encouraged many farmers to convert non-program acres to subsidized crops such as corn and soybean at the expense of livestock. History since has tended to confirm that theory as pasture, forage, and small grain acres have declined dramatically in Iowa over the past two decades along with the number of farms raising livestock.

Today, most federal and state programs require farmers to follow rigid guidelines or use specific practices in order to qualify for payments. Yet, Clark is not convinced that government agencies know better than individual farmers what is best for their land, their farming operations, or their communities. One size does not fit all, he maintains, and feels it is critical that farmers choose wisely the technologies and funding sources they will accept, even if it sometimes affects revenue.

The BreDahls have used funding available through the Environmental Quality Incentives Program (EQIP) to assist with some of their pasture renovation, but have largely self-financed farm improvements through private lenders. They began their shift to managed grazing in the late 1970s, long before most current government grazing incentive programs existed. Since then, they have invested heavily, at their own expense, in electric fencing, watering equipment, and solar-powered energizers that allow them to pasture more animals per acre in an environmentally sound manner.

Clark believes farm commodity programs have encouraged many producers to substitute the external resources of technology and capital for the internal resources of labor and management. As production has shifted heavily to corn, soybean, and other program crops, farms have lost diversity, become more open to risk, and, ironically, more reliant on future government support. He also feels the security afforded by commodity programs has dampened farmers' interest in trying new alternative approaches.

TRIAL OF ORGANIC PRODUCTION

The BreDahls' attitude of using internal resources to the fullest ultimately led them to experiment with organic production in the mid-1990s. They were considering organics at the time of their original interview for *Alternative Agriculture* in the late 1980s. They began the transition process on cropland in 1995 and maintained all of their home farm in a certified-organic regimen for 10 years. Although both Clark and Linda embraced the idea of eliminating synthetic fertilizers and pesticides from their operation, the results of their experiment did not meet expectations. Clark's biggest concern was soil loss from the tillage associated with organic production. "I knew that erosion on our farm previously had been extremely low and tillage just seemed like a giant step backwards," Clark said. "Besides that, soil tests showed organic matter starting to decline in some of the fields receiving multiple tillage passes."

Other shortcomings with organic production included marketing contracts that were sometimes deceptive, price premiums that were less than expected, burdensome paperwork, and, in the case of direct-sale meats, customers reluctant to pay the added costs. While organics aligned well with their philosophical beliefs, bottom-line costs and returns did not point to long-term sustainability.

BENEFITS FROM THE BIOFUEL INDUSTRY

In recent years, Iowa has become a national leader in the production of biofuels, and this new industry has had a positive effect on the BreDahls' farming operation. While providing an additional cash market for excess grain, the industry has also helped them lower feed costs for their cattle and sheep. They use both wet and dry forms of corn gluten and distillers grains to stretch summer pastures, extend the fall grazing season, and utilize low-quality crop residues for feed that otherwise would have little value. Ethanol co-products have especially improved the efficiency of their pasture and forage-based diets because, with starch removed, those feeds do not upset the balance of roughage-digesting microbes populating the rumen. The BreDahls have relied heavily on research done at Iowa State University to help them turn this new energy source into an asset for food production as well.

SUMMARY AND FUTURE OUTLOOK

The BreDahl Farm's management strategy of internalizing operating costs and minimizing purchased inputs while maintaining soil quality and animal health is a key factor of the farm's success. Their experience shows it is not necessary to farm thousands of acres or rely heavily on government safety nets to be profitable. An active management style, committed family involvement, and diverse mix of enterprises have helped the BreDahls make a comfortable living while improving soil health and contributing to their community. Clark's outlook on farming remains positive. He still very much enjoys the challenges associated with production agriculture and looks forward to implementing more new ideas as he "retires" to full-time farming again in the near future. Although they recognize change as a certainty, Clark and Linda believe the farm has potential to provide both an adequate living and satisfying life style to another generation of the family. Talks are currently under way, but no specific plans have been made.

Ferrari Farms, Inc.

Ferrari Farms, Inc. is a family-owned and family-operated farm located in California's San Joaquin Valley. The farm specializes in certified-organic fruits and nuts. George Ferrari and his son Wayne established the farm in 1963. It is currently operated by Wayne, his wife Irene, and their sons Jeff and Greg. Wayne and Irene have started to make the transition of ownership to their sons. Ferrari Farms, Inc., includes 450 acres of farmed land, 340 of which the Ferraris own and 110 of which are leased. Wayne and Irene have built the farm to provide opportunities for their sons to participate and have continued to expand.

FARMING PHILOSOPHY

When Wayne Ferrari was attending a cherry growers' meeting, he concluded that he needed to figure out a different way to farm. The customers at a farmers' market were asking for organic produce, and he determined that organic production would not require a lot of changes to the practices and approaches he already used on his farm. Wayne has continued to fine-tune his organic production system. As stated on the farm's website (www.ferrarifarms.com), the Ferraris "hope to not only provide you with the most naturally wholesome and delicious fruits and nuts around, but also to help restore the land to its natural balance" by using organic methods.

MANAGEMENT FEATURES

Crops

Ferrari Farms currently produces 20 acres of cherries, 20 acres of plums, 100 acres of peaches and nectarines, 20 acres of apples, and 190 acres of walnuts (including some leased land). Transition to organic takes about three years and all newly acquired production acreage becomes certified organic after three years, except for walnuts. The new acreages for walnut are grown using conventional production for the first three years. Building a walnut orchard requires substantial investment in the trees. The Ferraris use conventional fertilizers to obtain high yields for the first three years so that those acreages are economically viable. The orchard then undergoes transition to organic production during the fourth to sixth year and becomes certified organic thereafter.

Pest Management

Pest management practices on Ferrari Farms have expanded from using biocontrol initially to now including more dependable organic pest control methods and products that were developed in recent years. A wider variety of commercial products is now available for use in organic systems and provide the Ferraris with more options. They use bat houses, owl houses, and selected plantings to attract natural predators. They planted sunflowers to attract beneficial insects to counter peach borers and peach aphids. Greg's expertise in

ILLUSTRATIVE CASE STUDIES

disease and insect management is a valuable asset, and the Ferraris find scouting for pests a valuable practice. Selecting the most effective and efficient practices is critical because the production costs of the crops are so high that they cannot afford to miss a crop. In addition, they find value in crop insurance, especially for cherries, to protect their investments.

Fertility Management

Purple vetch is used as a cover crop, and composted steer manure is incorporated to help build the soil organic matter. Local regulations mandate that orchard waste be shredded and not burned. When Ferrari Farms renovates orchards, they pull trees, grind them, and put the material back on the land. Walnut shells are also returned to the land. That approach of fertility management has costs and benefits; spreading compost uses fossil fuels and creates dust, but those additions help to build soil organic matter.

Labor Management

Both sons have joined Wayne in the farming operation. Greg, the older son, who had always been interested in the farm, earned a bachelor's degree in plant science from Fresno State University and a master's degree with a double major in integrated pest management and in horticulture and agronomy from University of California, Davis. He handles pruning, picking, and spraying. Jeff, the younger son, had not originally planned to be involved in agriculture. He studied finance at Santa Clara University and earned a law degree at University of California, Davis. He now handles packing and sales. Irene is the book keeper and Wayne fills in any jobs that need help and works on big projects.

Ferrari Farms employs a labor force of 25–30 workers on a year-round or nearly year-round basis. It is important to the Ferraris that they employ these workers without exposing them to potentially harmful chemicals. Winter pruning and walnut cracking balances the growing season field work. They have not experienced problems in finding labor.

Equipment

Specialized machinery and equipment is essential to the Ferrari operation. Harvesting machinery is used just three weeks each year. During that time, machinery use is intense. They built a cracking plant to add value to their walnuts and installed forced-air coolers to upgrade postharvest handling. Hydrocooling was the previously used method, and if brown rot mold spores were present, the problem could spread and spoil everything in the cooler. Now they pack the products and cool with forced air. Excellent field management is important to getting fruit in and cooled right away. Precise temperature control is essential.

The purchase of a $240,000 laser infrared walnut shell sorter encouraged the Ferraris to reduce fruit production and expand walnut production to maximize the use of the equipment purchased. The sorter ensures that no metals are inadvertently mixed with the walnuts. Investments in larger and more effective equipment and machinery are an important part of the Ferraris' efforts to stay competitive in their market.

Marketing

Ferrari Farms has been selling its produce at the original San Francisco farmers' market for the past 37 years. Its goal is to provide consumers a tasty product at a reasonable price. The farmers' market is a good venue because it is not a high-end market, and consumers there are willing to purchase small or blemished fruits. Currently, a workman goes to the San Francisco market, and Jeff goes to the weekend market in Sacramento. The percentage of their products sold through farmers' markets is lower now than previously, but the markets still provide an outlet for their products and important source of income.

The farm sells wholesale through Veritable Vegetable, a company that focuses on its relationships with the farmers from whom it buys. It also sells to Whole Foods and Wild Oats, and used to sell to Albert's until it was bought out. Wholesalers are bigger and fewer today, with only about 18–20 wholesalers that might be interested in Ferrari Farms products. Much of the Ferraris' products go to Los Angeles where they are consolidated with other goods and shipped across the country. A small proportion of their products are shipped into Canada. The National Organic Program has added credibility to organic goods, especially for exporting. Within the United States, any organic certification is acceptable.

LEARNING NETWORKS

Wayne considers farming magazines an important source of ideas and information. He used to attend and speak at the Asilomar Eco-Farm Conference. However, he felt it had too much "story telling and sugar coating," with too little honest discussion of problems, so he stopped attending. He characterizes himself as an independent thinker who reads to gain valuable information and proceeds in his own way.

The University of California, Davis, is now doing quite a bit of organic work. Ferrari Farms hosted some on-farm trials with them several years ago. Wayne noted that extension farm advisors influence the farmers with whom they interact. If the extension agents are pro-organic, farmers begin shifting to organic; if not, farmers do not change. Presently, most of the Ferraris' neighbors still farm conventionally.

PERFORMANCE INDICATORS

The Ferraris manage the farm in a way not to threaten the natural resources. Wayne has observed that the soil organic matter is building up slowly.

Ferrari Farms, Inc. is not in any subsidy programs, and it generates enough income to provide for three families. The knowledge and expertise of Greg and Jeff contribute to the success of the farm.

KEY CHANGES

Ferrari Farms, Inc. has increased from 223 acres of farmed land in the 1980s to 450 acres in 2009. Because of increasing production expenses, Wayne feels that the farm has to increase total sales by increasing acreage and improving yield to earn enough income to support three families.

When Wayne first started using organic practices, the risk was high. There were few shipping channels, and the Ferraris had to grow many different types of produce to entice truckers to ship their products. Up until the late 1990s, the Ferraris grew walnuts, fruits, and vegetables (corn, tomatoes, cauliflower, broccoli, onions, red onions, and garlic). The diversity became unfeasible because of the increasing competition in the organic market. The bigger organic farms were getting better at what they did, but it was difficult for the Ferraris to improve their knowledge on the yields of all the different crops they raised to compete with the large growers. Moreover, the small acreage allocated for each crop made it difficult to use machinery that would improve efficiency. About nine years prior to the

2009 interview, Ferrari Farms dropped vegetables from its production and specialized in fruits and nuts. That way, it could also get larger accounts.

Another apparent change in the Ferrari operation over the past 20 years has been its vertical integration. Machinery investments have allowed it to add value to the crops it grows and have forced it to focus its efforts.

The local farming support network has started to change, with the agriculture commissioner becoming more supportive of organic farming. Agribusiness and service providers are also incorporating some alternative approaches.

CHALLENGES

Although the farm has an adequate water supply, the water table is dropping and is now at 135–140 feet. Where the farm is located, there are no places to put a small dam to collect rainwater. The Ferraris are learning to grow crops with less water by using drip irrigation. In their organic system, however, it is not feasible to control the weeds that grow around the irrigation emitters with limited flaming options.

Managing finances on Ferrari Farms is a challenge. They have to spend money to make money. Land rental requires upfront investment, which causes financial strain. Input costs for Ferrari Farms are about $3,500/day, with $1.3 million or more in annual expenses. Wayne feels caught in a cost/price squeeze, even with organic prices. He has observed the organic premium greatly decreasing as time passes.

The fuel needed to operate critical machinery also presents a challenge. There is some evidence of fuel additives contaminating area ground water, and the cost of diesel fuel was close to an all-time high at the time of the interview.

SUMMARY AND FUTURE OUTLOOK

Wayne Ferrari sometimes finds it a challenge to stay motivated after 63 years of farming. He sees that Ferrari Farms needs to be big enough to stay competitive, but not become too leveraged. Farming is hard work, six days a week, year-round. Today's farms have become so big that farmers cannot do all the work themselves and pay as they go. He and his wife are happy that their sons are involved in the farm. It has been expensive to expand, but they are finding ways to make it work.

Jeff and Greg Ferrari are farming because they enjoy the way of life. As Wayne said, it is not about the money but about being happy and gaining satisfaction from the work. They are willing to make a decent living in a way that they can feel good about how they manage the land.

Wayne believes that many farmers are choosing organic or sustainable farming nowadays for the wrong reasons (for example, because loss of pesticides and new government rules are driving them out of conventional farming). He urges efforts to raise awareness of such benefits of organic and sustainable farming to farmers as increased profits and better personal and family health. Wayne came from a conventional farming background, and he sees the good and the bad in both approaches. For him, the middle of the road is a good place to be.

Brookview Farm

Brookview Farm (known as Sabot Hill Farm in NRC, 1989) is located in Manakin-Sabot, northwest of Richmond, Virginia. Sandy and Rossie Fisher own and manage the farm, which has been developed to balance nature with profitable livestock production. It is a destination for people interested in good, healthful food. The Fishers own 480 acres and rent about 500 additional acres from their neighbors at Sabot Hill.

FARMING PHILOSOPHY

Brookview Farm's mission is "to sustainably produce the highest quality food and farm products, in a manner that preserves and enhances our community and natural resource." Sandy Fisher's farming philosophy was shaped in part following his Peace Corps experience in Colombia. He worked for nine years growing grass-fed beef. Years later, he drew on that experience when he shifted the Brookview Farm beef operation to become grass-based. Rossie Fisher's background in conservation and gardening has also helped to shape the farm.

The Fishers began shifting to organic practices in the early 1990s and have been certified organic for about 12 years. Certification has forced them to improve their record keeping, a practice that has validated their farming decisions. They enjoy having organic inspectors visit their farm. The inspectors appreciate the balance achieved by the Fishers among the beef, the chickens, and nature.

MANAGEMENT FEATURES

Crop Management

When the Fishers first took over management of Sabot Hill Farm in the late 1970s, the farm was planted with nearly 300 acres of corn and 200 acres of soybean. The soils on the farm were poor, and Sandy had to fertilize heavily. He also spent a lot of money on herbicides to control weeds. Realizing that heavy use of fertilizers and chemicals was not sustainable, he sought alternatives. In the 1980s, he reduced the corn and soybean acreage to 175 and 150 acres, respectively. The reduced acreage of corn and soybean decreased agrichemical expenses from $20,000 to $6,000. They focused on getting as much feed value as possible from pasture for the cattle and selling the hay crops to the farm's neighbors.

The Fishers bought Brookview Farm in 1982. They grow hay for cattle on that farm, but not crops. They added chickens and a compost operation to diversify the farm. Both enterprises have contributed to soil fertility, added income and profit, and complemented the beef enterprise.

The Fishers never tilled intensively even when they were growing corn and soybean. Feeding and finishing cattle entirely on grass has meant that they no longer needed to grow corn and they can concentrate on growing excellent pastures. The number of pastures has

Lundberg Family Farms

Lundberg Family Farms is located in Butte County in northern California. Albert and Frances Lundberg started the farm in 1937. They passed the farm onto their four sons, Eldon, Wendell, Harlan, and Homer. Although the four men are actively involved in the farm, the third-generation Lundbergs now run most of it. When the interview was conducted in 2008, Grant Lundberg was the chief executive officer and Jessica Lundberg was the chair of the board of directors. Lundberg Family Farms specializes in rice products that they grow, mill, process, and pack. It has expanded from a 3,100-acre farm in 1989 to an operation that includes 14,000 acres of farmed land, but the actual area farmed varies from year to year. The Lundbergs farm 5,000 of the 14,000 acres. The other 9,000 acres are contracted to 35–40 growers. Of the 5,000 acres they farm, 3,000 acres are managed organically and 2,000 acres are what the Lundbergs refer to as eco-farmed. Of the 9,000 acres managed by contracted growers, 7,000 acres are managed organically. More people are involved now in the operation with varying commitment to labor than before. Four farms are operated by the Lundberg family under the names WEHAH-Lundberg, Organic Rice Partners (which is run by a subset of the family), B&E Lundberg, and Lundberg, Lundberg, and Schultz.

FARMING PHILOSOPHY

Albert and Frances Lundberg left Nebraska in the 1930s after they saw the devastation of the Dust Bowl caused by poor soil management and poor farming techniques. Their philosophy of farming was to care for the soil by rotating crops and resting the land. Lundberg Family Farms have maintained the same family values in their business throughout the years and three generations of farming, despite two economic downturns. The operation is run with a tight business standard for efficiency and productivity. The owners are committed to maintaining a strong brand reputation for quality, environmental stewardship, and farm worker welfare.

MANAGEMENT FEATURES

Operations

Farming and food processing used to run seamlessly as one operation, but they were separated into two organizations since 1989. The farming business is called WEHAH-Lundberg, Inc. and overseen by a four-member board of directors. The food processing farm is called Lundberg Family Farms and is overseen by an eight-member board of directors. The two operations collectively are better known to the public as Lundberg Family Farms, and this case study uses that name throughout to refer to the farming operation.

Crop Management

The Lundbergs pioneered organic rice farming in America. Lundberg Family Farms grow mostly the same mix of crops as they did in 1989—rice, vetch, oats, clover, and some beans (for example, fava beans). They use a two-year rotation that alternates rice with purple vetch and fallow. They, however, have increased the number of varieties of rice they grow from 6 in 1989 to 16 varieties plus wild rice. Wild rice, which grows aggressively like a weed, is an integral part of the crop rotation. The Lundbergs also grow rice that is not organically certified, but it is grown with farming practices and approaches that aim to minimize the environmental impact of agriculture and to maintain the balances of nature.

Weed, Pest, and Disease Management

As in the 1980s, the Lundbergs' weed and pest management strategy involves first drowning the grasses with water and then drying the field. The continuous flooding to a depth of 8–12 inches for 21 to 28 days after planting the rice helps to control watergrass (*Echinochloa phyllopogon* and *E. oryzoides*). The watergrass drowns while the strong rice survives. After the watergrass drowns, the soil is allowed to dry to control for water weeds (for example, bulrush, small-flower umbrella plant, and duck salad *heteranthera limosa*). When the rice becomes fully established, the field is flooded until harvest. In a field with deep spots, the grass will grow and the weed population will stay in the field for years. Such fields are laser-leveled and alternately flushed and shallow-tilled to control weeds. The purple vetch in the rotation is used as a green manure or mulch. The mulch was thought to inhibit weed seed germination. As in the 1980s, diseases have not been a severe problem in the Lundbergs' rice fields because they emphasize the maintenance of soil health using such strategies as crop rotations and rolling the rice straw down to expedite the decomposition of the straw and sclerotia.[5]

In the 1970s and 1980s, the Lundbergs only had one crop of rice each year. Insects would infest the organic rice stored in bins because their eggs attached to the harvested rice. The Lundbergs had to sell the rice before the weather got warm and the insects hatched. Since then, they have had breakthroughs in postharvest storage, such as food grade carbon dioxide (CO_2) with sealed bins. The bins were painted white to keep the temperature inside the bin 10–15°C cooler and the rice fresher. They also purchased chillers from Sweden to keep the core of the bins cool and fresh, and they installed fans with temperature and moisture sensors in the silo. Fans are turned on automatically if needed to keep the rice fresh. When the bins and buildings are empty, the Lundbergs use heaters to raise temperature to 140°F to kill insects without chemicals.

Fertility Management

As discussed earlier, purple vetch is used as green manure at the Lundberg Family Farms. Until the 1980s, the Lundbergs applied chicken manure to what was called "the experimental field," which was the field they used to develop methods for rice production without chemical pesticides and fertilizers. In the late 1980s, the Lundbergs tried to provide their own fertilizer in an attempt to close the nutrient cycle. They grew two to three years of cover crops for every year of organic rice, but they still struggled with maintaining soil fertility. Since then, they found that chicken litter works well in their systems and have been using a combination of chicken litter, organic feather meal, or pelletized fertilizers that are purchased off farm.

---

[5]Sclerotia are compact masses of mycelia produced by the fungus *Magnaporthe salvinii* that causes stem rot in rice.

Energy Use

The farms have been investing in alternative energy and obtain 380–385 KW from alternative sources, including two solar arrays. The farms use about 1.2 MW of electricity. The Lundbergs support the development and use of green energy and purchased wind energy credits even though the cost per unit energy is higher than fossil energy. The farms, however, still uses diesel-fueled combines and other equipment. Bryce Lundberg explained that some old equipment is just as energy efficient and incurs little soil compaction. The Lundbergs evaluate the efficiency of their equipment periodically and assess how they could reduce the farms' carbon footprint. They are concerned that nonpoint source emissions are not well modeled and would like to know how they could manage those sources of emissions better. They would like to have tools for evaluating the carbon footprint of their farms.

Environmental Management

The Lundbergs have been keenly aware of the wildlife in fields and are concerned about the animals' well-being. They noticed that the cover crops attract ducks to nest in spring. They have been providing "egg aid" whereby they collect thousands of duck eggs, incubate and hatch them, and release the ducklings. They only clear half of the ditches at a time on the farm and leave the other half as undisturbed habitats for snakes. They have placed owl and bat boxes to provide habitats. They have been putting in ponds and planting trees in targeted areas in recognition of their ecological and aesthetic values. The Lundbergs aim to enhance the environmental value without compromising production.

Labor Management

Farm labor at the Lundberg Family Farms has been fairly consistent. The Lundbergs employed 6.5 full-time equivalent people year-round in the 1980s. Today, they hire 10 laborers. Because the farms' labor need is consistent throughout the year, they are able to offer job security. The good employees tend to stay with the Lundberg Family Farms because the farms offer a benefits package and a profit-sharing plan. The labor practices of the Lundberg Family Farms have earned them the honor of being named California Workplace of the Year by the Employer Resource Institute and the Top Small Workplace by Winning Workplaces™ (a nonprofit group) and by *The Wall Street Journal*.

Marketing

In the 1960s, Eldon, Wendell, Harlan, and Homer Lundberg started the Lundberg Family Farms® brand to provide consumers with the choice of purchasing rice that was not grown with conventional methods. They market their products under the "organic" and "eco-farmed" labels. The eco-farmed products are not certified organic, but they are produced with management practices that aim to improve agricultural sustainability.

In the 1980s and early 1990s, organic business seemed to flatten. At that time, the Lundbergs were uncertain how much organic rice they could sell or how much land they should transition to organic production. Although the eco-farmed business has flattened, the organic market has been growing, as indicated by customers' purchases. Having the California Certified Organic Farmers' seal on their products helps to promote sales.

The Lundberg Family Farms have mostly U.S. and Canadian customers; therefore international trade does not affect them much. The West Coast and northeastern United States are large markets for them. As energy costs increase, they might consider selling a larger proportion to nearby markets (for example, San Francisco and Sacramento, which are 150 miles and 100 miles away from the farm, respectively).

LEARNING NETWORKS

Lundberg Family Farms benefits from interactions with its group of contract farmers. The group meets 3–4 times each year to discuss production challenges, many of which are resolved as a result of the grower-to-grower interaction. Before the group was formed, some organic farmers felt isolated and did not know where to seek help.

University extension has helped the Lundbergs better understand the composition of their soil. The Lundbergs use many of the rice varieties developed in the Rice Experimental Station. The emphasis of the university's weed control division is not organic weed control and, hence, not as helpful to the Lundbergs. Faculty at the University of California, Davis, has helped the Lundbergs with fertility management so that the Lundbergs can time the application of pelletized organic fertilizer better.

The Ecological Farming Association provides inspiration to the Lundberg Family Farms. It holds a conference in January each year and provides opportunities for likeminded farmers and researchers to exchange ideas. In the Lundbergs' opinion, Acres USA also holds good conferences and provides useful information.

The Lundbergs do not find local agribusiness as helpful as the other groups mentioned above because the local agribusinesses tend to encourage the use of herbicides to control weeds and are less likely to provide much help to organic farmers. However, a growing number of companies market organic products.

PERFORMANCE INDICATORS

Yields per acre at the Lundberg Family Farms have increased. The Lundbergs have maintained the high quality of their soil despite farming more intensively. The organic production was low in the 1980s, and revenue from the sale of products grown on their "experimental fields" was well below expenses. In the 1980s, the yield from organic rice production was much lower compared to conventional production. Organic-rice production has to produce at least half the yield of conventional production to be economically viable. From experimentation and experience, the Lundbergs found that some varieties grown organically produce yields closer to conventional production than others. For example, vigorous varieties yield 9,000 lbs/acre with conventional production, and the Lundbergs might get 5,000–6,000 lbs/acre with organic production. With less vigorous varieties, such as California basmati rice, organic production might produce one-third to one-half the yield of conventional production. Basmati rice, however, has a much higher product value than short grain rice, and the Lundbergs have managed to increase their yields over time. Since the 1980s, their organic rice has been sold at a premium and is profitable. The Lundbergs were putting 200 acres of land in transition to organic in 2008. Their knowledge and experience in organic farming and the improved predictability of the organic market give the Lundbergs the confidence to shift a significant portion of land to organic management. The Lundbergs, however, have not put all land in organic production because they try to respond to customers' needs and some customers would like the nonorganic option.

ILLUSTRATIVE CASE STUDIES

Lundberg Family Farms participated in traditional commodity programs. Those programs pushed farmers to set aside acreage and were complementary to the Lundbergs' approaches as they put land into fallow anyway. Only 60 percent of the family's land, however, is recognized by those programs. The Conservation Security Program recognizes much of their organic production systems as sustainable production systems. Other programs are not as financially helpful to the Lundberg Family Farms for different reasons. First, the Lundbergs' production baseline is low. Second, they did not benefit from the burning cessation program, which offered a credit for cessation of residue burning, because they had always rolled down the rice straw instead of burning it.

KEY CHANGES

The organizational structure of the farming and food processing operations changed, but the Lundberg family still runs them both. The Lundberg Family Farms has a much larger proportion of land under organic production for two reasons. First, there have been some important breakthroughs in weed management for organic farming. In 1989, organic rice farming was significantly riskier than conventional rice farming. Although still riskier compared to conventional rice farming, growing organic rice now has more predictable results. Scientific research has increased the options for managing weeds in organic systems, and the Lundbergs' experience has improved their understanding of how those options work. Likewise, the options for managing soil fertility organically have increased. Second, the organic market has expanded substantially since 1989. Nowadays, customers who prefer natural and organic foods can purchase them in many locations. The market expansion also contributed to driving the increase in scientific research and tools developed for organic farming discussed above.

The Lundbergs have found that their neighbors have become more receptive to the idea of organic farming. In the 1980s, the Lundbergs had to build a mill for their organic rice because not many organic mills were readily accessible. That caused some friction with their neighbors, who viewed the Lundbergs' trial with organic production as a likely failure. Nowadays, neighboring farmers are more understanding and accepting of the diversity of the agricultural market. The Lundbergs were not trying to take sales from their neighbors by marketing a different product. In fact, more of their neighbors are moving toward organic production.

CHALLENGES

The price of land has increased substantially. Land costs about $5,000–$6,000 per acre, which could pose a challenge for new farmers. Land rental cost has increased, but not as much as land acquisition cost. The Lundberg Family Farms expanded by contracting more growers rather than purchasing more properties. The third-generation Lundbergs are concerned that passing the farm onto the next generation (which has more members) will dilute the resources. The farm was started by their grandparents, who passed it onto their four sons. The third generation includes 11 members, 7 of whom work daily on the farm. The fourth generation includes 20 members aged 12 months to 20 years and is expanding. The high costs for land acquisition could be a barrier to expanding the size of the farms. Yet, a full-time farmer needs at least 250 acres of land to be economically viable.

The Lundbergs have not observed much difference in weather patterns over time, but they consider water availability a looming challenge. Water is available in northern California, but needed in southern California. The Sacramento delta, which features many unique plants and wildlife, lies in between. Moving water from north to south across the delta will be challenging.

With regard to crop production, the Lundberg Family Farms recognizes customers' preferences for natural products and is concerned about the potential contamination of genetically modified organisms[6] (GMOs) in their fields from neighboring farms. The Lundbergs hope that the government regulates the use of GMOs and better protects organic farmers. The Lundbergs are also concerned about the potential spread of new weed species (for example, blueflower duck salad) in the future that cannot be controlled by current organic practices. Although herbicides can eliminate the weed, organic growers cannot use herbicides, and they need researchers to develop a solution quickly to control its spread. The Lundbergs believe that solutions to the challenges of organic farming result mostly from on-farm research.

SUMMARY AND FUTURE OUTLOOK

Lundberg Family Farms has been farming with practices and approaches that care for the soil, water, air, and wildlife for more than 70 years. The Lundbergs plan to continue the tradition and hope that the farm will be passed onto the fourth and fifth generations. They have a program designed to let family members experience different jobs on the farm. The smooth transition in the past depended largely on good planning and the absence of financial stress at the time of transition.

They feel that the future of organic agriculture is bright, because an increasing number of consumers are looking for high-quality foods that support their principles of protecting the environment. People realize that their values and concerns can be expressed through food choices. The Lundberg Family Farms not only have a larger consumer base than they did 20 years ago, but also they know their consumers' preferences better. They have their own brand, which sells at a premium. They plan to continue to expand the variety of rice products, with a continued focus on rice. They would like to communicate to their consumers that when they purchase the Lundberg brand, instead of a store brand of organic products, they are contributing to improving environmental and social sustainability (for example, fair labor treatment).

---

[6]The term "genetically modified organisms (GMOs)" was used by the farmers interviewed to refer to genetically engineered crops (see Chapter 3); hence, the term GMOs is used in this chapter.

Pavich Family Farms

Pavich Family Farms operated in the Harquahala Valley, Maricopa County, Arizona, and in Delano, Kern County, and Tulare County, California. Stephen Pavich, Sr., and his sons, Stephen Pavich, Jr., and Tom Pavich, operated the farm. At the time of the 1989 interview, Pavich Family Farms was in a growing phase following the 1986 and 1987 Alar (daminozide)[7] scare, which brought organic farming into the public's view and helped to drive its growth for many years. Pavich Family Farms took advantage of changing consumer desires and increased its sales in the organic market from 20 percent to 50 percent of their table grape crop. Pavich Family Farms ceased operation in 2001 because of several unfortunate conditions.

FARMING PHILOSOPHY

The Pavich family wanted to ensure that their produce was nutritionally superior. They conducted nutritional studies on their produce and believed that they were higher in quality of appearance and flavor than nonorganic produce. They worked on organic legislation with California and with the U.S. Department of Agriculture.

LEARNING NETWORKS

The Pavich family felt they had to stay 10–15 years ahead of the universities or they would be behind in the market. The Paviches experienced frustration in their interactions with local universities. Stephen Pavich, Jr., recalls being treated as though the family had no science behind its practices and found that the faculty were interested only in their own narrow disciplines (for example, insecticides or soil chemistry). In contrast, Pavich Family Farms took a holistic approach to organic farming and looked at the interactions of all of the farming systems. In his opinion, the universities' research did not adequately involve applied science or address agricultural applications.

KEY CHANGES

Pavich Family Farms expanded its farmed acreage in Arizona and California to 4,700 acres, up from the original 1,432 acres when the Paviches were interviewed in 1989. In the 1980s, they mostly grew and sold table grapes. The farms expanded their portfolio and marketed 65 crops, including imported grapes and berries from Chile, bananas and pineapples from Costa Rica, bell peppers from Mexico, and grapes from South Africa. The farms were

---

[7]Alar is the trade name of the growth-regulating chemical daminozide. Alar used to be sprayed on fruit trees to regulate their growth. The U.S. Environmental Protection Agency halted the use of Alar in food uses on the basis of evidence that it causes tumors in laboratory animals and that lifetime dietary exposure to this product might result in an unacceptable risk to public health.

selling about 1 million packages of vegetables each year under the Pavich Family Farms Organic label.

The Paviches began producing raisins from some of their grape crop in 1991 to further diversify their offerings. They also began to work with other growers in California and Arizona and were the biggest organic table grape growers in the country. By 1998, they were producing 2.5 million boxes of table grapes per year. They rapidly became the fifth largest raisin producer in the United States and coined the term "jumbo raisin."

The Paviches recognized the organic market as a growing market and wanted their farm to provide quality organic produce to consumers. They were the first producers to sell organic produce to Wal-Mart and Raleys supermarkets. Eventually, they were selling produce to 19 of the top 20 supermarket chains in the United States.

CHALLENGES

In 1997 and 1998, El Niño presented a major challenge. Due to unpredictable and unprecedented rainfall, the farm experienced massive crop losses in the form of damage from flooding and intensive rain, as well as rot and plant diseases caused by the excessive moisture. During the time leading up to El Niño, Pavich Family Farms had been expanding at a rate of about 10–20 percent per year. Its debt-to-equity ratio was marginal. With little competition in the organic market, Pavich Family Farms expanded rapidly to meet the demand. Stephen Pavich, Jr., stated that the family would not take such chances in today's market because there is more competition in the organic market compared to 10 years ago.

In his view, the farms could have financially withstood the crop losses if they were the only challenge the farm faced at the time. However, Pavich Family Farms was involved in three insurance claims. One claim was to recoup losses after a grower sprayed herbicide on its organic vineyard in Arizona. Another one was for flood loss. The third claim was for a malfunctioning box in cold storage that caused a large harvest of grapes to rot.

The substantial financial losses, the slow court process (which took seven years), and the expense of hiring lawyers created a "perfect storm" that was the beginning of the end for Pavich Family Farms. These chance events occurred at a time when the business was experiencing rapid growth and stretched financially. A significant upfront investment was required to initiate international expansion, and the business plan projected that the venture would break even after two years. The Paviches were importing 100,000 boxes of produce per year to start, and Stephen said they would have been importing more than 1 million boxes per year had the farm stayed in operation. The growers with which Pavich Family Farms had contracted returned to conventional growing practices.

The financial losses due to weather and the court costs were too much for the farm business. The bank owned everything at the farms by the time the family received its insurance money. In 2001, the farms were foreclosed upon, and the Paviches soon filed for bankruptcy. The Paviches could not find willing buyers to help save their farms; the losses had been too great for a company of their size. Stephen Pavich, Jr., remarked that similar farms (with different crops) that did not experience losses of this nature have been growing and have become multi-thousand acre organic farms.

SUMMARY AND FUTURE OUTLOOK

The farmland was auctioned off by the bank. The acreage in California is all being farmed organically, largely producing grapes and blueberries by five different organic producers. An organic vegetable grower bought the 1,300 acres in Arizona. The remaining acreage was purchased by a melon grower; half of his production is organic, the other half is marketed as conventional produce but its production incorporates many organic practices.

In addition, 160 acres of family land remains in California and is operated as an organic vineyard by Frances Mary Pavich, Stephen Pavich, Sr.'s daughter, under the name FMP Vineyards. She also produces 1 million pounds of organic raisins each year and markets them under the label Franny's Organic. Tom Pavich, who has an M.B.A., continues to farm organic raisins at the San Joaquin Valley farm for his sister. His farming practices include foliar sprays, cover crops, composting, and incorporating humic acid. It is an intense operation, albeit much smaller than what the Pavich family used to farm. In total, nearly all the 4,700 acres are still in agriculture (with different owners) and nearly all are being farmed organically.

Steven Pavich, Jr., who has a B.S. in viticulture, has become an agricultural consultant and represents a major organic agricultural fertilizer input company, Global Organics. He works mostly with large companies who supply food to grocery chains. He also works with nonorganic farms that want to grow high-quality produce with greater yields. He has found that if organic practices can do that, farmers will use them (or a combination of organic and conventional practices). He is very supportive of small-scale and family farms, but focuses his effort where he believes he will have the greatest impact—large organic producers that sell their produce in chain stores and farm in a manner that benefits the environment and worker safety.

Thompson Farm

Richard and Sharon Thompson's farm in Boone County, Iowa, is a 300-acre diverse crop and livestock farm. They started farming in 1958, and after 10 years of conventional operations they began experimenting with alternative approaches to growing crops and livestock.

FARMING PHILOSOPHY

As in the 1980s, the working philosophy of the Thompson Farm is to limit or find substitutes for off-farm inputs whenever possible to reduce costs and promote the health of livestock and people. The Thompsons take a "middle-of-the-road" approach to farming and continue to perfect a system that works well for them. They do not farm with conventional approaches because they found these approaches were not effective or profitable for their size of operation. They decided not to have the farm certified as organic because they wanted the freedom to experiment—sometimes with materials or approaches that are not permitted by the National Organic Program. Also, the Thompsons believe that they cannot totally protect their farm from the effects of neighbors who use synthetic chemicals around them.

MANAGEMENT FEATURES

Crops

Current crops on the Thompson Farm include 100 acres of corn, 50 acres of soybean, 50 acres of oats, 50 acres of hay, and 55 acres of pasture. The five-year crop rotation of corn, soybean, corn, oats, and hay, and the ridge-till system, have remained fairly consistent over the past 20 years, with small adjustments to fine-tune the system. The Thompsons plant rye cover crops in the fall to reduce erosion, help manage weeds, and reduce nitrogen loss. They use livestock manure plus bio-solids from the nearby city of Boone, which allows them to eliminate purchased fertilizer.

Although they do not use Roundup-Ready® corn seeds, they suspect that they have some Roundup-Ready corn in their field as a result of cross-pollination from neighboring fields. Typically, volunteer corn in rotation crops, which follow corn, is killed when Dick spot-sprays Roundup, but the corn did not die after a Roundup application in 2008.

Weed and Pest Management

The Thompsons use an integrated pest management approach and try to minimize their use of herbicides and pesticides. As mentioned, the crop rotations help control weeds. Additional weed control is achieved using a rotary hoe and cultivation and without use of broadcast herbicides. The Thompsons' use of cover crops, hoes, and cultivators is detailed in *"Steel in the Field": A Farmer's Guide to Weed Management Tools*.[8] They occasionally spray pesticides to control aphids

Livestock

Current livestock enterprises are 50 beef cattle and 90 sows raised for the "natural" meat market, without hormones or antibiotics. The beef cows are fed on pasture using a rotational grazing system and with farm-raised hay. The farm-raised feed is a mix of ground ear corn and oats to provide roughage for the cattle. A bunker, built during the 1980s, stores manure.

The Thompsons converted 30 hog farrowing houses to A-frame, winterized structures. The hogs are farmers' breed, which are not lean but flavorful. The breeding animals are selected for meatiness with some fat for flavor. The hogs are fed the farm-raised ground ear corn, shelled corn, and oats, which provides more fiber and less protein than conventional feed. The ration works well even for young pigs.

Equipment

Dick believes that a scale is a critical tool for success. Any farmer who wants to track the impact of farming experiments needs to weigh harvested grain, compare treatments, and keep records. The value of a scale becomes clear when one envisions a farmer conducting one experiment with six replications and two comparisons. To measure the yield from that one trial, the farmer would need to drive to the elevator 12 times to weigh the results. Dick believes that if USDA really wanted to encourage farmer research, it would subsidize purchases of farm scales. A platform scale that weighs one axle at a time works fine to weigh crops, animals, and manure.

Labor

The Thompson Farm is completely operated with family labor—Dick and Sharon plus their son Rex and his family. The Rodale Institute and Wallace Institute supported an extra hired person to fill in while Dick and Sharon traveled around the country to share their farming approaches and experience with others.

Marketing

Hogs from the Thompson Farm are sold as "natural" through Niman Ranch, which provides a price floor. The Thompsons used to sell their "natural" beef cattle—both black and red—individually, but have discontinued that. They purchased an additional livestock trailer that allows them to divide the livestock that they take to market, selling heifers and steers separately and dividing by grades.

LEARNING NETWORKS

One of the Thompsons' legacies was to begin the Practical Farmers of Iowa (PFI). Dick Thompson found it "a godsend" to spend time with other people interested in and trying to

---

[8] Bowman, G., ed. 1997. Steel in the field: a farmer's guide to weed management tools. Available at http://www.sare.org/publications/steel/steel.pdf. Accessed on March 3, 2010.

make sense of agriculture. Inspired by a similar organization elsewhere, he and a few like-minded people started an organization with farmers who are trying similar farming approaches across Iowa so that they can compare experiences. Early PFI board meetings were filled with discussion about farming problems and practices. For example, someone may have asked how farmers could control Canada thistles, and another farmer may have suggested spring plowing. As PFI encouraged more replicated trials, the results were printed and shared. PFI encourages its members to adapt, not adopt. It emphasizes cooperation among members and encourages them to develop solutions together.

The Thompsons get along well with Iowa State University and its Extension Service. Like PFI, the Thompsons' philosophy on research and extension is "to adapt, but not adopt, ideas." They believe that agricultural knowledge and practice can be greatly advanced if researchers work closely with operating farms and curious farmers, who could try out ideas as they and other PFI members have done. The Thompsons also maintain positive relationships with their neighbors and community despite their different farming approach. They, however, do not receive much help from their large (3,000–4,000 acre) neighboring farms.

Dick, a pioneer among "limited input" farmers, has always promoted farmer-led on-farm research. A compilation of Thompson Farm research is available online on PFI's website (www.practicalfarmers.org). In one year, the report was downloaded 25,000 times. Dick continues to conduct trials and comparisons on the farm and "tinker" with his production system because he gains satisfaction from small successes and enjoys problem-solving. The inspiration for some of Dick's innovative ideas came from his mentor, Dr. Warren Sahs, formerly at the University of Nebraska.

PERFORMANCE INDICATORS

Organic matter and soil loss on the Thompson Farm have been measured by the National Tilth Lab, which found that the Thompsons have 6 percent organic matter compared to 2.9 percent on the next farm. The Thompson's soil loss was just 4 tons/acre compared to 11 tons/acre next door. Earthworms average 19,000/acre in Boone County, but 1,269,000 earthworms/acre were found on the Thompson farm. The positive effects of the increased organic matter, low soil erosion, and high earthworm populations complement each other and provide positive feedback.

The Thompsons' farming system benefits the environment. Of the 300 acres of their land, 135 acres are in grass to protect water quality. Farm windbreaks and an ear corn crib provide habitat for wildlife. Fall manure application has improved air quality. The rotations include pasture, but after about six years in pasture, Dick finds that the soils harden. He plants alfalfa at that time because its roots penetrate deeply and provide opportunities for water to percolate through the soil.

Early in their farming lives, the Thompsons tried ridge tilling, using compost on the ridges and broadcasting rye cover. Hauling manure in the spring proved to be difficult and impractical. They only had a yield of 45 bushels of corn per acre, substantially below the county average at the time. They discontinued that practice and started to apply manure in the fall, plant hay, and then plow under. Initial corn yields were still only 45 bushels of corn per acre. Yields started to increase around 1979. By 2003 and 2005, they achieved yields of 231 and 210 bushels/acre, respectively, which were 39 and 38 bushels over the county average in those years. Records from 1988 to 2006 reveal that the Thompsons on average maintained a net income of $119 per acre for crops, mostly from increased crop and residue income. From 2002 to 2006, net income for crops averaged $172 per acre, an amount that does not include any government payments or premiums. Those numbers do not include profit or loss from livestock.

The Thompsons do not participate in any federal farm programs. They say that the programs for conserving natural resources lack flexibility to accommodate their approach to farming. Dick plows once every five years because he observes that no-till, over time, reduces water infiltration on his soil even though it reduces erosion. If the programs were more flexible, the Thompsons say they would like to participate. They do not participate in commodity programs.

Dick believes that the farm provides comfortable support for his family. Their yields and net income have improved since 1989. They attribute the success of the farm to good record keeping and constant trial and error. He and Sharon have always been careful not to spend too much money. He credits one of his neighbors as his "frugality mentor." They were also careful not to borrow too much money and to pay off debts as quickly as they could. Dick believes his ability and willingness to fix things (for example, electrical, machinery, or plumbing problems) around the farm, as well as their minimal off-farm purchases, contribute to controlling expenses on the farm. He explained that the rule at the Thompson Farm is "try to fix it ourselves first," because a trip to the dealer for a repair is usually 40 or 50 miles and, hence, costs time and money. They also built diversity into the farm to improve the buffer against potential loss or reduction of some crops.

KEY CHANGES

The acreage of the Thompson Farm has remained steady. Richard and Sharon Thompson's farm has stayed on the trajectory that they had established when interviewed for the 1989 assessment. They try to minimize chemical input and continue to fine tune the practices and approaches used on their farm. When the *Alternative Agriculture* report was published, the Thompsons were heavily involved with on-farm research, much of it focused on finding and correcting approaches that did not work.

CHALLENGES

Changing weather patterns are among the challenges that the Thompsons face on their farm. They use a flail machine (rotary scythe) to mow and process hay and to dry it faster between rain events. The wetter weather, plus a neighbor's shift to no-till, has worsened water runoff from a neighbor's farm onto the Thompsons' farm and created worsening gullies.

The cost of farming inputs has increased sharply. The Thompsons feel that the increased costs of equipment parts affect them the most. Because they have worked to create a largely closed loop system that allows them to avoid purchasing fertilizers and pesticides, the price increase of those inputs have not affected them much.

The rising land cost is a barrier to expanding the farm. Dick purchased land for $440 to $600 sometime ago. At the time of the interview, the land for sale around his farm cost about $6,000/acre.

Another challenge that the Thompsons are working to master is cattle size. Dick uses his animal science educational background to find the best size and mix of breeds—Holsteins, long horns, Herefords. Their goal is to balance an animal's sale price with the cost to keep and feed the animal.

SUMMARY AND FUTURE OUTLOOK

The family has established a trust to ensure the farm's 300 acres will remain intact for at least the next two generations. The Thompsons' family includes four children. One son and his family (including a grandson) are involved in farm work and are deepening that involvement. At least two Thompson generations beyond Dick and Sharon will operate the farm.

Dick Thompson feels that conventional agriculture has become less sustainable now than it was 10 years ago. Organic is an alternative, but it might not be suitable for all farms. He believes that the key to improving the sustainability of farming and farming communities lies in farmers' environmental stewardship and their taking responsibility to figure what works and what does not through trials and record keeping.

Green Cay Farm and Green Cay Produce

Green Cay Farm, in Palm Beach County, Florida, was established in 1957 as a 350-acre family vegetable-growing operation by Ted and Trudy Winsberg (as it was at the time of the 1989 report). As land use shifted in Palm Beach County from agriculture to residential and Ted prepared to retire, they sold much of the land. The remaining land was not large enough for commercial farming. Ted and Trudy began renting out plots to several growers to earn income. Their daughter and son-in-law run a small native tree operation on 20 acres. Another producer grows herbs on 10 acres. Charlie and Nancy Roe, of Farming Systems Research, Inc., rent 10 acres and sell mostly to local residents, restaurants, and resorts through subscription to their Community Supported Agriculture (CSA) program. Nancy Roe first began working with Green Cay Farm during her Ph.D. research at the University of Florida when the Winsbergs wanted to investigate alternatives for plastic mulch. The Winsbergs approached Nancy and her husband, Charlie, after her dissertation was complete, about starting a small farming operation. Nancy and Charlie returned to Green Cay Farm and started Green Cay Produce.

FARMING PHILOSOPHY

Ted Winsberg worked to improve the sustainability of his farm. He was a leader in and supported research and extension programs among South Florida vegetable growers. When a new technology was on the horizon, his peers often said, "Let Ted try it first to see how it works." The Winsbergs supported Nancy and Charlie selling locally as Green Cay Produce because they believe "people are far too removed from their food supply; they should know who grows their food and how it is grown." They also feel that customers are willing to pay a fair price for fresh, healthful, and tasteful vegetables. Since the transition, they have found that many people in the community agree with those principles and are willing to support Green Cay Produce.

MANAGEMENT FEATURES

Crops

Green Cay Farm changed from a commercial peppers and cucumber farm to become a diverse vegetable farm that grows more than 40 crops plus herbs year round (via the multiple growers using land from Green Cay Farm). Cover crops are planted in the summer, and drip irrigation is used. Green Cay Produce is not certified organic because it would not be profitable in its location. However, many methods and materials are used that are recommended for organic production.

Pest Management

Unlike other local growers, Green Cay Produce is selective about pesticide use. The Roes only spray pesticides when they observe a problem. They use mostly organic pesticides. They purchased beneficial insects for pest control at one time, but found that encouraging the natural populations of beneficial insects provided more effective pest control than introduced insects.

Fertility Management

The Roes use compost made from horse manure and bedding, which they obtain at no cost from nearby farms where many horses over-winter. The cover crops are tilled in the soil to provide nutrients. They also add potassium and nitrogen through the drip system and use a phosphorus starter solution when they first plant or transplant a crop.

Marketing

Before transitioning to becoming a small farm, Green Cay Farm sold most of its produce through conventional marketing channels and shipped to East Coast markets. In contrast, Nancy and Charlie operate their CSA on 10 acres they rent from Ted and Trudy. The CSA, which began operating in 2000 with a 26-week season, now has 400 members plus restaurant sales and operates for 34 weeks each year, from October through May. In both the CSA and restaurant markets, demand far exceeds the supply. It is the only CSA in the area, but several new ones will be starting soon.

The CSA operation includes the original vegetable packing house. The Winsbergs provide the land and equipment for the CSA at no cost to ensure that it can generate adequate income. Prices were raised by 10 percent in 2008. Some members thought the prices were high, but most proved to be loyal customers and stayed as members despite price increase and even when hurricanes wiped out crops. Of greatest importance to CSA members is that the produce is local and fresh. Only a few have left because they preferred certified-organic produce.

Labor

Labor for area commercial farms is from Central America or Mexico, contracted for the harvest. The Winsbergs' labor was initially from Puerto Rico, then Haiti. Ted once employed 70 people, 60 of them Haitians and mostly women, who did the field harvesting. Most lived on the farm or were assisted by the Winsbergs to buy houses. Most of the crew retired in 2000, and their children did not farm. Nancy and Charlie, plus five full-time and two part-time semi-retired employees, handle the CSA production. They also have seven part-time drivers and one part-time office staff person. Nancy views the current labor as "ridiculously stable" and would like to pay them more. The crew is aging, with only one person under 40 years old. One big challenge is how to find work and funds to keep the crew working through summer months when crop output is small.

LEARNING NETWORKS

The farm has cooperated with university and industry researchers to develop innovative, environmentally friendly systems for many years. Currently, it has contracts to perform research through Farming Systems Research, Incorporated. The Roes work with seed companies to grow out seed varieties and with universities to conduct on-farm research. They have participated in research about the effects of compost on postharvest quality, tropical cover crops or cover crops as alternatives to fumigation, and measuring yields of squash blossom production. They have a new project with USDA Fort Pierce to examine the ability of organic hay mulches to control nutsedge.

Both Ted Winsberg and Nancy Roe regard themselves as products of land-grant universities. Ted earned B.S. and M.S. degrees in soil science from Cornell University, and Nancy earned B.S. and M.S. degrees in horticulture from the University of Arizona and a Ph.D. from the University of Florida in horticultural sciences. She has managed vegetable crop operations for 30 years and spent five years in extension and research with Texas A&M University. Their backgrounds have made it easy and natural for them to cooperate with the University of Florida, with extension, and with industry groups.

Ted Winsberg has observed that large-scale farming operations do not share information with one another in the same way that smaller ones often do. The Winsbergs and Roes are starting to see others try to farm on a small scale and market locally. Nancy works with those farmers, and the local farm-credit office holds meetings and sponsors small farmers. Ted Winsberg sees that their farm could become a center to help small farmers. He would welcome more opportunities to share his mechanical experience and expertise.

KEY CHANGES

When Ted Winsberg was interviewed in 1989 for the *Alternative Agriculture* report, he was growing 350 acres of irrigated fresh market peppers. Although some of that acreage is still used for agriculture, it is very different today. In addition to the 350 acres of their home farm in Palm Beach County, they also rented a 400-acre farm near Martin, Florida, for 27 years.

Development approached on all sides of the farm, which was located in Palm Beach County. Ted and Trudy wanted to stay in the county. They explored alternatives with the American Farmland Trust, but believed that they did not have enough resources to defend the land against development. In 1985, Ted discussed his concern with Bob Wiseman (Palm Beach County's manager), the local water utility head, and a Palm Beach County commissioner. Wiseman's contributions to the community and the positive personal relationships that he had developed with key leaders helped them to work out options that could benefit the county and the Winsbergs. In the mid-1990s, the Winsbergs sold 176 acres of land to the county at one-third of its appraised value to create the Green Cay Wetlands and Nature Center. The interpretive nature center and 1.5-mile boardwalk provide suburban green space while the land filters 35 million gallons of water from Palm Beach County's Southern Region Water Reclamation Facility. The wetlands also incorporate 86 different species of trees, shrubs, grasses, and aquatic vegetation that help to recharge ground water resources.

A plant constructed in 1992 produces dried biosolids. The county is using 100 acres of its purchased land and the remaining 76 acres that are not in wetlands are rented back to Ted at $1/year. Another 40 acres were sold for an affordable housing development that has 100 townhouses and 320 condominium and rental units. Hazen Ranch Road, where Green Cay Farm is located, was once rural but is now filled with gated residential communities, with many retirees living in the area. There is little interaction between the farm and its neighbors.

CHALLENGES

Ted Winsberg considered using the biosolids produced nearby as a soil amendment and to boost fertility on the farm, but Environmental Protection Agency and Florida Department of Environmental Protection regulations made it impossible. Ted appreciates the need to safeguard both human health and the environment, but he questions regulations that prohibit any discharge of reclaimed water into surface water. He also questions the regulations regarding any composition of compost within organic systems. He believes some food safety and surface water regulations are not practical. Third-party inspection requirements have slowed conservation initiatives and efforts to recycle and re-use water.

As the area urbanized and commercial and residential development crowded out farms, the local agricultural infrastructure has severely declined. No other farms operate within five miles. Extension in Florida is being defunded, and 10 staff positions are unfilled in Palm Beach County Extension alone. Now the Winsbergs may need to drive 70–80 miles to attend an extension meeting. Ted Winsberg and Nancy Roe believe that small-scale farmers stand to lose the most if extension is lost. The farmers need research and extension specialists to help them solve pest problems, and they are concerned that large agrichemical companies drive much of the current research agenda.

Changing weather and climate present additional challenges. The winters of 2007 and 2008 were warm and caused problems for cool-weather crops and led to worsening insect problems. Palm Beach County is the warmest winter-growing area in the United States. With generally warmer weather, farmers can now grow winter products as far as 100 miles north of Palm Beach County. The extended range of growth removes the competitive advantage they once enjoyed. Winter vegetables are now grown as far north as the Carolinas. Palm Beach County's unique winter season is short or gone. Summer is nearly 30 days longer, as documented by Winsberg (not related to Ted) and Simmons (http://coaps.fsu.edu/climate_center/docs/flhotseason.php).

The farm has experienced some air quality issues with Vapam escaping from land used by an herb grower. Ted Winsberg anticipates new fumigant regulations.

The farm was cleared in the late 1940s. Its soils have been highly productive, but alternatives to methyl bromide will not sustain the high yields achieved. Ted's average price for peppers over the years of his farming operation was 25 to 50 cents/pound, which is low compared to prices sold in other areas. For example, the Winsberg's oldest son farms and grows peppers in Palo Alto, California, and sells them for $6/pound in farmers' markets. It took the Winsbergs 20 years to pay for their farm. While they farmed, they needed production loans of up to $1 million/year. At one point, they were offered $300,000/acre for their land and the opportunity to sell it for development rights for 950 houses. Ted thinks it is almost impossible for a small farming operation to be financially viable with increasing input costs. In addition, he sees that distribution of the products from small farms is energy inefficient. He has been working hard to create a fuel-efficient distribution system for his vegetables.

SUMMARY AND FUTURE OUTLOOK

At nearly age 80, Ted Winsberg's outlook has not shifted, and he is upbeat about the future. He wants to continue being involved in agriculture even in an urban area. Whether the farm will be passed on to the next generation is uncertain. The Winsbergs' daughter and son-in-law, Sylvia and Michael, established a 20-acre wholesale native tree nursery as a part of the farm. Their business is linked to construction. Their teenaged son is not currently interested in carrying on the operation, so its future is uncertain.

Ted suggested that municipalities and developers should set aside land for small farms. He believes that more consumer education is needed so that people better understand why products are grown a certain way, why food costs what it does, and why there are not more organic products. Most Americans are not concerned at all about vegetables, but they should be. If people put more thought into what they eat, they would likely be willing to pay more for good food.

New Case Studies

In addition to reviewing the current status and changes taking place among the original farms included in the 1989 *Alternative Agriculture* report, this committee commissioned a new series of case studies to provide in-depth examples of 21st-century successful sustainable farming systems. To increase the depth of the new case studies, the committee selected three farms from each of three important U.S. commodity sectors: dairy farms, specialty crop farms, and field crop farms. The nine new case-study farms were selected from a list of candidate farms that included recent winners of regional Sustainable Agriculture Research and Extension (SARE) program awards, farms known to members of the committee, and farms nominated by committee members' professional colleagues and networks. Within each commodity type, three farms from different geographic regions were selected to highlight the diversity of biophysical and socioeconomic environments surrounding U.S. agriculture.

Two experienced consultants—Lawrence Elworth (executive director of the Center for Agricultural Partnerships, a nonprofit organization) and Clare Hinrichs (an associate professor of rural sociology at Pennsylvania State University)—were contracted to conduct face-to-face interviews with the owners of the selected case-study farms. The consultants used a semi-structured interview schedule (see Appendixes E, F, and G) to ensure that similar topics and questions were used across all of the case-study farms. However, the flexible nature of the interview schedule allowed the consultants to probe for details on topics of particular relevance on each farm. Key topics covered in all interviews included:

- An overview of the size and scope of the cropping and livestock enterprises.
- A detailed understanding of innovations and creative solutions used by producers to address some common systemic and specific production challenges facing the sustainable production of dairy products, grains, and specialty crops in the United States.
- A detailed understanding of how the farmers manage risks associated with production, marketing, and family and worker well-being.
- A sense of how their farming systems have developed over time and what practices the farmers have tried and rejected.
- Information about what the farmers see as their major challenges in the future and how they plan to address them.
- Information about the farms' current sources of information, the role of public and private sector science and extension networks, and possible remaining information gaps that might be addressed by renewed investments in the science of sustainable farming systems.
- Information about the role of public policies that might have facilitated or thwarted the producers' success.

ILLUSTRATIVE CASE STUDIES

The case studies illustrate different farmer-designed approaches to sustainability within the context of their unique socioeconomic and biophysical environments. A profile of the new case-study farms is listed in Table 7-2.

The committee recognizes that many other farms have outstanding attributes that contribute to improved sustainability, but it selected farms that highlight important approaches to producing milk, specialty crops, and grains in ways that balance and enhance productivity, environmental quality, economic viability, and social acceptability. The methods used in the new case studies are best suited to provide in-depth descriptions of various aspects of each individual farm's production and marketing practices. The small number of cases, nonrandom selection, and diversity across the case-study farms preclude any formal statistical analysis or generalization to the broader population of U.S. farms. The existence of these farms in and of themselves, therefore, is not unequivocal evidence of either the success, viability, or desirability of these systems. Rather, they provide insights into issues,

TABLE 7-2 Summary of Characteristics of Farms Featured as Case Studies

Farm Name	Location	Primary Products	Size of Operation	Certified Organic	Non-certified Organic
Dairy					
Bragger Family Dairy	Montana, Wisc.	Milk, beef, poultry	285 milking cows, 100 beef cattle, 64,000 pullets	No	No
Radiance Dairy	Center, Iowa	Milk, poultry	75 milking cows, 25 chickens	Yes	No
Straus Family Creamery, Inc.	Marshall, Calif.	Milk	215 milking cows	Yes	No
Specialty Crops					
Full Belly Farm	Yolo County, Calif.	Vegetables, herbs, nuts, flowers, small fruit, milk, poultry	250 acres, 90 ewes, 400–500 chickens, a few goats, 1 milking cow	Yes	No
Peregrine Farm	Newlin, N.C.	Vegetables, ornamentals	4 acres	No	Yes
Stahlbush Island Farms, Inc.	Linn County, Ore.	Vegetables, small fruits	4,000 acres	30%	Other certificate program—Food Alliance
Grain					
Goldmine Farm	Shelby and Christian Counties, Ill.	Corn, small grains, soybean, hay, pasture, beef	2,200 acres (1,100–1,300 acres of row crops annually), 260 beef cattle	Yes	No
Rosmann Family Farms	Lincoln, Iowa	Corn, small grains, soybean, beef, poultry, swine	600 acres	Yes	Only broiler chickens
Zenner Farm	Latah County, Idaho	Wheat, small grains, peas, lentils, garbanzos	3,100 acres	No	Other certificate program—Food Alliance

barriers, and opportunities for innovative structure and technologies that could enhance sustainability alternatives.

Some common themes and issues that emerge from the full set of cases are discussed before the detailed narrative descriptions of the nine new case-study farms presented in this chapter. In the following sections, the committee provides a summary of ways in which producers have responded to common production challenges, addressed socioeconomic goals and concerns, and interacted with markets, policies, and knowledge institutions. Information from the case-study farms illustrate some of the important principles for improving sustainability of farming systems discussed in Chapter 2.

PRODUCTION CHALLENGES

Soil Management

All the new case-study farms recognize the importance of protecting soil resources and use either no-till or conservation tillage. A few farmers expressed that no-till is not practical for them because of concerns about weed pressure, particularly in organic production systems. Strategies used to reduce soil erosion and compaction include no-till or conservation tillage, winter cover crops, application of poultry litter and dairy manure to hillsides to reduce compaction, and leaving crop residue in the field. One of the dairy farms developed surfaced cow lanes, in which surfacing consists of a base of large rocks and a surface of fine rocks. The cow lanes allow the animals to access the paddock without inducing erosion.

Most of the farms interviewed use soil tests, and some even conduct plant tissue tests, to guide their soil fertility management. They all recognized the importance of recycling nutrients or using internal inputs to the extent possible to minimize costs and losses to the environment. Other than Stahlbush and Zenner Farms, most of the new case-study farms (even the grain and specialty crop farms) keep livestock. Manure or poultry litter are often used to manage soil quality and fertility. Although Stahlbush Island Farms does not have animals on site, it purchases manure and litter from a nearby layer operation. Cover crops are also widely used among the nine farms to maintain soil organic matter and nutrients. Cover crops and livestock manure, however, are not always sufficient to provide the nutrients required by crops. As a result, some case-study farms use purchased fertilizers to complement internal inputs. Three farms (Bragger, a dairy farm, and grain farms Goldmine and Rosmann) have nutrient management plans. Goldmine Farm's nutrient management plan is mostly for erosion control.

In addition to ensuring nutrient adequacy of crops and pasture, the dairy farms have to manage nutrient output from their operations. Bragger Farm tests the manure quality of its animals periodically and adjusts their livestock rations to reduce phosphorus excretion. Radiance Dairy (75 milking cows) and Straus Family Creamery (215 milking cows) emphasized that they maintain the ratio of about 1 cow/acre or less to ensure that they have sufficient land and crops to utilize nutrients generated by their livestock.

Straus Family Creamery separates the solids from liquids in their dairy manure, then applies the solids to fields and uses the liquids in an anaerobic digester. Although Bragger Farm contemplated getting a methane digester, Joe Bragger, who has 285 milking cows, felt his operation is not large enough to benefit from it. The use of anaerobic digestion is not limited to dairy operations. Stahlbush Island Farms recently completed an onsite anaerobic digestion and biogas recovery plant that uses the farms' vegetable byproducts. The size of the Stahlbush Island Farms (4,000 acres farmed) and the vertical integration of production and processing make anaerobic digestion and biogas recovery particularly feasible for this specialty crop farm.

- Alternative medicines, animal husbandry, and disease prevention practices appropriate for organic dairies.
- New precision-agriculture technologies, including sensor technology to distinguish between weeds and crops in cultivation equipment.
- Development of small, scale-appropriate equipment for vegetable enterprises.
- Studies on the implementation, impacts, and compliance problems associated with the USDA organic standard.

Although some of these topics might best be addressed by formal research conducted by public sector or private sector scientists, two farmers expressed a specific desire to see increases in on-farm research (echoing the conclusions of the committee discussed in Chapter 5).

GOVERNMENT PROGRAMS AND POLICIES

In general, federal farm commodity and conservation programs do not appear to be important drivers of (or obstacles to) the improving sustainability of the farming systems in the nine case-study farms. Among the three dairy farms, Radiance and Straus did not identify any government programs that provide important resources or opportunities for their farms. Albert Straus, however, mentioned that California requires farms with animals on pasture and dairy farms to have a nutrient management plan. The Bragger Farm also has a state-mandated nutrient management plan and participates in federal and state conservation programs and federal commodity crop programs. Although two of the grain farms, Rosmann and Zenner, participate in government programs, Russ Zenner believes that farm programs are flawed in terms of ensuring vitality of rural communities and maintaining or enhancing natural resources. Jack Erisman of Goldmine Farm avoids federal farm programs or crop insurance because his farm is economically viable without those programs and because of his own negative experience with crop insurance. None of the specialty crop farms interviewed reported participation in government farm programs. Although Full Belly Farm has developed several conservation initiatives, none occurred with government program support. Alex and Betsy Hitts said Peregrine Farm was too different from most farms in the United States to participate in government commodity or conservation programs. Bill Chambers cited bureaucracy, administrative costs, and reduced flexibility as the reasons for Stahlbush Farm's lack of participation in government programs.

LESSONS LEARNED

The new case-study farms use various combinations of farming practices and approaches discussed in Chapter 3 and marketing strategies discussed in Chapter 4 (Table 7-3). All the farms are reported to be economically viable, and most are optimistic about their future prospects. Each farm illustrates important characteristics of adaptability, robustness, and resilience—key components of sustainable farming systems discussed in Chapter 1.

Adaptability is evident in the fact that all the farmers, irrespective of the size of the farm they own, either conduct trials or experiments on their own farms or participate in experiments run by universities or other entities because they recognize the importance of adapting their farming approaches to local conditions. Although most farmers have identified the farming approach that works well for their system, they continue to test new approaches and cultivars to improve efficiencies, reduce costs, and enhance environmental quality. A key component of each farm's overall strategy is a conscious effort to adjust their

TABLE 7-3 Case-Study Farms' Practices or System Types That Move Agriculture Toward Sustainability Objectives (as discussed in Chapters 3 and 4)

General Indicator	Practice or Approach	Specifically Discussed in Cases
Soil management	Conservation tillage	Bragger Radiance Straus
	Cover cropping	Full Belly Peregrine Stahlbush Rosmann Zenner
Crop and vegetation management	Crop rotations	Bragger Radiance Full Belly Peregrine Stahlbush Goldmine Rosmann Zenner
Water use management	Gravity systems	Radiance
	Sprinkler irrigation	Full Belly
	Trickle or drip irrigation	Full Belly Peregrine Stahlbush
	Water reuse	Straus Stahlbush
	Small dams or ponds	Bragger Radiance Straus Peregrine
Water quality management	Buffers	Straus Rosmann Zenner
Nutrient management	Soil and tissue sufficiency tests	Bragger Radiance Straus Peregrine Stahlbush Goldmine Rosmann Zenner
	Nutrient management plans	Bragger Goldmine Rosmann
	Animal manure	Bragger Radiance Rosmann Straus
	Dietary modification to adjust manure composition	Bragger

TABLE 7-3 Continued

General Indicator	Practice or Approach	Specifically Discussed in Cases
	Compost	Radiance Straus Full Belly Stahlbush Goldmine Rosmann
	Precision agriculture	Stahlbush
	Anaerobic digestion	Straus
Weeds, pests, and disease management	Biocontrol	Stahlbush Goldmine
	Monitoring pests and use of threshold	Full Belly Stahlbush Zenner
Animal production management	Breeding	Radiance Goldmine
Business and marketing diversification	Value-trait marketing	Radiance Straus Full Belly Stahlbush Goldmine Rosmann Zenner
	Direct marketing	Radiance Straus Full Belly Peregrine
	Agritourism	Full Belly
Labor management	Best labor management practices	Radiance Straus Full Belly Stahlbush Goldmine Zenner
Systems type	Organic crop (certified)	Full Belly Stahlbush Goldmine Rosmann
	Low-confinement hog system	Rosmann
	Management-intensive rotational grazing	Radiance
	Integrated crop and livestock system	Goldmine

crop or livestock mix, process farm commodities into higher value products, or develop new products in response to emerging market demands. They strengthen their resilience by marketing through channels that reward farmers for using environmentally and socially responsible production practices. Direct sales are used in some cases to control price volatility and to eliminate loss of profits to processors, distributors, or other middlemen. Overall, the ability of the case-study farms to adapt their farming systems to changing climatic, market, and development conditions enhances their resilience and long-run viability.

The farms achieve robustness in different ways, including diversifying their mix of crops and livestock, selling to a variety of market outlets, and minimizing inputs from external sources. Diversification of the enterprise contributes to robustness because if one crop fails, other crops or product sales can support the farming enterprise. Minimizing external inputs insulates the farms from fluctuating costs and, hence, contributes to robustness. The farms also illustrate that they could use some of the best labor management practices discussed in Chapter 4 without compromising economic viability.

Although only a few case-study farmers described their farms as a "system," their approaches reflect key elements of systems thinking. The following descriptions of each farm highlights some of the "systems" elements of case-study farmers use in the overall management of their farms.

- Bragger Farm—Although its least profitable enterprise, the beef cattle play an important role as scavengers and help recycle unused feed and forages not suitable for dairy cattle consumption. The beef cattle are grazed on upland slopes that are not suitable for cropping. The poultry enterprise complements the beef and dairy enterprises by providing a diversified income, an important source of crop nutrients, and opportunities for the efficient use of available labor. In other words, the beef cattle and poultry enterprises contribute to maximizing the use of feed and forage and land on Bragger Farm.
- Radiance Dairy—Francis Thicke selects cows within his own herds to raise bulls for breeding. He selects for cows that provide a calf every year, maintain body condition and udder health, produce moderate amount of milk on pasture, and have a long productive life. Instead of focusing on short-term milk production, he balances long-term milk production with animal health.
- Straus Family Creamery—Albert Straus emphasized that "closing the loop" on resources is a driving concern for the Straus Family Creamery. The operation has an anaerobic digester with a biogas recovery system that generates 90 percent of the electricity used by the dairy and 50 percent of its propane needs. Manure solids are composted and reused on farmland.
- Full Belly Farm—A clover understory is typically planted in the orchards because it aids water infiltration, sequesters carbon and nitrogen, and poses little competition for water with crops. Paul Muller believes that a diversified organic system increases carbon sequestration and soil fertility via cover crop management, increases water infiltration by building soil organic matter, and enhances beneficial insect and soil microbiota.
- Peregrine Farm—Alex and Betsy Hitts use intensive rotations and cover cropping to manage their system. Their farm consists of two acres of vegetable crops (80 varieties) and blueberries, and two acres of cut flowers (more than 50 varieties). The rotations and cover cropping are key strategies for weeds, insect, and disease control, but they also contribute to soil quality and fertility and the farm's financial stability.

ILLUSTRATIVE CASE STUDIES

- Stahlbush Island Farms—Bill Chamber emphasized five concerns about pesticides: 1) pesticides often lower the yield of crops receiving applications, 2) pesticides can drift from where they are applied, 3) pesticides have negative effects on beneficial soil microorganisms, 4) pesticide application poses health risks, and 5) consumers are worried about potential pesticide residues in foods. These concerns illustrate the potential adverse effects of pesticide use on the farming system. His approach to pesticides is to balance productivity with profitability and long-term sustainability.
- Goldmine Farm—Integration of livestock into the cropping system contributes to the success of Goldmine Farm because it adds to and diversifies the farm's income streams. Keeping land in pasture helps to restore soil health and contributes to weed control when that land is planted with cash crops.
- Rosmann Farm—The diversified crop–livestock system is premised on internal cycling of nutrients and reduced purchased inputs. Six-year crop rotations include corn–beans–corn–oats or barley or succotash–alfalfa–alfalfa. The two years of corn in the rotation contribute to economic viability and the oats and barley are used to feed animals. The alfalfa in the rotation and animal manure provide nutrients for the cropland.
- Zenner Farm—Russ Zenner's approach is to improve soil microbiology by reducing use of chemical inputs, and thereby enhance crop health, quality, and yield. He direct-seeds his entire farm and has adapted and designed his planting and spraying equipment to better suit his operation.

Bragger Farm

BACKGROUND AND HISTORY

The Bragger Farm is located near the town of Independence at the head of a long valley in the Driftless Region of west central Wisconsin. The area is characterized by hilly terrain and narrow valleys, most of which contain streams. It receives 30–35 inches of rain yearly. Topography and rainfall make careful soil and nutrient management important for the sustainability of farms in this region, most of which have focused on dairy.

The Bragger farm was founded by Joe Bragger's father and mother, who each came from Switzerland in the 1960s and met in the United States. With little money in the early years, Joe's father started working on dairy farms near Waumandee, rented a farm, and then bought the current farm in 1968. The farm started with 400 acres and 40 dairy cows. Pasture predominated. There was a stanchion barn and a dilapidated shed, and a ram pump provided the water. Joe recalls from his childhood on the farm that "we'd march the cows out to pasture and we'd march the cows in." Joe's father slowly built up the herd from the original 40 head by buying cows at auctions.

Initially, Joe had no intention of working on the farm, having been trained as a diesel mechanic. He farmed for one year with his father, but "Dad and I didn't see eye to eye." In 1990, his father died in a tractor accident while spreading manure, and Joe assumed a major role on the farm. There were 63 cows and 660 acres of pasture, woods, and cropland at that time.

The farm is currently owned and operated jointly by Joe, his mother, Hildegard, and his brother, Dan. Hildegard lives in the main house on the central farm, Dan lives in a new modular home on a nearby hillside, and Joe and his family live on another farm parcel about a five-minute drive away. Joe refers to the organization of the farm as "a loose affiliation of tribes," one that works for the family members participating in the operation. Joe takes overall management responsibility for the farming operation, and he focuses on cropping and the farm machinery. Hildegard raised all the calves until 2007. Dan oversees the dairy, which now milks 285 cows on two farms. Noel, Joe's wife, is in charge of the contract pullet operation and the beef operation.

FARM PRODUCTION SYSTEM

Land

The Braggers currently own 884 acres of land. They supplement their holdings through an involved system of land rental arrangements with nearby property owners, many of whom live a long way from the operation. They rent approximately 400 acres from eight or nine different land owners that, with the land Braggers own, fit together like a parcel of mostly contiguous pieces. Many of the rented parcels were previously owned by farmers, but are now owned by hunters and sportsmen, who often live in Kenosha, Janesville, or Madison. Joe rents and farms those lands because he needs ground to spread his manure. The rental agreements include various prohibitions on practices, including specifications on when manure can be spread. However, he also rents some of his land to hunters, which provides income to pay off the newest piece of land he has purchased.

The upper parts of the hills are open and have a nice topography, but they support less production, and the risk of failure is higher on these somewhat shallower, rockier soils, some of which slope as much as 16–32 percent. Joe does not remove corn stover for bedding on the ridges because he prefers to leave more residue on the ground. In addition, he spreads poultry litter rather than dairy manure on the hillsides to reduce spring compaction by the truck traffic, which creates field roads that will later erode into ditches and gullies. The Braggers try to make use of all their land by keeping pastures or making hay on the edges of the woods at the tops of hills. Rents were in the $20–$25/acre range when Joe's father started farming. Now they are closer to $70/acre, and land in the Arcadia area to the south is in the $150/acre range.

Soils and Fertility

The soils on Bragger Farm include heavy clay soils on the ridge tops that require careful management to harvest as much rainfall as possible. Other areas of the farm have loess soils that range from 8–20 feet of topsoil. Although the soils on the ridges are mapped as only 40 inches deep, it is likely that the rocky soils impede probes from measuring the true depth of the ridge soils that are actually 8–13 feet deep.

Compaction is a major concern. Deep tillage is not an option because of all the rocks in the soil, so compaction has to be minimized by staying off the fields during high soil moisture conditions. In addition to watching when they work on their fields, the Braggers minimize vehicle traffic and the impact of equipment by maintaining tire pressures and using large-profile tires.

Joe takes soil samples each year on about 25 percent of the land to fit the four-year rotation on his hay fields. His soils often call for an application of potassium and phosphorus, but he uses phosphorus judiciously when needed. He is not convinced he needs to conduct soil testing at the level recommended for his farm (samples on a 5-acre basis) because of the relatively small-sized fields. For example, he does not split an 8- or 10-acre field into two and send two separate samples for each portion of the field. His opinion is that because he has managed these fields as one and has maintained the same rotation and management practices on them for more than 15 years, there is little or no value in dividing them into 5-acre grids. The results delivered tend to be similar over time and do not appreciably change his management strategies.

He has begun planting corn for a second consecutive year on some of his gentler slopes. If he plants a second year of corn on a piece of land, he will add 8,000–9,000 gallons of dairy manure per acre to achieve the desired levels of potassium. Joe manages his manure applications to the amount of phosphorous needed and balances out the nutrient requirements with commercial fertilizer. He applies manure in strips on his slopes, basically on the contour. He also foliar-feeds potassium on his alfalfa four times a year to produce "excellent" alfalfa. Poultry litter, cleaned out from the pullet house, is applied where needed to his pastures and new seedings.

Corn typically needs 55-60#/A of phosphorus and 180# of nitrogen. Based on an analysis of manure at 3-3-8, he applies roughly 20 tons of manure. The actual nitrogen requirements depend on soil residue levels to account for the no-till farming system. For corn following alfalfa, Joe credits 120 units of nitrogen from the alfalfa, and the remaining nitrogen requirements come from the 20 tons of manure application. The second year of corn following alfalfa receives a 50-lb nitrogen credit, and manure is applied at a rate of 6,000–8,000 gal/acre, which provides another 50 units. Some nitrogen is supplied in the starter fertilizer (10–20 units), and the remaining nitrogen needs are met through a commercial fertilizer application.

The liquid dairy manure application (6,000–8,000 gal/acre) provides a form of slow release nitrogen that breaks down and is available to the crop throughout the growing season, much like the nitrogen supplied through decaying alfalfa. It is important to note that in the first year, the Braggers are putting on a little extra nitrogen supplied by the manure in an attempt to build up soil potassium levels for the following alfalfa crop. The corn planted after soybean receives 50 units of nitrogen from manure, a 50-unit nitrogen credit from the soybean crop, 10–20 units in the starter, and around 80 units from commercial fertilizer. Fields planted to continuous corn receive about 50–80 units of nitrogen from manure and the remaining needs through commercial fertilizer. This operation does not try to meet the total nitrogen requirements on continuous corn through manure applications because doing so could increase the soil test phosphorus above acceptable levels.

Crops

The farm currently has 100 acres of pasture, which includes 68 acres of grassland that is hayed; 390 acres of corn, of which 150 acres are harvested as silage and the remaining 240 acres are harvested as grain for the dairy cattle; 264 acres of alfalfa; 2.5 acres in the Conservation Reserve Program; 120 acres of soybean; 21 acres of winter wheat; 10 acres fallow; 4.5 acres in a food plot for deer; and 1 acre of sunflowers for entertainment, public service, and a small experiment. The Braggers own 880 total acres, of which about 500 acres are tillable. The remaining 400 acres of cropland is rented.

The base rotation is a five to six year rotation of alfalfa and grass (fescue) or hay and then corn. Some fields are planted to corn, soybean, and occasionally wheat and do not have alfalfa or grass in the rotation. The land in the valley is mostly in the forage-based rotation (seeding, followed by two to three years of alfalfa–hay mix, followed by one to two years of corn harvested as either grain or silage). Occasionally a field or two will be planted with a barley cover crop that is harvested as baleage and fed to the beef cow herd. The fields on the ridge tops are in a rotation of corn and soybean or in a corn on corn followed by beans. A field might be planted to alfalfa following soybean, but this is limited because the stony soil damages the harvesting equipment.

According to Joe, corn has proven "a pretty economic crop." Because the on-farm nutrient sources improve soil fertility, yields have gone from 80 bushels up to 160 bushels on some fields. The farm is really operated under different management systems using some common resources such as manure and equipment. While Joe sells some of his corn and beans when he has excess, many area farmers sell cash grains. Joe prefers to plant corn, soybean, and wheat on the tops of the hills as cash crops. He puts forages on his bottom land.

Farming Practices

When planting corn into the previous year's hay or alfalfa, Joe will spring-kill the existing cover and no-till (or actually strip-till) the corn into high residue. Soybean residue is harvested and used as bedding or feed for dry cows because soybean residue is less effective at preventing erosion than corn stover. Despite the value of the residue, high levels of residue can lower soil temperature, and that factor has to be considered at planting. Joe initially went to no-till because of time considerations, but as he saw the advantages for his soils, he was won over. He would like to see crop insurance reward good practices that aim toward improved sustainability on the farm. For example, he suggests that insurance premiums could be reduced for those who use such practices as no-till or strip-till.

Pest Management

Leafhoppers are a big problem in alfalfa—Joe might spray chlorpyrifos for control up to three times a year with his potassium applications if infestations warrant it. He uses triple stacked corn varieties (Roundup-Ready®, corn borer resistance, rootworm resistance) and looks at Bt as an insurance against corn borer. He speaks of some difficulty in recognizing and responding appropriately to economic thresholds with insect pests: "I quit scouting. Scouting has always led to spraying for leaf hoppers." He now sprays for leafhoppers routinely one week after the first two cuttings and then evaluates for the need to spray after the third cutting. Soybean are routinely scouted not so much to determine if spraying for aphids is necessary, but to determine when to spray to get the most for the investment.

Weed Management

The key weed problems on the farm are lamb's quarter, nut sedge in the lower areas, and, increasingly, hemp dogbane on the ridge tops. Joe uses Roundup before no-till planting and an additional Roundup application after planting along with a pre-emergence herbicide. Price considerations strongly influence decisions about spraying on the farm. Joe tries to use the most cost-effective chemical program possible, and that means the chemical program used might change as prices change. Overall, however, he says, "I won't skimp on Roundup. Eventually that one weed that gets by will become 1,000." However, Joe does not rely exclusively on Roundup; he also uses other products.

Equipment and Buildings

Joe is now in the midst of constructing a large new machine shed on the central farm to replace a slightly smaller 40-year-old building that recently went down under last winter's snow load. As much as possible, he is incorporating energy-conserving features (including good insulation and windows to provide high levels of natural light) into the shed. A wood boiler will eventually heat the new machine shed using trees and brush that are removed from the edges of his fields.

Rather than owning all the equipment needed on the farm, Joe leases a tractor from a nearby grain farmer for chopping corn and alfalfa and another tractor for packing silages and snow removal as a way of keeping costs down. Their needs for the equipment are at opposite ends of the calendar, and the system works well for both. "There is very little machinery that we have that we use just twice a year."

Because of his background in diesel mechanics, Joe cares a lot about the condition and performance of farm machinery. Although he plants a lot of field trials and different crops and covers, if a particular cropping practice wrecks a machine, it might affect his decision to use it again.

Livestock Enterprises

Dairy

The dairy herd includes 180 cows on the Bragger Farm and 70 cows managed for the Braggers on a nearby farm. The latter arrangement, in its fifth year as of 2008, has enabled the Braggers to expand their dairy enterprise. The arrangement has also provided a stream of income for their neighbor, who uses an existing barn at her farm and appreciates being able to work from home, while caring for a special needs child. The Braggers own the cows and provide the feed and move the manure, and the neighbor milks the cows. Joe has modeled this arrangement on his own largely satisfactory experience as growers for Gold'n Plump. It generally works well, although the Braggers sometimes have to stretch to accommodate special requests by their neighbor to "take time off." Joe suspects that those cows might "some day come home."

The Bragger herd is primarily Holsteins, with a couple of Brown Swiss and a few Finnish Red–Holsteins and Brown Swiss–Holstein crosses. The Braggers raise their own breeding stock, and bulls are sold because of the current cost of grain for raising them. Prior to the high feed prices (March of 2008), the Braggers raised about 100 head of feeder and finished cattle. They keep only a few heads now to sell as "Bragger Beef" out of their freezer on the farm.

The milking cows are bedded with sawdust, which is a substantial expense ($450/load which lasts only a week). The sawdust acts as a replacement for the corn stover, which was previously harvested from fields. Leaving the corn stover in the fields helps reduce erosion and improves the soil health and structure. Fresh cows are segregated by the days they are in milk. In segregating them, the Braggers are aware of the "social" networks among cows and the stresses of moving them around and from one group to another. The Braggers are careful to make sure that there is ample light, ventilation, and sawdust bedding to keep the cows dry.

The cows are milked twice a day. The feed is alfalfa and corn silage, supplemented with high moisture corn, soybean meal, and a protein–mineral mix concentrate that the Braggers purchase. The Braggers have built storage areas for corn silage, protein supplements, and haylage. They continue to use their two upright silos for high moisture shelled corn. That saves them substantial corn-drying costs while providing palatable and high-energy feed sources for the cows.

Joe is paid based on the butter fat, protein, and quality of the milk, and he believes that healthy cows produce more protein, have higher butterfat content, and a lower somatic cell count. Much of their milk goes to a large butter maker in the area. The Braggers have considered the possibility of a value-added option such as cheese, but it would depend on the interest of the next generation and on market necessity.

Recent energy efficiency improvements to the dairy have included a plate cooler, which paid for itself in three years because of energy savings, a variable speed vacuum pump, insulation to the new shed, and wood heat for the machine shed.

One issue of concern for the dairy is that of stray voltage, which comes from an older inadequate distribution system. The voltage stresses the dairy cows and has an effect on milk production and can reduce production by 10–20 lbs/cow per day. It also greatly affects longevity of the dairy cows. The problem is particularly common in rural systems. The stray voltage problem on the main farm is one reason the Braggers do not use rBST on the herd at the main farm, although they use it on the herd tended by their neighbor.

Beef Cattle

One hundred beef cattle are pastured on the farm. The herd includes Limousin, Short Horn, and Holstein–Hereford crosses. The beef operation is the least profitable part of the farm, and Joe said he would drop it, particularly if they could put up another poultry barn. The priority of the beef cattle is evident in the "green cow" zone of the dairy barn. Joe's "green cows" are steers that get leftover feed from the dairy cows. The point here is to minimize inputs purchased directly for them.

The beef cattle are grazed on the upland slopes that border the woods. Those areas have not been cropped because of the slope and terrain restrictions. The cattle are over-wintered on a flat corn field on the home farm to minimize damage to slopes during the spring thaw. In this operation, the beef cows act as scavengers and recycle the unused feed and forages that are not suitable for dairy cattle consumption.

Heifers

The heifer calves are raised in hutches until they are about 2 months of age, at which time they are vaccinated and moved to a group pen. Heifers are raised in groups of about 10 heads per pen until they are about 4 months of age. They are, then, dehorned, revaccinated and moved off the milking operation to another farm. Joe and Noel live on an adjacent farm one to two miles from the milking operation. The heifers older than 4 months of age are raised on this farm until they reach breeding age, which is about 12–14 months. When they are ready to be bred, the heifers are moved to another neighboring farm where they are artificially bred. The heifers remain on this operation until they are within two months of calving, at which time they return to the dairy and are placed in a pen with the dry cows. The Bragger Dairy Farm raises all its heifer calves as replacements for its dairy operation.

Pullets

The poultry enterprise began at the instigation of Joe's wife, Noel, in 1995. She quit her teaching job, seeing contract poultry as something that would allow more time at home to raise their children. The farm produces 64,000 pullets annually under contract with Gold'n Plump, a poultry company based in the upper Midwest, which has 140–150 broiler barns in the area and a few pullet barns. The birds are raised from chicks brought in by the processor, which also provides the feed. The pullets spend 21 weeks in the two-story barn, and then go out to four large laying barns in the area. After paying off the initial costs of the barn (about $500,000 10 years ago), the Braggers have only insurance, taxes, and minor upkeep as current expenses. Joe estimates that Noel's work time in the pullet barn is approximately one hour a day, rarely more than two. The relatively low maintenance and profitability of the pullet operation has made it a good fit with farm needs and family responsibilities. The manure from this operation also has a high value that contributes greatly to the success of the operation. This manure replaces approximately $8,000–$12,000 of commercial fertilizer purchases and provides additional soil benefits from the micronutrients and other soil amendments.

Labor

The farm uses six full-time employees. Two of these men are in the H2A program, and they run equipment. The rest are in the United States on (one-year) J1 visas. They milk the cows and provide other herd care. Virtually no local people are available and willing to work on the farm, because the area has a competitive labor market with a major furniture manufacturer (Ashley Furniture) and a poultry industry (Gold'n Plump). Joe said "foreign workers are a nightmare" from a bureaucratic standpoint. However, he has had good experiences with many of the workers he has hired. Reliability has been valuable: "I can set my clock by Mario." He notes the quality of the workers on the farm is key to the farm's success. He expresses some frustration with regulations that can impede farmers' ability to support their workers. For example, Wisconsin state rules require workers' eating and sleeping quarters be 500 feet from where livestock are located (federal rules specify 250 feet). Hildegard's house is less than 500 feet from the cows, and its basement could house workers if allowed. Some of Joe's workers do not live on the farm; in fact the H2A workers must be housed off of the farm and have to arrange their own accommodations. The J1 visa holders are allowed to live in the main farm house with Joe's mother.

Manure and Nutrient Management Issues

A fundamental concern on the Bragger Farm is to balance phosphorus produced by livestock with the needs and availability of land on the farm. The Braggers believe that the current operation could support a maximum of about 300 cows. In 1999, Joe completed a nutrient management plan for the farm, covering 108 fields that averaged 8.3 acres. Joe looks at how to apply nutrients at certain phases of the rotation that will work well for long-term needs of the rotation.

Solid pen pack manure is always applied to land going into corn. If that manure is applied to alfalfa, the harvesting equipment would gather it up and contaminate the forage crop. Poultry litter is used on pasture and other areas that are difficult to access because of steeper slopes or long travel distances and on land that will eventually be seeded to alfalfa.

The Braggers evaluate manure quality through a frequent testing program. They are pleased to see that phosphorus levels in the manure have been decreasing because of the reductions in dietary phosphorus. The decrease in manure phosphorus has increased the level of nitrogen supplied to corn when manure is applied on the basis of phosphorus needs of the crop. The farm also tests the milk urea nitrogen levels to determine whether excess protein is being fed or if the cows would benefit from an increase in dietary protein.

Manure from the dairy barn flows by gravity into a manure pit; water from the dairy milk house is spread into a drain field that is checked and relocated regularly. The manure pit was constructed with cost share and technical assistance from the local Land Conservation Department. (In Wisconsin, each county has a Land Conservation Department funded through state funds from the Wisconsin Department of Agriculture and the Department of Natural Resources.) The liquid dairy manure is analyzed for nutrients each year, and all of it is applied on the farm's land. The manure application is contracted out twice a year. As noted previously, poultry litter from the pullet operation has been a valuable amendment for soils on the farm.

Other Land Enterprises

Joe has a deep passion for his woodlands. He cuts firewood and sells timber, such as white oak, red oak, and hickory, and builds hunting and horse riding trails for hunters and others to enjoy.

Radiance Dairy

BACKGROUND AND HISTORY OF THE FARM

Francis and Susan Thicke own and operate their organic, grass-based dairy just outside of Fairfield, a town of 10,000 in southeast Iowa. Their current herd includes about 80 milking cows, all unregistered Jerseys. They run an on-farm processing plant, where they produce cream-line milk and other dairy products, sold in the Fairfield area. The distinct marketing arrangement is a "feasible proposition" according to Francis, because of Fairfield's unusual market demographics. In the early 1970s, Fairfield became a magnet for a growing community of people associated with the practice of Transcendental Meditation. The "meditator population" has, from the beginning, been strongly committed to natural and organic foods, creating an excellent niche market for a dairy such as Radiance.

The dairy operation began in 1980, first at another location in the Fairfield area. Starting as a small cooperative effort among friends, the business milked two cows and sold raw milk at that time. Demand soon grew, and more cows were added each year. In 1987, the state determined that the dairy violated Iowa's raw-milk regulations, and the dairy was required to begin pasteurization. In 1992, the Thickes moved to Fairfield from Washington, D.C., and purchased the dairy, which had 22 cows at that time. In 1996, they moved the dairy to its present location, which had been a small, conventional farm. In moving their operation, the Thickes specifically sought to purchase less costly marginal land. It was rolling but well-suited for grazing and close to town on a paved road. Developing the dairy farm took some time. For four years, they lived upstairs in the barn, while also slowly constructing their present home.

Francis' movement into organic dairy farming follows from his family history of farming, academic training, and work for some years in Washington as a national program leader in the USDA Cooperative State Research, Education, and Extension Service. He grew up on a conventional dairy farm in Minnesota, milking Ayrshires. The farm converted to organic in 1976, a transition encouraged by Francis and his brothers, and assented to, at first somewhat skeptically, by their father. Francis' brother, Art, continues to farm there today and is widely recognized for his approach to grazing. Art has provided critical mentoring and advice as Francis developed his own grass-based dairy farm. After completing a B.A. degree in liberal arts, Francis completed an M.S. in soil science at the University of Minnesota and a Ph.D. in agronomy at the University of Illinois. After graduate school, Francis worked for USDA in a position that allowed him to travel around the United States, see diverse agricultural circumstances, and play a role in designing and implementing agricultural programs. He liked his work, but when the opportunity to begin dairy farming in Fairfield arose, he decided to return to a farm. At the time, Francis was concerned he might not remember how to do day-to-day farming. But, as Francis observed, "Farming is like riding a bicycle;" once you know how to do it, you don't forget.

FARM PRODUCTION SYSTEM

The Thickes' approach to farming centers foremost on thinking of the farm as an organism. They follow the evolving science of organic and sustainable farming, but not at the expense of their intuition. (Francis strongly advocates listening to one's "inner agronomist.") He sees a need to turn around recent assumptions about farming, including that "in Iowa, we tend to think the longer we farm, the more we'll wreck the environment. Instead we need to farm in ways that enhance the environment, rather than degrading it." His philosophical take on farming emphasizes "farming in the moment" and doing what needs to be done. His approach, he suggests, enables one to have more fun and lets important things rise to the surface.

The Thickes own 236 acres of land. They also rent some 150 acres at other farms in the area. As the Radiance herd has grown, they have needed more hay, which lines up well with local farmland owners, who will rent out land if farmed organically. In some cases, people seeking to rent out their farmland have approached Francis. The resulting rental arrangements are verbal and informal.

Pasture Management

With its emphasis on rotational grazing, balancing herd size to available pasture is central for Radiance Dairy. Francis aims for a maximum of one cow/acre, but would prefer a bit less, even one to two cows/acre to ensure an adequate buffer on the land. He discussed excess or buffer capacity—meaning more pasture forage than absolutely necessary in good weather conditions—in several contexts from the standpoint that poor or extreme weather could leave the farm without high-quality pasture for the cows. Excess pasture under good weather and growing conditions is harvested as hay.

Francis' goal is to make the farmland erosion proof. Healthy pasture helps prevent soil erosion and leaching of nutrients. The farm has 200 acres of pasture, divided into 60 paddocks of roughly two acres. The paddocks are divided into halves or thirds with portable fencing materials for grazing on any particular day. They have rented another 150 certified-organic acres on which they raise hay, small grains, and soybean.

Francis laid out the paddocks based on his scientific agronomy background and an intuitive sense of how best to configure the pastures given their varying characteristics. The pastures were established in 1997. Francis uses a no-till drill to interseed the pastures. The farm soils are a silty clay loam that has a geological clay hard pan about a foot down. When the farm was purchased in 1996, it had been in corn and soybean production for many years. Francis planted the whole farm to a variety of grasses and clovers. Although he conducts soil tests on the farm about every five years, he does not consider the tests the best or only way to monitor improvement of the soils in his pastures. He sees building of soils as the primary and critical task, which then leads to good forage. He applied three tons of lime per acre in 2006 to correct soil pH. He used calcitic lime because the soil has more than enough magnesium. Over time, Francis has seen an increase in the number of earthworms, which he takes as a clear indicator of increased soil health.

The Thickes do not have many empty spots in their pastures, although they do have Canadian thistle patches in a few places. They have found in the past two years that the patches have diminished in size and density, which they attribute to the increasing density of pasture forages interseeded with a no-till drill and to timely mowing of the thistles. They are not, at this point, concerned about importing nutrients onto the farm by importing feed because the farm had a very low nutrient status when they purchased it. They are now building soil fertility up to optimum levels. However, they began hauling composted bedding-pack manure to some of the rented land that is not contiguous to the dairy in 2008.

The cows are outwintered in paddocks selected for increased nutrient application. Hay bales (dry hay and baleage) are fed in a grid pattern in the paddocks to cause the cows to spread their manure at the desired rate of nutrient application. Residual hay from the bales serves as bedding for the cows. When needed, round bales of low-quality hay are unrolled in the wintering paddocks for additional bedding. When winter temperatures and wind are excessive, the cows are kept on a bedding pack in a pole barn near the milking facility. Paddocks selected for out-wintering are rotated each year. The following spring, residual bunches of hay, along with occluded manure, are pushed into piles, which are turned frequently to facilitate rapid composting.

Priority for outwintering cows is given to paddocks in which volunteer tall fescue (entophyte infected) has begun encroaching. In spring, after composting is completed on the outwintered paddocks, the paddocks are rotovated to a depth of 2.5 inches to kill the fescue and then planted to brown midrib sorghum–sudangrass, or some other annual forage crop. The following year, those paddocks are planted to a diverse mix of perennial grasses and legumes. Planting a summer annual like brown midrib sorghum–sudangrass for one year before reseeding perennial forages provides several advantages. First, it allows fescue seeds in the soil to germinate and then be killed the following year, and thus, depletes the fescue seed bank. Second, a summer annual will grow aggressively during midsummer and provide grazing forage during the time when the growth of cool-season grasses slows. Third, a summer annual requires warmer soil temperature for planting, so it is planted later in the season, which allows time to complete composting of residual material from outwintering. During the composting period, surviving cool-season forages in the paddocks are grazed at least once.

The manure and bedding from the barns and milking parlor are also composted. The perennial forages planted in paddocks include bromegrass, orchardgrass, timothy, red clover, white clover, Kura clover, alfalfa, and chicory.

In the spring and early summer, the pasture grows more rapidly than the cows can consume it, so some pasture areas are cut and baled for winter feeding. By midsummer the pasture growth slows and the areas that had been cut earlier for hay are brought back into the pasture rotation. During the rapid growth phase in spring and early summer, the rest period before a paddock is regrazed is 20–30 days. As the summer gets hotter and drier, forage regrowth slows, so rest periods are extended to 40 days or more, depending upon rate of forage regrowth.

The rented land is planted to a mix of alfalfa and grass. After four to five years, when the alfalfa stand diminishes, the land is rotated to soybean, then to small grains, then back to hay.

To reduce mud and erosion, Francis has surfaced cow lanes (with EQIP cost-share support) that provide access to the paddocks. Lane surfacing consists of a base of larger rock (1.5 inches in diameter) with fine rock on the surface. Francis is now contemplating planting trees along the pasture lanes to provide shade for the cows as they walk to and from the paddocks.

Livestock

The farm now has about 80 milking cows, with about 150 in total, including the dry cows and heifers of all ages. After each milking, the cows are moved onto new grass. The grazing animals are sorted into three groups: lactating cows, dry cows and bred heifers, and yearling heifers. The lactating cows are rotated through the paddock system first, so they can glean maximum nutrition from the forages. The dry cows and bred heifers come through next. The yearling heifers rotate through a separate group of paddocks.

The cows are also fed about 5 lbs/day of barley and wheat during the summer and about 10 lbs/day during the winter, although the Thickes would like to move away from feeding grain. They grow some of the needed grain on their rented land and buy the balance from other organic farmers. On average, the Thickes' lactating cows produce 32 lbs of milk per day. Some of their cows continue to be good milkers for 11, 12, or even 14 years.

From the beginning, Radiance Dairy has milked Jerseys, which produce milk high in protein, fat, and total solids. Cows generally produce one calf per year. Francis selects cows from within his own herd to raise bulls for breeding. He describes his approach as selecting for cows that are well adapted to his farming system, rather than focusing on milk production. He favors cow characteristics such as producing a calf every year, the ability to maintain body condition while producing a moderate amount of milk on pasture, maintaining a healthy udder, a good disposition, and living a long life. Recently he brought in a bull from a herd in Nebraska that feeds no grain, and through the use of artificial insemination has brought in selected genetics from Jerseys in New Zealand and elsewhere.

Francis manages the herd for two calving windows—one in spring and one in fall—in an effort to avoid calving during the hottest of summer and the coldest of winter. He needs to have milk year-round but aims to have more cows dry in the summer, when many of his customers are away on vacation and less milk is sold.

Francis gets a fair number of calls from people wanting his cull cows to serve as "family cows." When he has excess milk for his market, he sometimes sells cows that are older, or less than ideal for his herd, as a family cow. The roughly $300/cow received when such a cull cow is sold essentially matches what it would bring from selling it for slaughter, and the cow gets a new lease on life and extra coddling in its new home.

The farm also includes a small, noncommercial chicken flock of about 25 birds. The chickens free-range around the farmyard and help with fly control by eating fly larvae.

Herd Health

Animal health, as Francis sees it, is integrally tied to the robust ecological health of the farm. Problems associated with herd health have been diminishing on the farm over the years. Nonetheless, pinkeye has sometimes been a health issue for the Radiance herd, exacerbated by flies. Experimenting with a comprehensive approach to fly control has helped to manage that challenge. In summer, Francis buys weekly shipments of parasitic wasp larvae that he spreads around the farm, particularly in pasture areas where the cows have recently grazed. The tiny wasps hatch and then lay eggs in fly larvae, preventing flies from hatching. To capture flies that hatch in spite of the farmyard chickens and parasitic wasps, fly traps (using attractants) are used around the barnyard. In the milking parlor, he treats the cows with a spray of soybean oil and essential oils on their backs and legs to kill flies. Sticky tapes are strung along the ceiling of the milking parlor to catch flies that leave the cows and fly to the ceiling.

During milking, any cow that shows signs of pinkeye (for example, tearing up) receives a light eyewash spray. The careful monitoring approach, coupled with eye patches for cows that do become afflicted, has kept the problem in check. Nonetheless, Francis sees a need for more research on fly control systems suitable for organic (animal) agriculture.

The calves sometimes get scours, which is treated effectively with herbal boluses. In most cases, sick animals are brought back to good health with natural methods. When antibiotics or other materials prohibited in organic production are used on sick animals, the treated animals are removed from the herd.

Livestock Waste

Cow manure is not a waste at Radiance Dairy. Most of the manure produced on the farm is spread throughout the paddocks by the cows as they graze. Manure from the milking parlor, the barnyard where the cows wait for milking, and calves in a barn is composted near the barn. In the past, the compost has been spread on pasture areas that need additional fertilizer. Future plans are to haul the compost to the rented land off the farm, which is used to grow hay and small grains for the cows. Dead cows are composted on the farm.

On-Farm Dairy Processing

Milking takes place in a New Zealand swing-style milking parlor that Francis designed and built. It has a central operator's pit, with spaces for eight cows on either side of the pit. Francis knew what design he wanted after visiting grass-based dairies in Minnesota. What he now has for Radiance flows well for handling the cows. The dairy processing facility is immediately adjacent to the milking parlor. He had to work creatively with the inspector from the Iowa Department of Agriculture and Land Stewardship to keep milking and processing under the same roof, although in functionally separate spaces. The inspector's supportiveness and flexibility has been critical to implementing some of the innovative features of the dairy. Processing equipment was comparatively affordable, because some items were brought from the other farm to the present farm, and some items were bought secondhand, or adapted to purpose.

Francis sees the farm's movement into milk processing in the late 1980s as a key to its success. At that time, no dairy in Iowa was processing milk on farm, although a few others have since begun to do so. "We were ahead of the curve on that," he says, noting the current enthusiasm for local foods.

Radiance Dairy does not homogenize its milk; thus it has the "cream line." Most of the milk it sells is whole milk, but Francis notes that even Jerseys' skim milk tastes rich compared to conventional milk. Beyond whole, 2 percent, and skim milk, the product line includes whipping cream, yogurt, cheese (Monterey jack and ricotta), and paneer (a South Asian cheese popular with many in the meditator community). The Thickes are extremely responsive to customer feedback and concern. For example, their yogurt included a probiotic at one time. However, some customers became concerned about that, so they took it out. On the demand side, customers have recently asked if Radiance could produce fresh mozzarella, so they are now looking into developing this new product line.

At present, Radiance Dairy milk is sold in plastic jugs. Francis has contemplated and researched the implications of switching to reusable glass bottles, which are appealing on aesthetic, environmental, and possibly health grounds. He thinks many of his customers would like a reusable glass bottle. However, such a switch might not be cost-effective, given the capital costs required to purchase a new bottler and a bottle-washing machine, and to build an addition on the processing plant to house the bottle washer. For now, Francis plans to install a solar hot water heating system for the processing plant and milk house. With solar hot water, the economics of returnable glass bottles will be more favorable, and Francis hopes to make the switch to glass bottles within the next two years.

The Thickes deliver all their processed dairy products to grocery stores and restaurants within 24 hours of production. They have a cold storage in which they keep inventory until delivery and an incubation room for making yogurt. The dairy processing facility appears clean, orderly, and efficient. Francis says he does not obsess about "biosecurity" with the various visitors to the dairy. (For example, visitors are not required to use shoe covers.)

Labor

Radiance Dairy is a family labor farm, with a clear division of labor between Francis and Susan. He leads on farm and herd management, and Susan handles much of the cheese making, marketing, and business management. However, a small cadre of nonfamily workers is critical to the successful operation of the farm and dairy. A young person (in his 20s) has helped full-time with the cows for the last three years. He comes from a local conventional farm. Francis emphasizes the opportunity to help him develop as a grass-based organic dairy farmer. Working with an idea offered by his brother Art, Francis has moved to an arrangement where this worker can select two heifer calves for himself each year, as a way of increasing his stake in the herd. That arrangement would facilitate the possibility of his taking over Radiance in the future, if that becomes a viable option, or starting his own herd. The worker lives in a modular home on the farm.

Radiance Dairy has two additional part-time workers from the nearby area. Both work about three-quarters' time, processing and making deliveries. They have been working at Radiance fairly long term (one since before the Thickes bought the dairy). They are close friends (one recruited the other), and their friendship might contribute to workforce stability, as they enjoy working together. In addition, the daughter of one of the plant workers works a few hours a week assisting with cleanup in the plant. Also, a high school student who lives nearby assists with milking on weekends during the school year and more often during summer vacation. The longest-term processing worker at Radiance makes $17.50/hour.

Farm Equipment

The most frequently used equipment on the farm are two all-terrain vehicles to open and close paddock gates, move cows and fencing, and undertake general chores. The farm also has a full line of hay-making equipment (disc mower-conditioner, rake, round baler, and balage wrapper) and four tractors. The tractors are moderate-sized, and one has four-wheel drive. Francis would like to get a second four-wheel-drive tractor.

NATURAL RESOURCES, ENERGY, AND CLIMATE CHANGE

Water and Air Issues

Pond development (cost-shared through EQIP) has been critical for watering cows in the various pastures. Water quality concerns are generally minimal at Radiance Dairy. Although no swine concentrated animal feeding operations (CAFOs) are in the watershed, Francis is concerned about environmental threats posed by CAFOs in the larger region. He has been active in Jefferson County Farmers and Neighbors (JFAN) to fight CAFOs.

For watering cows in paddocks throughout the farm, Francis installed a solar-powered water system. He placed a 4,000-gallon tank on top of the highest hill on the farm. In a pond just below the hill, he placed a submersible pump and installed a solar panel on the edge of the pond. The solar panel powers the pump, which pumps water to the tank on top of the hill. Water then gravity-feeds from that tank to a water system that provides water to all 60 paddocks on the farm. The water system consists of a 1-inch polyethylene pipe injected under the ground about 8 inches deep and over a mile long. In the fence line between every other paddock is a 55-gallon water tank with a float valve connected to the underground pipeline. The water system is drained and blown out with an air compressor for the winter. Because ground water in southern Iowa is low in quantity and quality, water for the watering system came from Iowa's rural water system (water pumped into the countryside from a system of reservoirs) before the solar-powered water system was installed.

Francis strives to keep his cows out of the creeks running through the farm. One strategy is to allow grazing with access to the creek at night only, when, Francis says, they are less likely to go into the water. With flash grazing, cows do not damage streambanks.

In anticipation of possible local concern about wastewater processing from the dairy, Francis had the Department of Natural Resources (DNR) check the wetland it runs into for any problems. The DNR identified no environmental issues.

Energy and Carbon Concerns

Francis expressed concern about the emphasis on corn-based ethanol in current federal and state approaches to renewable fuels, both for the environmental impacts of more extensive row cropping and for trends towards centralization in biorenewables processing.

In general, Francis sees the current emphasis on greenhouse-gas emissions in agriculture as a new way of talking about a longstanding concern: efficiency in the operation. He sees grass-based dairies as a particularly energy-efficient model. Because of his own emphasis on grazing, he does not feel particularly worried about the rising energy and feed costs that now concern some agricultural sectors.

With rising energy costs and intensified public discussion about energy, Francis is now thinking more about other alternative energy options to incorporate into the farm. He is considering the possibility of a wind generator, as the area is good for wind energy. He thinks a solar hot water heating system could be a good addition to the dairy.

MARKETING, BUSINESS MANAGEMENT, AND FINANCIALS

Marketing

The Thickes made a conscious decision to sell all their dairy products locally, because they "didn't want to lose that connection." Radiance Dairy sells its products to two groceries and 12 restaurants in the Fairfield area. Thus, Radiance does not sell products directly to customers. Hy-Vee, an Iowa grocery chain, and Everybody's, a locally owned wholefoods store, are the retail outlets. Most of their buyers have standing orders. Products are distributed as they are produced, with bottled milk delivered to stores on Mondays and Thursdays. The processing schedule means that Radiance never keeps much inventory of milk or other products on hand at the dairy. The low inventory sometimes surprises and disappoints regular customers who expect to drop by and make special purchases.

The Thickes raised prices for Radiance products a few years ago, but Francis says their prices are lower than the prices for other organic milk brands. He and Susan are now considering whether they need to raise their prices again. They prefer to keep their prices as low as possible—because they are selling to friends and neighbors—and base their prices on the cash flow needed to pay bills and make a modest living.

Radiance does community service advertising (for example, an advertisement in the high school yearbook), but in general does little to no formal advertising. It relies on reputation in the community and word-of-mouth. The dairy does not have its own website.

Certifications

Francis and Susan certified Radiance Dairy organic in 1993 and became certified organic again in 1997–1998 when they moved to their new farm site. In accordance with organic certification requirements, Francis records what pasture every cow is in on every day. Francis noted, "I don't like the paperwork for organic. I'll be honest; I do the minimum to get by. I don't keep a lot of records. I mostly do it by observation."

An interesting benefit of being a certified-organic farmer, said Francis, is not only the premium it brings for dairy products but also the legitimacy it brings to his speaking about the organic sector at conferences and in public policy settings.

Radiance Dairy is affiliated with the Buy Fresh Buy Local Initiative in the Fairfield area (the distinctive logo is posted on the side of the Radiance Delivery truck), and Susan plays a leadership role in the initiative.

Finance and Business Management

The very purchase of the dairy back in 1992 entailed what Francis calls "creative financing." As the Thickes were unable to interest a bank in lending to them, Francis' brother Art helped to finance purchase of the dairy herd. The dairy owner rented them the farm and small processing plant and was willing to forgo payments on the farm and processing equipment for one year. By 1995, Francis and Susan had accumulated enough equity to secure loans to purchase a farm about four miles away and build the milking and processing facilities there. They moved the operation to the new site in 1996.

The farm had been organized as a sole proprietorship. Based on tax arguments offered by their tax accountant, Francis and Susan have recently made it a limited liability company (LLC). They are now budgeting and paying themselves salaries.

On the production side, they keep track of each cow, herd, and paddock. They also track purchased feeds and seeds and keep records of hay and other crops produced on the rented land. Francis is disinclined to calculate or dwell on measures such as net profit per cow, because he finds those measures do not capture what matters most to him about his farm, which is meeting the demand for dairy products in the local community.

On the marketing side, sales records (invoices) help in inventory management and adjustment. The federal milk marketing order also requires them to keep records of their sales of each product. Susan does all record keeping and accounting for the farm. Records for the farm are not computerized.

At one point, the farm carried $500,000 in debt, but that now stands at about $300,000. Four years ago the debt-to-asset ratio was 46 percent; it is now under 30 percent. The bank now considers Radiance Dairy a very good credit risk and would gladly make further loans. However, Susan, who handles the farm finances, is accelerating the payments. They now seek to buy things needed for the dairy on a cash flow basis.

In terms of risk management, Francis said, "I don't think too much about that. I don't do a lot of analysis, but I'm not the best business person." He quickly noted that Susan pays attention to the numbers and the business side of the operation, and looks for efficiencies. But Francis stressed that efficiencies do not always infer economic efficiencies.

SOCIAL AND COMMUNITY CONSIDERATIONS

The Thickes very much see Radiance Dairy as embedded in and serving their local community. The Thickes interact frequently with their customers. As Francis notes, "People just stop you on the street and thank you." While such connections are gratifying, they also bring the "little burden of informal tours"—those sometimes spur-of-the-moment requests by customers to show friends or visitors the farm.

Aside from immediate local social connections, Francis has long put a priority on participating in the broader education and policy arena. As his farm has flourished, he has used it as an example, and he believes this lends credibility to his positions. He has made presentations to Rotary, Kiwanis, and national organic conferences. He is a Food and Society Policy Fellow (Class of 2002–2004), which provided training and a venue for public writing and speaking on food and agriculture issues. He serves on the boards of the Organic Research Foundation, Iowa Food Policy Council, and Iowa State Technical Committee. He further noted, "With my extra time, I like to do political things." Francis has served on the Iowa Environmental Protection Commission, where he relished pressing for a more conservationist approach to environmental issues facing the state. Francis enjoys this work and has considered running for office. He explained his attraction to policy work: "I think farmers won't change on their own. Policy needs to lead to change."

Francis is also active in formal and informal educational efforts related to sustainable agriculture and food systems. He speaks to classes at Iowa universities and colleges. He has long been an active member of the Practical Farmers of Iowa (PFI). For example, in the summer of 2008, he hosted a PFI farm tour at his farm that focused on wildlife and agriculture.

RISKS, CHALLENGES, AND CHANGES

The Thickes have reached a place where experimentation has yielded to a well-established, diversified, and generally resilient system. Planning for the transition of the dairy when Francis and Susan are ready to retire is one issue on the horizon, in part, "because this is a community dairy." The Thickes have considered looking for someone to carry on the dairy in the future. They are exploring options that include working with present workers on the farm to help them build equity in the herd and explore an eventual management role, and creating a community board. Francis underscores that "To us, it's important that this dairy would continue as a community dairy."

OBSERVATIONS AND CONCLUSIONS

When other farmers come to him for advice on how to do similar dairy farming and processing, Francis offers the following advice: 1) First look at the market. While at one time, Fairfield seemed a uniquely promising market, Francis notes that the growing interest in local foods, particularly in many college towns and larger metropolitan areas, presents good opportunities for farmers interested in local marketing; 2) Look carefully at the costs of dairy processing equipment and consider working with a consultant to be sure these critical decisions are the best ones for the operation; 3) Make sure to have a specialty product, with unique features. An on-farm dairy processor selling ordinary conventional milk would probably be less successful than one selling milk with some unique qualities.

Several factors, then, help to explain the success of Radiance Dairy. First, its very loyal and distinctive customer base in Fairfield has been built over the 28 years the dairy has existed. Second, its overall product is unique: local, organic, not homogenized, grass-fed, and from Jersey cows. Third, the dairy product line is diversified and carefully and regularly finetuned in response to customers' requests.

Francis summarizes the approach of his rotational grazing system and local-market focused dairy as follows: "All the pieces work together. We try not to change any one thing without doing it in the context of the whole." He also stressed that he believes in producing "food for people, not a commodity for the market."

Straus Family Creamery

BACKGROUND AND HISTORY

The Straus Family Creamery is located in western Marin County, California. The dairy farm sits just on the east side of scenic Tomales Bay, while the creamery is located approximately five miles inland on a property leased from a local beef rancher.[9] The region has a long history of dairy farming. In the early 1960s, there were 150 dairies in Marin County. As of 2008, 27 dairy farms remained in the county.

Albert Straus' parents, Bill and Ellen, started the dairy farm in 1941 with 23 cows. Ellen was particularly influenced by the message of Rachel Carson's *Silent Spring* in the early 1960s, which would come to shape both their farm practices and wide involvement with agricultural land stewardship efforts in Marin County. The farm ceased using chemical herbicides by the mid-1970s and has not used chemical fertilizers since the 1980s.

One of four siblings, Albert received his B.S. degree in dairy science from California Polytechnic State University in San Luis Obispo, where he wrote his senior thesis on how to set up a processing plant for raw milk.[10] He became a partner in the family farm in the early 1980s. At that time, the Strauses were doing no-till for silage. They fenced off all the creeks through the 1980s and created riparian buffers. Albert estimates that they built a pond per decade from the 1950s on. In 1993, Albert began the formal conversion of the dairy to certified organic. In 1994, the farm became the first certified-organic dairy west of the Mississippi River. Up until then, the Straus family farm marketed its milk through the local dairy cooperative. When they decided to produce their own dairy products in 1994 (to gain more control over the prices they were receiving), they had to leave the cooperative.[11] Albert observed, "If we hadn't gone organic, we wouldn't be around as a farm. It got us out of having to keep getting bigger."

The decision to produce and market organic milk, yogurt, butter and ice cream under the family name was another critical juncture for the Straus family. Today, the dairy and creamery are two separate operations organizationally. The dairy is established as a C-corporation; the creamery is an S-corporation. Originally, the dairy was set up as a partnership, which later became a sole proprietorship before becoming a C-corporation. The creamery started as a sole proprietorship, became a C-corporation, and has recently become an S-corporation.[12] In a formal arrangement between the two distinct enterprises,

---

[9]Until abandoned around 1980, the creamery property had been one of several communal residence sites in California for the group Synanon (see http://en.wikipedia.org/wiki/Synanon). What became the Straus Family Creamery plant was at one time a commercial kitchen serving some 2,000 people.

[10]Albert now notes, "As a farmer, I grew up on raw milk. But that's not for this business. It requires constant battles."

[11]California dairies have 100 percent exemption from the federal milk marketing order. However, California has its own marketing order. It has to contribute into the federal marketing for marketing costs ($00.10/CWT).

[12]The income of a C corporation is taxed, whereas the income of an S corporation (with a few exceptions) is not taxed under the federal income tax laws. The income, or loss, is applied, pro rata, to each shareholder and appears on their tax return as Schedule E income /(loss).

the creamery contracts to buy milk from the dairy. The creamery also contracts to buy milk from two other nearby certified-organic dairies in Marin County, one of which has been supplying Straus Family Creamery since 1995, the other since 2007.

FARM PRODUCTION SYSTEM

Farm Production

The present Straus dairy farm consists of 660 acres of sometimes steeply rolling land that encompasses two farms—one of 166 acres purchased in 1941 and another of 493 acres acquired in 1956.[13] The dairy typically includes approximately 300 milking cows and 300 replacements, which allows for about 1 cow/acre.

Soils are mostly a clay–loam structure. Most of the animals are removed from the fields during the winter to minimize their impact on the pastures. The Strauses use no-till methods to plant grasses and for 25 to 33 percent of their silage crops. They favor no-till methods specifically to minimize soil erosion.

Herd Management

In the 1960s, the herd consisted of both Jerseys and Holsteins. As of 2008, the herd is about one-third Jersey, one-third Holstein, and one-third Jersey–Holstein crossbreed.

Since 1986, the farm has milked three times a day rather than only twice (at 4:30 AM, 12:30 PM, and 8:30 PM).[14] The currently used system has produced 10 to 15 percent more milk using shorter shifts, but it requires more labor than the old system. Instead of two seven–eight hour milking shifts, the milking takes about four hours, three times a day. Milking three times a day is somewhat more expensive than twice a day but yields benefits consistent with an organic sensibility. The cows are seen by workers more frequently. There is less stress on the cows from shorter milking times, and they spend shorter periods of time on concrete. The herd produces on average 65 lbs/day of milk per lactating cow.

An automatic transponder on the legs of each cow allows for individual identification and the ability to track volume of milk, temperature, amount of solids and fat, and other factors. The information allows the dairy to manage the individual cows more effectively, ensure higher-quality milk, and, ultimately, become more profitable.

In terms of genetics and breeding, 30–35 percent of the cows are artificially inseminated with other genetics, which is essential for achieving genetic diversity without the risks of bringing in outside animals into the closed herd.

The animal nutritionist who has for many years helped design feed rations for the cows now serves on the National Organic Standards Board. As the Straus dairy farm made the transition from conventional to organic, the task of developing a balanced feed formula became simpler because there were fewer options. At the same time, it became more difficult given some scarcity of suppliers. When Straus was a conventional dairy farm, the Strauses used anything they could find that would deliver needed nutrients—saki waste, tofu waste, orange peels—but the feed had to become consistent when they went organic. They currently use a computerized feeding system that can precisely account for the dry matter that the cows get from pasture.

---

[13] Albert is in the process of purchasing the 493-acre parcel from his siblings. He leases the 166-acre parcel, site of the original dairy farm and home place, from them.

[14] Albert was inspired by the thrice-daily regimen when working on a kibbutz in Israel after high school.

Albert stressed the need "to be diligent" about herd health and focus first on prevention of disease. The main health issues include mastitis, problems with feet and legs, and calf issues such as scours, pinkeye, and flies. The Strauses stopped de-worming calves six years ago with no ill effects. They focus on preventing diseases by addressing feed problems right away, keeping bedding clean and dry, and, when necessary, using alternatives. Specific actions include careful documenting cow health, dipping calves' navels, and keeping feed off the ground. Albert uses nongenetically modified vaccinations, treats hoofs with copper sulfate, and uses hydrated lime on the bedding. Flies are managed with solar-powered zappers and through efficient manure handling. Fly control aids in the reduction of pinkeye. Cows that need treatment or other cows that are culled from the herd are all sold as conventional. They sell some steers for organic beef. Albert notes that they do not really want to be in the organic beef market, but can rationalize doing some to support the local slaughterhouse, which might otherwise go out of business.

They have recognized that their high production times in the summer do not match well with the relatively low consumption that characterizes that season. As a result they try to manage their herd so that production matches the demand curve by drying cows and calving to match the consumption cycle.

Pasture and Silage

Most of the farmland is in pasture, although 180 acres are typically devoted to a silage mix (oats–bell beans–vetch) that is planted in October and harvested in April, chopped, and stored in a bunker silo (concrete on the bottom and dirt sides). The farm uses a no-till seeder and some discing to plant the silage, which is grown on the tops of the hills. The best pasture also is available during roughly the same time—November to March, after which it dries up. The area receives on average 30 inches of rain per year. Water and the climate of the farm are not suitable to raise crops or hay; therefore, making silage is important. As a result, 50 percent of the hay needed is brought onto the farm, mostly alfalfa and grains and concentrates fed to the cows. They figure that the cows get 30 to 40 percent of the dry matter in their diets from pasture. Eventually, Albert hopes to increase the production of forages and silages to use on the farm and, if possible, obtain feed from as close to his farm as possible. Regarding current debates about the pasture standard of the National Organic Program, he says, "It's not a simple issue. They try to paint it black and white. I can do 120 days on pasture. But I can't do 30 percent of cows' dry feed from grazing."

Fertility and Nutrient Management

Manure solids from the dairy are applied on the fields, after being separated from liquids that go into the anaerobic digester. They conduct soil tests regularly to ensure that a nutrient balance is maintained. They spread the manure in a widening set of concentric circles from the dairy. Since 1976, water quality tests in streams have indicated that nothing is getting into waterways from the confined areas.

Pest Management Concerns

The primary weed problems are hemlock and thistles, which are managed chiefly through mowing, grazing, and competition from other more desirable plants. In very wet years, they might have some mold problems in their silage. California Fish and Game introduced wild turkeys some years ago. The wild turkeys have become a problem because they eat seeds and grains and pose a small but possible *Salmonella* threat. The methane digester has helped with fly problems

Creamery

The creamery has been essential to the successful development of the farm's agricultural system. Albert observed, "If I didn't have the creamery, I would probably be selling to Horizon. I did talk to them." Instead, he now produces his own line of organic dairy products.

Production

Between its three local supplying farms, the creamery currently brings in and handles 9,000 gal/day of raw milk, a decrease from 12,000 gal/day of raw milk that was once brought in. Albert noted that too much organic milk is on the market, because many new organic dairy producers have recently come on.

Product Line

The milk products are not homogenized; thus, the whole milk is cream-top. Surveys were initially conducted at farmers' markets to determine what milk products customers prefer. Although the survey results indicated that most people would prefer reduced-fat milk, 60 percent of their sales are whole milk. Milk in glass bottles with the Straus Family Creamery label accounted for more than 30 percent of milk sales in 2007. The milk products include whole milk, 2 percent milk, 1 percent milk, nonfat milk, half and half, and cream. The Strauses have discontinued production of chocolate milk and buttermilk.

The farm also sells ice cream, yogurt, and butter. Butter is sold as lightly salted and sweet butter varieties, and in quarters, pounds, or a 20# and a 40# size for food service, such as bakeries and high-end restaurants. Ice cream is made with egg yolk stabilizer and is organic. The five flavors of ice cream are all super-premium and are the highest priced in their category.

Whole-milk yogurt is sold as a European-style yogurt. It is incubated in a vat and then condensed via reverse osmosis for packaging in quarts and pints. The yogurt, which contains only milk and culture, is the best-selling yogurt in its category. Sixty to seventy percent is sold under the Straus label with about 30 percent sold under private label as "thick and creamy" yogurt. The family would prefer to sell more yogurt branded as Straus product.

The Strauses attempt to seek sources for the flavorings in their products as close to California as possible to improve sustainability. They are considering using agave from Mexico as a sweetener, which has the added benefit of a low glycemic value. They are looking into developing some food service products such as a nonfat yogurt base and nonfat ice cream.

The Straus Family Creamery also innovates with some very specialized dairy products. For example, its recently produced special "barista milk" won second place for making cappuccinos in an international competition.

Packaging

Albert mentioned, "My notion of sustainability says that all packaging should be reusable, as much as possible." The creamery uses glass milk bottles, which are made in Canada from 40 to 50 percent recycled glass and sport the Straus Family Creamery label and logo.

The use of glass bottles was a deliberate decision that emerged from his personal view that "organic" should include minimizing the waste stream and having a very clean, inert product to contain the milk. Consumers did not drive this packaging choice, as most of their consumers do not necessarily remember the days when glass milk bottles were the norm. Delivery trucks pick up and return the used bottles. Nonetheless, the glass bottle system has its tradeoffs as it incurs extra work and expense.

Plant Procedures and Issues

Considerable attention has been paid to changing standard chemical use in the plant and moving to nontoxic cleaners. The plant uses a formulation of peracetic acid as the primary disinfectant in its equipment, because it does not leaves any residue.[15] Scientific staff members at the creamery are also experimenting with colloidal chemistry in developing nontoxic formulations.

Water from the creamery is currently hauled (12,000–15,000 gal/day) to the methane digester at the dairy. They are working to re-locate the dairy to the Petaluma–San Rafael area to include a demonstration farm and facilities for processing and retail sales on 80–100 acres. The new location would have the advantages of more available water, waste treatment, energy savings from reduced water trucking, proximity of labor, nearness to distribution networks, and options for alternative transportation compared to the current location.

Labor: Farm and Creamery

Albert oversees the dairy, but noted, "Right now, I am not hands-on there, although I'm managing certain aspects." Much of the dairy management falls to a full-time farm manager who has been there for 20 years. There is also a part-time book keeper. The dairy farm employs two milkers. The milkers tend to remain for three to five years, and an additional relief milker provides support. The farm also employs a full-time mechanic who works on dairy and farm equipment and brings the definite benefit of "being inventive." The family is seeking an additional full-time person to raise the calves; Albert believes a dedicated person is need for that position to ensure animal health. They provide "decent" housing at the dairy for employees. One structure is a mobile home, which they plan to improve.

The creamery has a larger and more complex staffing structure (70 employees), and Albert is highly involved. At the management and office end, the organization now includes a vice president for operations, vice president for marketing, chief financial officer, merchandising manager, and three to four part-time merchandisers whose territories includes southern California and the South Bay. As the Strauses consider food service options, they recognize that they will need a sales force.

Most of the creamery's processing and plant workers are from a particular dairying area of Mexico. They commute from Petaluma and Santa Rosa, because of a lack of housing in the immediate area. Employees are generally recruited by word-of-mouth. Albert seeks to recruit people with higher education (many have not much more than a third-grade education). "Bringing up the quality and experience level for workers is a goal and it will take time," he said. In order to retain employees, the Strauses offer competitive wages, health and dental insurance, a simple IRA option, one to two days off each week, paid vacation, and safety training. They also offer incentives for education for bilingual employees and for car-pooling.

---

[15] Albert noted that the National Organic Program currently allows chlorine-based cleaners.

Further Business, Marketing, and Financial Considerations

The price of feed has risen 80 percent since December 2007, which has increased prices for dairies, while the price of milk has inched up 3 to 5 percent and is unlikely to increase much more. Too much conventional milk and more than enough organic milk are already on the market. The dairy and creamery are looking to reduce costs and gain efficiencies. Given that they produce a high-end product, they also keep in close contact with buyers to ensure their continued market position. Albert noted, "If we had not gone organic, we would not be in business now." The commitment to organic is strong: "We're an organic company, not a conventional company with an organic line."

Organic Certification

Organic certification is central to the success of the Straus Family Creamery. They use two different certifiers—Marin Organic Certified Agriculture for the dairy and California Certified Organic Farmers (CCOF) for the creamery. Because the needs of the two enterprises are different, Albert recently switched the creamery certifier to CCOF, which he sees as "more proactive" in areas that matter for that business.

The certification process requires annual documentation of practices and inspections (now yearly, but the frequency might increase), but overall Albert finds the procedures manageable. He notes national organic standards and dairy industry standards sometimes conflict.

According to Albert, organic certification requires a different mindset for the farmer: "It changes the tools in what you can use and you have to think about things differently." That means different feeding; different animal care; and a more proactive approach to herd health, which includes keeping bedding clean and dry, avoiding feeding problems, and relying on alternative medicinal approaches.

A particular concern has been verification that his organic feeds are free of GMOs. While testing is not common in the United States, he said, "We're not the first to do this. The European Union, Japan, they do it." After finding that a quarter of the supposedly organic corn he had purchased showed GMO presence, he is now seeking 100 percent compliance for non-GMO feed on his farm and the other two dairies supplying the creamery. He contracted recently for organic triticale from the Central Valley. While he does his own milling, he also now works with a mill in the Central Valley that purchases grain from the Midwest and elsewhere and does the needed GMO testing. Albert sees non-GMO verification as a critical issue overlooked by many in the organic dairy industry: "If we don't show people it [non-GMO verification] can be done, it just gets buried . . . ." Milk from the Straus Family Creamery is also certified kosher. In addition, the dairy follows a Marin County grass-fed standard.

NATURAL RESOURCES, ENERGY, AND CLIMATE CHANGE

"Closing the loop" on resources is a driving concern for the Straus Family Creamery. Energy, water, and waste issues exemplify this concern.

Energy

A major project at the dairy is its anaerobic methane digester, which currently produces 90 percent of the electricity used by the dairy and 50 percent of its propane needs. Phase I of the project was a pond retrofit, supported by the U.S. Environmental Protection Agency.

Phase II was the generator and co-generation, done through a 50 percent cost-share with the California Energy Commission. Twice a day, the barn is flushed clean with recycled water and manure is scraped with a tractor toward a holding pond, where decomposition and methane digestion begin. Solids are separated from liquids. The liquids go into a second, covered pond where the anaerobic digestion takes place. The separated manure solids are composted and spread on farmland.

The methane digester system is designed to take advantage of regulations for "net metering," where meters can run in reverse, so any excess electricity feeds back into the grid. Albert believes that installation of the methane digester has led to reduced manure odors and lower fly populations. He notes that others in the area are now following his lead and installing on-farm methane digesters.

For the last five years, he has driven a plug-in electric car (RAV4), powered by the farm-generated electricity and used mostly for errands. The Strauses are now trying to develop a fully electric truck, including building it and testing it with various sizes and types of load. "Doing something no one else has done takes more money and more time," says Albert. He recognizes that the innovations entail significant costs and time upfront, but the energy savings can pay back those costs over time. He sees such technological development as exciting and necessary for the farm and creamery. He had some failed experiments with wind generation. His long-term goal is to become independent of imported energy sources.

Water

The dairy farm relies on 10 wells for water. The creamery hauls in its water from Petaluma (about 50 miles away) in its own trucks. Creamery waste water is hauled to the digester on the farm. Water is not that expensive, but the transport and labor to move it represent a notable cost.

Waste

There is considerable attention to recycling water from the dairy and the creamery. The Strauses have built a pond in each decade to handle waste water. They now have five ponds, one of which is aerobic. Another pond is anaerobic with a floating cover. It was built using Clean Water Act Section 319 funds from the Environmental Protection Agency ($70,000–$80,000 at 50 percent cost-share), which came through the Resource Conservation and Development Program for pollution reduction. The anaerobic pond is the digester with the floating cover to collect the gas. They have achieved a 99 percent reduction in fecal coliform and an 80 percent reduction in biological oxygen demand in the effluent that comes out of the system. No effluent is allowed to go into the streams.

The Straus Family Creamery is participating in a benchmarking and best practices effort with the State Energy Commission in conjunction with Lawrence Livermore Berkeley Laboratory. The effort aims to create tools for more efficient water and energy use in dairy plants. The Straus Family Creamery hopes to be one of two pilots, which will assess return on investment.

Local Environment

Local environmental quality issues and ensuing regulations have been important for development and change in the dairy industry in the Tomales Bay area. In 1976, California regulations came into effect stating that nothing could enter waterways. As Albert said, "That was the start of it." Following a study in 1998, Tomales Bay was designated as impaired from coliform bacteria. Among the concerns are the health and viability of oysters grown in the Bay. California is about to require nutrient management plans for animals on pasture. Dairies in the area are all required to have nutrient management plans. Farmers must report every autumn to a regional water quality board.

On the farm and in the surrounding area, Albert observes a lot of songbirds and takes this as one good indicator of biodiversity. Albert's mother, Ellen, played a leadership role in establishing the Marin Agricultural Land Trust (MALT), which has now attained easements on some 42,000 acres in the county. The California Coastal Commission has made it difficult to develop in this area. But land prices have escalated to $10,000/acre, even if they are worth but a fraction of that. Unfortunately, according to Albert, when very wealthy people buy MALT land, it might effectively go out of agricultural production, as their interest is often more in maintaining a "visual corridor" than in maintaining working farms. Management of nearby Point Reyes National Seashore has sometimes also been at odds with retention of farming activities, such as dairy. Thus, competing land use interests clearly influence the future possibilities for farming in this region.

DISTRIBUTION AND MARKETS

The Straus Family Creamery has seen double-digit sales increases for the past several years. The glass bottle milk is sold in Washington, Oregon, California, Nevada, Arizona, and New Mexico, with the majority sold in northern California. Sales are growing by 30 percent each year. The San Francisco Bay area is a large and important organic market.

Distributors move 99 percent of the milk for the Straus Family Creamery (maybe 96 percent for all packaged products). However, Straus has encountered problems associated with distributors, including ensuring a premium price and making sure the distributor provides priority to moving Straus' products. Priority for distribution can become an issue, as in the case when a distributor eventually launched its own organic line. The Straus Family Creamery was planning to increase its milk prices in September by 3 to 5 percent. Albert believes that a larger increase would cause them to lose market share. Therefore, they have to work on reducing costs and becoming more efficient.

Straus Family Creamery butter is marketed nationwide through Whole Foods and also goes to some food services, where it is used for making ghee and other specialty foods. Until five to six years ago, the Straus Family Creamery made all of Trader Joe's private-label milk, but the retailer first took the southern California and then the northern California stores away from Straus. (The farm still sells its own cream-top milk in some Trader Joe's.) Albert Straus commented on the challenges of working with a major company such as Trader Joe's, where contracts are not made and loyalties seem sometimes to be limited. The situation with the Straus Family Creamery's whole-milk yogurt at Trader Joe's illustrates this point. The creamery had developed a very successful and popular European-style yogurt sold at Trader Joe's under the Straus Family Creamery label. Trader Joe's then sought a shift so that the yogurt would be sold instead under the store label. At the moment, yogurt produced at the creamery is marketed 60 percent Straus Family Creamery-branded and 30 percent private label.

The creamery also produces premium organic ice cream that uses only egg yolk as a stabilizer. Flavors include vanilla bean, Dutch chocolate, mint chocolate, raspberry, and coffee. Ice cream has long been a special passion of Albert's since he won an ice-cream making competition as a student at California Polytechnic State University. He remains involved in the research and development of new products. In the past, products were developed to "our taste." Recently, more attention has been paid to customers' tastes and input, as in the case of a new nonfat frozen yogurt.

The company does little advertising, but it is in the process of developing a marketing plan, given new professional staff in this area. It had one bad experience with an advertising campaign in which the retailer did not put its product on the shelf until after the campaign was over. It conducts in-store demonstrations with retailers. Although the Straus Family Creamery participates with its distributors in advertisement and discount programs, it has so far been able to avoid slotting fees. The Straus Family Creamery has thus far not tapped the school milk market, but it sees that market as unlikely in the near term given the creamery's product costs and the current limitations of school food service budgets: "No way we could sell it. We're at 40 cents to break even, and it would have to be 20 cents."

The Straus Family Creamery markets at the San Rafael Farmers' Market and is considering the Palo Alto Farmers' Market as well. Overall, however, direct marketing plays a minor role in the marketing mix. The Straus Family Creamery also supplies milk for a nearby artisanal cheese-making enterprise, Cowgirl Creamery. The Straus Family Creamery participates in two large trade shows each year, including West Coast Fancy Food.

In terms of financial arrangements, the farm leases equipment to reduce debt. The dairy also takes out a silage loan each year.

SOCIAL AND COMMUNITY CONSIDERATIONS

Farming in California requires dealing with a wide range of public and governmental interests. Albert has been involved with a number of groups addressing the agriculture–environment interface. He has been involved in the Animal Waste Committee, a multiple stakeholder group in Sonoma and Marin counties that has been working for some 20 years on increasing dialogue on agriculture and environmental issues in the Bay. He has also been part of the Tomales Bay Agricultural Group, a nonprofit organization that engages all livestock farmers and works to collect data on successful nutrient management practices in the region. Albert is also part of a state and industry Dairy Quality Assurance program that includes the use of voluntary "good agricultural practices" (GAPs). He said, "In California, with more and more people, we're so populated and this puts the focus on agriculture. Parts of it are good, but it can be very draining. It can be a lot of time at meetings."

Albert was scheduled to participate in the Slow Food Nation event held in San Francisco late summer of 2008. He attended Terra Madre in Italy with Slow Food International in 2004 and 2006.

The Straus Family Creamery is still a producer member of Marin Organic. However, he expressed some concern to "get all the meanings into one label." In terms of research partnerships, the Straus Family Creamery has worked with California Polytechnic State University on energy issues and with Chico State University, which has an organic dairy program. It has had fewer interactions with the University of California, Davis. He commented, "Universities are sometimes manipulated in different directions than I would necessarily take."

Albert is trying to build a business that is appealing to others to become involved, including perhaps his 13-year-old son. He took his son with him to Terra Madre to begin to see the world context in which the dairy and creamery operate. The trip generated excitement and a clear sense of the viability of their enterprises, relative to other like-minded food and farming ventures.

He tries to share information with new dairymen and women interested in his approach. He has answered emails from Alaska, Russia, and other distant places, and he consulted once for a dairy operation in Virginia.

Albert observes how easy it is to become consumed by the business, with its accelerating demands. When he met his wife, he agreed to take one day off a week. Now he tries to make that two days off weekly, in some effort to achieve balance between work, family, and public service.

SUSTAINABILITY

Albert's definition of sustainability includes developing practices in anticipation of issues rather than in reaction to problems. His farm attempts to do that by figuring out ways for the cows to provide energy for the farm through the digester; developing electric vehicles; finding a viable market niche with premium organic products; and conducting on-farm processing.

Key challenges to sustainability have been the availability of water and the need to bring in feed from off-farm. A major asset for the Straus Family Creamery has been its proximity to and links with a burgeoning customer population that wants to support organic or sustainable farming and purchase quality, organic dairy products. The timing of the Straus Family Creamery's entrance into organics and the location of its dairy are relevant. He also tries to look forward, such as with the electric truck, and to differentiate the creamery from others through products, such as the ice cream and the high quality of the entire product line.

His stance is premised on leadership in experimentation: "I try to show that things work and then people follow." He also noted, "Being first is part of our story" (for example, the first organic dairy west of Mississippi and the first to use a methane digester in California). He strikes a strong note for innovation and risk-taking: "If I were to do the same thing day in–day out, I'd be stuck." Albert has begun to think about specific indicators to track and monitor the sustainability of his integrated enterprises.

RISKS, CHALLENGES, AND CHANGES

The Straus Family Creamery faces a decision about relocating the creamery, most likely to another location somewhere between Petaluma and San Rafael. Albert speaks of the importance of having a labor force living closer to the plant and being closer to major roads to enhance product distribution. Such a move will cost tens of millions of dollars. However, a new location would create an opportunity for a more publicly visible site, with public demonstration and education possibilities. Such public interactions on site are not really possible or desirable at the present location.

"Controlling the message" has become increasingly important as the business has developed. A *New York Times* article several years ago erroneously implied that he used antibiotics, and the article got him into trouble with others in the organic business. Given the rising public interest in food issues, many journalists are looking for stories, and Albert sees the media attention as something that needs to be carefully managed.

To pursue some of the research and development opportunities at his farm and creamery, Albert believes he would benefit from a dedicated grant writer on staff. New, innovative ventures such as the methane digester are worthwhile, but require outside money and support. A new idea for putting a fuel cell on the dairy will also require creative financing.

Albert believes that more research is needed in alternative medicines, animal husbandry, and disease prevention for organic dairies. He also cites the need for more and better bilingual materials for training the growing Spanish-speaking work force in dairy.

Full Belly Farm

BACKGROUND AND HISTORY

Full Belly Farm is a 250-acre certified organic operation raising vegetables, herbs, nuts, flowers, fruits, and livestock near Guinda in the Capay Valley of Northern California. Smaller, independent farms continue to operate in the Capay Valley, a narrow, picturesque area running parallel to the Sacramento Valley to the east and the Napa Valley to the west. The farm runs along Cache Creek, which supplies much of the farm's water. Full Belly Farm involves an active partnership among four farm owners who live in three households on or close to the farm.

Paul Muller,[16] one of the four farm owners, brings a Swiss Catholic family farming background to Full Belly Farm. He grew up in San Jose on a "drive-in" family dairy farm, which marketed directly to customers in the area. In 1968, in the face of urbanizing pressures on farms in the San Jose area, the Muller family (including Paul's four brothers and one sister) closed the dairy and moved to Woodland, located about 45 minutes south of Full Belly Farms, where the Muller family bought a 300-acre farm. Over time, Paul's parents, four brothers, and sister have grown that operation to 10,000 acres producing wheat, tomatoes, and peppers. The conventionally run farm continues to be operated by members of Paul's family.

Paul started farming on his own, separate from the family farm, in the early 1980s on small rented parcels in the area, while he also was working part-time. He raised seed crops at first and then moved into fresh vegetable crops, after he saw they could be viable. His wife, Dru Rivers, worked then as manager of the University of California, Davis, student farm. Having seen the environmental impacts of agricultural pesticide use, such as farmer cancers, farmworker illnesses, drift from agricultural chemicals such as paraquat, and out of concern when Dru was pregnant with their first child, the couple decided to move higher up in the watershed in 1984. An aging 100-acre almond orchard was for sale. As they could not then afford to buy the farm, they rented it for five years, along with Judith Redmond and her husband at the time. Rule #1 starting out at Full Belly Farm, according to Paul, was "no pesticides—we're going to figure it out."

For those first five years, the farm marketed its produce through a local farmers' cooperative. That arrangement ultimately proved to be a losing proposition for all of the co-op members due to difficulties in coordinating production and the overall small market share for their products. In 1989, when the farm's landlord died, his widow asked if the four Full Belly Farm partners would buy the farm. In 1989, a multiple family partnership was formed that, with some changes, has continued to the present.

The early years of Full Belly Farm coincided fortuitously with key developments in alternative and sustainable agriculture. First, CCOF, which had formed in the mid-1970s, was

---

[16]Paul Muller was the key contact and interviewee for this farm case study. The consultants met but spoke much more briefly with the other three partners at Full Belly Farm—Dru Rivers, Judith Redmond, and Andrew Brait.

evolving as an organizational presence and influence in California and beyond. Through its involvement with CCOF, Full Belly Farm contributed to the early institutionalization of organic in California, including the state's legal definition of organic. In those years, restaurateur Alice Waters had begun traveling around northern California, looking for and speaking about the value of fresh organically raised produce. Among her finds was one of Full Belly Farms's early crops of sweet corn. In that way, Full Belly Farm connected almost from its start with the nascent movements for both organic and local agriculture.

Full Belly Farm is based on a partnership among four people: Paul Muller, his wife Dru Rivers, Judith Redmond, and Andrew Brait, a former Full Belly Farm intern who bought Judith's former husband's share. The partnership owns 100 acres, of which 70 acres are farmed, leases another 200 acres, and also operates 70 acres owned by Andrew, of which 30 acres are farmed. The nonfarmed acreage is primarily riparian areas that offer critical wildlife and biodiversity value.

The current business structure allows the combination and integration of the individual skills of each partner. Paul observes that the arrangement is good for a management-intensive operation like Full Belly Farms. Currently, Paul oversees soil management, associated planning and design, and equipment maintenance and repair. Dru manages sales and oversees the animal and flower operations, and the interns who work on the farm. Judith handles financial management, book keeping, computers, and general business operations. Andrew manages farm operations for specific crops and orchards and handles sales to accounts that he has developed. Paul comments that "partnerships are interesting—like a marriage. You're not king of the heap. You're constantly working together." The arrangement has enabled, if not required, the partners to develop strong interpersonal and cooperative skills, which they consider important models for their children.

Their business model has also facilitated the farm partners, their family members, and some workers to introduce new activities and enterprises, continually diversifying the mix. There is room, and indeed encouragement, to experiment with new ventures that are then treated and evaluated as new enterprises that eventually have to survive on their own. New enterprises are assessed by how they fit with the existing farm, how profitable they are, and how well they serve the wider community.

A summer camp run by Paul and Dru's daughter, Hallie, who studies agricultural education at Chico State University, offers a good example of how a new enterprise becomes part of Full Belly Farm. She came up with the idea, hired her siblings to help, and developed a week-long camp that involves harvesting and cooking foods raised on the farm and participation in other farm activities. The camp has now grown to three one-week sessions, which serve about 80 children who are mostly from families who belong to the farm's CSA. The camp concept has further expanded to include visits from school children, teachers, and parents during the school year. The educational activities bring in additional income, build stronger relationships with the community and customers, and enrich public understanding of food and farming. Based on her experience developing the camp, their daughter has offered to become the farm education coordinator upon graduation.

FARM PRODUCTION SYSTEM

Full Belly Farm currently operates 370 acres in total. Of those, 176 acres are owned and 194 acres are rented. Some of the leases are for multiple years; others are shorter term and potentially less stable. The farm involves a complex multiplicity of crops and strategies, based on thoughtful efforts to "design a farm that is ecologically diverse; vertically integrated—from healthy soil to content consumers; has a stable, fairly compensated workforce; and has year-round cash flow for economic stability."

Planting and Rotations

Full Belly Farm grows some 80 crops, including a diverse range of vegetables, flowers, herbs, nuts, and tree fruit that create year-round activity. The orchards are interspersed with, and not distant or separated from, the other crops. A typical field size is 3–10 acres.

According to Paul, out of 100 acres at any one time, there are typically 30 acres to vegetables for the current season or being harvested, 30 acres to vegetables for the next season, 30 acres to cover crops or forage, and 10 acres in tree crops, roads, hedge rows and others. However, he stresses that the ratios are not fixed and could shift on the basis of the farm's needs and market outlook. The rotations, although part of the farming system, are somewhat fluid, with no formal or fixed master rotation plan. Generally, all cropland has a cover crop about once a year. The summer cover crop, drilled in with a 15-inch grain drill, is a buckwheat–cow peas–sudangrass mix. A vetch and oats combination (with some screenings from the wheat that Paul has experimented with and some clover) is sown in mid-September as a typical winter cover crop. The winter cover is managed like a crop; for example, it gets an early irrigation to encourage fall season growth to protect the soil from winter rains.

Paul and his partners typically plant a clover understory in the orchards. It aids water infiltration, sequesters both carbon and nitrogen, and only marginally competes with the crop for water in the summertime. Paul believes that a diversified organic system harnesses a wide range of energy for the system—from beneficial insects and soil microorganisms, to animals that convert complex carbon into meat and fertility, to increased solar gain and carbon sequestration through cover crop management, to increased water infiltration by building soil organic matter. He also stresses that a diversified system creates a more interesting work environment that engages the imagination and energy of partners and of farm laborers.

Full Belly Farm relies on cover crops and animals to provide nutrients for the fields. Winter crops include rape and safflower. Paul and his partners are considering the possibilities of using those two crops for biodiesel. At present, selling safflower oil for $6/qt through the CSA is a better option, although they like the idea of leasing the oil to users who would then return the used vegetable oil to the farm to convert to biofuel.

Tillage

Paul explained, "When we got here, the place had not been farmed that hard. Frankly, the soils may have been better then because of years of no discing. We started tilling, and learned from the results that we observed. With cover crops, compost, calcium and micronutrient additions, the tilth now is pretty good." He and his partners try to keep permanent beds, which they work less (in practice, these "permanent" beds go for two to three years). The pattern is to mow any excess material that is standing after crops or after the cover crop and sheep grazing, then turn the beds over with a chisel plow, beddisc, and shaper. They have done some laser leveling on occasion. Conservation tillage could work, but they would need the right cultivars as a cover and issues about how to do conservation tillage without pesticides arise. Tools and knowledge for conservation tillage in their system remain at an early stage, but he sees growing possibilities for conservation tillage on their farm.

Weed Management

A central strategy in managing weeds is to manage moisture, which is perhaps easier to do in California agriculture than in some other places. Basically, weed growth can be managed by first managing moisture. Paul typically discs a field before planting and then pre-irrigates to let the weeds germinate and come up. Then he works the beds lightly again before planting in order to kill germinated weeds. Depending on the crop and weed pressure, these operations might take place twice. During the growing season, Paul and his partners rely on close plantings and cultivation to reduce any subsequent weed problems. They also flame the beds on the carrots to reduce weed competition. In the tree crops and some of the vegetables (like broccoli or leeks), they establish an understory crop such as clover. All the weed control strategies are management intensive and require careful timing.

Pest Management

Paul emphasized that healthy plants generally experience less insect and disease pressure and that timing of planting can matter (neither too early nor too late). He and his partners have encountered leaf curl but rely on cultivar selection to reduce problems. They use lime sulfur and organic copper as a treatment if necessary. Apple blight pressure can be reduced with agromyacin. They have encountered tobacco mosaic on tomatoes, which can be transmitted by thrips and can be exacerbated by tobacco-smoking workers who tie up the tomato vines with hands that may bear tobacco traces. They use GC-Mite (cottonseed oil, garlic oil, clove oil) for thrips, mites, and aphids, if needed. For codling moth on apple, they use mating disruption. In corn, they have used trichogramma releases and *Bacillus thuringiensis* (Bt) formulations, as well as reduced the amount of corn they grow in the dead of summer. They hire a consultant to monitor their tree fruits and nuts (which results in a formal written report). But more informal, ongoing monitoring is also critical for this farming system. The labor crew, for example, is trained to monitor for insect pests: "We tell them they're an important part of this." They deal with navel orangeworm in almonds and walnuts by harvesting early and shelling their own nuts. In addition, they count on generalist beneficial predators and parasitoids to help in pest management. Overall, the farm's insect strategy connects to choices and options for markets and particularly knowing what is acceptable for a given market (for example, walnuts that are cosmetically damaged by walnut husk fly are not sold in the shell).

Animals

Animals have been a part of Full Belly Farm since the beginning. They are Dru's particular passion and something she pushed for (based on his dairy childhood, Paul's initial inclination was to get away from animals). Animals serve as an integral part of the rotation (for example, sheep "graze off" harvested fields) and a valuable part of the market mix for Full Belly Farm.

The farm keeps 90 ewes. It lambs out every year, and bring about 140 lambs per year to market. CSA customers generally buy lambs live and then pay for the slaughter. Full Belly Farm has been working with local growers to establish a USDA-inspected harvest facility, and it is soon to realize a local cut-and-wrap facility. However, the bureaucratic and regulatory hurdles are immense. The organic wool from the sheep is also a desirable and year-round marketable product. The farm raises a vetch–oats hay for the sheep. It is starting to establish four to five acre pastures for the sheep. One person is designated to manage the animals; it takes about half of each day to care for them. Netted wire and/or New Zealand fence protect the sheep and lambs from coyotes and dogs, which can be a problem.

Initially, Full Belly Farm raised chickens only for internal use. Then, Paul and his partners saw a neighbor's model with chicken houses and started producing chickens as a market enterprise. Starting out with a couple hundred chickens in 2007, the farm now has 400–500 layers on pasture per year.[17] They keep the chickens in mobile coops that they build on the farm. The coops are rotated through the orchards and along the field edges, because shade requirement for the chickens is an issue.

They also keep goats, which are milked for on-farm consumption, one milking cow all year for on-farm use, and a few pigs. Paul describes the full complement of four cows residing on the farm as "a utilitarian slash love relationship." Seeing and caring for those animals also form part of the camp experiences that Full Belly Farm offers to children.

Nutrient Management

The nutrient program is fairly crop specific. Paul and his partners use cover crops to provide nitrogen and maintain carbon and microorganisms in the soil. Given the hot and dry weather, the latter two factors require attention. They apply 8–10 tons/acre of purchased compost per crop cycle to each field. The compost comes from Sacramento green waste. It is monitored for pre-emergent herbicides and also heated to a specific temperature. At pre-plant and transplant, they also sidedress with seaweed or liquid organic fertilizers that are on the Organic Materials Review Institute's products list (that is, OMRI listed®). They have added gypsum to the soil, but are changing to adding lime at one to two tons/yr. In addition, they will use composted pelletized chicken manure when additional nutrients are needed and for additional nitrogen, based on the plants' need and field history. Paul notes that many current cultivars are "racehorse" feeders, designed to grow quickly and produce tasty vegetables, but they require a full complement of inputs, including nitrogen. He is looking for cultivars that are better suited to their system of farming and the specifics of their site, and they have done some seed-saving. He asks, "Can we breed plants for high yield on lower inputs or as better nutrient foragers in organic systems?" He sees appropriate cultivars as important to help reduce their current reliance on external (organic) inputs.

Equipment

Full Belly Farm has deliberately kept its equipment to the minimum needed. Paul commented, "A cutting torch and welder are the best tools for setting out and starting to farm." Because Paul and his partners got into farming at a time of consolidations, they were able to pick up equipment fairly easily and rebuild and modify it. They have tended toward smaller, cheaper cultivation equipment and do as many repairs as possible on the farm. The farm has eight 80–120 hp John Deere and Kubota tractors and two two-wheel-drive offset Kubotas for cultivation. The goal for the equipment is that it be scaled to the operation, can be serviced by the dealer on which they rely, and is affordable. Paul sees a need for smaller machinery. Although Italy and Asia make some great models, he said, for the most part, such equipment is not readily available in the United States.

Given the fruit and vegetable crops produced, Full Belly Farm needs and has various storage equipment and coolers. However, the farm's strategy of trying to harvest crops only when the market is secured (that is, picked to order) helps to reduce storage capacity needs.

---

[17] Full Belly Farm has recently been selling the eggs at farmers' markets for $6/dozen.

NATURAL RESOURCES, ENERGY, AND CLIMATE CHANGE

Energy and biodiversity provide organizing and integrating concerns on Full Belly Farm. Paul commented, "We harvest much more total energy in our system. A diversified, integrated system harvests more total energy on the farm." In discussing the logic of the clover understories, for example, he added, "We see the interface with pollinators as an underappreciated crop." Similarly, he pointed to the carbon sequestration possibilities on the farm as a benefit of the farm's system and approach. Paul stressed that such energy and biodiversity considerations need to be part of "the *total* economic bottom line."

Energy

Full Belly Farm has tried B-99 biodiesel to power their diesel irrigation pumps, but the fuel has caused problems for seals in fuel injector systems. Now it has made the transition to B-20, which it hopes will avoid similar problems. It has solar panels on its large shop, where important postharvest activities take place. The $100,000 system provides 23 kilowatts and has a grid tie. It powers about 60 percent of packing shed needs—the coolers, ice maker and root washer, and shop. In the long term, Paul and his partners hope to invest in enough solar to meet all of their power needs.

Beyond fuel and electricity concerns, Paul speaks frequently of energy in a broader sense, as in the way that plants growing and infiltrating the soil with their roots are also capturing carbon for restoring the soil.

Biodiversity

The riparian areas and hedgerows provide important habitat for native pollinators and wild spaces that add to the aesthetic values of Full Belly Farm. Native plant corridors have also been integrated into wide hedgerows between fields. They intend for those plantings to widen the range of pollen and nectar sources, which directly benefits many farm crops, but also contributes to the broad concerns of enhancing wildlife habitat, biodiversity, and environmental quality in the region.

Water

The farm irrigates with water from Cache Creek and wells if needed. The irrigation system uses drip tape in the tomatoes, which the farm reuses for other crops such as melons. Sprinklers are used in the orchards and other crop fields and have the ability to shift to micro-sprinklers (very little flood irrigation is used). Cache Creek and several wells supply the farm's water. There have been no significant recent concerns about quality or quantity of water.

MARKETING, BUSINESS MANAGEMENT, AND FINANCIALS

Marketing

Paul stressed that beyond the strong philosophical and ecological values that guide Full Belly Farm, ultimately "we have to grow something we feel tastes better and has a better shelf life." He also recognizes the importance of "growing what the market wants, rather than just what we like." In its marketing, Full Belly Farm emphasizes both the freshness and great variety of the products. The farm has a diverse mix of direct-market and wholesale outlets. It sells to 15 restaurants and at 3 farmers' markets in the region. It also sells 1,500 CSA boxes of fresh produce per week that are delivered to locations in the East Bay, South Bay, San Rafael, Sacramento, Davis, Woodland, and Esparto. (The Bay area is about 100 miles away.) In addition, it sells to Whole Foods stores in Mill Valley and San Rafael, Whole Foods Wholesale, and a number of other wholesale accounts.

Because the farm has a relatively large full-time labor force, it can grow and handle its very wide diversity of crops and products. For example, it can generate $50,000/acre gross on flowers, because of the availability of labor to cut, handle, and bunch the flowers. It is also able to service a large CSA clientele. As fruit and vegetables are packed, it can efficiently and systematically take culls and process them as sun-dried products—from peaches to onions. The labor force enables Full Belly Farm to reduce loss from cullage and also extend the market window with differentiated products.

Harvesting proceeds to custom-fill orders, which reduces the need for storage because only what the market needs and wants is picked. That strategy mostly eliminates the undesirable situation of being forced to sell what has been picked in less than profitable markets. The day's picking schedule is laid out on a large chart in the packing shed at the beginning of the day. The chart precisely organizes each specific customer order and which staff person is responsible, with the exact amounts of each crop identified for harvesting, cleaning, processing, and loading. For wholesale, the orders can be palletized for hauling in refrigerated trucks. Paul noted that refrigerated trucks have been a welcome addition, because they allow earlier loading, resolve some hazard analysis and critical control points concerns, and improve the work experience.

A recent innovative twist on the CSA began in the summer of 2008: CSA-style "wellness boxes" of food for 100 employees at a firm in Palo Alto. Paul is considering how to direct this market niche toward its potential to support sustainable agriculture development. Firms committed to wellness for their employees could be encouraged to appreciate the links to local and regional agricultural land use, possibly by investing in the land resources for new sustainable farmers in the Capay Valley.

The sheep have proven to be another profitable and multifaceted microenterprise. Wool is sent out for making into yarn, which is sold under the farm's own label. It costs $5/skein, and Full Belly Farm can sell the yarn for $8–$9/skein. Fleece hides are tanned in Bucks County, Pennsylvania; Full Belly Farm can sell the hides for $150 each.

Considering the farm's various food products, Paul indicated that focused attention on flavor and freshness relates to its marketing approach. He pointed to remarks from a very satisfied CSA member who said, "My kids didn't eat tomatoes, but they love your tomatoes." He noted the sophistication about wine, and speculated that similar things can be done for other crops, identifying superiority in taste and nutrition.

Pricing

Full Belly Farm is concerned about the "fair price" for its products. Although not an exact formula, arriving at a "fair price" involves consideration of the current market price, farmers' market price, and the conventional and organic wholesale price. Fair also means an accounting that considers the sustainability of the system, and necessary compensation to, and investment in, soil, farm biology, health care for farm employees, and the long-term vitality of the farm.

Finances

Initially, buying the farm required a co-signature from Paul Muller's father. Purchase of the farm was paid off in the first 10 years, because Paul and his partners were "aggressive in choosing to be out of debt as quickly as possible." Since then, the partnership has deliberately avoided taking on new debt, although it has not been averse to making investments in the farm. Paul observed that, "The financial end [of the farm] came together through good management."[18] After some initial hesitations, the partners have instituted a bookkeeping system that enables them to track cash flow and create an operating reserve sufficient to maintain the farm for multiple months into the future. Maintaining the reserve means they do not need outside production loans.

SOCIAL AND COMMUNITY CONSIDERATIONS

Labor

Workers are essential to the farm production system. They are also integral to the complex social fabric of this farm. The farm employs, at the peak of the summer season through October, about 55 full-time employees. Year round, 25–30 full-time workers are employed. They include entire family units; one extended family has worked at Full Belly Farm for some 25 years. Because of the year-round nature of the production system, the farm can provide stability for its employees and for many, full-time workers through the year. All employees are paid hourly wages with profit-sharing bonuses in August equal to one to three weeks pay.[19] Full-time workers receive health insurance with a co-payment. Workers take three weeks off around Christmas.

Many workers' extended experience at Full Belly Farm is an asset, enabling flexibility. The workers know how to handle the diversity of tasks that the year-round production and marketing requires. The many tasks in the packing shed provide ways to meet workers' families' needs, by creating jobs for workers' wives, for example. Paul said, "This is work that has some dignity if you give it some dignity." He added that "part of the reason for diversity on this farm is to create stability for workers." Multiple crops and enterprises and year-round activity enhance stability for Full Belly Farm workers.

Internships

Paul described education as a core commitment of Full Belly Farm: "We want to grow more farmers." Full Belly Farm conducts a thriving internship program, mostly coordinated by Dru. It began in 1986 when some Israelis came to intern. Now interns apply online and are selected on the basis of their seriousness about a career in agriculture and their ability as self-starters, among other factors. The interns are asked to make a one-year commitment to the farm so that they can see the whole production cycle. Five interns are employed at any one time, and they work on a rotation of five- and six-day weeks. They are paid a salary and provided room and board on the farm.[20] Full Belly Farm tries to include one international intern in each group (over the years, they have had interns from places such as Japan and

---

[18]He recalled challenges in the early days of generating cash flow for the farm, but noted that their starting "miracle crop" of mixed lettuce at $11/box made a big difference in getting a financial footing.

[19]A very few workers are on salary.

[20]An intern with little farm experience would be paid $600/month and then reviewed at one month for a possible raise.

Peregrine Farm

BACKGROUND AND HISTORY

Peregrine Farm, owned and operated by Alex and Betsy Hitt, is located about 15 miles from Chapel Hill, on the eastern edge of Alamance County, North Carolina. It is a region that remains rural, although under increasing subdivision pressure. The farm's location near the large, well-educated, and prosperous demographic of the Research Triangle area was critical to the Hitts' decision to acquire this farm and to their eventual farming success. Some 40 other small farms (three to five acres) operate in the area, all within a 50-mile radius of Chapel Hill.

The Hitts began farming in North Carolina in 1981, shortly after they both graduated from Utah State University, he with a degree in soils and she with a degree in forest recreation. They moved east specifically to begin a horticultural operation and chose that region of North Carolina for its demographics, climate, and proximity to family. Alex credits early mentors in business and in horticulture as instrumental in shaping their planning on how to start and organize their farm. His summer job as a college student at a profitable startup outdoor equipment and backpacking store in Houston exposed him to a successful business entrepreneur. That entrepreneur provided an inspiring model of "bootstrapping" and conveyed the value of developing a written business plan. While at Utah State University, Alex also benefited from training with a horticulture professor who maintained his own sideline U-pick enterprise. The professor's farm enterprise gave Alex a realistic first-hand look at small acreage and small capital farming.

From the outset, the Hitts planned to create a self-sustaining farm operation. Not a lot of land was available in that part of North Carolina in 1980, and interest rates were high. They found the farm based on a set of criteria that they had established in creating the business plan. It is a 26-acre parcel that, until 1975, had been part of a larger 108-acre farm that had been pasture and some mixed cropping.

The Hitts worked with an accountant to establish a subchapter S corporation to enable them to finance purchase of the land. They developed a prospectus based on their business plan that eventually reached 200 people (family members, friends, friends of friends). Eighteen individuals invested in the farm, with amounts ranging from $3,000 to $10,000; most investors resided outside of North Carolina.

The Hitts had read many studies (including those of Booker T. Whatley) about self-sustaining operations that, at the time (early 1980s), focused on pick-your-own produce as a particularly appealing and profitable farming option. They began growing 22 varieties of raspberries and thornless blackberries, which they eventually switched over to thorned varieties (for their superior taste) on 4 of the 26 acres. In the process, they also learned the vagaries of the pick-your-own business, which for them proved not to meet the economic projections in the popular media or extension publications. For example, although the Hitts achieved the "car count" they projected for their location, customers generally picked far less than expected (rather than 10 lbs/car, perhaps only 1 gal/car).[23] The Hitts gradually transitioned to selling their berries wholesale by the flat to nearby shops. They had to learn all facets of wholesale production and marketing—handling appearance, shelf-life, scheduling—that were challenging and, as they determined, still insufficiently profitable for their small operation. They eventually stopped growing all berries except for blueberries, noting with hindsight that they had also brought to North Carolina assumptions from another climate. Over time they moved into vegetable greens and began to rely primarily on the Carrboro Farmers' Market (Carrboro is a town adjacent to the west side of Chapel Hill) and local restaurant and co-op sales to market the now four acres of vegetables and flowers that they manage.

In the early years of their operation, the Hitts say, "we were willing to live very close to the ground." They initially lived in a tent on the property. For the first eight years, Alex also worked off the farm as a painting contractor. Running the farm was all consuming: "We didn't go anywhere or do anything else for 10–12 years." They gradually built their current house, which sits in the woods a short walk from their four production acres. The investors were all paid off by the end of 12 years.

FARM PRODUCTION SYSTEM

At present, crop production at Peregrine Farm consists of two acres of vegetable crops (80 varieties) and blueberries and two acres of cut flowers (more than 50 varieties). Alex takes responsibility for the vegetables, the blueberries, and the seasonal work crew of two hired people. Betsy is in charge of the flower business, including cutting every single stem herself.

The farm is divided into quarter-acre sections (100' × 100') with 24 planting beds in each section. The division provides the ability for the intensive rotations and cover cropping that are important to the system. Low tunnels occupy one-quarter of an acre, although one-eighth of an acre is planted at any one time—the tunnels are on slides so that they can be moved during the growing season and plantings rotated. High tunnels occupy three-quarters of an acre with only half of that space planted in tomatoes at any one time. They also have greenhouses in which they start raising their transplants in mid-December.

They plant every week, half transplants, half direct-seeded. The small tunnels are started in February. They start in the fields with sugar snap peas and spinach–kohlrabi–beets, and with lettuce (transplants). The busiest planting time is in March and April. They raise all their own transplants in their greenhouse. They then plant tomatoes and peppers, and the high tunnels are for tomatoes. Vegetable planting is usually completed by May because they stopped growing fall greens in 2000. The way they decided to discontinue growing fall greens was typical of the way they make decisions—the greens offered minimal marginal income and, with the additional time in the fall, they were able to spend more effort on preparing land for the next season.

Varieties are selected and retained based on their flavor and not the units of production. New varieties are tried based on suggestions from chefs and customers at the farmers' market. The Hitts say that it takes three years to learn how to grow new varieties well; they often raise new varieties within the crops that they are already growing successfully.

The entire key to their operation is a tightly designed system of rotations—for fertility pest and crop management. Their system provides the farm with a diversity of crops and the ability to alternate heavy and light feeding crops, and cool- and warm-season crops.

---

[23]Possible explanations include a decline in home canning and food preservation, due in part to household time constraints as more women entered the formal labor market in the 1980s.

Small tunnel rotations are done on a 12-year cycle; field rotations are done on a 5-year rotation; high tunnels are on a 3-year rotation (tomatoes–flowers–cover crop).

Another key crop management practice that they have instituted is a series of uniform blocks (100′ × 100′) that use the same dimension of landscape cloth, irrigation lines to provide for economies of management, standardization of materials, and the ability to compare the performance of blocks of varieties over time.

Soil and Fertility Management

The soil on the majority of the farm is a Cecil sandy loam with a porous clay subsoil that is productive and retains moisture—it was formerly in pasture. Some additional bottom land tends to flood; hence, that land is not always in production.

They test the soil in each quarter-acre rotation unit annually. They typically apply dolomitic lime, rock phosphate (for phosphorus and minerals), and potassium sulfate as indicated from the tests. They see phosphorus as their "biggest nutrient to manage." They also rotate the farm's turkeys from block to block (blueberries, flowers) to provide nitrogen and phosphorus.

Early on they used composted horse manure; they had tried dairy manure but had trouble with the seeds. They currently apply feather meal on the day of planting their cash crops. They do not use side dressing and apply nitrogen maybe one out of five years. They also add purchased worm castings to the transplant mix.

Regular use of cover crops restores soil fertility and organic matter. They rely on cover crops—two sets of winter cover crops and two sets of summer cover crops, rye–hairy vetch and oats–crimson clover—chosen to maximize organic matter and nitrogen. The cover crops need to be timed correctly. Tomatoes and peppers are raised on a no-till system that relies on crimping the cover.

Weed, Pest, and Disease Management

The Hitts' biggest pest challenges are vertebrates—rabbits, deer, birds, and coyotes that have become established in the area and can threaten the turkeys they raise. Those animals are the hardest to control. They rely on electric fences around all their fields. They have relatively few insect problems, and they rely on the diversity of the system to retain generalist predators. The main pests are fruitworm on tomatoes, for which they use Bt, and western flower thrips.

To control weeds, they rely on a combination of rotation, cover crops, well-timed cultivation, some hand thinning, and some flame-weeding. They also use landscape cloth on early tomatoes and hot peppers.

The disease management strategies are to break up disease cycles with rotations; use high tunnels for tomato diseases; avoid excess irrigation; and maintain careful field sanitation. The primary issues are foliar anthracnose, bacterial leaf spot, and southern stem blight, which they manage with an organic formulation of Kocide, resistant variety selection, and field sanitation (removing diseased plants). Because of the range of crops on the farm and the amount of flowers grown, they have ample pollinator habitat so they do not have any separate planting to increase diversity.

Animals

The Hitts raise about 100 turkeys, half of which are Bourbon Reds and the other half of which are Broad Breasted Bronzes. They began raising turkeys in part to address an insect problem. They get a premium for the birds at Thanksgiving, but there have been complications in the past couple of years in getting the birds processed. The Hitts have learned a lot about the business, and they believe that they now have a stable processor in the area to handle their needs.

Labor

The Hitts' operation makes use of two "seasonal" employees who each work from March through October about 30 hrs/wk. They work under Alex's direction and take turns helping with the two weekly farmers' markets. The workers are provided benefits, but they live off the farm. The housing arrangement differs from other small farms that take on and house "interns" and is a strong preference of the Hitts. All of their employees, who tend to change from year to year, are interested in sustainable agriculture. Many have come from the sustainable agriculture program at nearby Chatham (County) Community College and live in the area. Employees assist with the full range of production and handling tasks. Getting workers to understand and practice an "economy of motion" in those tasks is sometimes an issue.

Equipment

The Hitts have one small Ford tractor, chisel plow, flail mower, cultivator, and hand rototillers and mowers. They have two small walk-in coolers for the cut flowers and a cool room for handling tomatoes to sort by ripeness and prepare for farmers' markets. They also have two trucks for delivery to markets.

MARKETING, BUSINESS MANAGEMENT, AND FINANCIALS

Marketing

Given the history of their operation, the Hitts feel that they "were at the beginning of the curve on local foods." The majority of production from Peregrine Farm is now sold at the Carrboro Farmers' Market (Wednesday and Saturday) in which they have been involved since 1986. They are required to be at the farmers' market 27 weeks of the year and note that 75 percent of their gross income comes from those two markets. The rest of their production is sold through Weaver Street Cooperative—a local store with three sites with which Peregrine Farm has had a longstanding relationship—and to a handful of local, chef-owned restaurants. Customers, including the chefs, can and do order items to pick up at farmers' market, thereby simplifying delivery. The key to their marketing has been to create direct and solid long-term relationships with their customers. Because they can track crops and sales over time, they can make decisions to alter the crop and marketing mix.

Certifications

The farm was certified organic for about five to six years in the 1990s. Over time they have let the certification lapse, even though they have not changed their practices. The administrative time and expense of certification have small marginal value now that they have a well-established customer base and reputation. However, they can see the value of the certification process in that it provides an incentive for planning, for keeping records, and for tracking inputs, such as manure, more carefully.

Business Management

Peregrine Farm is very much a partnership between Alex and Betsy Hitt, and the partnership is critical to the success of the overall enterprise. The size of the farm is based on what they can manage on their own with modest hired seasonal labor.

Attention to record keeping has been key to this farm's operation from the outset. They began with extensive records on Quicken software and later on QuickBooks as the need for more sophisticated records increased. Alex also took a course at North Carolina State University on farm-management record keeping that yielded useful information to adapt for their type of operation. Their records have included, from the beginning, planting, production, and financial records. Record keeping was important initially to inform investors about progress in the business but has become a foundation for the farm's ongoing monitoring and success.

Record keeping is essential given the need for each production unit to be maximally productive and to manage the biological and marketing diversity on which the operation is based. They use Excel software for recording crop production and planning. In the Excel workbook, they track more than 200 varieties of vegetables and flowers. Turkeys and blueberries are handled as separate enterprises. The recordkeeping system is particularly valuable as a management tool in planning for orders, as an assessment of the value of individual varieties and plots, and in marketing. They are very systematic in their record keeping, and they input data weekly.

Alex noted that their ability to build their own sheds and outbuildings has been particularly important in keeping expenses down and allowing them to add buildings as needed. Alex and Betsy are technically employees of the S-corporation.

Finance

Peregrine Farm typically carries very little debt. The Hitts do not take out operating loans but occasionally take short-term loans ($3,000–$5,000) for capital or cash flow in the spring. They had no debt at the time of the interview.

They keep extensive records to make sure that their minimum target of $20,000 gross per acre is met. They have done progressively better each year, recently getting $27,000–28,000/acre. They actively chart trend lines to monitor the economics of each part of their operation. For example, they track labor expenses and try to keep those expenses to 15 to 19 percent of their gross sales.

The Hitts are unsentimental about crops that don't yield adequate income per acre. They eliminate poor performers and try something else or shift the season, if the strategy can increase income.

SOCIAL AND COMMUNITY CONSIDERATIONS

Quality of life is an explicit and ongoing priority and consideration underlying many of the Hitts' decisions about elements and practices of Peregrine Farm.

Markets as Community

The Hitts gravitated to the Carrboro Farmers' Market in part for what they cite as quality-of-life reasons. They noted, "The farmers' market is our community.... To us, local is the community in which you live, shop, go to the doctor, and it's basically 15 miles away."

Both Alex and Betsy have been active in governance of the market that is critical to their business and their community.

Outreach

For the past five years, they have sent a weekly electronic newsletter to about 300 people. Through the newsletter, they let people know about crops coming in, keep in regular contact with their customer base, and share seasonal pictures of the farm and news from the farm. They do not currently have a website, but are considering creating one, in part "to manage information about them and their farm that is already on the web" at other sites.

They also attend and speak at sustainable agriculture, "slow food," and other meetings. Alex has served on the board of directors for Weaver Street Market Cooperative and is currently serving on the board of Carrboro Farmers' Market.

Government Programs

They have not participated in government programs, but they went to the Small Business Administration for help to rebuild their greenhouse after Hurricane Francis. In general, they said, "We were too early, too weird, too different" for government agriculture programs.

Learning and Obtaining Information

The Hitts have many contacts and "good working relationships" with university and extension people from North Carolina and other parts of the country. They are cooperating on variety–disease susceptibility trials with North Carolina State University. Alex remarked, "We go straight to the specialists," but they also turn to fellow growers at their farmers' market with questions, "since they are most like us." Networking in general is important to the Hitts. Betsy has become very active in the National Association of Specialty Cut Flowers. Alex and Betsy attend conferences related to their specializations on the farm.

Food Safety

They have not experienced food safety problems. The scale and care with which they farm, their precise record keeping, the direct control they have over their crops, and their direct marketing to long-term consumers minimize their exposure to food safety concerns. They note that, across the board, food safety protocols such as a national animal identification system would have a disproportionate and negative effect on small farms such as their own.

Labor Practices and Mentoring Workers

Half of their previous employees now own farms. The Hitts are very open in sharing information on farming and business management with their employees so that, in a way, the employees are apprentices. They are concerned about where help is going to come from in the future, as fewer people seem available. A gas allowance was provided to workers when fuel costs increased dramatically in 2008.

The Hitts have become more actively involved in the local and regional food movements, including the Southern Foodways Alliance and Slow Food (they have attended Terra Madre in Italy several times since 2004).

NATURAL RESOURCE ISSUES, ENERGY, AND CLIMATE CHANGE

Water

Two ponds and an adjacent stream, without much development upstream, provide water for the irrigation system. With careful management, the Hitts seem to have enough water, although they have expressed some concern about quantity. Everything is on drip irrigation—although the drought in 2007 affected establishment of cover crops. May is the month with the highest demand for water.

Energy and Recycling

With the exception of transportation to markets, the Hitts' fossil fuel demands are minimal. They have been using durable flats for raising transplants and have reusable landscape cloth. They had been using drip lines for about three years at the time of interview.

Climate Change

The Hitts have noted some evidence of climate change in terms of temperature and rainfall. They are thinking about how climate change might shift their planting calendar. They might do less production in high summer, when it is very hot.

RISKS, CHALLENGES, AND CHANGES

Flooding in the bottomlands provided a lesson. They moved that site out of their regular production rotation to reduce risk, and "We got terrible new populations of weed seeds there after that." Farm transition issues are a major preoccupation for the Hitts, who are in their early 50s now. They are not yet ready to stop farming, but are beginning to think about succession and transition issues.

SUSTAINABILITY

"Being as diverse as we are has been making us sustainable," the Hitts said. This idea is key to sustainability and to risk management. "Our system is established or carefully balanced." Curiously, their very success creates new challenges; they share a sense that farming is not quite as exciting anymore, because in a way they have the system down. The day-to-day operation now is fairly straightforward in terms of knowing what to do.

The Hitts identify labor—and especially quality labor—as the chief barrier to their farm's sustainability. This is a new time for small-scale farming as a second generation of farmers is coming along. The Hitts and other farmers are facing the issue of farm succession, and many of them do not have children or children who are likely to enter the business.

They feel that society has turned a corner on improving food quality. They have concerns about the economic downturn, which leads them to want to develop even stronger relationships with their customers.

OBSERVATIONS AND CONCLUSIONS

Peregrine Farm has always been a business that was also a farm. Although the Hitts are skilled and attentive farmers, their development of and adherence to a well-designed business plan along with their ability to keep, analyze, and apply an extensive record-keeping system has been very important. From the outset, their habit of careful and regular record keeping helped in their overall success. Their innovative financing system and their willingness to be frugal in the early years allowed them to build a successful operation and to learn from their mistakes. They are proudest of the fact that they have done better financially every year they have been in business; that they have both been able to have full-time jobs on the farm; that half of their employees now have their own farms; and that they are able to contribute back to the sustainable agriculture movement.

Their willingness to do all the work on the farm and be personally involved in the marketing, along with their decision to focus on high-quality projects, has enabled them to develop strong relationships with their customers. They take care to maintain those relationships. The value of their location in proximity to Chapel Hill with its well-educated, prosperous, and socially active population cannot be overstated.

Some advantages in being first-generation farmers and not from the area in which they are farming are that "we always look at things with fresh eyes, and we ask questions." Rotations and the subdivisions of land into quarter-acre sections with 100' × 100' plots makes possible the diversity of crops. The diversification reduces risk from overdependence on one crop, allows for direct marketing of a range of crops over an entire season, and offers opportunities to try small amounts of new crops in response to demand. Rotation is the key to the fertility and soil health programs and also to the pest management regime. A key contribution is the development of a well-defined system of plantings and cover crops along with the systematization and record keeping that is integral to its functioning. This has provided for important efficiencies and, as a result, income. A testament to the value of the system is its adoption by many of the people who previously worked with them and now own farms.

predators. With their blackberries, they are particularly concerned about orange tortrix; they monitor carefully and end up spraying about one out of three years. The farm has no calendar spray program, but use monitoring and measuring of problems to make appropriate cost-effective decisions.

The Willamette Valley is dry enough that disease problems are not great. Disease management is probably most intensive on cane berries, on which they use lime sulfur and Bordeaux mixture.

Rainfall and Irrigation

Rainfall varies considerably across the three farm locations, with the Corvallis farm area receiving 40 inches/year while the Medford farm area receives 19 inches/year. Therefore, Stahlbush Island Farms uses every form of irrigation except flood irrigation, given the wide range of crops raised at the multiple locations. The preferred form is drip, used on all perennials except rhubarb and strawberries. The Chambers are trying to eliminate big guns because of the energy costs and land lines because of the labor involved. They currently use pivot and liner irrigation on their other acreage and are just starting to experiment with annual drip tape. Water continues to be plentiful; they have had only one year where water rights were an issue in the regions of their farms. Even so, water in the Willamette is fully allocated, and developing additional irrigated land would be difficult because of concerns about salmon. Two-thirds of the water used in the Corvallis area comes from the river.

Equipment

Crop-specific and efficient equipment is critical to the functioning of this vertically integrated operation. Stahlbush Island Farms now fabricates most of its own plant and field equipment and has done so for a number of years. The operation has 13 fabricators, a draftsman, and an engineer on staff. For example, employees developed innovative in-row cultivation equipment with sensors for crops like pumpkins. Precision and custom equipment is designed to enhance the productivity of the operation overall. Some new and specially designed equipment, especially for harvesting, reduces hired labor requirements.

Because of the size of the operation and the amount of equipment required, the Chambers can order equipment to their specification from dealers who are eager for their business. For example, at the time of interview, they had just taken receipt of several new tractors, specifically built for their operations.

NATURAL RESOURCES, ENERGY, AND CLIMATE CHANGE

Water

Water is visually prominent, especially at the original farm site in Corvallis, which is an island in the river. A braided water system, constantly changing, can flood in winter. Rains can be significant at some times of year. Before World War II, the original farm site flooded frequently and dramatically, such that people could not really live there. When the Army Corps of Engineers built dams, the land could be more readily farmed.

For Stahlbush Island Farms, water for irrigation is a more significant concern than water for the processing operation. Crops are irrigated from surface water. Well water is used in the processing plant. All water in the processing plant is used three times: 1) as a coolant; 2) for washing fruits and vegetables; and 3) recycled again for use in irrigation.

Any organic residue picked up in the processing plant through that cycle thus goes back onto fields. This sequence conforms to Bill's view: "There's no such thing as garbage, just underutilized resources. The question is how do we make use of them to create value?"

Wildlife and Biodiversity

Bill noted that they have virtually lost wild bees in the areas where they farm, although he recently found two wild hives. They hire bees to pollinate their cucurbits and berries.

Energy

At the time of the interview, Stahlbush Island Farms was actively addressing ways to use waste products from processing and turn them into biogas for the production of electricity. Construction was underway on an on-site digester at the main Corvallis farm.[27] With the processing plant running 24 hrs/day, 365 days/year, it uses about 750 kwh/day. The Chambers hope to produce two times the electricity they currently use. Bill refers to the energy cycle on the farm as solar energy processed through plants and then through the digester. He sees the potential for "electricity becoming another crop of the farm." The goal for the farm is to eventually be better than carbon neutral.

When Bill purchased seven tractors in the summer of 2008, his decision was guided, not by lowest capital cost, but by the tractors with the best fuel economy. He is also interested in compressed natural-gas vehicles, and he is beginning to look into them for the farm.

MARKETING, BUSINESS MANAGEMENT, AND FINANCIALS

Stahlbush Island Farms is structured as a privately held corporation, with Bill and Karla as the only stockholders. The partnership between them is at the very heart of the enterprise. As Bill says, "My wife and I are business partners and life partners. Her skills and attributes are almost a perfect complement to mine. She does marketing and short-term administration. I do farming, operations, and long-term administration."

The guiding principle for the company is to provide quality for the customer by finding out what is important to the buyer. Bill and Karla used to visit customers face to face once a year. Stahlbush Island Farms now has six full-time staff in sales and marketing, each with a portfolio of customers to remain in contact. Karla oversees the establishing and maintaining of strong working relationships with customers through regular communications. Bill notes that Stahlbush Island Farms will grow different varieties as requested by its customers or process with more or less solids as the customer specifies. The Chambers, however, will also tell a customer when something cannot be done and still be a quality product. The business is very demand driven, yet guided by the Chambers' knowledge and expertise. The Chambers have found that giving customers choices can produce better business decisions. For example, they always try to give two samples to customers and then ask which they like better and why. That approach is part of working toward customer commitments and contracts: "We don't grow it if we don't have a market for it," Bill said. "First we get the contracts, and then we plant."

---

[27] According to a May 31, 2009, article in the *Corvallis Gazette-Times*, the $10 million biogas plant generating methane and electricity from processing plant residues was to start up in June 2009 and is projected to be capable of generating enough electricity to power 1,100 homes.

Stahlbush Island Farms sells into three food market segments: 1) industrial food ingredients; 2) retail; and 3) food service, but it has targeted the first two the longest and most actively. Its industrial food ingredients segment, which began with the frozen pumpkin, has now expanded to processing 42 different crops into more than 300 different SKU's sold domestically and to export markets. The retail segment is a smaller, but growing part of the business. Retail includes the frozen sustainable fruit and vegetables "Stahlbush Island" brand, canned organic "Farmers Market" label (including canned pumpkin, canned sweet potato, and canned pie filling), and, most recently, a canned organic pet food brand, "Nummy-Tum-Tum."[28] The development of the retail brands creates the greatest value for the business. The food service segment has been the last to be developed, but it is now gaining attention with focus on local restaurant chains and food service distributors.

Twenty percent of the farm's business is exported, going to about 20 countries. Japan is its major export market, and China is one of growing importance. It also does some spot marketing through brokers but generally finds that using brokers inhibits communication with customers. Stahlbush Island Farms attends and exhibits at food-trade and natural-food shows.

The processing plant, which operates year-round, functions, according to Bill, as a "job shop processor." He said, "We are not usually the low-cost producer. We seek the customer where price is not the key deciding factor." As a result, Stahlbush Island Farms only plants what it has sold under contract, which eliminates sales and price risks. Stahlbush Island Farms will accept a waiting list for orders of its products. Despite their positive experience with contracts for industrial food ingredients, the Chambers also moved into the retail market "where we could be our own customer." Bill noted that with industrial buyers, the average life span of a product is only three to four years before the product is discontinued. Developing its own retail lines provided a way for Stahlbush Island Farms to gain some control over the costs of that constant product churning on the industrial side of the business.

Audits and Certifications

Stahlbush Island Farms occupies an ultra high-quality market niche, which has become important to avoid potential recalls of its food products. The company brings in outside auditors to assess operations for food safety, and its industrial customers can also (and do) send in auditors. The company goal is 98 percent or better out of 100 points for good food-safety performance. It typically achieves that level or better when audited by a private food-safety auditing firm. The Chambers' approach is to "take food safety and push it back down on the farm." They also regularly audit the farm with their own internal quality-control group, ensuring that they meet Good Agricultural Practices.

Sustainable and organic certifications have been important for Stahlbush Island Farms as ways to increase the overall competitiveness of the operation by addressing the demand of quality-oriented market segments.

---

[28] In the fall of 2008, Stahlbush Island Farms introduced Nummy-Tum-Tum, its new line of canned organic pumpkin dog food. The Chambers developed this product after learning that some customers of their Farmers Market line of canned organic pumpkin were feeding it to their pets as a healthful diet addition.

SOCIAL AND COMMUNITY CONSIDERATIONS

Labor and Staffing

Stahlbush Island Farms is a significant employer, with 120 full-time year-round employees and another 340 either part-time or seasonal workers. Bill's goal is to provide 10 months' full-time employment for most of the workers, which involves continually looking for things that can be done year-round.[29] Hispanics make up the majority of seasonal labor at Stahlbush Island Farms. In Oregon, about 70 percent of employees in natural resource industries, including agriculture, are Hispanic, according to a recent study. The diversity of the crops raised enables Stahlbush Island Farms to provide work over a large portion of the growing season and to make best use of the human resources engaged in the operation: "I've tried to select crops that allow us to have employment as long through the season as possible and a diversity of things for those people to do." In perennials, workers might prune blueberries in winter, then harvest rhubarb, then harvest strawberries, then harvest blueberries, then train blackberries, for example. Employees often go back and forth between work on the farm and in the processing plant, although some workers specialize on particular tasks. Bill mentioned, "A lot of business people think of labor as an expense. I think of them [laborers] as an asset. People are important to me."

The Chambers recruit nationally for skilled and professional positions, but they are able to heavily rely on the Corvallis area, which, with nearby OSU and other companies, is a hub for scientific and technological expertise. The farm also employs two in-house crop consultants as technical specialists for the farm operation.

Community Support, Service, and Recognition

Among other community activities, the business supports youth efforts like 4-H/FFA. Bill noted approvingly, "4-H supports entrepreneurship and kids learning responsibility." In 2008, Karla bought 45 animals from 4-H/FFA members at the county fair, and she used her influence to encourage others in the community to purchase the animals as well.

The Chambers receive many queries from other farmers and prospective farmers and have been happy to share information about their farming practices with anyone who asks. They view information about the processing plant as more proprietary. In addition, both Bill and Karla have served in many different faculty, leadership, and training roles at OSU, which permits them to reach others about entrepreneurship in agriculture and the business of Stahlbush Island Farms. Karla currently serves on the Federal Reserve Board of San Francisco. She has served on the boards of various regional and local foundations and organizations as well. Bill has served on local school and several corporate and district improvement boards. The couple's work and service have been recognized with awards such as the 2000 U.S. Presidential Award for Leadership in Sustainable Agriculture and 2001 Agribusiness of the Year.

SUSTAINABILITY

The Chambers' experience of supplying ingredients for the baby food market in the early 1990s helped motivate their turn to more sustainable farming practices. Rethink-

---

[29] At the time of interview, the Chambers were looking at individually quick frozen products, including wild rice, as another option for extending work through the year.

ing his received assumptions about pesticide use, Bill saw externalities for which society was not paying. He emphasizes five things about pesticide use that came to concern him: 1) pesticides often lower the yield of crops receiving applications; 2) pesticides don't always stay where you put them; 3) pesticides have negative effects on beneficial soil microorganisms; 4) applying pesticides poses health risks, and Bill does not like asking someone else to do something that he does not like doing himself; and 5) pesticide residues in foods worry consumers, and therefore "less is more." In 1992, the Chambers set a goal to eliminate pesticides from their operation in five years, although, Bill noted, "We failed in this." The cost of organic nutrients was a factor, but he also decided complete elimination of pesticides was perhaps an inappropriate and nonsustainable goal.

According to Bill, Stahlbush Island Farms sees sustainability in terms of three main criteria: profitability, responsibility with resources (such as soil, water, air, wildlife, and people), and taking care of customers. The Chambers' goal is to continually make improvements in all three areas.

Bill noted the importance of finding the balance between profitability and the unsustainability of pesticides. He has developed a rule of thumb he calls the $2 \times 4 \times$ rule, which requires thinking through alternatives. For example, if the alternative to a pesticide is no more than twice the direct cost of a pesticide application, he will use the alternative. If the alternative is four times the cost of the pesticide application, he will use the pesticide, carefully choosing the one that is least harmful to the environment.[30]

RESEARCH AND POLICY CONCERNS

Research

Stahlbush Island Farms has sought to create a culture of continual innovation to reduce costs, improve quality, and enhance the environmental and resource conservation aspects of its operations. The Chambers helped found the Willamette Farm Improvement Association (this group no longer exists), which was funded by USDA-Western SARE. They work extensively with OSU, participating in plant breeding trials for berries and studies on insects and cover crops. They also conduct their own internal research on the farms, although this is more informal, through deliberate experimentation, rather than formal, structured scientific research. Bill sees two categories where further research is especially important: 1) soil biology, and especially nutrient availability; and 2) the application of electronics to agriculture (for example, sensor technology to distinguish between weeds and crops in cultivation equipment).

Policy Concerns

Stahlbush Island Farms has not availed itself of government conservation programs or farmland protection programs. Bill cited the transaction costs ("the cost of dealing with the paperwork creates friction greater than the value from participating") and reduced flexibility (program participation could limit future options) as reasons why not. The Chambers support country-of-origin labeling, in the belief that consumers deserve to

---

[30]Bill said that an alternative that costs three times what a pesticide costs usually requires some debate to decide which way to go. He further notes that they never use some pesticides under any circumstances, such as fumigants.

know where their food is coming from and can make better choices for themselves with that information.

OBSERVATIONS AND CONCLUSIONS

Perhaps the most distinctive feature of Stahlbush Island Farms is its tight integration of production, processing, and marketing within the operation. The Chambers see their model of "vertical integration" as a strategy for better managing the risks inherent in agricultural operations. The stability allows Stahlbush Island Farms to produce a consistent quality product for distinct sets of valued customers. It also allows for continual innovation on the production side to increase the farm's efficiency and profitability. Ultimately, Bill thinks of their sustainable farm operation as a farm that does processing and not as a processor that happens to farm.

This vertical integration strategy distinguishes Stahlbush Island Farms to some degree from other sustainable farming enterprises in the region. It certainly reflects the vision and evolving business plan of Bill and Karla Chambers. At the same time, their vertically integrated enterprise has been facilitated by particular circumstances of time and place. For example, the site of the original Corvallis Farm was subject to an exception in land use rules that allowed them to locate a plant for processing their own product. Without that provision, it might have been more challenging to start, develop, and expand their complex "vertical integration" model.

The strength of the company and the key to its business sustainability lie in having carved out a very stable market in which they are a contract supplier of high-quality products. The company's strategy enables the Chambers to ensure a price and market for their crops as they are planted, helps them know how to modify their production to meet a specific market, and provides them the security and flexibility to carry out environmentally sound production practices and long-term improvements to the farm.

From his experience, Bill has distilled three pieces of advice for others who want to build on the lessons of Stahlbush Island Farms:

- Know your real customer.
- Grow what your customer wants.
- Understand costs.

such as Red Brangus, Black Simmental, and Polled Herefords. All his cattle were initially grain-finished. Now he pasture-finishes his cattle and is trying to develop a phenotype for smaller frame animals that are better adapted to his operation and more efficient. At the moment, the transition involves some line breeding, with six lines of females on the farm breeding (AI) with seed stock from a top Murray Grey bull from New Zealand. The intention is locally adapted genetics that will include desired traits—specifically being able to finish well on grass, because it has proven too costly to feed his cattle organic grain, from his farm or elsewhere.

The cattle might be divided into several herds (three to five) that rotate among the pastures, depending on water and on pasture availability and quality. The majority of the animals are rarely brought in from the pasture. Supplemental hay is fed as necessary in the winter months, but pastures are managed for winter grazing.

Jack normally calves in June, which he sees as optimal. The cattle-raising cycle has generally lengthened "more in keeping with nature's cycles," according to Jack. For example, he does not tend to wean calves until the following spring when they can be put on fresh grass. That schedule gives the calf 9–10 months with its mother, which Jack believes is more healthful for the animals. With a forage-finished system, the calf then goes another year. The system is flexible year-to-year, gradually harmonizing across phenotype of cattle, season, and mineralization in pastures.

The Erismans provide kelp and Redmond salt plus a little selenium as a free-choice supplement for their cattle. If an animal develops an infection or presents a disease that they are unable to resolve homeopathically, they will treat it with an antibiotic. Such animals then come out of organic certification. They are generally taken to the sale barn. Sometimes a local beef customer is happy to buy such an animal. They do not seem to care about antibiotic treatment in the context of what they know about the overall quality of Jack's animals and his beef cattle system.

Adequate shade is an issue for the beef cattle. Therefore, Jack is considering planting trees or building shelters for the cattle in the newer permanent pasture areas. Although he notes that good veterinary services are available in the area and address most of the needs for their livestock, he observed that large animal veterinary services are not as prominent or available as they were in the past.

Equipment

Jack owns an extensive assortment of large farm equipment particularly for crop production, management, and harvest. He said, "I'm a machinery buff, but I do have some pink elephants around." Much of the equipment used on the farm is adapted to its specific conditions and needs.[34] He has equipped an older tractor with flotation tires to reduce compaction for harrowing and rotary-hoeing solid seeded soybean. The air drill can plant grain at a desired depth while also incorporating micronutrients or overseeding a grass–legume mix. Jack noted that the machine has 144 monitoring points to ensure precise delivery of the seeds and materials. Such equipment can present not only mechanical but also electronic problems, but he notes, "We thrive on the challenge of what can we use or adapt to make things work better." A driving concern is improving precision of application and minimiz-

---

[34]Jack notes that the area has a lot of people who do fabricating, which can be useful for customizing equipment. The grain-bin manufacturing out of nearby Assumption and the legacy of farm manufacturing (including Caterpillar, Inc.) in Decatur, a 30-minute drive to the north, have probably contributed to a level of fabricating skills and innovative capacity in the local population that benefits those who farm in the area.

ing trips over the field. Given his own considerable range of equipment, he does not hire custom operations.

NATURAL RESOURCES, ENERGY, AND CLIMATE CHANGE

Soils on the farm can sometimes be wet, although flooding seems to be controlled, probably because of the drainage system and benefits from the rotation system. Jack sees some resilience from his organic farming system: "I'll do better than my [nonorganic] neighbors in a droughty year." The resilience is a result of the water-holding capacity of his soils. However, he notes that the reverse is also true: "Wet years become harder for us."

Jack believes that increasing no-till on the farm would help reduce its carbon footprint, but no-till presents challenges in an organic system. Because of high fuel costs in 2008, he has made some changes that include making fewer passes with equipment, less mowing of roadsides, and less cutting of some pasture. He has looked into wind energy and has explored its possibilities on his land. He pays more attention now to farm-related travel and tries to consolidate trips into town.

MARKETING, BUSINESS MANAGEMENT, AND FINANCIALS

The farm business includes three different business structures: one property is a limited partnership of Jack, his wife, and sons; another property is a sole proprietorship; and the rest of the farm operates as a C corporation. The different fiscal year closing for the C corporation provides a helpful way to manage some income and tax liabilities, although it has potential negatives, for example, with estate planning.

All marketing is done through the corporation, which is a certified-organic entity. The operation keeps meticulous records on production and income, which Jack sees as useful for finetuning and adjustment that ensures success. For example, the Erismans receive numbers from their buyer on the "cut-outs" when their animals are slaughtered. That data help to inform decisions on herd genetics and selecting animals.

Marketing and Organic Certification

The Erismans received their organic crop certification before USDA set and administered federal organic standards. Organic certification has been important for Goldmine Farm's access to profitable national and international organic grain markets (for example, much of its white corn has recently gone to the European market). Although the market premium for organic certification is important to the bottom line, the decision to transition to organic was initially also about reducing production costs. "In the first year of transition [to organic], I saved $100,000 in fertilizers and chemicals," Jack said, "but I didn't get good yields." Over time, his yields in organic production have, of course, improved. Non-GMO verification, however, is becoming a new area of concern. Even with his longstanding organic production system, he sees a risk that there might be "no zero anymore," given more widespread cultivation in the region and country of transgenic crops.

Jack did not organic-certify the cattle operation until 2003. He noted, "You can be successfully organic without the beef line. For us, the organic beef is a complement, a supplement. Most of our income is in the organic commodities." There is only one organic-certified kill floor in Illinois. The majority of finished animals (roughly 20 head annually) are sold through a grass-finished market. A few head are sold into the Chicago organic market (about two to four head annually) and into the St. Louis organic market (about six to nine head annually). Some of his beef (perhaps six head annually) is marketed locally, more as a community courtesy than a focused market niche.

Jack has been willing to try new varieties targeted for new markets. He is trying an experiment with red corn in addition to his production of blue corn. One critical issue with the development of new crops, in addition to getting the production system right, is the uncertainty of working with a buyer who is unfamiliar with the dimensions of the new market. Those challenges often increase the risk to a producer and can cause vexing delays in delivery and payment.

Although he had a contract for his first blue corn production, Jack views contracts cautiously, noting "I've been burned." Now, when he is setting up a contract, he tries to sit down with the other party, and ideally they write it together. He cites the importance of having trust and being trusted but emphasizes that he tells them, "Here's what I want." Ultimately, however, he said, "My preference is no contract. I'm a gambler." Therefore, he plans each year with the idea of storing all his grain, which allows him the flexibility of not having to sell his crop on the market at a specific and perhaps unprofitable time. Such storage capability is, in his view, essential for a farmer to be successful in organics. Over the years, he has shipped corn and beans directly to international markets.

Financial

Goldmine Farm's income derives overwhelmingly from crops and commodities rather than from the beef. Jack was able to pay off the debt on his farm, even before he started the farm's transition to organic around 1990. He has now reached a stage of financial comfort and success in farming. In 2008–2009, his financial concerns paralleled those of other Americans, whether or not they were farming, who have questions about how best to invest and protect their financial resources.

Risk Management and Insurance

For personal and philosophical reasons, the Erismans had for a long time chosen not to buy medical or life insurance. At Jack's wife's request, they now carry supplemental medical insurance. Although Jack had crop insurance on 320 acres farmed the first year he farmed with his father, he has not taken out crop insurance since then. As he said, "I don't believe in it [crop insurance]. What am I going to do? Sit around and pray for hail? I just never take it."

SOCIAL AND COMMUNITY CONSIDERATIONS

Labor

Despite farming more than 2,000 acres and managing livestock, Goldmine Farm operates with a minimal amount of hired labor. Jack currently has one full-time employee, who helps in all areas and is "as good with a wrench as he is with a sick animal." This employee grew up about an hour away and has a background in construction. He began working on the farm in 1999. The employee has been building equity in the operation by taking ownership of one cow in the herd each year plus her offspring. (In late 2008, the employee was up to 26 head total.) Jack developed similar arrangements with previous workers that helped to retain them for extended periods of time and that also recognized and rewarded the importance of hired labor as a resource contributing to the farm's success. In 1987, at the end of 10 years work at the farm, one worker went away with a $10,000 bonus. Jack described his full-time workers: "All those people have value. I don't want to be a user. I'm not talking about altruism. It's good logic for making systems work." Aside from having a full-time, year-round employee, he also hires college students to help on the farm intermittently in the summer.

Community Involvement

Jack has been active in his local community (for example, he has served as a volunteer fireman for some 30 years, and he has served on the board for his local conservation district). Particularly within his state and region, he has been a leader and spokesperson for what many describe as "sustainable agriculture." Although Jack started "doing" sustainable agriculture in the 1970s, it was in the mid-1980s that he became involved with efforts in Illinois to "bring those concepts into the mainstream." To that end, he played a part in the Illinois Sustainable Agriculture Society, an organization that is now mostly defunct because, as Jack puts it, "We've done what we set out to do, bringing the concepts into the mainstream." He continues to speak to groups of farmers and others on his approach to organic production of crops and beef cattle. Although few of his immediate neighbors farm organically, he has supported or informally advised other people in his region as they have transitioned about 1,500 acres to organic production.

Observations on Access to Organic Food

Jack sees the farmers' need to develop strong organic markets as linked through production innovation to the consumer's ability to access those markets. He says, "If you want the organic market to grow, you need to make it more affordable for more people. With forage-finished [beef], you can maybe compete." He has had the opportunity to work on these issues as a member of the Illinois Local and Organic Food and Farm Task Force, which was established by the Illinois Food, Farm and Jobs Act of 2007.

RESEARCH AND POLICY CONCERNS

Farm Programs

Jack serves on the board of the Organic Trade Association (OTA). He has assiduously avoided the use of federal farm programs or crop insurance. His aversion is based on personal experience in the failure of insurance and a belief, borne out by the success of his farm, that he can thrive economically without those programs. He firmly believes, "You don't have to have a government program to survive." The farm has received some technical assistance with conservation measures, such as terraces and waterways, but not financial assistance.

Research Participation and Needs

Jack served on the original elected board of the Illinois Council on Food and Agricultural Research (C-FAR), an entity with a broad base of stakeholders that in the past 15 years has directed more than $100 million of state money for needed research at four public Illinois universities in such disciplines as agricultural production systems, water quality and conservation, rural development, and food safety. For organic production in particular, he believes that research is best done at the farm level.

In terms of research needs, he is concerned about development of and access to good organic varieties. He believes that the organic seed supply system has become too proprietary and sees a need for more public plant breeding targeting organics. More work by land-grant universities on organic soybean is especially needed. Not many companies supply good lines of non-GMO seed, and some have unilaterally dropped lines that Jack considers good. Growing organic on contract is a way to get organic seed, but contracts have drawbacks, in his view. He feels that contracting should not be the only way to farm organically. He sees a need for farmers to adapt varieties to local conditions over time. More research at the farm level could help develop new organic varieties that are appropriate for specific sites, regions, and markets. He also sees a need for additional research investigation into the federal organic standard, its implementation, impact, and problems of compliance.

SUSTAINABILITY

The challenge to the gradual accomplishment of sustainability on this farm has been understanding and optimizing the system: "The most important thing we do here is understanding compatibilities and synergisms of different crops and plants in the system. It's knowing what crop do I need to follow with." He stresses the need to work with and reinforce the natural order of things, creating the lightest carbon footprint, achieving the greatest return from nature "without having to do much," and making the best use of the natural topography and soil resource.

Jack believes that his system is transferable because it is flexible. It involves knowing the site's soils and the topography, having good drainage, and understanding what steps nature will take in combination with a realistic sense of markets. A realistic sense of the market requires, he said, knowing what you are good at and what markets you can afford to supply.

Farm Transition Issues

The question of farm transition has become more salient lately. His two sons have gone away for college and for military service, and he is unsure whether either would be interested in the farm, given interests and obligations in their adult lives as they marry and start families. Jack muses, "The biggest question for me: How do we transcend my presence, if it's my presence that has been important. The farm may be more my personality than it should be." He recognizes that for someone to take on this farm requires a special combination of will and skill. He notes the depletion of young people from central rural Illinois, especially compared to generations past, but he also insists that he is "open to some bright young person who might come in."

Labor as a Limiting Factor

Jack identifies labor as a factor that could constrain what activities or enterprises can be taken on and developed at the farm. For example, value-added initiatives (such as packaging some of their specialty grains) could be considered if additional qualified labor is available. In that sense, limited access to sufficient labor might circumscribe some entrepreneurial options for his farm.

OBSERVATIONS AND CONCLUSIONS

Two factors appear central to the success of Goldmine Farm. First, the incorporation of livestock into the cropping system not only adds an important income stream, but also figures centrally in establishing a sound, highly integrated rotation system. Keeping land in pasture helps restore soil health and minimizes weed problems when cash crops are planted. Second, the ability to access a premium (for organic crops and beef) and to reduce and even eliminate costly chemical inputs of conventional farming enhances economic profitability. Those factors contribute to the success of the enterprise overall.

As Jack Erisman reflects on his life's work at Goldmine Farm, he is proudest of not having used government programs, educating his sons, and having created an organic agricultural system that has survived and thrived, even when at some junctures many people (including some family members) thought it was not possible.

Rosmann Family Farms

BACKGROUND AND HISTORY

The Rosmann Family Farm is deeply rooted in the rural Iowa community in which it is located. The original ancestor, George Rosmann, came to Westphalia around 1875, part of a wave of German Catholic immigrants brought to work and settle the area by the railroads. Present-day Westphalia is a small and still largely German Catholic hamlet located on a ridge a few miles north of the current Rosmann Farm. Westphalia remains an important community focus for the Rosmann family.[35] Ron Rosmann's great grandfather began farming in the area in 1883, at the original farm and homestead that the family retains and where Ron's middle son now lives. Ron's grandfather acquired additional farmland in the area. Born in 1907, Ron's father, as the eldest son, had to farm. He moved to Ron and Maria's present farm, after this farm was lost by another family during the Depression. It was on this farm and in this farmhouse that Ron grew up.

Ron went to Iowa State University in 1968, beginning as a farm operations major, shifting to distributed studies, and finally graduating with a degree in biology. He began college with no particular intention of going back to the farm, and while there became influenced by the anti-Vietnam War movements and other social issues of the time. Upon graduating, he worked briefly in a youth home in Ames and considered going into psychiatric social work. Ron's two older brothers were by then pursuing non-farm career paths. When Ron's father said, "You're the last one. Try it or I'll have to rent the farm," Ron decided to return to the farm, then 320 acres, for a year. He found he liked the independence of farming and the opportunity to be a leader in a small community. As he says, "I never looked back."

When Ron returned in 1973, his father was raising cattle and hogs, doing crop rotations, and using pesticides and anhydrous. His father was locally regarded as an innovative farmer. He had the first combine and first corn dryer in the area. He also kept a team of Belgian draft horses (until 1969) to work on the farm, even though they had three tractors. Ron and his father worked together from 1973 to 1980, when his father passed away "far too young." Ron regards those years of farming with his father as invaluable for their father-son relationship and crucial in his own development as a farmer.

The current farm of 600 acres (200 of which are rented) is in Shelby County, in a region of fertile rolling hills, roughly 35 miles from the Missouri River. All the land is classified as highly erodible, with slopes of 8 to 11 percent. The soils are loess soils that include Marshall silt loam on the high flats, Monona on the hillsides, and Judson in the bottoms, which were all tiled by hand by Ron's father and grandfather. The tilth of the soils is, in Ron's words, "incredible"—they dry out well but they also hold moisture. The farm has no ponds and no major streams, although Keg Creek starts on the Rosmann land. Ron and his wife Maria

---

[35]The Rosmanns are active in the Westphalia Catholic Church, which leads or coordinates a number of community and service programs, including recent construction of a new community center.

have three sons. The eldest, David, works for an agriculture-focused nongovernmental organization in the Twin Cities. The middle son, Daniel, has returned to farm with Ron and Maria, after completing a B.S. in agronomy at Iowa State University. The youngest, Mark, just finished a double major in agronomy and history at Iowa State University. He works on the farm summers and vacations and is still determining his post-college plans.

FARM PRODUCTION SYSTEM

Rosmann Family Farms is a diversified crop–livestock system, premised on internal cycling of nutrients, reduced purchased inputs, and organic certification, which provides formal recognition of its sustainable farming practices and a market premium.

Crops

The Rosmanns currently follow a six-year crop rotation of corn–beans–corn–oats or barley or succotash–alfalfa–alfalfa, with crops certified organic since 1994. The two years of corn represent a concession to economic reality, as corn is raised for feed and for cash; its production contributes to the profitability of the farm. The Rosmanns feed their barley and oats to their hogs and cattle. After harvest, cattle are run in the fields for gleaning; otherwise "we are throwing feed away." It is hard to do more than two years of alfalfa in a field, before pocket gophers begin to damage roots and make the terrain more difficult for farm equipment.

Planting

Their seed corn is 50 percent Blue River and 50 percent an untreated and non-GMO Pioneer® hybrid; their soybean for planting comes from Blue River. The Rosmanns also raise some soybean for Blue River. The farm is divided into 40 fields, between 5 and 40 acres each, depending on the slope.

They ridge-till corn and beans, planting the rows 38 inches apart. A key to their system is planting on time, which is roughly May 1 for corn and May 20 for beans, depending on the weather and soil. Ridge tillage works best for their system, which has four to seven times less weeds than other systems (for example, no-till) and provides them with the best yields. As required for organic certification, the Rosmanns keep a 30-foot buffer around their fields from plantings on neighboring farms of GM crops.

Yields

The Rosmanns' crop yields are at the county averages, which are 135–145 bu/acre for corn, 45 bu/acre for soybean, 80 bu/acre for oats, 65 bu/acre for barley, and 5–6 tons/acre for hay. (In 2007, the farm averaged 55 bu/acre for beans and 160 bu/acre for corn.) The farm does not currently produce enough manure to get higher yields, and the Rosmanns do not haul any additional manure onto the farm.

Inputs

Ron stopped using pesticides in 1983, but the farm did not become certified organic until 1994, when organic markets were strengthening. He notes that when he first moved toward organic farming practices, purchasing off-farm organic inputs was the emphasis.

ILLUSTRATIVE CASE STUDIES

The Rosmanns do not purchase nitrogen inputs for their farm and have never used fish meal. They have not applied potassium since 1983. They use soft rock phosphate every third year at 250–300#/acre; they also add micronutrients (zinc, copper, and sulphur) based on soil tests, which they conduct every third year. They often bring home from the mill the cleanings from processing of their organic soybeans into meal to feed their own hogs to maximize economic value from that crop.

Livestock

Hogs

The Rosmanns have 50 sows at any one time, and raise about 600–800 head a year. Their hogs are a Berkshire–Chester White–Duroc cross (Berkshire and Duroc for the meat quality). Organic Valley requires half-Berkshire for the red color of the meat and the marbling. The hogs are housed in small groups in barns with access to the outdoors for disease and parasite reduction. Indoors, Ron keeps the straw bedding sufficiently deep so the hogs stay dry, and he notes that bedding application levels need to be coordinated and agreed on among everyone working on the farm to achieve that. Pyrethrins are not added to the feed; a chrysanthemum-based spray is used on the pigs for lice and mange. The Rosmanns use vinegar in the feed rations to manage scours to keep the hogs' guts somewhat acidic. Sick hogs are isolated, treated, and then sold as nonorganic (not through Organic Valley).

The hogs are raised in a farrowing house, which has individual units, for six weeks after they are born. They remain in the farrowing house for at least two additional weeks after they are weaned and the sows are removed. The Rosmanns no longer use farrowing crates for the hogs. When they reach 30 pounds, the hogs are moved to a small hog barn, where they are kept segregated. At 50 pounds, the hogs are moved to the grower unit where they are raised to 100–120 pounds. When the hogs reach 120 pounds, they are moved to a finishing barn. All of the barns are treated in the same way as hoop houses, with a constant cover of bedding and a cleaning once the hogs are removed. The Rosmanns try to allow the farrowing house to sit empty between groups of young, but find that they often cannot do so for very long. They rely on good sanitation practices, using a power and chlorine wash on the building after it is emptied.

They seed peas in with the barley and harvest them together as feed for their hogs, which the hogs love. Changes over time in the hog operation include discontinuing docking tails and taking out needle teeth. Those changes, which Ron supports, were his son Daniel's choices as he assumed more control of the hog enterprise.

Cattle

The Rosmanns raise 70 feeder steers and heifers and about 85 cows and calves, all a Red Angus breed for which they are continually try to select for individuals suited to their farm location. Ron notes that "every breed has good individuals, and it's a matter of selecting within the breed" to improve their own internal system on the farm. They also raise their own bulls. Their cattle do not have as many health problems as their hogs. They have had only a few isolated outbreaks of pneumonia over the past 10 years, for which they isolate the cattle in the same way they do sick hogs. Per requirements of Organic Valley, their buyer, they feed corn to their cattle, because that helps them make "choice" rather than "select" grade.

The Rosmanns are considering a transition away from such heavy corn feeding, but they recognize the need to find the right genetics to get choice grade meat with grass-fed cattle. Their hope is to do some selection for individuals with the right characteristics for their farm and location that would thrive on a pasture regime with some grain. However, those considerations are tempered by their knowledge that it typically takes at least two to three months longer to raise a pasture-raised animal for market.

They rent a tub grinder to grind hay that they pile for the cattle when the snow is too deep (from December to February). They spread the manure out from the cattle feeding areas before they are ready to plow.

Poultry

Maria manages another small enterprise of about 150 non-certified organic broilers once a year. She starts the birds in late summer and butchers them in the fall before the cold of winter, a system that is the reverse of many farmers, but more efficient in her view. The birds are used on-farm and sold informally to local extended family.

Pest Management

The Rosmanns now have minimal insect problems, although they stopped raising food-grade soybean (for the Japanese market) in part due to an infestation of bean leaf beetles that discolored the beans and downgraded their market value. They have since moved entirely to feed-grade organic beans that yield better and pose less risk. On occasion they have problems with corn borers for which they use Bt, but they have only used Bt once in 25 years.

Their weed problems include a variety of thistles, which they believe they are managing through hand-digging from their pastures. They have velvet leaf and giant ragweed in the crops. Velvet leaf was at one time much worse. Giant ragweed has become a more recent concern, spreading up into field areas adjacent to small creek bottoms. Their primary tools for dealing with weeds are ridge tilling and crop rotations.

Pocket gophers and deer are serious vertebrate problems. As mentioned, the gopher mounds damage crops and equipment and prevent the Rosmanns from going more than two years in alfalfa if they decided that was desirable. They also have to consider gopher infestations as they think through their rotations. The deer tear up fences and crops at the perimeter of fields.

Pasture Management

The Rosmanns have 120 acres of pasture, which they like to keep for 7–10 years (one pasture has now been in place for 14 years). They do not rotate pasture as much as they used to, and they see a need for more nitrogen on their pasture lands. Ron notes, "There's a weak link. We recognize we need more nitrogen [in the pastures] for optimal growth." They fence their field with high tensile wire on the outside perimeter and move cattle from paddock to paddock every three to four days, based on a visual assessment of the condition of the forage in the pasture.

They have added shrubbery to some of their terraces even though doing so has not matched recommendations from NRCS. The shrubbery provides beneficial insect and bird habitat. Despite concerns from NRCS, the terraces have remained stable and intact.

Equipment

The Rosmanns have a large diversity of equipment, including an old Case IH tractor. Most of their equipment is bought used, except for the manure spreader, which was bought new. Over the past five years or so, they have purchased about one new piece of machinery each year. They bought a new wheel rake in 2007 and a new manure spreader in 2008. They also purchased a new disc mower for cutting hay in 2007. They rarely get rid of machinery, noting that one never knows what might be needed. It has become harder to get parts for some of their older machinery, partly because it is old and partly because of the decline in nearby farm equipment dealerships. The Case IH dealership in nearby Harlan went out of business in 1983, so they go to Avoca for Case, which is 15 miles away, even though Ron acknowledged that 15 miles is not so far considering what many farmers now have to travel. Their ridge tillage and planting equipment is smaller than typical, which allowed them to get it used for less expense and then adapt it to their farm. They do some sharing of farm equipment with neighbors.

New technology, such as cell phones, has been useful for the farm, enabling Ron, Maria, their sons, and a hired man to check in with each other, answer questions, and reduce truck miles around the farm or into town.

Labor

The Rosmann Family Farms are very much a family operation, with Ron managing most of the production side until recently, Maria handling most of the marketing, and their sons providing critical labor and input. Daniel, in his mid-20s, plays a particularly important role and has joined the farm full-time, assuming greater responsibilities. Among other things, he has taken over feed management. With his younger brother, he has taken the lead on the hog enterprise. Currently, the Rosmanns also employ a nearby farming neighbor three days a week year-round. This worker has exceptional mechanical skills. He is critical for "making us more efficient for production" particularly given some of the ongoing challenges in operating, repairing, and maintaining equipment.

Nutrient Management

The Rosmanns have a Comprehensive Nutrient Management Plan, which they completed with help from NRCS. They use composted manure from their cattle and swine. They compost the manure in windrows with the bedding—barley and oat straw—along with round and square bales of stubble hay that are used for farrowing. The compost is mechanically turned to achieve optimal temperatures and then applied to fields with a manure spreader.

The alfalfa in their rotation is also used to provide nitrogen for the soil. Being on a two-year rotation maximizes the tonnage of hay and the nitrogen, which peaks at the second year. Manure is also distributed in the field through the feeding system in the winter for the cows. The system uses a hay grinder to create feeding piles that are moved around the fields.

NATURAL RESOURCES, ENERGY, AND CLIMATE CHANGE

Water

The farm's water comes from wells and the rural water system. The wells frequently go dry because they are less than 100 feet deep. The typical rainfall is 28–30 inches/year. The farm is on the drier side of Iowa and was not adversely affected by the torrential rain and flooding in the spring and summer of 2008.

Energy and Carbon Concerns

The Rosmanns have long been interested in alternative energy and in reducing their energy use. Ron and Maria organized an alternative energy conference in nearby Harlan some years ago, which was attended by 300 people. They built a solar nursery and farrowing house (from a design of Iowa State University Extension), which saves significantly on energy costs. They have focused on purchasing smaller, more efficient equipment appropriate to their size operation. They have also cut back on some operations, such as mowing. They are considering growing beans specifically to make biodiesel and would like to explore wind power. Alternative energy is something Ron says he would like to "key in on."

They look at their system as an economically productive ecosystem in its use of carbon. The size of their operation allows for a rotation of crops to restore nitrogen and carbon to the system (without importation of nitrogen). The livestock provides a means to create a value-added product that also returns nutrients to the biological system.

MARKETING, BUSINESS MANAGEMENT, AND FINANCIAL

Marketing

The Rosmanns have been marketing organic beef for 10 years and organic pork for 4 years through Organic Valley. They thought it would be more difficult to meet Organic Valley's requirements for pork than was the case. Organic Valley frequently shifts where it has the beef cattle sent for slaughter. The slaughter site for pork, which is in Sioux Center, Iowa, has not changed since they started. Ron would like to see Organic Valley move toward standards for more pasture-feeding of animals. His interest in how markets for grass-fed meat develop has been reinforced by emerging information about the potential human (heart) health benefits of grass-fed meat. Current debates about U.S. agriculture and the American diet resonate with him, considering his own personal and family health histories.

The Rosmanns also process some organic beef at the Amend plant in Des Moines, a longstanding arrangement that has worked well. They pay for the organic certification at the Amend plant. For 10 years, they have marketed that beef under their own Rosmann Family Farms label. Most of it goes to central Iowa retail establishments, notably Wheatsfield Cooperative in Ames and Campbell's Nutrition, which has three stores in Des Moines. Those retail shops have been good outlets for their label. Maria has held cooking demonstrations and tastings at Wheatsfield.

They also process some organic pork at a different meat processor to sell under their own label. That pork is not certified organic (instead "natural") because they are not sure they can get the processor certified. It is sold to the same shops as their beef.

They were the first to market organic meat in the area. For a brief time, they had a mini-CSA in Des Moines through the support of an interested acupuncturist. It involved both Ron and Maria being in Des Moines once a month (12 days a year). As their children reached the more activity-laden years of middle school, they found the arrangement worked less well, and they focused on retail outlets. In the late 1990s and early 2000s, they frequently supplied meat for special events at Iowa State University that sought to serve "Iowa-sourced meals." Although that arrangement provided good visibility for their farm and product, they have moved away from it. Ames is not close to their farm, and "we've streamlined now for our needs."

They keep seven freezers on the farm to ensure sufficient supply of their farm-raised meat to service their central Iowa accounts and customers. Even with the economic downturn in 2008, sales remain steady. Although they might now be selling "less steak, it's consistent for ground beef."

They have never had to advertise their own label meats, as word-of-mouth recognition has spread. They recently bought their first advertisement, in the Wheatsfield Co-op newsletter, but this was more to support the cooperative than because they actually need to advertise.

They market the grain that they do not use on farm for their own animals to a variety of buyers. In general, they "can't meet the demand" for organic grain and have no problem selling it. They have marketed their grain through Scoular Grain in Omaha, Grain Millers in Minnesota, and Heartland Organic Coop. They were involved with a marketing cooperative that focused on organic soybean for export to Japan. It went under five years ago, because, Ron asserts, it became a victim of poor timing and perhaps too narrow of a focus relying on Japanese sales. Everybody in it lost money.

Certifications

The cropping operation has been certified organic since 1994. The Rosmanns' beef operation has been certified organic since 1998.

Finance and Business Management

The farm is organized as a sole proprietorship, although the Rosmanns have considered an LLC or another corporate structure. They carry production loans each year because of the crop cycle. Their accountant works with them to use the tax system effectively (paying attention to depreciation and timing of sales) for their farm. They continue to be comfortable with the debt load they carry and look at it as a management tool that is key to the success of their operation and their peace of mind. Ron noted that "debt is not a bad thing necessarily. You have to manage it."

They carry catastrophic insurance. They do not carry revenue insurance even though they believe it is better than hail insurance. (It hails often in their area.) The primary reason is that they typically receive higher than average prices for their products so the insurance would not adequately compensate them. With Daniel joining the farm now, they feel greater concern about farm income to support both generations.

SOCIAL AND COMMUNITY CONSIDERATIONS

The Rosmanns' orientation to farming is undergirded by a strong sense of social responsibility. Both Ron and Maria are active in church and local activities, and they have also been active in state and national policy debates. Both have provided testimony before Congress on agricultural issues, and Ron was running for a seat in the Iowa House of Representatives at the time of the interview. He had been thinking of entering state politics for a while, and he saw the moment as a good time to launch a bid, given issues in the state and Daniel's developing role on the farm.

The Rosmanns have long relied on and learned from a network of other Iowa farmers also interested in low-input and organic farming. They have been members since the beginning of the Practical Farmers of Iowa (PFI), and Ron has served as its president. Sustained connections with other PFI farmers over the years have allowed informal back-and-forth about ridge tilling, crop rotations, composting, animal rations, and other topics. The Rosmanns have held numerous field days for PFI farmers and others at their farm.

They have also participated in considerable research on their farm. Some of their participation has been through field and farm experiments coordinated and managed by PFI. The Rosmanns execute the trials and collect the data. Many Iowa State University researchers have also conducted research on the Rosmann farm.[36] When asked why they participate in research, Ron notes he is "enthralled with it," and sees involvement with farming research as a way to be more involved with science. At the same time, he is bothered when he sees grants for sustainable agriculture research that budget little or no compensation to the participating farmers. Both Ron and Maria have served on various grant review panels related to sustainable agriculture, including for the USDA Cooperative State Research, Extension and Education Services' organic program, Value-Added grants, USDA Sustainable Agriculture and Research Program, and the Organic Research Foundation.

More recently, Ron has become involved in and served as treasurer of the Iowa Organics Association, a group formed in part to provide a collective voice for the interest of Iowa organic farmers relative to issues of genetic drift from GM production. He has some concerns about the viability of another new organization and how it fits within the larger landscape of organizations addressing sustainable agricultural concerns.

Informally, the Rosmanns often confer and compare notes on farming or specific challenges with a neighboring farmer who also farms organically. He does not have livestock, but this connection is still an important source of information and support.

Federal Farm Programs

The Rosmanns receive commodity payments for corn and soybean (they also receive some small payments for oats and barley) but have a comparatively small base relative to their neighboring farmers.

They have been involved with Environmental Quality Incentive Program for their comprehensive nutrient management plan for windbreaks, pasture management, and buffer strips. They also have 2.5 acres of land along Keg Creek in the Conservation Reserve Program. They would have been glad to be involved in the Conservation Security Program, but their farm is not in a priority watershed. They hope to participate in the Conservation Stewardship Program coming out of the 2008 Farm Bill.

They will not put any more land into terraces because the terraces take too much out of production and are difficult to maintain. Instead they will rely on buffer strips that are more flexible. They currently make good use of their headlands, keeping them organic and cutting hay from them.

---

[36]Other studies, beyond the many with Iowa State University researchers, in which the Rosmanns have participated are Michigan State University examining ridge till vs. conventional cultivation and the University of Iowa investigating methicillin-resistant *Staphylococcus aureus* in the animals and workers on sustainable versus conventional farms.

Ron stresses the importance of helping beginning farmers in the community. He notes his own father's approach, which involved renting out two 160-acre farms to young men wishing to get started, rather than his father farming that land himself. Daniel is participating in the USDA Beginning Farmer Program for a low-interest loan (a 30-year loan at an interest rate of 2 to 3 percent) for the purchase of an additional 70 acres (all in CRP), which will bring the Rosmann Family Farms to a total 670 acres. Daniel has spoken publicly in Washington about the importance of that program.

RISKS, CHALLENGES, AND CHANGES

While diversity is a linchpin for the Rosmann Family Farms, it is also time consuming. Ron is quick to underscore the advantages of diversity on the farm—its beauty, productivity, associated ecosystem services, and energy processes. However, he and Maria quip about the possibility of being "overly diversified," pointing to challenges in identifying "optimal diversification" on the farm, itself a condition that is different over time in response to the various elements of the system and needs and resources of the people farming. Nonetheless, they feel considerable pride and pleasure about the system they have developed.

The Rosmanns are beginning to think through transition issues, especially now that Daniel is farming full-time. Ron's experience in working well with his father in that farm transition has provided a good model for involving his son's ideas and energy in the farm. Although there is some day-to-day operational learning in the transition, there are also issues of calibrating the farm's enterprises overall and ensuring they can support all family members who look to the farm for livelihood.

OBSERVATIONS AND CONCLUSIONS

The keys to the success of the Rosmann Family Farms, as they see it, have been their longstanding commitment to crop rotation, finding a ridge-till system that works well for them, including ruminants to recycle nutrients, growing a diversity of crops, and pursuing value-added (organic) meat production. Having an excellent location for farming, including good soils, provided a good base. The involvement by all their sons and a commitment to farming as an occupation by at least one of them have also been critical to the vitality and prospects of their farm.

Sheer zest for farming is also important for understanding Rosmann Family Farms. Ron says of farming: "I love it. There is always something to learn." They see research as an ongoing and pressing need to ensure the future of their type of farming, particularly for development of more and better organic crop varieties (for example, barley). Although they have been very involved with sustainable and organic farming research, they see the need for much more in this area.

Zenner Farm

BACKGROUND AND HISTORY

The Zenner Farm is a family-operated sole proprietorship run by Russ and Kathy Zenner. It is located in Latah and Nez Perce Counties, Idaho, in the southeast corner of the Palouse region. Dryland farming predominates on the rolling terrain of the Palouse. The Zenners currently farm 2,800 acres of land (640 owned, 2,160 leased) and produce wheat, small grains, lentils, peas, and garbanzos. Russ has been a regional pioneer in adopting, learning about, and promoting direct seeding (a method of planting and fertilizing done without prior tillage to prepare the soil). He is a longstanding member of what is now the Pacific Northwest Farmers' Cooperative, which handles a large portion of the pulse market in the United States. More recently, he has marketed a portion of his crops through Shepherd's Grain, a new regional value supply chain certified by the Food Alliance.

Russ's family came to the area in the 1890s from Luxembourg. His grandfather managed to get six boys started farming, beginning in 1936. Russ reflects, "We've been blessed with where my granddad settled. We've got good dirt here." He says that the first generations of Zenners who farmed had the goal of living conservatively to make future opportunities possible for themselves and their children. By the time Russ was growing up, the family farm included cropland and a cow-calf operation that Russ's father and uncle ran in a partnership. Russ and Kathy married after high school, and she worked at a bank and helped put him through the University of Idaho. After finishing his B.S. degree in agricultural economics, Russ worked for a while for the Farm Credit System. The couple came back to the farm in 1970. Russ was the oldest of his generation in the extended family who returned for some involvement with the farm.

By the early to mid-1970s, the farm was structured as a corporation that involved members of the extended family, including Russ's uncle, his father, his brother, and various cousins. At that time, the farm involved an extensive (1,500 head) livestock-finishing operation and about 4,000 acres of cropland and 8,000 acres of rangeland. In 1984, the farm split into two separate partnerships, one with Russ and his brother (which included livestock finishing and 2,200 acres of cropland) and the other run by his cousins. In 1993, Russ and his brother split their partnership in two, with Russ's brother getting the finishing operation and Russ and Kathy taking over the cropland. Because Russ observed consistent profits in livestock farming and had developed his passion for the cropping side, he preferred to turn his full attention to crop farming. His approach and philosophy were influenced heavily by what had been his first volunteer experience: serving for 12 years as Latah County Soil Conservation supervisor. That work opened Russ's eyes to the dramatic impacts of soil erosion in the region and the role of agriculture in that problem. He says, "I was on a mission early in my farming career to reduce the detrimental influence of tillage on our cropland. I didn't want to go broke doing it, so we went slow with changes on our farm." In that way, Russ developed a clear goal early on to question and change long-accepted practices of farming in his region.

ILLUSTRATIVE CASE STUDIES

Over the years, Russ and Kathy expanded their land holdings through purchase and developed some long-term relationships to rent farmland.[37] For about 20 years, Kathy augmented their farm income by working at the local cooperative. In time, the Zenners had three children—a son and two daughters. The couple also made some sound investments that provided security and enhanced their ability to develop their farming system in accordance with their values and priorities. Russ now enjoys reading about and discussing food, agricultural, and environmental issues. In the summer of 2008, he spoke avidly of ideas he had encountered in Michael Pollan's *The Omnivore's Dilemma*, William McDonough and Michael Braungart's *Cradle to Cradle*, and Andrew Duffin's agricultural and environmental history of the Palouse, *Plowed Under*.

FARM PRODUCTION SYSTEM

Russ's approach to his farm and development of a farming system emerge from his attunement to topography and the health and productivity of his soils. He says, "My main motivation is to farm this ground in a manner that we can build topsoil. That means we can't do much tillage." His concerns regarding the sustainability of his farming system thus center first on tillage and the need to reduce disruptions, and subsequently on the impacts of manmade chemistry (that is, how agricultural chemicals might affect soil biology). He describes himself as becoming increasingly interested in "biological farming": an approach that focuses less on chemical inputs and more on improving the microbiology of soils and plants as a way of enhancing crop health, quality, and yield.

Soils and Growing Conditions

The soils in the region are unique, young, wind-blown loess soils that are easily affected by erosion on the rolling hills. Tillage erosion, over the roughly 120 years during which the area has been farmed, has exposed clay ridges. As the land has been tilled with moldboard plows initially, and chisel plows and disks today, the soil has been moved further down the slopes so that the topsoil on the upper part of the hills has thinned. The evidence of erosion and knowledge of its impacts have compelled Russ to move to direct seeding.

Between Russ's farmlands and up into Canada, little irrigation is done. However, the conditions are more desert-like southward. Annual rainfall on Russ's farm is about 22 inches. The weather can affect the quality of the pulse crops he grows. There is good moisture in May and June, and weather can be cool even in the summer. At the same time, moisture has critical effects on the quality of lentils and garbanzos; it can negatively affect the color or can discolor the crop, which reduces its quality. The region can have wet autumns, which can make late-season harvesting difficult. In addition, harvest times differ across the rolling topography. A week's difference in maturity can affect the quality of the crop with frequent differences between north- and south-facing slopes.

Crops and Rotations

In this region, according to Russ, winter wheat has been king for a long time. With the shift toward planting more pulse crops, wheat on Zenner Farm now has a place in a more

---

[37] Russ notes that farm expansion opportunities have become very competitive in his area. Today they can be more driven by money, than by long-term relationships. However, in some cases, being a direct-seed farmer can help, as when an older landowner, taking his land out of CRP, specifically sought a direct-seed farmer and leased to Russ.

complex rotation of cool-weather crops. The rotation typically involves winter wheat, followed by a spring grain (wheat or barley) or certified seed, and then followed by a broad leaf crop (such as lentils, peas, garbanzos, or a brassica). Within a category, the Zenners try to diversify at least every three years; hence, they rotate garbanzos with peas or lentils to break up disease cycles. At times, Russ has also planted grass for seed. He is exploring the possibility of other rotations, such as sweet clover–alfalfa as a nurse crop for late-season planting, which could be a means to include green manure in rotations and still minimize the need for tillage.

In the process of considering rotation options, Russ has looked at the possibility of raising livestock, which was a more typical production option in previous decades than at present. He has considered sheep, which could pulverize the stubble and eat volunteers (and possibly allow less glyphosate use). However, most livestock in this region require an over-wintering facility, which could create concentration-related problems, because deep mud can limit grazing in winter and early spring.

In terms of current crop allocations, about 1,200 acres are to winter wheat and winter peas, 800 acres to spring grains, and 800 acres to pulses; another 30 acres are in CRP. Russ averages 1700#/acre production on garbanzos. He averages 95 bu/acre on winter wheat, and 60–65 bu/acre on spring wheat. For winter peas, he averages 2,500#/acre; for spring barley, 5,000#/acre; and for lentils, 2,000#/acre.

Fertility Program

Russ conducts regular soil tests before every grain crop and applies fertilizer according to Washington State University and University of Idaho recommendations. The requirements vary dramatically by crop. For dark-red winter wheat, he typically applies 120#N/20S/10–20# per acre. He will then top-dress soft white wheat (20#/A of N) and high protein wheat varieties (40#/A of N). He does not put fertilizer on his pulse crops.

Russ has seen a general decline in soil pH over time on his farm, but he has no cost-effective source of lime to apply. He has seen a decline in yield for pulse crops, perhaps due to the change in pH and the absence of new genetics for those crops. He has not observed as much change in the grain yield. There appears to be nothing in the literature on work to adapt crop genetics to deal with declines in soil pH. He is particularly frustrated that current management practices are not sustainable because of factors such as declining soil pH. He does not believe that what has worked in the past will necessarily work in the future.

Russ monitors pH and Brix in the sap of the plants while they are growing. He has looked at some research into the response of plants to biological agents such as the application of molasses at 1 pt/acre. He has also tried biological foliar sprays, but with equivocal outcomes: "We thought our discovery process would happen quicker."

Direct Seeding

As noted earlier, Russ's service with the Latah County Soil Conservation District after returning to the farm helped him see firsthand the need to create a more sustainable agricultural system and prompted his interest in direct seeding (no-till) as a viable conservation option. The late 1970s had seen a push for no-till, but various challenges from disease problems, inappropriate rotations, poor yields, residue management problems, and the cost of glyphosate all served to create new risks for farmers interested in the transition. Russ observes that in those days, they did not understand, for example, the "green bridge" and failed to anticipate how heavy residues could pose a problem for planting. In a flurry of activity, the result was that a lot of farmers jumped into no-till too quickly and then did not do it successfully. Russ says that "no-till has struggled to be acceptable, to have a good name. That's actually been another motivation for PNDSA [the Pacific Northwest Direct Seed Association]."

As Russ looked at direct seeding for his own farm, he did not want to lose his certified seed capability (and the associated income) and realized that he would have to figure out how to reduce risk in the system. The transition took seven or eight years and involved considerable experimentation, inquiry, and informal education. For example, he learned a lot from experience of others, including Dr. Dwayne Beck at the South Dakota State no-till research station. Russ sees the transition to direct seeding as key to his long-term profitability.

As the Zenners have become established in direct-seed production, they have also adapted and designed planting and spraying equipment that is more efficient and particularly suited to the region and their operations. Russ comments, "We're so much more labor and equipment efficient than we were 20 years ago. It's dramatic." They also provide custom services with their direct-seeding equipment through ViCo (see below and see also a case study on the Zenner Farm by Washington State University Extension[38] for a good description of the equipment).

Disease and Pest Management Issues

Although this area of the country is fairly dry, plant diseases can still pose problems. Garbanzos can be infected with ascochyta blight, which damages the plant (stem, seed, and pod) and is exacerbated by cool wet weather. Solutions have included finding and using resistant varieties (which the USDA Agricultural Research Service has been working on at Washington State University). Nonetheless, managing the problem is challenging, as evidenced by the production moratorium on garbanzos in Idaho from 1988–1991 as an effort to break the disease cycle.

Crop rotations, which contribute to more diversified production, also help with disease management. Russ has found a two-year or ideally three-year interval between planting a particular crop helpful in managing diseases. He sees rotation management and incorporation of diversity as critical for his farming system. The use of certified seed also seems to have reduced disease problems. Finally, Russ has made prophylactic applications of some fungicides when grass herbicides are applied to reduce the potential for infection. Overall, the blight has not posed much problem in the past two years. Russ attributes the scarcity of blight problems to farmers in the region commonly using one fungicide application as a preventive measure, having better seed sources, and attending more carefully to rotations. Russ is fairly satisfied with his own current ability to address plant diseases. He also sees his own generally successful disease management as a function of monitoring his fields. Nonetheless, he says, "this issue [of pest and disease management] does bother me. It relates to our dependence on manmade chemistry to manage these problems." He thinks more knowledge about soil and plant health could be useful for devising other management options for disease.

Another important disease issue Russ identified on his farm is "green bridge." Under minimum tillage or direct seeding of spring crops, the volunteer grain and weeds grow-

---

[38]Mallory, E.B., R.J. Veseth, T. Fiez, R.D. Roe, and D.J. Wysocki. 2001. Direct seeding in inland northwest. Zenner Farm case study. Available online at http://pnwsteep.wsu.edu/dscases/ext_pubs/pnw0542.pdf. Accessed on December 6, 2009.

ing between crop harvest and spring seeding can serve as a "green bridge" host for root diseases and other pests. A good fall weed control program has seemed to contain that potential problem. Nonetheless, Russ feels that insufficient research has been done, particularly with regard to the role and function of soil and plant health in minimizing or even suppressing "green bridge" problems.

With respect to insects, wireworms can pose a problem in lentils. Russ has responded with early seed treatment and some insecticides. He has also had problems with slugs on Austrian winter peas and has used some baiting. Aphids can be a significant problem in pulse crops; Russ has in the past had to spray as often as twice a year, although that is not the case now. He believes that his overall farming system is healthier now, as his rotation diversity has increased. He has noticed that weaker plants are more likely to be infested and speculates that aphids are more attracted to those plants, which reinforces his belief that plant and soil health needs to be a priority.

The direct-seed system, which is at the core of the Zenner farm, is intended to minimize weed competition. However, changes in weed species problems are observable during the transition process. The use of glyphosate as burn down at planting is typical. While in general, Russ would like to see less glyphosate in his farming system, it plays an important role at present. "We're not close yet to zero [glyphosate] use," he says. With the use of the direct-seed system, Russ has observed some shift in the types of weed problems he faces. It is more common now to have problems with bedstraw, china lettuce, and rattail fescue than in the past.

NATURAL RESOURCES AND WILDLIFE CONCERNS

Russ has observed changes in wildlife populations in the years he has been farming. He notes it was rare to see elk when he was a child. But elk, moose, and deer populations have increased in this area, as wolf reintroductions in the high country have chased them southward. As well, farmers see a lot of evidence that the elk and deer love the garbanzos and the Austrian winter peas. However, Russ did not frame predation on crops by wildlife as a major problem.

Russ has the impression that bird populations have declined in the area and wonders if the decline is related to bigger fields and fewer fencerows. He also wonders how pesticides are affecting bird populations, but he stresses he does not have the answers to those questions.

Russ has participated in an Idaho Fish and Game program promoting buffer strips for wildlife. The agency pays farmers $20/acre, for up to a total of 100 acres, to leave a foot of stubble on lands along existing bird habitat. He and other farmers also participate in the state's Fish and Game's Access Yes program, which provides public hunting opportunities on private lands.

MARKETING, BUSINESS MANAGEMENT, AND FINANCIALS

Preserving identity and adding value to products are central to the marketing strategy for this farm. Russ observes that identity-preserved crops can involve more work (for example, meticulous cleanout of combines), but their greater profitability makes it worthwhile.

The Zenner Farm markets about 80 percent of its crops through the Pacific Northwest (PNW) Farmers Cooperative and the remaining 20 percent through Shepherd's Grain. The Pacific Northwest Farmers Cooperative emerged in June 2008 from the union of two preexisting cooperatives in Colfax, Washington, and Genesee, Idaho. The new PNW Cooperative now includes 600 farmer members who raise 500,000 acres of cool-season crops: peas, lentils, wheat, and garbanzos. The cooperative has $100 million in gross sales and 13 million bushels of storage capacity. Forty to 45 percent of the sales are domestic, a fair amount under contract. For the past 25 years, PNW Cooperative has focused on providing valued-added products sorted by size, color, and quality, with the goal of ensuring high product uniformity to meet customers' expectations. The strong orientation toward adding value to products has enabled the cooperative (and its immediate predecessors) to grow even in tough times. The cooperative's products, depending on destination for export or domestic markets, are either loaded on barges in Lewiston and transported down the Columbia River to Portland, trucked directly to Seattle, or trucked to one of two PNW Cooperative rail-loading facilities in the growing region.[39] As a farmer-owned institution, the cooperative, Russ stresses, is an integral part of the local community. Russ has agreed to serve on the board of directors for the cooperative.

Russ is also a member of Shepherd's Grain (http://www.shepherdsgrain.com/index.htm), a marketing label and alliance of farmers in the Pacific Northwest, who use sustainable production practices and market differentiated wheat products together. Shepherd's Grain consists of 28 farmer-members, all of whom are certified by the Food Alliance.[40] It has drawn growing attention from agrifood researchers and activists as an example of new "value chains" that can help support an "agriculture of the middle." Shepherd's Grain has emphasized wheat varieties with special flour functionality desired by artisanal and quality markets. It supplies flour, for example, to family-owned Hot Lips Pizza, which has four restaurants in Portland, Oregon. It also supplies to Bon Appétit, a food service company that has become very engaged in regional sourcing. Most of Shepherd's Grain's distribution occurs within the Pacific Northwest or northern California.

Russ views Shepherd's Grain as "a very fun project." It has brought him into greater contact with the Portland food market, which he sees as currently one of the most innovative and sophisticated in the country. Experiences and insights from his participation in Shepherd's Grain, in turn, are useful for his involvement with PNW Cooperative, especially in terms of how to anticipate and respond to the challenges facing value-added agricultural products and the possible impacts of economic downturn.

The Zenner Farm has been certified by the Food Alliance since 2004, the first farm in Idaho to receive this certification. It was certified on the basis of its direct-seeding practice and additional criteria, such as worker safety and chemical storage. The Food Alliance inspects the farm every three years. Russ notes that the certification compelled him to make some changes in areas such as chemical storage. He approvingly notes that the Food Alliance now has a cropping *system* certification option, rather than only a focus on certified crops. Russ thinks that the certification offers helpful differentiation in the market place, is the most recognizable of the certification programs, and elevates awareness and commitment for sound growing practices. "For what we're [Shepherd's Grain] doing, Food Alli-

---

[39]The consultants learned during their field visit that PNW Cooperative now supplies garbanzos to food manufacturer Sabra, which recently entered the U.S. market and now makes hummus on the East Coast. The hummus is then shipped back to the western United States, where it can be purchased in the Lewiston, Idaho, Costco.

[40]Many links and overlaps exist between groups such as Food Alliance, Shepherd's Grain, and the Pacific Northwest Direct Seed Association (PNDSA). Some direct-seeding farmers saw the potential of Food Alliance certification to provide a value-added marketing opportunity that could reduce the risk of transitioning to direct seeding. From its start, Shepherd's Grain, comprised solely of direct-seeding farmers, worked with Food Alliance. Furthermore, a significant core group within the Shepherd's Grain alliance is the Columbia Plateau Producers. Columbia Plateau Producers (CPP) is an LLC with about 14 farmer members, including Russ Zenner. CPP farmers constitute about half the farmers participating in Shepherd's Grain.

ance is the best match," he says. His long-term goal would be no-till organic, but "we're not close to it at all."

Financials

The Zenners took out production loans regularly in the past, but have not done so for the past four years. Russ notes that the farm incurred considerable debt to get where they are today, but that over the years, he was still able to "push the envelope and have consistent profitability." His adoption of no-till farming had to meet the test of being "sustainable financially," which led to a measured and cautious approach and "doing a lot of homework." In 2008, the Zenners made a significant pay-down of long-term farm debt, so that farm debt is now approaching zero

SOCIAL AND COMMUNITY CONSIDERATIONS

Labor

Despite the considerable number of acres farmed, Zenner Farm has only one full-time, one part-time, and two seasonal workers. Good mechanical skills and an ability to recognize and respond to timing issues in getting critical jobs done have been especially important attributes of the full-time worker. Russ sees a strong technical skill set and reliability as essential for worker productivity. Although communication and social skills are desirable, they might not be as critical as technical skills and reliability. In general, fewer individuals with the needed technical and mechanical skills and interest are available in the surrounding community to hire, in part because fewer farm children grow up in the area. A pending dearth of local labor to work on the farm could become a problem in the future.

For the full-time worker, the farm provides health insurance, a retirement plan, a house to live in, and a crop bonus share. The part-time and seasonal workers play important roles during the busy season, but overall their hours are limited. Part-time and seasonal workers tend to be older, retired people, often with rural and farming roots, who have had nonfarming occupations (in some cases, professional occupations) for much of their adult lives. Some of them, Russ notes, "maybe would have preferred to farm."

Learning

Russ's approach to farming is premised on active learning and experimentation: "I'm constantly trying to glean information from someone else's experience. I've attended no-till conferences nationally and internationally." He has connected with and visited direct-seeding farmers in Australia, and he believes he learned a lot from them. Russ says Australian farmers generally have much tougher weed control issues than farmers in Idaho. He also believes they are much farther along with the "biological farming approach" than most American farmers.

Russ's personal interest in continual learning has spilled over into auxiliary enterprises with others. For example, he is involved with ViCo (stands for "visions cooperatively"), a small LLC he founded in 1998 with three fellow growers in the region to provide innovative farm management services.[41] A relatively new company partner is a former extension agent

---

[41] Three of the four ViCo grower partners are Columbia Plateau Producer members, and hence also members of Shepherd's Grain.

who manages "information discovery" on subjects the other partners want to learn more about—essentially more about practices that reduce impacts from chemical and fertilizer applications and other farming practices that can enhance soil health.[42] The research-driven work of ViCo is done in addition to managing custom equipment operations. Regarding the motivation for ViCo, Russ said: "To a man, we're concerned that current management practices we have are not sustainable. We're not getting answers from traditional research resources and what we've done in the past is not going to carry us into the future." ViCo received an NRCS Conservation Innovation grant for "technology innovation" focused on how precision agriculture can reduce chemical fertilizer applications. It has also partnered with Shepherd's Grain on a grant project to study the soil health and human nutrition link. They intend to apply for a larger grant. The members of ViCo have shared their farm employees and also sensitive personal financial information. Overall, ViCo emphasizes finding new farming practices and approaches to try, first on a small scale, with a priority to maintain profitability.

Russ also interacts frequently with other farmers, which is often a learning exchange: "I don't mind sharing information. I've been blessed. I've had some opportunities most people will never have." His stance on sharing information and learning follows consistently from his admission that "this [farming] is my main passion in life."

Russ's orientation to learning includes attention to the consumption side of the food system. Among the things he has enjoyed with Shepherd's Grain is getting into cities like Portland (for promotional events, for instance) to meet and interact with consumers of the Shepherd Grain's product. "It's fun to talk to people who really understand how food is produced," he says. Those events allow Russ to provide information about the realities of farming in the Palouse to the customers of his products and to learn about their preferences.

Community Relations and Service

Russ suggests that models from the past have affected his views on the importance of good community relationships for farmers. He notes that in his father's generation, farming neighbors did not always get along well. Observing those social dynamics "has had a profound impact on how I get along with my neighbors" and made him aware of the long-term implications of social interactions in the community.

In addition to his involvement with the Cooperative and the Conservation District, Russ has been active in organizations that conduct research, provide consultation, and support direct seeding. He spoke at a South Australian no-till farming conference and more recently at the first no-till conference held in Finland. Russ has also been involved with the Pacific Northwest Direct Seed Association, which "was formed in 2000 to provide information exchange and advocacy on conservation policy issues and research coordination that will assure adoption of economically-viable and environmentally-sustainable direct seed cropping systems" (from website www.directseed.org).

Russ goes so far as to suggest that as his time has been freed up by direct-seed farming practices, he has more discretionary time for volunteer and public service. He served as a director on a regional (Oregon, Washington, Idaho) bank board. That role afforded numerous regional contacts and insights on the local economy. His role ended recently when the bank was sold. He views his participation on various volunteer boards as an opportunity,

---

[42]Work on Brix measures for garbanzos is an example of research undertaken by ViCo that, thus far, is not being done at the land-grant universities.

not only for service, but also for learning: "There's some self-motivation in all this volunteering, you know. You can ask the right questions, keep your ears open, listen to different people from different places."

RISKS, CHALLENGES, AND CHANGES

Research Needs

Russ has become very interested in the possible connection between organic matter, soil health, the nutritional value of food, and their impacts on human health. He believes that connection is under-researched, and the knowledge gained would be very beneficial. He also expresses concern about the sources of funding for research: "Some research efforts and their funding come from the chemical companies, like for glyphosate. I don't think enough research is being done to monitor the effect on the soil biology of repeated applications of glyphosate." Such information is important for designing and improving direct-seeding systems.

He worked with other farmers and STEEP (Solution to Environmental and Economic Problems), a joint program of the University of Idaho, Oregon State University, and Washington State University, which was an innovative interdisciplinary research and education initiative focused on developing profitable cropping systems technologies for controlling cropland soil erosion and protecting environmental quality. His views about the contribution of public research to his farming enterprise are ultimately somewhat mixed. On the one hand, he recognizes some definite advantages in his location near two land-grant universities (Washington State University and University of Idaho) and has personally experienced benefits, particularly from USDA-ARS work on green bridge management and from STEEP's work on cropping systems rotation research. On the other hand, he notes the constraints now facing public agricultural research. He says, "Generally speaking, the land-grant universities are not always doing the kind of work we're looking for [to answer the questions we have]." This, in part, motivates his involvement with ViCo, as discussed above.

Russ sees a need for much more research investment in the genetics of pulse crops, where knowledge has lagged the extensive work on corn and soybean. He identifies a continuing technical challenge that research could address—how to avoid the "yield hit" in the early stages of transition to direct seeding. Managing the heavy residue common in this region (which can depress yield) is another area that needs research. Better information and resources for weed control in no- or minimum-till systems would be very helpful, in his view.

Transportation

A big issue for growers in this somewhat remote region, and of concern to Russ, is dependable and efficient transportation infrastructure. Rail access is particularly important for the cooperative, which is looking to be more strategically positioned in terms of its rail access. The cooperative is also very concerned about the river system on which it relies to move grain from Lewiston to Portland. Environmentalists and sportsmen are pushing to breach the dams on the upper Snake River, but that would make the barge transportation on which the cooperative now depends no longer viable.

Farm Transition Concerns

Russ and Kathy encouraged their three children to obtain college educations. Given that their children, now adults, are established in professional careers and living in Boise or Seattle, and, as Russ puts it, "none had the passion (for farming) I did," the Zenners are beginning to think about other options for continuing the farm operation, including hired management. A year before the interview, when Russ had back problems, he felt ready to make the transition. Resolution of that issue made the transition question less urgent, although it has not gone away. Russ notes, "I feel I have some obligation to what my father and grandfather did." Their hope is to set the farm up with top-level management that can mentor any eventual family members in the succeeding generation. At the same time, they concede that it is difficult to find people who can fulfill all their expectations as well as those of their children. Russ and Kathy have been to a Farm Credit's succession program on family business transition. They have held several all-family meetings about the future of the farm. Their children say they are not interested in selling the farm and express some desire to keep the farm so that their own children (Russ and Kathy's grandchildren) could come back to it and know that work ethic. Reconciling the various internal family interests with maintaining profitable farm operations remains a challenge.

GOVERNMENT PROGRAMS AND POLICY INVOLVEMENT

Russ observes, "I've been a significant recipient of farm program benefits over the years, but I think the system is very flawed in terms of ensuring rural communities and sustainable resource management." In general, he believes that regions reliant on program crops experience a stifling of innovation and diversity. Those regions are likely to find their economic opportunities restricted to those associated with niche or specialty crops. Russ is interested in seeing policies that are more sustainable and that encourage resource conservation and more value-added options at the local level.

Russ has been involved with the Dry Pea and Lentil Council, serving as chairman of its research committee in the 1990s, at which time he pushed for sustainable cropping systems research and links to the work of STEEP. The Dry Pea and Lentil Council later sought to address federal policies, but Russ was not involved in that effort.[43] He underscored that the system as currently structured does not adequately support sustainable resource management or rural economic health and does not support crop diversity. For example, if a grower has a diverse rotation (grows a crop one out of every three years), it is extremely difficult to develop the yield history required to participate in crop insurance—even though such a rotation would involve less risk from yield loss.

Russ has been involved in the Conservation Security Program (CSP) in the Clearwater watershed (2007, his first year, and 2008). He likes that type of incentive program, which he sees as promoting sustainable resource management. He says the CSP application was geared to no-till, so it was fairly easy for him to apply. Zenner Farm is getting full CSP funding as the Zenners are addressing many of the issues that CSP is concerned with, notably water quality. Russ articulates some concern that, at present, CSP does not reward the new biological farming approaches that he believes hold promise for the future. Zenner Farm has also participated in EQIP to develop buffers around streams.

---

[43]In 2002, the pulse marketing assistance loan program came in, but peas and lentils do not have program crop status nor the associated direct payments.

OBSERVATIONS AND CONCLUSIONS

Russ Zenner attributes the success on the farm to good soil, his opportunity to exercise responsibility at a young age, involvement on the conservation district board, and his gut feelings about what is strategically important. He has been particularly glad to share his experience and the information he has gained with others. The Zenner Farm has four specific features, which together distinguish its sustainability approach from many other farms:

- Conversion of the entire farm to direct seeding.
- Involvement in value-added marketing efforts at the commodity level through the cooperative.
- Extensive involvement in research and education efforts to increase the use of direct seeding and other environmentally sound practices.
- Involvement in innovative marketing efforts that connect with discerning local and regional consumers through Shepherd's Grain.

In addition, as true with many of the farmers at the farms studied for this report, Russ Zenner has a very active mind, such that he is continuously looking for new ways to pursue his interests and passions related to farming and to learn more. As Russ says, "The farther I've got in my farming career, the less I know. We remain so far from sustainability."

in some regions of Africa can potentially be expanded (FAO, 2009a). Projections indicate that a number of African countries could make much progress toward poverty reduction and food and nutrition security over the next 15–20 years by targeting policies and investment strategies that raise average crop yields by 50 percent, increase livestock numbers by 50 percent, and accelerate overall gross domestic product growth rates to 6.5–8.0 percent and the agricultural sector growth rate to 6 percent. Several experts agree that to achieve such a level of growth would require a commitment among African governments to reallocate up to 10 percent of their national budgets to agriculture, up from an average of 5 percent over the past decade continent-wide and only 4 percent in sub-Saharan Africa (African Union Report, 2008; World Bank, 2008). Although the growth performance implied above is high by historical standards, it is within the range of recent economy-wide and agricultural growth rates observed across Africa since the late 1990s (Runge et al., 2004; African Union Report, 2008; World Bank, 2008). Recent data also show that even agricultural production in sub-Saharan Africa grew at a rate of 3.5 percent in 2008 (FAO, 2009a). The Comprehensive Africa Agriculture Development Programme (CAADP) and the Sirte Declaration on Agriculture and Water are at the heart of efforts by African governments under the African Union to accelerate growth and eliminate poverty and hunger. The main goal of CAADP is to help African countries to reach a higher path of economic growth through agricultural-led development that eliminates hunger, reduces poverty and food insecurity, and enables expansion of exports. As a program of the African Union, it emanates from and is fully owned and led by African governments (African Union Report, 2008).

CONSIDERATIONS OF U.S. "LESSONS" LEARNED

Transferability of Agricultural Practices for Improving Sustainability

A large number of scientific-based issues relating to agricultural sustainability have been discussed throughout this report. Most, if not all, of the findings could be argued to have relevance to nearly every country. However, the specific methods chosen and priorities for their use in Africa need to be determined primarily by local and regional contexts and needs, as well as costs, potential and timing for impact, national R&D capacity, and the ability to attract resources from development assistance agencies.

The committee recognizes that many of the findings and conclusions in this report concur with recommendations made in recent reports that include *Realizing the Promise and Potential of African Agriculture* (InterAcademy Council, 2004); *Emerging Technologies to Benefit Farmers in Sub-Saharan Africa and South Asia* (NRC, 2008); *Science and Technology for Development* (IAASTD, 2009); and *The World Report 2008, Agriculture for Development* (World Bank, 2008). The commonalities among reports demonstrate that some sustainability principles and approaches are widely relevant, although, as discussed below, the details of implementation on the ground will be highly context specific. A series of science and technology recommendations to increase food security in Africa recommended by the InterAcademy Council (see Box 8-1) illustrate many of the commonalities in sustainability principles and the specific needs for the African context.

Further discussion and explanation of the recommendations in Box 8-1 can be found in the relevant sections below. The International Assessment of Agricultural Knowledge, Science and Technology for Development (IAASTD) reached many similar conclusions in its 2009 report (IAASTD, 2009). IAASTD is a multidisciplinary and multistakeholder effort that was initiated by the World Bank and the Food and Agriculture Organization of the United Nations in 2002. It evaluates the relevance, quality, and effectiveness of agri-

> **BOX 8-1**
> **Science and Technology Recommendations to Increase Food Security in Africa Proposed by the InterAcademy Council of the United Nations (2004)**
>
> **Near-Term Impact:**
>
> - Adopt a production ecological approach with a primary focus on identified continental priority farming systems.
> - Pursue a strategy of integrated sustainable intensification.
> - Use a blend of knowledge-intensive and technology-driven approaches that integrate with indigenous knowledge.
> - Adopt a market-led productivity improvement strategy to strengthen the competitive ability of smallholder farmers.
> - Recognize the potential of rain-fed agriculture and accord it priority.
> - Reduce land degradation and replenish soil fertility.
> - Explore higher-scale integrated catchment strategies for natural resource management.
> - Enhance the use of mechanical power.
> - Embrace information and communication technology at all levels.
>
> **Intermediate-Term Impact:**
>
> - Bridge the genetic divide.
> - Improve the coping strategies of farmers in response to environmental variability and climate change.
>
> **Long-Term Impact:**
>
> - Promote the conservation and the sustainable and equitable use of biodiversity management.

cultural knowledge, science, and technology on hunger, poverty, nutrition, human health, and environmental and social sustainability, and the effectiveness of public and private sector policies and institutional arrangements that focus on smallholder agriculturists. The assessment addressed how agricultural knowledge, science, and technology could reduce hunger and poverty, improve rural livelihoods, and facilitate equitable environmentally, socially, and economically sustainable development. It also proposed that new priorities and shifts in agricultural knowledge, science, and technology recognize and give increased importance to the *multifunctionality* of agriculture, which encompasses multi-output activity producing not only commodities (food, feed, fibers, biofuels, medical products, and ornamentals), but also noncommodity outputs such as environmental services, landscape amenities, and cultural heritages. It proposed, as well, that new institutional arrangements and policy changes be directed primarily at resource-poor farmers, women, and ethnic minorities. Fifty-eight countries approved the executive summary of the IAASTD synthesis report, but three countries (Australia, United States, and Canada) had reservations about some parts of the report, particularly the findings concerning the role of genetically engineered (GE) crops in sustainable agriculture development. The use of GE crops was not rejected in principle; rather, the report found that GE crops were appropriate in some contexts, but as of yet, the potential of GE crops to serve the needs of resource-poor farmers remains unfulfilled. There is no conclusive evidence so far that GE crops offer solutions to the broader socioeconomic dilemmas faced by developing countries (Kiers et al., 2008).

The next section first discusses the relevance of conclusions from earlier chapters of this report at the whole-system level, and then discusses component technologies that could be appropriate for the African context. The committee identified 12 major areas of agricultural science and technology, agricultural-supporting infrastructure, policy, and development process that are critical for the United States and have relevance, with appropriate adaptation, to African sustainable agricultural development.

1. **Sustainability is ultimately defined by the goals and objectives determined through an inherently political process and are highly context dependent.**

   Sustainability is a process of moving toward identified goals, but progress can be made in many different ways or by using a combination of different strategies. The four sustainability goals[1] outlined in Chapter 1 of this report are sufficiently broad to apply to the African context, although specific objectives within each goal, and the priority given to each objective, need to be determined through a political process (informed by scientific principles and knowledge) by people in the different regions of Africa. The importance of reflecting the priorities of African countries is strongly stated in the United Nation's InterAcademy report (InterAcademy Council, 2004) and in the report from the African Union (African Union Report, 2008). The need for African ownership of development efforts to improve food production and sustainability will require building a stronger indigenous research and education capacity. Increasing the involvement of farmers, especially women farmers, in research, policy discussions, and activities is critical to pursue appropriate goals and strategies (IAASTD, 2009).

   Throughout this report, the importance of understanding the biophysical, socioeconomic, and political context within which a farming system operates when seeking strategies to increase productivity sustainably has been discussed at length. That understanding is critical in a highly diverse continent such as Africa. The strategies for achieving different sustainability objectives will be specific to particular regions of the continent, and as such will require creation of interdisciplinary research and education institutions at multiple levels, from regional and national to local, with effective mechanisms to exchange information and knowledge among them.

2. **Sustainable systems need to be productive, efficient in resource use, and robust.**

   System attributes that are important for sustainability—productivity, system efficiency, and robustness (that is, have a combination of resilience, resistance, and adaptability to stress and changing conditions; see Chapter 1)—are emphasized in this report. In other words, a system needs to have the ability to continue meeting identified goals in the face of unpredictable weather and fluctuations in cost and availability of inputs to be sustainable (see Chapter 1). These points are also made in other reports (InterAcademy Council, 2004; NRC, 2008; World Bank, 2008; IAASTD, 2009) that argue for specifically focusing on strategies and technologies to improve productivity and increase efficient use of resources, most notably water, and to address the ability to adapt to climate change.

   The importance of building resilient and adaptable systems cannot be overstated. Predictions are that under climate change, there will be higher rainfall variability and uncertainty than at present, especially in arid and semiarid areas; extreme events like floods and droughts will become more frequent; and temperatures will increase in sub-Saharan Africa (NRC, 2008; IAASTD, 2009). Given that only 4 percent of agricultural land in sub-

---

[1]Satisfy human food, feed, and fiber needs, and contribute to biofuel needs; enhance environmental quality and the resource base; sustain the economic viability of agriculture; enhance the quality of life for farmers, farm workers, and society as a whole.

Saharan Africa is irrigated, unpredictable weather patterns will greatly affect the majority of rain-fed systems. As discussed in the report *Emerging Technologies to Benefit Farmers in Sub-Saharan Africa and South Asia* (NRC, 2008), developing strategies to alleviate both the agroecological and economic impacts of climate change will be necessary. Farmers will need tools to have the flexibility to adapt to changing conditions. Adapting to changing climate conditions will involve agroecosystem design, such as use of multiple cropping instead of monoculture, and use of varieties bred to incorporate adaptation to multiple stresses such as drought, high temperatures and flooding, and landscape diversification. Systems that take advantage of natural processes, complementarities, and efficiencies can often reduce the need for external inputs, and thus reduce vulnerability to changes in input availability and cost. A diversity of products and markets would also help buffer farmers from fluctuating weather and prices, and reduce the risk of food shortage in bad years.

In addition, strengthening social and institutional networks (Turner et al., 2003; Nelson et al., 2007) and building appropriate infrastructure can also help buffer against fluctuating conditions. High capital investment (especially in infrastructure) would need, however, to be well planned, cost effective, and seek to improve both the productivity and adaptive capacity of farmers in the region.

3. **Criteria and indicators are needed to assess progress toward achieving sustainability goals.**

In addition to goals and objectives, criteria for assessment and well-designed indicators of progress toward sustainability are needed at each level from the global to regional, national, and community levels. Much attention is given to that notion in the Millennium Report, which discusses goals and indicators from the level of the Millennium Development Goals (United Nations, 2008) downward to goals for nations and for community-level civil society groups. In defining indicators of sustainability, the development and testing process has to be decentralized at the national, state, and community levels if the indicators are to be relevant and have broad ownership.

"Sustainability" has particular priority objectives and time frames when very poor farmers are striving to move toward greater productivity, quality of life, and resource stabilization, which indicators need to reflect. For example, ensuring adequate productivity for short-term survival is critical, as is sufficient system robustness to prevent yields falling below critical levels over the longer term. In addition, resource stabilization, such as building soil organic matter and inherent fertility, is a long-term but critical component. "Improved" systems need to address all these priorities simultaneously to effectively move toward sustainability, and therefore need to be evaluated against appropriate indicators for each component (see Chapter 1).

Well-constructed indicators can be highly relevant as guides for agricultural development agencies and groups at all levels. The process for their identification could be informal, but the indicators and the assumptions upon which they are based would have to be made clear by all development groups as interventions are made.

4. **Priority should be given to an integrated systems approach to R&D that encompasses ecological, technological, and socioeconomic elements.**

If the four sustainability goals are to be addressed, then efforts to develop new technologies need to use integrated systems approaches to assess performance characteristics and the agroecological, environmental, and socioeconomic drivers operating in the farming system in question. Integrated studies of performance and the various drivers are particularly important to identify synergies among different management practices or barriers to research and policy evolution that are designed to reduce tradeoffs and enhance synergies between the four goals and to manage risks and uncertainties associated with their pursuit.

Measuring Progress Toward Sustainability

Sustainability is best evaluated not as a particular end state, but rather as a process that moves farming systems along a trajectory toward greater sustainability on each of the four goals. For this report, the committee's definition of sustainable agriculture does not make a sharp dichotomy between conventional and sustainable farming systems, not only because farming enterprises reflect many combinations of farming practices, organization forms, and management strategies, but also because most types of farming systems can potentially contribute to achieving various sustainability goals and objectives. Pursuit of sustainability is not a matter of defining sustainable or unsustainable agriculture, but rather of assessing whether choices of farming practices and farming systems would lead to a more or less sustainable system as measured by the four goals.

Finding ways to measure progress along a sustainability trajectory is an important part of the experimentation and adaptive management process. Environmental, economic, and social indicators can be used to describe the performance of agriculture and to provide information on whether a farm, a farming system type, or agriculture at any scale is on a trajectory toward improved sustainability. Many indicators are means-based and others are outcome-based; both types have limitations and strengths. Efforts to develop indicators to assess social dimensions of agricultural sustainability are sparse. Some of the indicators being used, such as production energy costs and levels of implementation of best management practices, are useful at many levels of aggregation from farm-level assessments to regional and national accounting. Yet, there are no agreed-upon standards regarding which indicators to use under different conditions. Few indicators have been validated by scientists, farmers, and the public. Developing consistent and effective indicators would facilitate assessment of the sustainability of farming practices or systems. Understanding the relationships between sustainability indicators and the outcomes they are meant to represent is a priority for future research.

Farming systems that move toward greater sustainability on most, if not all, of the four goals generally strive to work with ecological and biogeochemical processes and cycles to maximize synergistic interactions and the beneficial use of internal resources, minimize dependence on external inputs, and use added inputs efficiently. Through those efforts, they potentially reduce discharges to the environment and additional waste disposal activities, provide economic resilience, and enhance social well-being. As exemplified in the case studies, many farmers who work toward improved agricultural sustainability manage their operations to encourage social and economic synergistic relationships on-farm and throughout the food chain. The overall sustainability or robustness of a farming system—the ability to adapt to stresses, pressures, and changes in circumstances over time—is a result of some mixture of resistance, resilience, and adaptability of the coupled biophysical and socioeconomic system.

TOWARD AGRICULTURAL SUSTAINABILITY IN THE 21ST CENTURY

Although all farms have the potential (and responsibility) to contribute to different aspects of sustainability, the scale, organization, enterprise diversity, and forms of market integration associated with different individual farms provide unique opportunities or barriers to improving their ability to contribute to global or local food production, ecosystem integrity, economic viability, and social well-being. Transformation of the agriculture sector will require long-term research, education, outreach, and experimentation by the public and private sectors in partnership with farmers and will not occur overnight.

If U.S. agriculture is to address the challenges outlined in Chapters 1 and 2, both incremental and transformative changes will be necessary. Therefore, **the committee proposes two parallel and overlapping efforts to ensure continuous improvement in the sustainability performance of U.S. agriculture: incremental and transformative. The incremental approach is an expansion and enhancement of many ongoing efforts that would be directed toward improving the sustainability performance of all farms, irrespective of size or farming systems type, through development and implementation of specific sustainability-focused practices, many of which are the focus of ongoing research and with varying levels of adoption. The transformative approach aims for major improvement in sustainability performance by approaching 21st century agriculture from a systems perspective that considers a multiplicity of interacting factors.** The transformative approach would involve:

- Developing collaborative efforts between disciplinary experts and civil society to construct a collective and integrated vision for a future of U.S. agriculture that balances and enhances the four sustainability goals.
- Encouraging and accelerating the development of new markets and legal frameworks that embody and pursue the collective vision of the sustainable future of U.S. agriculture.
- Pursuing research and extension that integrate multiple disciplines relevant to all four goals of agricultural sustainability.
- Identifying and researching the potential of new forms of production systems that represent a dramatic departure from (rather than incremental improvement of) the dominant systems of present-day American agriculture.
- Identifying and researching system characteristics that increase resilience and adaptability in the face of changing conditions.
- Adjusting the mix of farming system types and the practices used in them at the landscape level to address major regional problems such as water overdraft and environmental contamination.

INCREMENTAL APPROACH TO IMPROVING U.S. AGRICULTURAL SUSTAINABILITY

The proposed expanded incremental approach would include focused disciplinary research on production, environmental, economic, and social topics, and policies (such as expanded agricultural conservation and environmental programs) to improve the sustainability performance of mainstream agriculture. For example, large livestock farms in the United States produce the majority of the nation's meat and dairy products. Similarly, a large portion of corn and soybean are produced on highly mechanized grain farms that specialize in the production of a small number of crops and rely heavily on purchased farm inputs to provide crop nutrients and to manage pest, disease, and weed problems. Most, if not all, farms have adopted some practices for improving sustainability, and some farms, including large farms illustrated in the report's case studies, are highly integrated, but such methods have not been adapted to all environments, and none of the practices have reached their full potential for adoption. Each of these production systems has fostered high productivity and low costs, but many have led to serious negative social and environmental outcomes (or externalized production costs) that could hinder agriculture's progress toward improved sustainability. The negative outcomes have led to policy changes and publicly funded research programs explicitly designed to address those concerns. Efforts to improve the sustainability outcomes associated with mainstream production systems might be incremental in nature, but could have significant benefits given the dominance of those production systems in U.S. agriculture.

Science—including biophysical and social science—is essential to understand agricultural sustainability. Science generates the knowledge needed to predict outcomes likely to result from different management systems, and it also expands the range of farming system alternatives that farmers and policy makers can consider. Science is critical for informing the political process. Research on an array of farming practices and farming systems has led to increased understanding of how each practice (including production practices and marketing strategies) can contribute to improving environmental, social, and economic sustainability of farms under different conditions. Examples of practices that have advanced the sustainability of U.S. agriculture toward some environmental, economic, and social goals are summarized in Box 9-1. Many practices listed in Box 9-1 have been implemented to different degrees and most serve as key components for fully integrated, sustainable farming systems.

Although the research conducted to date has led to development of many farming practices that enhance environmental quality and the natural resource base, **continuous research, extension, and experimentation by researchers and farmers are necessary to provide the toolkit necessary for farmers to adapt their systems to the changing environmental, social, market, and policy conditions to ensure long-term sustainability.** The committee also notes that much of the research to date focuses on developing an approach or a practice to enhance a specific environmental quality (such as increasing soil organic matter) or solve a specific environmental problem (such as reducing or preventing soil salinization). **Research on the economic and social dimensions of agricultural sustainability complementary to research on productivity and environmental sustainability is scarce despite its importance in providing farmers with knowledge to design systems that balance different sustainability goals and improve overall sustainability.** Studies on economic and social sustainability are complicated by the fact that economic viability is influenced by market and policy conditions and that social acceptability of farms is influenced by the behavior of key actors (including farmers and consumers) and the values of community members. The lack of information on the economic viability of practices and approaches to improving environmental and social sustainability and on how market and policies influence the economics of those practices could be a barrier to wide adoption of those practices. Examples of research priorities aimed at understanding and devising best management practices for agriculture are listed in Box 9-2.

Because research to develop practices and approaches for improving environmental sustainability and to qualify or quantify their economic and social impacts does not result in a marketable product for industry, this type of research is generally not attractive for private sector investment. Therefore, such research would have to rely on public funding and institutions, farmer organizations, and civil society sectors.

> **RECOMMENDATION:** The U.S. Department of Agriculture and state agricultural institutions and agencies should continue publicly funded research and development (R&D) of key farming practices for improving sustainability to assure that R&D keeps pace with the needs and challenges of modern agriculture. They should increase support for research that clarifies the economic and

> **BOX 9-1**
> **Examples of Practices That Contribute to Sustainability**
>
> **Production Practices**
>
> - **Conservation (or reduced) tillage systems** have become common for many crops and soil types. As of 2004, 41 percent of planted crop acreage was managed with conservation tillage. Water-caused soil erosion and surface runoff of nutrients, chemicals, and crop residues have been greatly reduced. Although no-till leads to savings on fossil fuel and labor, it could result in lower yields and greater difficulty with weed control than conventional till. Thus, the economic effect of no-till versus conventional till is unclear.
> - **Cover cropping** provides ground cover to protect soil. Cover crops can also be used to provide other services, including maintenance of soil organic matter and provision of nutrients to subsequent crops (green manures), trapping excess nutrients in the soil profile following harvest of the primary crop, and preventing leaching losses (catch crops). However, cover crops are not widely planted because they require complex management skills and their seeding costs could be high.
> - **Crop diversity, including rotations, intercropping, and using different genetic varieties** can contribute to improving soil quality, enhancing ecosystem function, and managing pests and diseases. Although the use of diverse cropping systems has increased, it fluctuates widely with commodity prices. Diverse cropping systems require extensive knowledge and management skills to identify the right combination of crops to achieve multiple sustainability goals. Comparative economic studies reported economic advantages for diversified rotation in some cases and disadvantages in others. The variation in results is partly attributable to market and policy conditions.
> - **Traditional plant breeding and modern genetic engineering techniques** will continue to be used to develop crop varieties with increased yields, pest and disease resistance, enhanced water-use and nutrient-use efficiencies, and other important traits. Genetic engineering (GE) has the potential to contribute novel solutions for problems that could not be addressed with natural plant genetic resources or traditional plant breeding methods. New GE varieties would have to be tested rigorously and monitored carefully by objective third parties to ensure environmental, economic, and social acceptability and sustainability before release for planting.
> - Many technologies for **efficient water use** such as metering, improved distribution of high-pressure water, and low-pressure, directed-use systems offer promise to address water scarcity. **Water reuse** is another strategy for addressing water scarcity, but the biological and chemical quality of the reclaimed water would have to be monitored carefully.
> - **Best management practices (BMPs)**, including nutrient management planning, field buffer strips, riparian area management, surface and subsurface drainage water management, and livestock manure management, have been developed to mitigate the runoff of agricultural nutrients and chemicals into the nation's surface and ground waters. Effectiveness of BMPs at the watershed scale has been difficult to prove, in part because actions by individual farms might not be visible at the landscape scale. The benefits of BMPs can vary widely depending on characteristics of the landscape, weather events, and time lags between BMP adoption and physical changes in the dynamics of nutrient and chemical cycling on farm fields.

social aspects of the many current and potential technologies and management practices and that addresses issues of resilience and vulnerability in biophysical and socioeconomic terms.

TRANSFORMATIVE APPROACH TO IMPROVING U.S. AGRICULTURAL SUSTAINABILITY

If major farming systems and aggregations of systems within key production regions have gradually evolved toward meeting some sustainability goals while moving toward unacceptable ends of the others, as indicated by scientific knowledge accumulated over

People and organizations in developed countries and in developing countries can exchange useful information and ideas to solve problems related to sustainability of agriculture. Likewise, scientists and policy makers can learn from farmers and vice versa. Researcher and farmer partnerships and peer-to-peer exchanges among farmers could facilitate incorporation of local knowledge, making use of the best-available scientific process-level understanding, and enabling learning and developing knowledge systems to build the local capacity for improving agricultural sustainability.

IN CLOSING

This report identifies what is known about farming practices and systems that can improve sustainability. It discusses the potential benefits and risks if those practices are used and the potential synergies and tradeoffs that might present themselves if the practices are used in combination in a farm system. The report also identifies knowledge gaps and areas where greater research is needed to help inform future decisions and to move agriculture along the sustainability trajectory. Filling those gaps will require some innovative new approaches in the realms of resilience thinking, complex systems science and management as applied to agroecosystems, and a better understanding of the economic and social drivers and outcomes of various farming approaches. The report findings show positive and promising outcomes among the production systems, farming businesses, and communities that are pursuing improved sustainability. It also reveals the importance of government agencies, farmers, food industry companies, communities, and consumers to support research, policies, programs, and institutions that help U.S. agriculture move along the sustainability trajectory.

Appendixes

C

Presentations to the Committee on 21st Century Systems Agriculture

DECEMBER 10, 2007

Julia Kornegay, North Carolina State University
The Socioeconomic Context for Farm Systems and Sustainability

Curt Reynolds, U.S. Department of Agriculture (USDA) Foreign Agricultural Service
Trends and Forces in Global Agriculture Production

Jim MacDonald, USDA Economic Research Service
Trends in the Structure of Agricultural Markets

Mitch Morehart, USDA Economic Research Service
Structure and Finances of U.S. Farms

Richard Harwood, Michigan State University, Emeritus
Where Have We Come Since 1989?

Robbin Shoemaker, USDA Economic Research Service
Changing Trends in Agricultural Labor and Energy, and Agriculture's Relation to the Environment

Jeffrey Steiner, USDA Agricultural Research Service
Agricultural Systems Research: How Will We Know When Alternative Has Become Conventional?

MARCH 27, 2008

Otto Doering, Purdue University
Impact of Energy Crop Production on U.S. Agricultural Economics
Impact of the Farm Bill and the Energy Bill on Agriculture to Date

Thomas Dobbs, Food and Society Policy Fellows Program
Economics and Policy Conditions to Foster Sustainable Farm Systems

Gerald Bange, The World Agricultural Outlook Board
Update on the USDA Agricultural Outlook

Seth Meyer, University of Missouri, Columbia
Impacts of Overseas Demand (over the next 15–20 years) on U.S. Agriculture

MAY 2, 2008

Roger Claassen, USDA Economic Research Service
Impact of Conservation Programs on U.S. Agriculture

Douglas Lawrence, USDA Natural Resource Conservation Service
Impact of Agroenvironmental Policies on the Environmental Sustainability of U.S. Agriculture

Sally Shaver, U.S. Environmental Protection Agency (EPA)
EPA Regulations That Could Impact Agricultural Producers

Connie Musgrove, University of Maryland
State Policies and Programs for Improving the Environmental Performance of Agriculture

Cathy Kling, Iowa State University
Cost and Benefit of Developing Policies to Improve Water Quality

Jerry Skees, University of Kentucky
Impact of Crop Insurance on Farmers' Decisions to Adopt Agronomic Practices

AUGUST 5–6, 2008

Joan Nassauer, University of Michigan
Landscape Perspectives on Agricultural Intensification and Biodiversity

Laurie Drinkwater, Cornell University
Research and Understanding of Complex Farming Systems

Ariena van Bruggen, Wageningen University
Ecological Principles Underlying the Functioning of Farming Systems

Kathy Soder, USDA Agricultural Research Service
Opportunities and Challenges in Management-Intensive Grazing Systems

Alan Franzluebbers, USDA Agricultural Research Service
The Science Behind Integrated Crop/Livestock Systems

Lynne Carpenter-Boggs, Washington State University
Role of Composting on Soil Health and Other Aspects

Ariena van Bruggen, Wageningen University
Interdependence Between Soil Health, and Plant and Animal Health

Tony Grift, University of Illinois, Urbana-Champaign
Automation in Complex Farming Systems

Clay Mitchell, The Mitchell Farm
Precision Agriculture

Steve Evett, USDA Agricultural Research Service
Use of Water, Water Processing, Water Reclamation

Steve Naranjo, USDA Agricultural Research Service
Biocontrol and Transgenic Crops

Pamela Ronald, University of California, Davis
Marker-Assisted Breeding and Genetic Engineering to Enhance Crop Stress Tolerance

Eric Sachs, Monsanto
Germplasm Improvement and Disease and Pest Management

Mark Allan, USDA Agricultural Research Service
Animal Genetic Improvement

Joy Mench, University of California, Davis
Scientific Basis for Improving Animal Welfare

JANUARY 12, 2009

Hans Herren, Millennium Institute
International Assessment of Agricultural Knowledge, Science, and Technology
  Development Report

William Settle, Food and Agriculture Organization
Role of Integrated Pest Management in Improving Agricultural Sustainability in
  Developing Countries

Gary Toenniessen, Rockefeller Foundation
Role of Biotechnoloy in Improving Agricultural Sustainability in Developing Countries

Joyce Turk, U.S. Agency for International Development
Yesterday's Future: Sustaining Livestock Production Systems in Developing Countries

Amir Kassam, University of Reading
Conservation Agriculture as a Foundation for Sustainable Production Intensification

Jules Pretty, University of Essex
Recent Evidence on Improving the Environmental, Social, and Economic Sustainability of Agriculture in Developing Countries

Ruth Meinzen-Dick, The International Food Policy Research Institute
Rural Development and Institutions

Keith Moore, Virginia Polytechnic and State University
Networking Technology

Joan Fulton, Purdue University
Marketing and Trade Factors That Could Affect Adoption of Sustainable Practices

William Settle, Food and Agriculture Organization
Community-based, Discovery Learning: Farmer Field Schools as a Pragmatic Approach to Agricultural Extension

APPENDIX F

2. Relate soil characteristics to specific enterprises/crop rotations, vegetation management
3. Highlight any limitations (for example, are soils erodible, subject to flooding, poorly structured, sloped, etc.)?
4. What do you like best about your soil? How do you respond to challenges to soil quality (compaction, erosion, periodic flooding, poor structure, slope, etc.)?

Soil Management Practices

1. Fertility and nutrient management practices
    1.1. Do you have a nutrient management plan on file with NRCS?
    1.2. Do you use manure on your operation?
    1.3. What are your tillage management practices
    1.4. Do you have an overall erosion control program?
    1.5. Is your erosion risk primarily from wind or water?
    1.6. What are your off-season soil management practices?
    1.7. Any other conservation practices or benefits?
    1.8. Do you use protected agriculture—high tunnels, greenhouses, plasticulture, row covers, hoop houses?

WATER RESOURCES

Quantity

1. What is the average rainfall?
2. Do you irrigate?
3. How adequate is the water supply for your operation?
4. Have you made significant changes in your operation to adapt to water scarcity?

Quality

1. Have water quality concerns been raised in your area?
2. How have these concerns impacted your farming operation?
3. Have you made significant changes in your operation to adapt to water quality concerns?
4. What specific steps have you taken to protect water quality in your area?

CLIMATE

1. Have there been any major droughts, storms, or other weather events that have impacted your operation in recent years?
2. How do you feel your operation has been impacted by climate change or global warming?
3. How do you think this might impact your operation in the future?

CARBON FOOTPRINT ISSUES

1. To what degree are you aware of greenhouse-gas emissions (CO_2 or equivalents) resulting from farm production (fuel usage of machinery, application of fertilizers and the use of synthetics, including plastics)?
2. What specific steps have you taken to address potential concerns?

ALTERNATIVE ENERGY

1. Do you use any other forms of renewable energy (wind, solar, etc.) in your operation?
2. How well have these systems worked for you?
3. What are the best aspects of these systems?
4. What are the worst aspects of these systems?
5. Do you have an energy management plan?

EQUIPMENT

1. How has your equipment changed over time? Has your energy use increased or decreased over time?
2. Do you have any specialty equipment? Where do you get this equipment?
3. Do you use any contract equipment operations?
4. What portion of your equipment maintenance do you perform on-farm? (Do you consider maintenance as a major part of your operation for cost saving/income?)
5. Solid waste management?
6. Other features or comments?

2. Economic/Market Questions

SALES

1. How have the gross sales of your farm changed over the last 5 years?
2. What do you expect the change to be in the next 5 years?

MARKETING

1. Where do you sell your farm products?
2. What proportion of your produce is marketed locally?
3. What would motivate you to seek out local markets for your products (farm profit, local economic development, local food security, farm-to-school or farm-to-cafeteria)?
4. Do any of your products go outside the United States?
5. How do you determine what price you are paid for your product?
6. Do you sell any of your farm products on contract?

Organic Products

1. Do your farm productions have any other official "certifications" that you use in marketing your product?
2. If you are a certified-organic grower, please answer the following questions

2.1. Do you get organic price premiums for your products?
2.2. Have these premiums increased or declined in recent years?
2.3. How important are these premiums to your ability to use organic practices?
2.4. Are organic premiums important to your decision to use sustainable practices?

FARM ENTERPRISE FINANCIALS

Income

1. What are your approximate annual gross sales of farm products?
2. What is your approximate income from government program payments?
3. Are there other sources of farm income?

Net Income

1. In how many of the last 5 years has your enterprise made a profit?
2. Has your net income increased over the last 5 years?

Farm versus Off-farm Income

1. Does anyone in your family work off the farm (known as public work in the South)?
2. To what degree does your household depend on the farm for income and benefits?

Debt

1. Roughly what would you estimate is the current ratio of your farm debts to farm assets? (no debt, debts less than 10 percent assets, debts 10 to 40 percent assets, debts exceed 40 percent of assets)
2. Are you comfortable with your current debt levels?

LABOR

1. Do you have workers assisting you on the farm?
2. Have you been able to find adequate labor to sustain operation and quality of life?
3. What specific practices do you use to ensure labor is treated fairly?
4. How do labor issues affect your decisions about which production practices to use?

RISK MANAGEMENT

1. Do you have crop insurance?
2. What would cause you to purchase crop insurance?
3. If you currently have it, how could the policy be improved?
4. How important is crop insurance to your farm's long-term financial security?

3. Social and Community Aspects

1. What motivated you to choose this approach to farming? What keeps you enthusiastic about your approach?
2. Do your customers/buyers have an impact on how you farm? Do they impact what you produce? If so, what kind of impact?
3. How do you get information and advice for your farming? (main sources)
4. Who comes to you for advice about farming?
5. Do you share information with other farmers?
6. Do you make an effort to share information about your operation with the public?
7. Are you involved with any farmers' groups or other organizations?
8. Do you work with any government agencies?
9. What government programs have the largest impacts on your operation?
10. Are you involved in farm policy issues? If so, how?
11. How does the way you farm influence the way your community relates to you?

DEVELOPMENT PRESSURE

1. How much pressure does your operation feel from growth and development in your area?
2. What opportunities and challenges does urban growth and development present to your farm?
3. Farm transition challenges—how do you ensure viability of operation across generations/time?
    3.1. How to ensure sustainability (values) across generations/time?
    3.2. How to mitigate start-up costs for new farmers by removing land (and perhaps some infrastructure) costs through long-term lease and trust arrangements (to replace fee-simple purchase and/or inheritance models by community-based stewardship models)?
    3.3. How to connect successive generations to pass on knowledge and practical experience?
    3.4. How to secure the older generation while enabling the future of the next generation?

G

Specialty-Crop Farms

Topics of Discussion During On-Farm Interview

Elements of "Success" and Barriers and Broader Assessments of the Sustainability of the Farming Systems

1. What factors have contributed to progress or success for your farm/company?
2. In what important ways has this operation changed over the last 5 years?
3. Overall, how would you assess the performance of your farming systems on some broader indicators of sustainability—
   3.1. Local ecosystem sustainability (for example, water, soil, air, biodiversity and critical species)
   3.2. Global ecosystem sustainability (for example, energy use, climate)
   3.3. Social sustainability (for example, quality of life, labor, economics, community)
   3.4. Food system sustainability (for example, food quality, nutrition, affordability and access)
4. What are the main opportunities you have to improve your farming operation? What are the main risks?
5. How have you overcome problems or decreased risks? (provide examples)
6. Do you do research or experimentation on new methods? If so, what kind? Do you work with other organizations/people that do this work? If so, who?
7. Do you have any specific problems that might benefit from more (or better) scientific research?
8. What do you think is the future outlook for your own farming operation/enterprise?
9. What factors will most influence your long-term viability? What do you think is the future outlook for sustainable systems/organic farming?
10. Do you foresee any barriers to continuing to farm sustainably, or to developing additional strategies to enhance sustainability?
11. What are you proudest about when you think of your farming operation?

1. Production Practices

PRACTICES

1. What practices do you implement on your farm and what percentage?
   Conventional_____
   Integrated pest management_____
   Low input_____
   Certified organic_____
   Mixed crop and animal systems_____
   Biodynamic_____
   Other_____
2. Do you rotate crops?
3. Do you use cover crops?
4. Do you control flowering, fruit set, or growth of your crops?
5. How do you accomplish pollination?
6. Do you manage bees or other pollinators?
7. If you grow annual crops, do you use hybrid varieties?
8. Do you grow perennial, woody, or tree crops?
9. Do you have a strategy for managing biodiversity on your farm?

WEED, DISEASE AND INSECT CONTROL

1. What control practices do you use for weeds?
2. What control practices do you use for diseases?
3. What control practices do you use for insects?
4. If you use integrated pest management (IPM), what practices do you employ?
5. If you use IPM, do you hire or contract with IPM management firms or individuals?

SOIL RESOURCES—General Soil characteristics (for the major soil types only)

1. Please describe the soil types on your farm.
2. Relate soil characteristics to specific enterprises/crop rotations, vegetation management
3. Highlight any limitations (for example, are soils erodible, subject to flooding, poorly structured, sloped, etc.?)
4. What do you like best about your soil? How do you to respond to challenges to soil quality (compaction, erosion, periodic flooding, poor structure, slope, etc.)?
5. What soil management program do you use?
6. How do you assess soil health and fertility? Do you do bioassays to evaluate microbial activity?
7. Do you use fertilizers? __Yes or __No
8. Do you use compost? __Yes or __No
9. Do you use green manures/cover crops. If so, what crops?
10. Do you use soil amendments? If so, what kind?
11. Do you use soil conservation/prevention of erosion methods? If so, what kind?
12. Do you know the level of soil organic matter in your fields? If so, how do you measure this?

APPENDIX G

WATER RESOURCES

Quantity

1. What is the average rainfall?
2. Do you irrigate?
3. How adequate is the water supply for your operation?
4. Have you made significant changes in your operation to adapt to water scarcity?
5. Has water availability changed in recent years?

Water Quality

1. Have water quality concerns been raised in your area?
2. How have these concerns impacted your farming operation?
3. Have you made significant changes in your operation to adapt to water quality concerns?
4. What specific steps have you taken to protect water quality in your area?
5. Do you test your water source?
6. Do you have any drinking water health issues?

CLIMATE

1. Have there been any major droughts, storms, or other weather events that have impacted your operation in recent years?
2. How do you feel your operation has been impacted by climate change or global warming?
3. How do you think this might impact your operation in the future?

AIR QUALITY ISSUES

1. Are there any air quality or odor issues that currently affect your farming operation?
2. What specific steps have you taken to address potential concerns?

CARBON FOOTPRINT ISSUES

1. To what degree are you aware of greenhouse-gas emissions (CO_2 or equivalents) resulting from farm production (fuel usage of machinery, application of fertilizers, and the use of synthetics, including plastics)?
2. What specific steps have you taken to address potential concerns?

ENERGY

1. What are your main sources of energy?
2. Do you use any other forms of renewable energy (wind, solar, etc.) in your operation?
3. How well have these systems worked for you?
4. What are the best aspects of these systems?
5. What are the worst aspects of these systems
6. Do you have an energy management plan?

EQUIPMENT AND STORAGE

1. What kind of specialized farm equipment do you use (i.e., precision seeders, seedling transplanters, plastic layers)?
2. Do you use animals to farm?
3. Do you have postharvest storage capabilities? If so, what type, what capacity?
4. Do you have postharvest cooling capabilities for fruits and vegetables? If so, what type, what capacity?

2. Socio-Economic/Market Questions

SALES

1. How have your gross farm sales changed over the last 5–10 years
2. What do you expect the change to be in the next 5 years?

MARKETING

1. Where do you sell your farm products?
2. What proportion of your produce is marketed locally?
3. Do any of your products go outside the United States?
4. Do you sell any of your farm products on contract?
5. Do you get farm credit?
6. How have your marketing approaches changed over the last 5–10 years?
7. Do you provide value-added packaging or branding for your harvested products (i.e., clam shells, special boxes)?

Organic Products

1. Do your farm productions have any other official "certifications" that you use in marketing your product?
2. If you are a certified-organic grower, please answer the following questions
    2.1. Do you get price premiums for your products?
    2.2. Have these premiums increased or declined in recent years?
    2.3. How important are these premiums to your ability to use organic practices?
    2.4. Are organic premiums important to your decision to use sustainable practices?

FARM ENTERPRISE FINANCIALS

Although the committee does not need complete details regarding your farming operation finances, it is interested in the economic opportunities and challenges facing different kinds of farm enterprises. Can you provide us with rough estimates of the following economic information?

INCOME

Gross Income

1. What are your approximate annual gross sales of farm products?
2. What is your approximate income from government program payments?
3. Are there other sources of farm income?

Net Income

1. In how many of the last 5 years has your enterprise made a profit?
2. Has your net income increased over the last 5 years?

Farm versus Off-farm Income

1. Does anyone in your family work off the farm (known as public work in the South)?
2. To what degree does your household depend on the farm for income and benefits?

Debt

1. Roughly what would you estimate is the current ratio of your farm debts to farm assets? (no debt, debts less than 10 percent assets, debts 10–40 percent assets, debts exceed 40 percent of assets)
2. Are you comfortable with your current debt levels?

LABOR

1. Do you have workers assisting you on the farm?
2. Do you provide them with housing on-farm or off-farm?
3. Have you been able to find adequate labor to sustain operation and quality of life?
4. What specific practices do you use to ensure labor is treated fairly?
5. How do labor issues affect your decisions about which production practices to use?

RISK MANAGEMENT

1. What are the major sources of risk on your farm?
2. What strategies have you developed on your farm to deal with these sources of risk?
3. Given recent changes in energy and farm commodity markets, how well do you feel your operation can compete in the current high-energy and high-feed-cost environment?
4. What alternatives do you see in managing future (expected) shortages in energy, fertilizer, land and water resources, and concomitant increases in production costs?
6. Do you have crop insurance?
7. What would cause you to purchase crop insurance?
8. If you currently have it, how could the policy be improved?
9. How important is crop insurance to your farm's long-term financial security?

3. Social and Community Aspects

SOCIAL AND COMMUNITY

1. What motivated you to choose this approach to farming? What keeps you enthusiastic about your approach?
    1.1. Could you give us a thumbnail description of what you mean by sustainability?
2. Do your customers/buyers have an impact on how you farm? Do they impact what you produce? If so, what kind of impact?
3. How do you get information and advice for your farming? (i.e., main sources)
4. Who comes to you for advice about farming?
5. Do you share information with other farmers?
6. Do you make an effort to share information about your operation with the public?
    6.1. Other?
7. Are you involved with any farmers' groups or other organizations?
8. Do you work with any government agencies?
9. What government programs have the largest impacts on your operation?
    9.1. Positive?
    9.2. Negative?
10. Are you involved in farm policy issues? If so, how and at which level: federal, state, county, or town?
11. How does the way you farm influence they way your community relates to you?

DEVELOPMENT PRESSURE

1. How much pressure does your operation feel from growth and development in your area?
2. What opportunities and challenges does urban growth and development present to your farm?
3. Farm transition challenges—how do you ensure viability of operation across generations/time?
    3.1. How to ensure sustainability (values) across generations/time?
    3.2. How to mitigate start-up costs for new farmers by removing land (and perhaps some infrastructure) costs through long-term lease and trust arrangements (to replace fee-simple purchase and/or inheritance models by community-based stewardship models)?
    3.3. How to connect successive generations to pass on knowledge and practical experience?
    3.4. How to secure the older generation while enabling the future of the next generation?

FOOD SAFETY

1. Do you have a Food Safety Plan?